责任编辑：库流平　陈魁英
文字校对：冯　辰　王金龙
版面设计：谭晓林　张洪仨

涉密资料　列入移交

空军"五级"主官示范培训资料汇编

空军政治部组织部

徐滨士　院士

2004 年 2 月，在全国科技大会上，徐滨士院士获得国家技术发明二等奖和国家科技进步二等奖

国家自然科学奖
证 书
为表彰国家自然科学奖获得者，特颁发此证书。
项目名称：面向再制造的表面工程技术基础
获 奖 者：徐滨士（中国人民解放军装甲兵工程学院）
奖励等级：二等
证书号：2013-Z-109-2-05-R01

2013 年徐滨士院士荣获国家自然科学二等奖

2010 年波兰驻华大使为徐滨士院士颁发波兰科学院外籍院士聘书

2006 年 12 月，实验室获得军委四总部表彰的首届“中国人民解放军科技创新群体奖”；2008 年 4 月，实验室荣立“集体一等功”。实验室现已成为我国、我军再制造的技术创新中心与战略思想智库。图为徐滨士院士目前的学术团队

發展表面工程 為先
進制造技術做貢献

师昌绪 一九九七年十月一日

鐵甲流金
鑄軍魂

書敬徐濱士院士

遲浩田 二〇〇四年六月廿八日

中国工程院 院士文集

Collections from Members of the Chinese Academy of Engineering

徐滨士文集

A Collection from Xu Binshi

本书编委会 编

北 京
冶金工业出版社
2014

内容提要

本书是为祝贺中国工程院成立20周年和徐滨士院士参军执教60周年而编撰的，选取了徐滨士院士及其科研团队在不同时期所撰写的代表性论文，反映了徐滨士院士为我国国防事业所做出的贡献。本书共分四部分，包括表面工程，再制造工程设计及质量控制，再制造关键技术，再制造工程管理。

本书可供从事材料工程、机械工程、环境工程等研究的科技人员及高等院校相关专业的师生阅读。

图书在版编目(CIP)数据

徐滨士文集/《徐滨士文集》编委会编．—北京：冶金工业出版社，2014.5

(中国工程院院士文集)

ISBN 978-7-5024-6634-3

Ⅰ.①徐…　Ⅱ.①徐…　Ⅲ.①武器装备—机械工程—文集　Ⅳ.①TJ05-53

中国版本图书馆CIP数据核字(2014)第099818号

出 版 人　谭学余

地　　址　北京北河沿大街嵩祝院北巷39号，邮编100009

电　　话　(010)64027926　电子信箱　yjcbs@cnmip.com.cn

责任编辑　李　臻　卢　敏　美术编辑　彭子赫　版式设计　孙跃红

责任校对　禹　蕊　刘　倩　责任印制　牛晓波

ISBN 978-7-5024-6634-3

冶金工业出版社出版发行；各地新华书店经销；三河市双峰印刷装订有限公司印刷

2014年5月第1版，2014年5月第1次印刷

787mm×1092mm　1/16；29.5印张；2彩页；681千字；453页

175.00 元

冶金工业出版社投稿电话：(010)64027932　投稿信箱：tougao@cnmip.com.cn

冶金工业出版社发行部　电话：(010)64044283　传真：(010)64027893

冶金书店　地址：北京东四西大街46号(100010)　电话：(010)65289081(兼传真)

(本书如有印装质量问题，本社发行部负责退换)

《徐滨士文集》编委会

《中国工程院院士文集》总序

2012年暮秋，中国工程院开始组织并陆续出版《中国工程院院士文集》系列丛书。《中国工程院院士文集》收录了院士的传略、学术论著、中外论文及其目录、讲话文稿与科普作品等。其中，既有院士们早年初涉工程科技领域的学术论文，亦有其成为学科领军人物后，学术观点日趋成熟的思想硕果。卷卷文集在手，众多院士数十载辛勤耕耘的学术人生跃然纸上，透过严谨的工程科技论文，院士笑谈宏论的生动形象历历在目。

中国工程院是中国工程科学技术界的最高荣誉性、咨询性学术机构，由院士组成，致力于促进工程科学技术事业的发展。作为工程科学技术方面的领军人物，院士们在各自的研究领域具有极高的学术造诣，为我国工程科技事业发展做出了重大的、创造性的成就和贡献。《中国工程院院士文集》既是院士们一生事业成果的凝炼，也是他们高尚人格情操的写照。工程院出版史上能够留下这样丰富深刻的一笔，余有荣焉。

我向来认为，为中国工程院院士们组织出版院士文集之意义，贵在“真、善、美”三字。他们脚踏实地，放眼未来，自朴实的工程技术升华至引领学术前沿的至高境界，此谓其“真”；他们热爱祖国，提携后进，具有坚定的理想信念和高尚的人格魅力，此谓其“善”；他们治学严谨，著作等身，求真务实，科学创新，此谓其“美”。《中国工程院院士文集》集真、善、美于一体，辩而不华，质而不俚，既有“居高声自远”之澹泊意蕴，又有“大济于苍生”之战略胸怀，斯人斯事，斯情斯志，令人阅后难忘。

读一本文集，犹如阅读一段院士的“攀登”高峰的人生。让我们翻开《中国工程院院士文集》，进入院士们的学术世界。愿后之览者，亦有感于斯文，体味院士们的学术历程。

2012年7月

序

1931年我出生时，正值日本帝国主义发动九一八事变，大规模武装入侵我国东北。当时，中国老百姓被奴役，被宰割，过着牛马不如的悲惨生活。东北人民只准吃掺着大量橡子面的杂粮，吃点大米就是犯罪。上了小学，学伪满洲国语和日语，不让知道自己是炎黄子孙，只让知道日本的天照大神、日本的历史。当时不准讲自己是中国人，父母私下里告诉我：老家在山东。我知道了自己是中国人，更加痛恨日本侵略者和伪满洲国。

"八·一五"苏联红军解放哈尔滨以后，尚未公开的党组织已经开始活跃在各个阶层。我念书的四中校长和不少老师就是共产党员，他们宣传新民主主义能够救中国，给我们带来了希望。在共产党的引导下，我知道了共产党和国民党的不同，知道了以蒋介石为首的蒋、宋、孔、陈四大家族的腐败，蒋介石的不抵抗造成的东北沦陷，使我们饱尝了14年亡国奴之苦，而且当时蒋介石又要将中国卖给美国，因此，我坚决拥护共产党，决心跟着共产党把落后的旧中国建设成为富强民主的新中国。当时有少数人盲目相信"正统"，跑到长春国民党统治区去了。我坚定地留下来，并考入了共产党领导的哈尔滨工业大学。那时家里供不起我上大学，但党实行的供给制，使我读完了大学。我家祖祖辈辈没有大学生，我是家里第一个大学生，是党培养了我。

在哈工大，我学习十分刻苦。学校请来的苏联教授介绍了苏联的情况，特别提到建设社会主义需要大量的科技人员。回想起日本帝国主义铁蹄的蹂躏，看到国家落后的状况，我深深地感到学习的重要、科学技术的重要，因此发奋苦读。由于我在日伪统治时期打下的文化基础差，为了能够跟上他人学习的步伐，我只有加班加点弥补差距。

1954年从哈尔滨工业大学毕业后，我服从组织安排，来到哈尔滨军事工程学院装甲兵工程系，从事坦克维修专业的教学工作。开始我对这一工作认识不足，认为只有搞设计、制造才是高水平，才大有作为。后来有两件事对我震动很大。一件是我给苏联坦克维修专家当翻译时得知，苏联红军坦克、机械化部队的维修部门在卫国战争中抢修了43万辆次坦克、装甲车，相当于苏联战时最高年产量的15倍，这对保证装甲部队的持续战斗力、战胜法西斯起到了极其重要的作用。另一件是下部队调查时，看到因维修设备和技术落后，许多局部磨损的坦克零部件不能修复，整机只好报废，浪费很大，严重影响正常训练和战备任务的完成，干部战士们忙得团团转。这让我的心情久久不能平静，作为一名从事维修专业的教员，我暗下决心，一定要用所学的知识改变部队维修技术落后的现状，首先攻克坦克磨损件的修复难关。

20世纪50～60年代，我军坦克薄壁零件的修复技术相当落后，被称为坦克修理中的顽症。当时我既缺乏资料，又没有经验，仅在一本苏联杂志上找到了一条关于振动电弧堆焊修复薄壁零件的简要报道，而这一设备是什么样子，我们谁也没有见过。为了将资料变成实际的设备，我和助手们按照杂志上介绍的原理，一步一步地摸索着干。每次失败后，我和我的助手们就总结经验，相互鼓励，再重新开始。经过100多个日日夜夜的苦干，终于在国内首次研制成功了振动电弧堆焊设备，摸索出了新工艺，解决了薄壁零件修复的难题，突破了部分坦克薄壁零件不能修复的禁区。1958年该成果参加了在北京举行的全国科技成果展览会，受到党中央和中央军委的称赞，组织上给我记了二等功。不久，我的课题组又研制成功了“水蒸气保护振动电弧堆焊”，薄壁零件的修复质量明显提高。在部队推广中，我们又改造老设备，研制出了两种新的振动头（该设备获得了全国科学大会奖），使坦克零件的修复范围逐步扩大，看到一辆辆“趴窝”的坦克又驰骋训练场，心里感到无比的欣慰。

学习无止境，科研无尽头，我常常以此激励自己。振动电弧堆焊虽然解决了坦克薄壁零件的修复问题，但修复后的零件质量只能接近新品，为了提高易损零件的耐磨性，我四处寻找解决的办法。1973年，我利用休假机会到哈尔滨锅炉厂参观，受到了采用等离子堆焊技术制造高压阀门来提高零件耐磨性的启发，返院后，当我提出将等离子喷涂技术用于坦克零件维修时，有的同志持怀疑态度，有的同志不断提醒我，一旦试验失败，将会造成坦克车辆损坏的大事故，会给自己造成严重后果。个别领导也要我深刻检查，把试验停下来。我想，搞科研是部队建设的需要，试验不能停，顶着压力也要上。经过教研室全体同志的不懈努力，终于试验成功了等离子喷涂修复坦克零件的技术。在装甲兵首长及兄弟单位的支持下，六辆坦克各行驶12000km的实车考核证明，应用等离子喷涂技术修复的坦克零件，耐磨性比新品提高1.4~8.3倍，而成本只有新品价格的1/8，有效地延长了被修零件的使用寿命，大大提高了坦克装甲车辆的持续作战能力，节约了大量的装备维修经费。该项目获得了全军科技成果一等奖和国家科技进步二等奖。

通过在设备维修中应用各种先进的表面工程技术的实践，我深感各类表面工程技术在维修、制造领域的重要地位。20世纪80年代初期，英国伯明翰大学T. Bell教授建立了世界上第一个表面工程研究所，1985年创办国际《表面工程》杂志，同时将国际热处理联合会改名为国际热处理与表面工程联合会。我们根据多年的研究和实践，在1986年提出了具有中国特色的表面工程学科的设想，在中国机械工程学会和同行专家的支持下，于1987年建立了我国第一个学会性质的表面工程研究所，1988年创办了我国第一本《表面工程》杂志，1997年批准为向国内外正式发行的《中国表面工程》，受到国内外同行的重视和欢迎。1993年正式成立中国机械工程学会表面工程分会。从1988年至今已召开了九届全国表面工程学术会议。中国机械工程学会于1997年11月在上海举办首届国际表面工程学术会议。根据我军装备维修现代化发展的需要，我们于1991年建立了全军表面工程推广网，军委刘华清副主席亲笔为中心题词：表面工程出效益出战斗力。他的题词高度概括了表面工程的地位、作用和发展方向。1996年批准建立全军装备表面工程重点实验室。表面工程在我国的发展，从维修表面技术开始，逐步发展到成为系统的综合性的表面工程学科，并成为先进制造技术的组成部分，在国民经济和社会生活的各个领域发挥了日益重要的作用。这一特点生动地体现了科学技术是第一生产力的关键作用。

进入21世纪，针对我国资源能源匮乏、环境污染较重的问题，我提出了“尺寸恢复和性能提升”的中国特色再制造模式，在许许多多以发展我国再制造事业为己任的政府领导、专家学者、企业家及工程技术人员的共同支持与帮助下，再制造逐渐被大众所认同，被国家所接受，现已被写入代表国家意志的国务院文件及国家法律。2005 年 6 月国务院有关文件中指出：国家将“支持废旧机电产品再制造”，并把“绿色再制造技术”列为“国务院有关部门和地方各级人民政府要加大经费支持力度的关键、共性项目之一”；2009 年 1 月颁布生效的《中华人民共和国循环经济促进法》在第 2 条、第 40 条和第 56 条中六次阐述再制造，明确表示国家大力支持再制造。2009 年 4 月，我受邀向中共中央政治局常委、国务院副总理李克强汇报再制造产业的发展现状与对策建议，受到李克强副总理的高度重视。2009 年 12 月，我和中国工程院时任院长徐匡迪联名起草的中国工程院建议书——《我国再制造产业发展现状与对策建议的报告》得到温家宝总理的高度评价，他做出重要批示：“再制造产业非常重要。它不仅关系到循环经济的发展，而且关系到扩大内需（如家电、汽车以旧换新）和环境保护。再制造产业链条长，涉及政策、法规、标准、技术和组织，是一项比较复杂的系统工程。工程院的建议请发改委会同工信部、商务部、财政部等有关部门认真研究并提出意见”。到 2013 年底，我国已经在法律、行政法规和部门规章等不同层面上制定了 30 余项法律法规，国家发展和改革委员会与工业和信息化部发布的再制造试点单位已有 77 家，涵盖工程机械、矿山机械、石油机械、轨道车辆、办公设备、机床、内燃机、汽车零部件等，再制造产品达到 70 余种，我国再制造产业的发展已经初具规模。

目前，再制造已成为我国构建循环经济、建设“资源节约型、环境友好型”社会、创建创新型国家、实施节能减排战略，以及推动我军新时期装备建设、装备保障的重要技术手段。

回顾走过的路，我深感：没有国就没有家，更没有自己的生存和发展。只有时刻想着部队和国家建设，群策群力，不为困难压倒，不为挫折屈服，不为名利诱惑，奋勇攀登，不断进取，才能有所作为和贡献。无论干哪一行，只要人民需要，国家需要，就要深钻进去，前途将会无限广阔。

徐滨士

2014 年 4 月

编者说明

中国工程院院士徐滨士是我国表面工程学科和再制造工程学科的倡导者和开拓者，在国内外学术界和产业界享有盛誉。在装备维修工程、材料表面工程、再制造工程学科方面的研究取得了一系列重大突破，开创了一脉相承、特色鲜明的新型学科发展之路，培养了一大批优秀科技人才，构建了国家战略性新兴产业，为国家和军队现代化建设做出了重要贡献。

徐滨士是新中国培养的第一代大学生，他参军执教60年来，始终把党和国家、军队的利益放在第一位，把自己的青春、智慧和心血无私地献给了我国教育科研事业，他积极策划构建学科平台，坚持以学科发展引领教学和科研，20世纪70年代末提出并发展了我国的维修工程学科，80年代中期率先倡导在国内建立表面工程学科，90年代发展了具有中国特色的再制造工程，开创了一脉相承、特色鲜明的学科发展之路。

徐滨士作为人民教师，正心修身，笃学尚行，授业解惑，诲人不倦，用自己特有的人格魅力熏陶着一代代学生，培养了一批批优秀人才，成就了一支支教学团队，构建了一个个科研创新群体，从20世纪80年代起共培养了百余名博士、硕士研究生。他先后创建了国家级重点实验室、国家工程研究中心和全军重点实验室，为广大教员进行科研实践，开展学术交流创造了良好的研究平台。

徐滨士主持国家和军队百余项科研任务，从最初攻克装备零部件的修复难题，到研究维修表面技术，再到开创具有中国特色的装备再制造工程，始终坚持通过实践积累知识和经验，在实践的基础上理性思考，不断向更深更高的层次迈进，在平凡的岗位上做出了不平凡的成绩，他以科学家的远见和气魄把个人的学术思想转化为国家战略，以创新成果为武器装备发展和国家经济建设服务。他共获得国家级科技奖励10项，军队及省部级科技奖项19项，出版专著20余部，获得授权国家专利50余项，发表学术论文千余篇，先后荣获“何梁何利”基金技术科学奖、“光华科技工程奖”、国际热处理与表面工程联合会“最高学术成就奖”、中国机械工程学会“科技成就奖”、中国焊接学会“终身成就奖”、中国表面工程学会“最高成就奖”、中国摩擦学学会“最高成就奖”、中国当代发明家等荣誉称号。

徐滨士虽已年逾八十高龄，仍坚持战斗在教学科研岗位上，呕心沥血，奋斗不息，立足本职，辛勤耕耘，紧跟时代，拼搏创新，为国家循环经济发展和武器装备建设操劳奔波，始终没有停歇研究探索的脚步。

徐滨士院士的精神是我们不断前进的强大动力，为弘扬徐滨士院士的学术思想，展示先生坚持不懈的创新精神，激励青年学子，我们摘选先生的部分论文集结成《徐滨士文集》以供学术交流与参考。

谨以此书向中国工程院20周年院庆表示祝贺！

本书编委会

2014年4月

目 录

表面工程

再制造工程设计及质量控制

再制造关键技术

再制造工程管理

附　录

表面工程

中国表面工程的发展*

摘　要　表面工程是由多个学科交叉、综合、发展起来的新兴学科，是综合运用各种表面技术提高材料表面性能的系统工程，其研究和推广应用将有力地推动我国经济，特别是先进制造技术和高新技术的发展。

关键词　表面工程　表面技术　应用　发展

1　表面工程的发展日新月异、应用广泛

表面工程，就是经表面预处理后，通过表面涂覆、表面改性或多种表面技术复合处理，改变金属表面或非金属表面的形态、化学成分和组织结构，以获得所需要表面性能的系统工程[1]。表面工程的概念始于20世纪80年代，近年来，它的发展异常迅速，作为体现表面工程学科重要组成部分的表面技术，已成为当今世界的关键技术之一，中国机械工程学会1987年建立了学会性质的表面工程研究所，1988年出版了《表面工程》杂志（后改为《中国表面工程》），现已出版30期。1989年召开了第一届全国表面工程学术交流会，1991年召开了第二届全国表面工程学术会，同时举办了首届中日表面工程学术研讨会，1993年在日本召开了第二届中日表面工程学术会议。1993年成立了中国机械工程学会表面工程分会，1995年在北京召开了国际表面科学与工程学术会议，1996年7月召开第三届全国表面工程学术会。

材料的性能往往通过表面表现出来。材料的破坏也往往自表面开始，诸如磨损、腐蚀、高温氧化等，改善材料的表面性能，会大幅度地提高材料的性能，有效地延长其使用寿命，节约资源，提高生产力，减少环境污染。表面工程的最大优势是能够以多种方法制备出优于本体材料性能的表面功能薄层，该薄层厚度一般从几十微米到几毫米，仅占工件整体厚度的几百分之一到几十分之一，却使工件具有了比本体材料更高的耐磨性、抗腐蚀性和耐高温性等。

表面工程综合了材料科学、冶金学、机械学、电子学、物理学、化学等领域的基础理论、技术和最新成果，它的发展不仅在学术上丰富了上述学科，而且开辟了新的研究领域。如高能束冶金学、等离子体物理学、动态金属学、摩擦化学、纳米摩擦学等，并且以其优质、高效、低耗等特点在国民经济和社会生活的各个领域发挥着重要作用[2,3]。

（1）化学：催化——贵重金属催化剂以薄膜的形式沉积在反应媒体上；耐用传感器——传感器的发展与薄膜技术的发展密切相关。

（2）建筑：轻型材料——通过应用涂层技术来满足对材料的耐腐蚀与磨损的要求。

（3）金融服务：信用卡——薄膜技术将能满足信用卡发展的需要。

---

* 本文合作者：马世宁、时小军、朱有利。原发表于《中国机械工程》，1996，7(5)：243～248。

（4）健康与生命科学：药剂释放系统，例如，通过涂层来控制药剂的慢速释放。

（5）运输：燃料效率，低排放动力工厂，能量回收系统——耐腐蚀及磨损涂层；精确定位系统——传感器；急救系统——传感器。

（6）通讯、数据传输及电子仪器：技术支持——薄膜技术。

（7）制造、生产及商业过程：改进加工设备的效率——表面技术；先进传感器的发展——薄膜技术，有效利用新材料的工艺改进。

（8）农业、自然资源和环境：遥感——传感器。

（9）能源：光电能时代——薄膜技术；"净煤"能时代——耐腐蚀及磨损涂层；更高的效率——耐高温腐蚀涂层；低排放动力装置——发动机设计及燃料质量传感器——耐腐蚀和磨损涂层及薄膜技术；更高效率的建筑，例如反红外涂层。

（10）零售及销售：智能包装——传感器。

（11）材料：表面工程——一个优先发展的重要领域，有利于环境的加工技术——无污染的表面加工技术；生物材料——用于移植的涂层技术；特殊应用的高温材料——耐腐蚀涂层；特殊应用的减重材料技术——耐腐蚀、磨损涂层；用于高温超导的加工技术——薄膜技术。

目前，表面技术的发展异常迅速，制造业普遍意识到表面工程将是未来制造技术的一个重大举措，是通过不断符合环境要求的系统设计而获得稳定增长利润的重要手段，加强与表面工程科研机构的合作极为重要。采用表面技术的费用一般虽然只占产品价格的5%～10%，却可以大幅度提高产品的性能及附加值（见表1），从而获得更高的利润[4]，采用表面工程措施平均可提高效益高达5～20倍以上。根据英国科技开发中心的调查报告，英国主要依靠表面工程而获得的产值每年超过50亿～100亿英镑，其他工业国家的情况也基本相当[4]。我国表面工程的研究与应用多从维修入手，并已逐步扩展到了新设备与新产品的设计和制造过程中，经过十几年的发展，在研究水平与规模方面，与国际水平相比也不逊色，并有自己的独创和特色，在重大工程中的部分应用已达到国际先进水平。采用表面工程在设备制造与维修中取得重大经济效益的实例不胜枚举，在节能、节材、降耗和提高经济效益方面发挥了巨大的作用，仅国家重点推广的热喷涂和刷镀两项表面技术在"六五"和"七五"期间已创经济效益32亿元以上。在国家的节能、节材"九五"规划中，拟建议将表面工程作为重大措施之一，并列为节能、节材示范项目。材料表面改性作为传统材料性能优化的基础研究也被列入国家自然科学基金"九五"优先资助领域。

表1　表面处理提高工具性能的实例[5]

应用实例	方法	寿命提高倍数
钻头	物理气相沉积	6～7
拉丝模	化学气相沉积	75
线切割高速钢	离子注入	9

表面工程不仅被用于一般设备零部件的防护、强化和修复，而且还为高新技术的发展提供了工艺及材料支持。例如：微电子工业就是以电镀、化学镀、离子注入、气相沉积等表面技术作为技术基础的。在制备高TC超导膜、金刚石膜、纳米多层粉和纳

米晶体材料、多孔硅、碳 60 等新材料中表面技术起到了关键作用。

2 表面工程学科体系与特色

2.1 表面工程的产生

表面工程的形成和发展，与其在生产中的重要作用是分不开的，它是发展生产，提高产品质量和经济效益的需要；它为高新技术提供了特殊的材料；它是设备技术改造与维修的有效手段；它是节约能源和资源的重要途径；它还是装饰与美化人民生活的得力措施。正是由于这些重要作用，表面工程发展成为日益受人瞩目的新兴技术领域。

2.2 表面工程的学科特色

表面工程是由多个学科交叉、综合、发展起来的新兴学科，它以“表面”为研究对象，在应用与综合有关学科理论的基础上，根据零件表面的失效机制，以应用各种表面技术及其复合表面技术为特色，逐步形成了与其他学科密切相关的表面工程基础理论，主要有：表面层失效分析理论；表面摩擦与磨损理论；表面腐蚀与防护理论；表面（界面）结合与复合理论等。表面工程的发展不仅在学术上丰富了材料科学、冶金学、机械学、电子学等学科，而且开辟了新的研究领域，如高能束冶金学、等离子体物理学、动态金属学、摩擦化学、纳米摩擦学等。

我们将表面工程的学科体系概括在图 1 中。

按照作用原理[5]，表面技术分 4 种基本类型：

（1）原子沉积：沉积物以原子、离子、分子和粒子基团等原子尺度为粒子形态在材料表面上沉积形成外加覆盖层，如电镀、物理气相沉积、化学气相沉积。

（2）颗粒沉积：沉积物以宏观尺度的颗粒形态在材料表面上形成覆盖层，如热喷涂等。

（3）整体覆盖：如包箔、贴片、热浸镀、涂刷、堆焊。

（4）改性处理：如磷化、离子注入、离子渗、扩散渗、激光表面处理。

2.3 表面工程的技术特色

单一的表面技术由于其局限性，往往不能满足日益苛刻的工况要求。随着科学技术和设备的进步，发展出了综合运用两种或多种表面技术的复合表面技术，或称为第二代表面技术。目前，表面技术的复合技术的研究和应用已取得了重大进展，如热喷涂与激光重熔的复合、热喷涂与刷镀的复合、化学热处理与电镀的复合、表面强化与喷丸强化的复合、表面强化与固体润滑层的复合、多层薄膜技术的复合、金属材料基体与非金属材料涂层的复合等。实践证明，复合技术使本体材料的表面薄层具有更加卓越的性能，解决了一系列高新技术发展中特殊的工程技术问题。

2.4 表面工程的技术设计

包括表面技术与涂层材料的科学合理选用，表面层成分、组织结构及力学性能的确定，工艺参数选择，以及各种表面层的性能检测方法等。

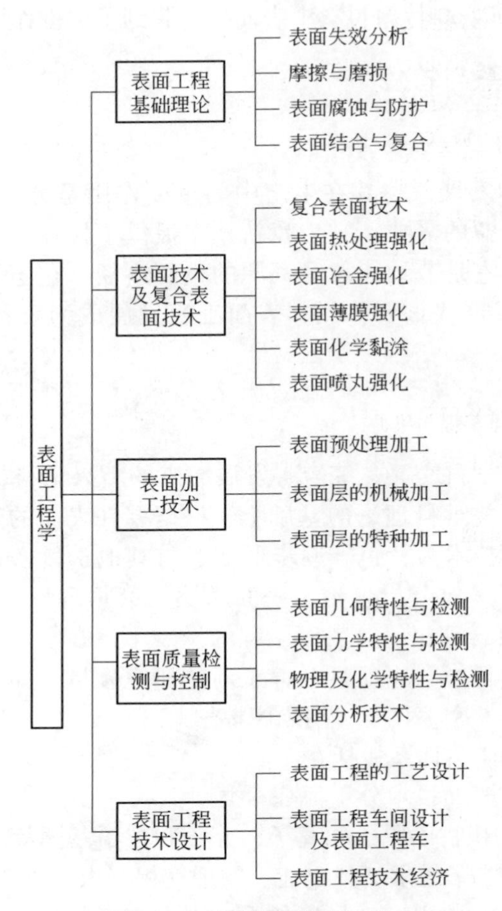

图1　表面工程的学科体系

3　开拓表面工程，推动国民经济发展

目前，许多发达国家都在努力研究和应用各种提高零件表面性能的新技术、新工艺，使得诸多表面技术成为了现代制造技术和高新技术中的重要工艺方法，并且在设备的技术改造和维修方面发挥了重要作用。作为一个发展中国家，我们的工业水平，特别是制造业的水平还比较落后，发展表面工程，研究开发和推广应用表面新技术，使其迅速向现实生产力转化，并推动产品制造和设备维修技术的进步，为促进我国的工业生产由粗放型向集约型转变做出贡献。为此提出如下建议：

（1）加强表面工程学科的建设，深入开展表面工程基础理论的研究。

（2）重点研究和发展复合表面技术，充分发挥各种工艺和材料的综合优势，以适应制造业对产品性能越来越高的要求。

（3）大力开发维修表面技术的应用及研究，为其在设计新设备和制造新产品过程中的应用积累经验。

（4）加强表面工程在设备制造和新产品制造过程中的应用与研究，特别是在提高产品的质量、品种、性能、效益等方面发挥作用。

（5）逐步创立表面工程技术设计，全面、系统、综合应用、优化各种表面技术，以提高产品质量和效益为中心，进行产品的表面工程技术设计。

（6）重视表面工程设备的研制工作，不断提高表面工程的装备水平。

（7）建立国家表面工程研究中心、产学研中心，以加强表面工程的研究开发与推广应用，加速实验室技术向工业应用的转化，为企业提供表面工程方面的技术、设备及建议等，以推动企业的发展。

（8）努力办好《中国表面工程》杂志，开展表面工程的学术交流，积极开拓表面工程的国际交流。

参考文献

[1] 徐滨士，马世宁．表面工程与维修．北京：机械工业出版社，1996.

[2] 徐滨士，马世宁．表面工程及其未来发展．机械热加工科学的未来论文集，国家自然科学基金委员会，1988.

[3] Bell T. Technology Foresight and Surface Engineering Sector. Surface Engineering，1995，11(3)：177 ~ 178.

[4] Bell T. Design Awareness in Surface Engineering：Surface Engineering，1993，9(4)：247 ~ 248.

[5] Hondros E D. Surface Engineering – The New Challenge in Materials Technologies. Surface Engineering Volume Ⅰ（Fundamentals of Coatings），1993(1)：1 ~ 9.

National Technology Strategy and Surface Engineering in China*

Abstract During the celebration of the sixtieth anniversary of the Chinese Mechanical Engineering Society held in Beijing in September 1996, I was present as an honorary member of CMES to hear the society honour Professor Xu Binshi, Director of the Surface Engineering Research Institute in China. In this guest editorial, Professor Xu has abstracted his award acceptance speech to the 7th National Assembly marking the CMES sixtieth anniversary and his technical paper on the Development of Surface Engineering in China.

Key words surface engineering, strategy

At the turn of the century, all countries in the world are busily engaged in the restructuring of their scientific and economic strategies so as to strengthen their economic strategies so as to strengthen their economic and scientific position. In accordance with the situation in China, the Party and State Council have drawn up a plan to invigorate China through science and education, which means that science and technology are the primary production forces and that science and education should be put at the forefront of economic and scientific development, so as to speed up the increase of prosperity in China. The technical level and scale of its machine building industry are important indicators of a country's overall economic strength. Rapid developments in the fields of communications, electronics, energy, and materials have put higher demands on the machine building sector of manufacturing industry. The Ministry of Machine Building has set three major goals over the next 5 year period:

(1) To improve and control the quality of engineering components;

(2) To optimize manufacturing methods;

(3) To enhance the national capacity for the design of new products.

These goals fully recognize both the importance of surface engineering as the next big step forward in manufacturing technology and the crucial nature of surface engineering in the design of engineering components. In the area of international exchange, CMES, in accordance with the further development of the machine - building sector in China, has signed cooperative agreements with a number of international scientific and academic organizations, including the International Federation for Heat Treatment and Surface Engineering (IFHT). The IFHT was, in fact, the first international organization to hold a scientific congress in China, in 1983, since the initiation of the "open door" policy by Deng Xiaoping in 1978, thus promoting a high level of recognition of surface engineering and related technologies of heat treatment and tribology in

* Reprinted from *Surface Engineering*, 1996, 12(4): 265 - 266.

Chinese manufacturing industry. With the development of a modern manufacturing industry, the demands on the properties of a component's surface are becoming more and more severe, especially for those components running at high speed, at elevated temperature, under heavy loads, in corrosive environments, etc. A local surface failure usually causes catastrophic failure of the component, which, in turn, may eventually stop the equipment itself running. The improvements in surface properties of materials, through surface engineering design, can effectively extend the surface life of components, reduce overall energy consumption, increase productivity, and reduce environmental pollution. It has been shown in China that, in many situations, surface engineering allows 5-20 μm its of economic benefit to be achieved through preproduction of components. Over the past 5 years it has been reliably estimated that more than £ 100 of economic benefit has been obtained by using surface engineering technologies for the manufacturing and renovation of machine parts in China. The level of this economic benefit to China is expected to 490 grow rapidly as propagation of the discipline of surface engineering increases over the next 5 years. In recognizing surface engineering as one of the most important measures for energy and materials conservation, the State Planning Committee is now considering the formation of a National Surface Engineering Research Center and surface engineering will receive a preferential subsidy from the National Natural Science Foundation.

References

[1] T. BELL. J. Surf. Eng. ,1989,(5):1 -2(in Chinese).

[2] Investigation report on the perspectives of energy and material saving technologies and major technical innovations in the 9th Five Year Plan, Beijing, 1995:72.

[3] Preferential subsidising projects of the China National Science Foundation in the 5th Five Year Plan, Beijing, 1996:54.

纳米表面工程*

摘　要　阐述纳米表面工程的产生背景、内涵和科学问题，介绍纳米表面工程相关材料的特性和制备方法，总结纳米表面工程在表面精加工、制备功能涂层、功能复合镀层、功能薄膜材料、热喷涂加工和抗磨减摩领域的应用现状。

关键词　纳米表面工程　表面工程　纳米材料　纳米表面工程应用

随着纳米科技的发展，微机电系统的设计、制造日益增多，制造技术与加工技术已由亚微米层次进入到原子、分子级的纳米层次。如日本已研制成功直径只有1～2mm的静电发动机、米粒大小的汽车。美国已研制成功微型光调器，并计划研制微电机化坦克、纳米航天飞机、微型机器人。在光电子领域，日本NEC公司在GaAs基体表面上，利用分子外延技术，把所需的原子喷射到一块半导体表面上，形成特定的岛状晶体而成功制作出具有开关功能的量子点阵列。美国已制造出可容纳单个电子的量子点，而量子点小到可在一个针尖上容纳几亿个。这些技术都是在特定表面上实现的，属表面工程范畴。但随着尺度的减少，表面积与体积之比相对增大，表面效应增强，表面影响加大，传统的表面设计和加工方法已不再适用。为适应纳米科技发展带来的变化，需建立与之相适应的表面工程——纳米表面工程。

1　纳米表面工程的内涵

纳米表面工程是以纳米材料和其他低维非平衡材料为基础，通过特定的加工技术、加工手段，对固体表面进行强化、改性、超精细加工，或赋予表面新功能的系统工程。纳米表面工程是在纳米科技产生和发展的背景下，对固体表面性能、功能和加工精度要求越来越高的条件下产生的。因纳米表面工程以具有许多特质的低维非平衡材料为基础，它的研究和发展将产生具有力、热、声、光、电、磁等性能的许多低维度、小尺寸、功能化表面。与传统表面工程相比，纳米表面工程取决于基体性能和功能的因素被弱化，表面处理、改性和加工的自由度扩大，表面加工技术的作用将更加突出。

2　纳米表面工程的科学问题

纳米表面工程的主要科学问题有两个：（1）材料的表面改性、界面及非平衡条件下低维材料的结构和行为。如纳米等低维非平衡材料结构的形成演化及表征，以及对结构、物理性能、化学性能、力学性能等基本问题进行深入研究，有助于材料表面的优化设计和加工控制。（2）宏观、介观和微观的一体化研究，从而揭示出两个新的科学问题：一是“尺度问题”，即怎么进行不同尺度层次——宏观、介观及微观下的过渡

* 本文合作者：欧忠文、马世宁、乔玉林、张伟。原发表于《中国机械工程》，2000(6)：707～712。

及其相应的内在联系，如体相材料表面原子排布对单晶格、超晶格和纳米超薄膜的生长、力学性能等有何影响；二是“群体演化问题”，即如何描述介观、微观结构和缺陷作为群体所体现的交互作用和演化问题。

3　低维材料是现代表面工程研究和发展的基础

低维材料包括薄膜材料和纳米材料，薄膜材料可分为表面工程意义上的薄膜、纳米超薄膜和原子尺度上的薄膜。纳米表面工程中用得较多的是纳米超薄膜。纳米超薄膜实际上也是二维纳米材料。

纳米粒子的表面效应使界面杂质浓度大大降低，从而改善了材料的力学性能。同一材料，当尺寸减小到纳米级时，由于位错的滑动受到限制，表现出比体相材料高得多的硬度，其强度和硬度可提高4～5倍。如n-Fe晶断裂强度比普通铁高12倍；纳米碳管[1,2]的密度仅为钢的1/6，但其强度比钢高100倍，杨氏模量估计可高达5TPa，这是目前可制备的具有最高比强度的材料。研究发现，骨、牙、珍珠和贝壳之所以具有很高的强度，是因为它们是由纳米羟基磷酸钙、纳米磷酸三钙与少量的生物高分子复合组装而成的[3]。

纳米材料界面量大，界面原子排列混乱，原子在外力作用下产生变形时，很容易迁移、扩散，表现出甚佳的塑性、韧性、延展性和比粗晶高10^{16}～10^{19}倍的扩散系数。如28nm的n-20Ni-P在280℃时的伸长极限比257nm的20Ni-P高3.7倍；n-CuF_2和n-TiO_2室温下的塑性变性也有类似现象。n-CaF_2在80～180℃范围内可产生100%的变形，且在弯曲时，材料表面的裂纹可不扩大。这些高强度、高塑性甚至超塑性的纳米材料对材料表面改性具有特殊意义。

小尺寸效应使纳米材料的热容和散射率比同类其他材料大，其熔点和烧结温度显著下降，使在常温和低温条件下加工陶瓷和合金成为可能。如2nm的Au熔点仅为330℃，比通常Au的熔点低700℃，而n-Ag熔点竟低至100℃，而在钨颗粒中附加0.1%～0.5%的n-Ni可使烧结温度降近2000℃。这些特性将为传统材料表面的合金化改造和陶瓷功能化改造带来新机遇。

另外，由于纳米材料的电磁性能发生改变及具有极高的光吸收率（大于99%），可用于制作红外敏感元件、雷达波纳米吸收涂层等，这在军事上具有特殊的应用前景。纳米材料在力、热、声、光、电、磁等方面的许多特性为纳米表面工程技术的发展奠定了基础。

4　低维材料的制备

传统材料表面的低维化材料生长、组装，以及利用低维化材料对传统材料进行表面超精加工是纳米表面工程的主体技术。

4.1　零维纳米质点（0-D）的制备

0-D的制备方法有固相法、气相法[3]和液相法。固相法可分为物理粉碎法、机械球磨法[4]、固相化学反应法[5～7]和爆炸法等。机械球磨法是将金属粉体或非晶态金属箔膜置于高能球磨机或行星式球磨机中，在惰性气体保护下通过研磨来制备0-D。此法制备的0-D一般粒径较大，但易批量生产。固相反应法是将两种或两种以上的固体在研磨过程中发生相间化学反应制备0-D。气相法可用于0-D的制备，但更多的是用于2D的制备。液相法是制备0-D普遍采用的方法，其关键是防止粒子团聚（特别是硬团聚）和粒度控制。团聚常发生在干燥过程中，防止团聚的措施有溶剂置换[8]、共沸蒸馏、冷冻干燥、喷射干燥、超临界干燥等。为达到控制0-D粒度的目的，最重要的是控制粒子成核率和生长率，为获得小粒度的0-D，应增加成核、控制生长。后期干燥温度对粒度影响也较大。一般而言，适当降低干燥温度，有利于获得小尺寸的0-D。另外，在产生原级粒子的溶液中加入适当的表面活性剂或聚合物，对获得团聚少、粒度适当、分布较窄的0-D是有利的。固相法制备中加入表面活性剂或介质对获得小粒径0-D也有好处。

4.2 二维纳米薄膜（2-D）的制备

4.2.1 单层纳米薄膜的制备

2-D的制备常采用PVD和CVD。PVD是在真空或超高真空下将金属加热气化，通过气态原子之间或与作保护气的惰性原子之间碰撞，并被捕集于基材上来制备2-D。PVD包括高频溅射（RFS）、离子束溅射（IBS）[9]、分子束外延（MBE）、原子层外延（ALE）、迁移增长（MEE）、离子束混合沉积（IBM）、离子束增强沉积（IBED）、离子束辅助沉积（IBAD）、离子束支持沉积（IAD）、电子束蒸发（EB）等。CVD将金属氯化物、醇盐、羰基化合物等挥发性化合物的蒸气在气相中进行热分解或水解，直接制得或与氧、氨、甲烷等气体反应制得2-D。CVD包括金属有机物的化学气相沉积（MOCVD）和金属有机物等离子体化学气相沉积（MOPCVD）[10]等。

PVD、CVD法工艺复杂、成本昂贵，不宜大面积制备。近两年，电沉积法[11]因设备简单、工艺成熟而逐渐受到重视。经过发展，利用电沉积技术已能制备厚度小于1nm的2-D。Sol-Gel法因化学计量比易控、成分均匀、成膜面积大等优点亦广泛用于2-D的制备。用模仿动物牙齿和骨骼生长的方法、激光剥蚀和真空气量控制的方法也可制作2-D。将单晶硅片上的聚酰亚胺LB薄膜真空热解可制得β-SiC单晶超薄膜。

现在制备的n-2D有Ti(N，C，CN，N)、(V，Al，Nb)N、Al_2O_3、SiC及Cu、Ni、Al、Ag、Au、金刚石等。其中TN、Al_2O_3、TiC是较典型的超硬膜，其显微硬度HV分别为1950、3000、3200，抗磨顺序是TiC > TiCN > TN > Al_2O_3。这些膜在微机械、微电子领域制作耐磨、耐腐蚀涂层及其他功能涂层将会有重要应用。

C_3N_4是另一有应用前景的硬质膜[12,13]。自LIU等首次用理论预测存在一种比金刚石还硬的碳氮化合物（β-C_3N_4）后，已可用多种方法合成它[14,15]，但现在还不能获得纯的β-C_3N_4晶体。在Si、Pt上合成的含有一定量α-C_3N_4的β-C_3N_4，其体积弹性模量可达349GPa。

4.2.2 纳米多层叠膜的制备[16~20]

叠层膜是广义上的金属超晶格，因二维表面上形成的特殊纳米界面的二元协同作用，表现出既不同于各组元也不同于均匀混合态薄膜的异常特性——超模量、超硬度现象、巨磁阻效应，其他独特的机械、电、光及磁学性能等，在表面改性、表面强化、表面功能化改造、表面超精度加工等领域极具潜力[21]。现在通过PVD、CVD和电沉积技术已制备出 Cu/Ni、Cu/Pd、Cu/Al、Ni/Mo、TN/VN、TiC /W、TiN/AN 等几十种纳米多层叠膜。

4.3 有序分子膜的制备

有序分子膜包括 LB（Langmuir-Blodgett）膜、SA（self-assembled monolayers）膜和 MD（molecular depositionfilm）膜。

LB 膜是将气液界面上的分子单层膜通过物理机械过程转移到固体基片上来构筑有序单层或多层膜的。膜中分子与基片仅靠物理吸附发生作用，同层分子间靠范德华力相连，因此膜的稳定性差。

SA 膜是让液相中的活性分子，通过固液界面具有反应活性的不同分子头尾基的化学吸附或化学反应，在基片上形成化学键连接、取向紧密排列的二维有序单层膜或多层膜。SA 膜比 LB 膜更稳定，具有更好的摩擦学应用前景。目前已研究过的 SA 膜有烷基硫醇、二硫化物在 Au、Ag、Cu 表面，羧酸在 Ag、Cu、Al、Fe、玻璃表面等。

MD 膜是利用有机或无机阴阳离子之间的静电相互作用为成膜驱动力，通过相反离子体系的交替分子沉积制备成的一种层状有序超薄膜。MD 膜及其技术已引起国际学术界的注意。目前，MD 膜的表面改性技术已成功应用于光化学、酶和纳米粒子的组装等领域[22,23]。

5 纳米表面工程应用现状

纳米表面工程是极具应用前景和市场潜力的。据德国科技部预测，到 2000 年材料表面的纳米薄膜器件组装和超精度加工的市场容量将达近 6000 亿美元。

5.1 超高精度表面加工

用分布很窄的 0-D 作磨光材料，可加工表面粗糙度（R_{max}）为 0. 1 ~ 1nm 的超光表面，如高级光学玻璃、晶体、宝石、金相表面等，其加工精度比传统的磨光加工提高了一个数量级。使用的抛光液是含 n-MoS_2、n-Al_2O_3、n-SO_2、n-Cr_2O_3 等纳米微粒的润滑油。据报道，英、美等国纳米抛光液的生产已商业化，其 $R_{max} \leq 2$nm。采用沉积超薄膜的方法也可加工超光表面。Takenka 等人用射频溅射技术，先在基体表面沉积一层可减少基体粗糙度、厚 3. 2nm 的 C 层作缓冲膜，然后在-20℃下沉积一层 2. 3nm 厚的 Ni，这样可以得到 $R_{max} \leq 0.25$nm 的、由纳米多层膜组成的超光表面。

表面超精加工的另一方面是对材料表面进行纳米尺度超微细图形加工，这是制备纳米结构和器件的关键。如 Sohn 等[24]用 AFM 针尖对 PMMA/MMA 超薄膜进行机械刻蚀，制得 40nm 宽的金属铬线；Magno 等[25]用 AFM 针尖直接在半导体上刻划出沟槽，最细可达宽 20nm、深 2nm。最著名的例子是用 SPM 针尖在镍金属表面用 35 个原子摆出的 IBM 字样。材料纳米尺度表面加工通常采用电场诱导局域物理化学变化和机械刻划等[26]。

5.2 制备功能涂料

纳米粒子添加的静电屏蔽材料比炭黑添加的静电屏蔽材料用于电器具有更好的静

电屏蔽作用。为了改善静电屏蔽涂料性能，日本松下公司已用 n-Fe_2O_3、n-TiO_2、n-Cr_2O_3、n-ZnO 等研制成功具有优良静电屏蔽作用的纳米涂料，且不同粒度、不同种类的纳米材料其颜色各不相同，因此可调配出不同颜色的静电屏蔽涂层，不会像炭黑屏蔽涂层那样颜色单一。

纳米粒子的光反射率低，加之其电磁性能的改变，可用于制造红外线吸收涂料和雷达波吸收涂料。美国花上亿美元研制了一项顶极绝密技术——纳米雷达吸波涂料，每辆坦克只需花5000美元就可获得涂层薄、吸波率高、吸收波带宽的隐身涂层，具有极高的军事利用价值。采用金属、铁氧体等纳米微粒与聚合物形成的0~3型复合涂层和采用多层结构的2~3型复合涂层，能吸收和衰减电磁波与声波，从而达到电磁隐身和声隐身的作用，这在潜艇等军事领域有广阔的应用前景。

0-D涂层还有促进成核作用，如用PVD、CVD生长金刚石薄膜时，在基体表面先沉积一层纳米金刚石（DNP）或巴基管作为金刚石成核的“晶粒”，从而有利于金刚石的快速成核，并提高薄膜质量。另外某些纳米微粒还有杀菌、阻燃、导电、绝缘等作用，用这些纳米粒子可制成仿生物涂料、阻燃涂料、导电涂料和绝缘涂料。

5.3 制作功能复合镀层

在镀液中加入0-D或1-D可形成纳米复合镀层。如在45钢上镀Ni-P-巴基管复合镀层，可极大地改善镀层的摩擦学性能。化学镀Ti-P镀层的磁盘基板表面若采用DNP复合镀后，可减少磨损50%。用来生产磁头和磁性记忆储存器磁膜的Co-P化学镀液中添加DNP形成复合镀层，其耐磨能力提高2~3倍。用于模具镀铬的DNP复合镀层，可使寿命延长，精度持久不变，长时间使用镀层光滑无裂纹。用于钻头镀铬的DNP复合镀层，寿命成倍提高。汽车、摩托车汽缸体（套）的铬-金刚石纳米复合镀，可使汽缸体寿命提高数倍，见表1。

表1　铬/金刚石复合镀层模具的使用测试结果

压制材料	铁及不锈钢粉末	无线电工业陶瓷粉末	塑料粉末
提高倍数	9~15	4~5	2~3

纳米材料还可用于耐高温的耐磨复合镀层。如将n-ZrO_2加入Ni-W-B非晶态复合镀层，可提高镀层在550~850℃的高温抗氧化性能，使镀层的耐蚀性提高2~3倍，耐磨性能和硬度也都明显提高。金刚石的导热性比金银高得多，DNP与金银形成复合镀层，能在保持金银良好导电性的同时，大大增强镀层的强度、耐磨性、导热性，可使电接触材料的寿命提高2倍以上。Ni、Ni-P、Cr镀层一般只能在400℃以下工作，钴基复合镀层，如Co-Cr_3C_2、Co-ZrB_2、Co-SiC的出现大大提高了高温耐磨性能。但采用Co-DNP复合镀更具明显优势，因能承受500℃以上的高温，机件使用寿命延长。若镀层采用短杆DNP，由于同镀层金属的接合面积大，摩擦时不易剥落，则效果更好。

将DNP和纳米陶瓷微粒为代表的、具有耐高温性能的纳米硬质粉应用于刷镀层，能较大幅度地提高电刷镀层的力学性能。笔者用电刷镀技术制备了含DNP的复合镀镍层。研究表明，DNP的弥散强化作用可有效改善镀层的生长，减少内应力，提高镀层的显微硬度，并使镀层在室温、高负荷下具有优良的抗疲劳、抗磨损性能，其耐磨性是纯镍镀层的4倍。在快速镀镍液中加入n-SiC，刷镀层的耐磨性和硬度均有较大幅度的提高，摩擦系数减小。加入的n-SiC微粒主要分布在镀层缺陷处和镀层中镍晶粒之间。

5.4 制作功能性薄膜材料

纳米单层膜、叠层膜具有许多特异性，在特定基材沉积、组装纳米超薄膜将会产生表面功能化的许多新材料，这对功能器件、微型机电产品的开发具有特别重要的意义，同时也是表面工程迈向先进制造技术和高新技术的重要方面。预计纳米薄膜将在以下领域得到应用，见表2。

表2 纳米超薄膜应用领域一览表

应用领域	具体产品
信息与通讯	磁性存储器（硬盘、磁带、软盘、读/写头）、光学存储器（CD存储器、磁-光存储器）、液晶显示
光学	制作功能性薄膜（反光镜、过滤器、防锈、X射线、HL激光器）
电子	开发电子元件（电子薄膜、绝缘薄膜、异质薄膜、衍射薄膜、聚合薄膜）
能源	太阳能电池、热敏绝缘体、大容量储能电池
测量与控制技术	制作传感器（气体传感器、生物传感器、pH值传感器、分离膜器）

另外，法国的汤姆逊公司正在利用纳米多层膜的巨磁阻效应开发用于汽车制动系统的产品。巨磁阻效应可使磁盘的磁记录密度增加许多倍，因而IBM公司和其他磁盘驱动器制造商正在生产巨磁阻磁头产品。利用巨磁阻纳米叠层膜存储芯在计算机开断时保持“记忆”的特性，制成了低噪声、快速、长寿命的MRAM[21]。为解决工具薄层硬度问题，德国Linkopin大学和美国西北大学成功地制作了二重交替的薄膜，如TiN/VN或TiN/NbN，尽管薄层厚度仅为5～10nm，但其硬度却超过了50GPa。在有机树脂光学镜头表面沉积10nm厚纳米薄膜，可改善其抗划痕能力和反射指数。

5.5 组装纳米结构涂层

与传统热喷涂涂层相比，纳米结构（ns）涂层在强度、韧性、抗蚀、耐磨、热障、抗热疲劳等方面会有显著改善，且一种涂层可同时具有上述多种性能。但0-D要用于热喷涂，需解决两个问题：一是0-D质量太小，不能直接喷涂；二是喷涂过程中怎样保证0-D不被烧结。解决第一个问题的办法是将0-D组装为可直接喷涂的ns喂料。研究表明，只要控制好条件，ns喂料在喷涂过程中，0-D是不会烧结的[27]，因为粒子从加热到接触冷基体的时间非常短暂，原子来不及扩散，0-D生长和氧化同时受限。组装ns喂料的方法主要有两种：机械法和溶液合成法[28]。Stutt和Xiao将n-Al_2O_3和n-TiO_2用超声波分散在溶液中后再加入黏结剂，制成浆状物，然后喷雾干燥，将0-D黏结重组为直径为15～100μm的ns喂料。喷涂过程中，颗粒从室温加热到1200℃，黏结剂被燃烧去除，从而在基体上得到由几十纳米到亚微米粒子构筑成的、涂层致密度为95%～98%的涂层。与商用粉末涂层相比，该涂层结合强度、抗磨能力均提高2～3倍[29]。Stewart等用ns喂料，采用HVOF方法也得到了抗磨性能和抗电化学腐蚀性能优异的纳米结构涂层[30,31]。Jiang 等[32]用机械碾磨法制备的 Ni，Inconel 718 和 316 不锈钢作 HVOF 的 ns 喂料，获得了相应的 ns 涂层，与普通涂层相比，其显微硬度分别提高 20%、60% 和 36%。

5.6 制作抗磨减摩润滑涂层

近年来，以零磨损、超滑[33]为目标的纳米粒子在表面改性技术、表面分子工程领域也取得了进展。如 DLC 膜、Ni-P 非晶膜和非晶碳膜等作为磁盘表面保护膜，以及利用 LB 膜技术组装有序分子润滑膜，获得了优异的减摩抗磨性能，使软磁盘表面每运行 10～100km 的磨损量小于一层原子，硬磁盘磨损率为零。通过研究纳米薄膜微观磨损特性，了解材料表面的物理化学状态[34]，优化薄膜设计、完善膜的制备方法，则纳米超薄膜，特别是硬度高、耐磨能力强的 BN、Al_2O_3、TiN、Si_3N_4 膜，具有超模量、超硬度的纳米叠层膜，以及 SA 膜、MD 膜等在摩擦学领域将有更大的实用价值。

6 结束语

笔者提出了建立纳米表面工程学科的必要性，并阐述了纳米表面工程产生的背景、内涵及所包含的科学问题。纳米表面工程的最终建立和被大家所承认尚需时日，但随着纳米科技的发展和在其他学科对其需要产生的动力推动下，纳米表面工程一定会迅速发展壮大。

参 考 文 献

[1] Iijima. Helical Micro tubes of Graphic Carbon. Nature，1991，354：56～58.

[2] Ruoff R S，Lorent D C. Mechanical and Thermal Properties of Carbon Nanotubes Carbon. 1995，33：95～98.

[3] Hahn H. Gas Synthesis of Nanocrystalline Materials. Nanostructured Mater，1996，19(1～8)：3～12.

[4] GeriA K，Julian C D，Gonzalez J M. Conductivity of Fe－SiO_2 Nanocomposite Materials Prepared by Ball Milling. J. Appl Phys，1994，76(10)：6573～6575.

[5] XN X Q，ZHAN G L M. Solid State Reaction of Coordination Compounds at Low Heating Temperature. J. Solid State Chem，1993，106：451～460.

[6] LANG J P，XIN X Q. Solid State Synthesis of Mo(W)－S C luster Compounds at Low Heating Temperature. J. Solid State Chem，1994，107：108～ 127.

[7] Treece R E，Gillan E G. Rap id Materials Synthesis by Solid State Metathesis Reactions. Mat Res Soc Symp Proc，1992，271：169～174.

[8] HU Z S，DONG J X，CHEN G X. Replacing Solvent Drying Technique for Nanometer Particle Preparation. J. Colloid Interface Sci，1998，208：367～372.

[9] LU T R，KUO C T，YANG J R. High Purity Nanocrystalline Carbon Nitride Films Prepare at Ambient Temperature by Ion Beam Sputtering. Surf and Coat Tech，1999，115：116～122.

[10] Taschner C H. Plasma－assisted Deposition of Hard Material Layers from Organom etallic Precursors. Surf and Coat Tech，1993，59：207～213.

[11] Ei－sherik，Abdelmounam M，Cheung，Cedric K S. Nanocrystalline Metals film. U S 5433797，1994.

[12] LUA Y，Cohen M L. Prediction of New Compressibility Solid Science，1989，254：841.

[13] Sjostrom H，Stafsrom M，Boman M. Superhard and Elastic Carbon Nitride Thin Films Having Fullerene-

like Microstructure. Phys Rev. Lett, 1995, 75(7): 1336 ~ 1339.

[14] NIU C, LU Y Z, Lieber C M. Experiment Realization of the Covalent Solid Carbon Nitride. Science, 1993, 261: 334 ~ 336.

[15] LID, CHU Z, CHENG S C. Synthesis of Superhard Carbon Nitride Composite Coating. Appl Phys Lett, 1995, 67(2): 203 ~ 205.

[16] LIOU S H, CHEN C L. GranularMetal Films as Recording Media. Applyphys Lett, 1998, 52(6): 512 ~ 514.

[17] WANG Y, XU H Z, ZHANG F Q. Study on Optical Properties of Reactive – Sputtering a – Si: H/a – Ge: H SuperLattices (in Chinese). J. Serniconductors, 1991, 12(12): 755 ~ 758.

[18] TestardiL R. Williens R H, Krause J T. Enhanced Elastic Model in Cu – Ni Film with Compositional Mediation. J. Apply Phys, 1981, 52(1): 510 ~ 511.

[19] Masil J, Leper I, Kolega M. Nanocrystalline and Nano composite CrCu and CrCu – N Films Prepared by Magnetron Sputtering. Surf and Coat Tech, 1999, 115: 32 ~ 37.

[20] Shin K K, DoveD B. Ti/TiN Hf/HF – N and W/W – N Multplayer Films with High Mechanical Hardnees. Appl Phys Lett, 1992, 61: 654 ~ 659.

[21] Falicov L M. Metallic Magnetic Super lattices Physics Today, 1992(10): 46 ~ 50.

[22] Lvov Y, Ariga K, Ichinose I. Assembly of Multi – component Protein Films by Means of Electrostatics Layer – by – layer Adsorption. J. Am Chem Soc, 1995, 117, 6117 ~ 6123.

[23] Komvopoulos K. Surface Engineering and Microtlibology for Microelectromechanical Systems. Wear, 1996, 200: 305 ~ 327.

[24] Sohn L L, Willett R L. Fabrication of Nano structures Using Atomic – microscope – based Lithography. Appl Phys Lett, 1995, 67(11): 1552 ~ 1554.

[25] Magno R, Bennett B R. Nano stuctures Patterns Written in Ⅲ – Ⅴ A Semiconductor by An Atomic Force Microscope. Appl Phys Lett, 1997, 70(14): 1855 ~ 1857.

[26] Mamin H J. Thermal Writing Using a Heated Atomic Force Microscope Tip. Appl Phys Lett, 1996, 69(3): 433 ~ 435.

[27] Rawers J C. Thermal Stability of Nano structured Metal Alloys. J. Thermal Spray Tech, 1998, 7(3): 427 ~ 428.

[28] Rawers J, Korth G. Microstructure of Attrition Ball – milled and Explosively CoMn Pacted Iron Powder Alloy. Nanostructured Mater, 1996, 7(1/2): 2 ~ 4.

[29] Stutt P R. XIAO. Thermal Spray of Nanostructured Alumina/Titania Coating with Improved Mechanical Properties. 99' Wuhan International Conference on Surface Engineering Wuhan, 1999: 142 ~ 146.

[30] XIAO T D, JIANG S, WANG D M. Thermal Spray of Nano – ceramic Coating for Improved Mechanical Properties. 12th Intl. Surface Modified Conf. Proc ASM Intl, 1998, 32 ~ 36.

[31] Stewart D A, Dent A H, Harris S J. Novel Engineering Coating with Nanocrystalline and Nanocomposite Structure by HVOF Spraying. J. Thermal Spray Technology, 1998, 7(3): 422 ~ 423.

[32] JIANG H G, LAU M L, Lavemia E J. Synthesis of Nano structured Engineering Coating by High Velocity oxygen Furel (HOVF) Thermal Spraying. J. of Thermal Spray Technology, 1998, 7(3): 412 ~ 413.

[33] Shinjo K. Dynamics of Friction: Super – lubric State Surface Science, 1993, 283: 473 ~ 478.

[34] Yoshizaw a H, Mcguiggan P, Israelachvili J. Identification of Second Dynamic State During Stickslip Motion. Science, 1997, 7(1): 1 ~ 6.

Nano Surface Engineering in the 21st Century*

Abstract Nano surface engineering is the new development of surface engineering, and is the typical representation that the advanced nano technology improves the traditional surface engineering. The connotation of nano surface engineering is profound. The initial stage of nano surface engineering is realized at present day. The key technologies of nano surface engineering are the support to the equipment remanufacturing. Today the relatively mature key technologies are: nano thermal spraying technology, nano electric-brush plating technology, nano self-repairing anti-friction technology and metal surface nanocrystallization, etc. Many scientific issues have been continuously discovered. Meanwhile they have been applied in the practice more and more, and have achieved the excellent remanufacturing effect.

Key words nano surface engineering, key technology, equipment, remanufacturing

Since the beginning of 1980's, surface engineering has continuously developed and been more and more perfect, becoming one of the key technologies in manufacturing and remanufacturing. Surface engineering is the important composition of the advanced manufacturing and remanufacturing, at mean time, it provides a technology support to them.

Surface engineering is a system engineering to obtain desired surface properties through surface coating, surface modification or duplex surface treatments on a pretreated metallic or non-metallic surface to alter its morphology, chemical composition, microstructure and stress condition[1]. The basic character of surface engineering is synthesis, intercross, compounding, and optimization. Surface engineering takes the "surface" as core, under the foundation of correlative subject theories, basing on the surface failure mechanisms, characterizing multi surface engineering technologies and their compounding, to create gradually the basic theory of surface engineering. The major advantage of surface engineering is that the surface functional coating, being superior to the substrate materials, can be prepared on the components by a lot of surface methods, to endow the properties such as temperature-resistance, anti-corrosion, wear-resistance and anti-fatigue with the components. Comparing with the substrate materials, the coatings are thin, small in area, but hand on the main shoulders of components.

By the end of 20th century, the nano materials, such as 0-dimension particles and 2-dimension films, have been applied perpetually in surface engineering[2,3]. Under the condition, "Nano Surface Engineering" is vividly portrayed. In June, 2000, Professor Xu binshi presented firstly the concept of nano surface engineering in Chinese Mechanical Engineering[4]. The conception

---

* Copartner: Wang Haidou, Dong Shiyun, Shi Peijing, Xu Yi. Reprinted from *Transactions of Materials and Heat Treatment*, 2004, 25(5): 8 - 12.

was rapidly acknowledged by professor T. Bell, the international founder of surface engineering, academician alien of Chinese Academy of Engineering, academician of Britain Royal Academy of Engineering and professor in Birmingham, and then a project named "Nano composite coating and composite surface engineering used in advanced car components" was rated as the Sino-Britain governmental collaborative science and technology project.

1 Nano Surface Engineering

Nano surface engineering is a system engineering that combining the nano materials and nano technology with traditional surface engineering, through particular processing techniques or methods to alter farther morphologies, compositions, microstructures of surfaces and impart them new properties much more.

The connotation of nano surface engineering is profound, at least five aspects should be included: first, the preparation of macro-coating possessing partial nano character. To the coating with macro size, when preparing, the nano materials, mainly pointing to nano particles with minor content, are added evenly into the coating. The coating actually is a composite coating that is strengthened by nano particles. Its properties of wear-resistance and anti-friction are further improved. Second, the preparation of nano functional films. To the functional film, with the decrease of device size, the thickness of film is thinner and thinner. When the thickness is down to several nanometer, the thickness of a single crystal is more than that of the film, namely a crystal grain runs through the whole film. At this condition, the performance of the film is determined by the performance of the crystal grain. Third, metal surface nanocrystallization. All kinds of metals, under common condition, were composed of crystal grains with micrometers or submicrometers. Through special processing methods, the size of crystal grains below the metal surface several centimeters will be changed from microscale to nanoscale. The physics and chemistry properties and processing performance of the metal surface will be enhanced remarkably. Fourth, precision processing within nanoscale on the surface of materials or films. To the surface operated at the nanoscale tolerance, it is very important for the precise processing and special processing on the surface, such as nanoscale rubbing, nanoscale etching, nanoscale nucleation, etc. Fifth, the preparation of macro-coating composed entirely of nano materials. The macro-coating is the most important coating form. Through the special methods, the coating with micrometers even millimeters thick is composed entirely nano crystal grains or nano materials. If that, the performance of the coating could be extremely excellent.

The order of the above five aspects of nano surface engineering is just the order of the difficult degree to carry out industrially. In view of the development state of the nano materials and surface engineering, the preceding three aspects of the nano surface engineering contents, namely "the preparation of macro-coating possessing the partial nano character", "the preparation of nano functional films" and "metal surface nanocrystallization", can be completed or have been achieved. They can be called as the primary stage of the nano surface engineering. Whereas the latter two aspects of the nano surface engineering contents, namely "precision processing within nanoscale on the surface of materials or films" and "the preparation of macro-coating composed entirely of nano materials", due to their difficulties in technology, their industrial application has been a long way to go. They can be called as the senior stage, and they are the aim to struggle. In the preceding three aspects, the second aspect is "the preparation of nano functional films", it has a relatively long distance to the engineering application with the characters of high velocity, heavy load and severe wear. So the emphases the paper discussed is the first aspect "the preparation of macro-coating possessing the partial nano character" and the third aspect "metal surface nanocrystallization".

2 The Key Technologies of Nano Surface Engineering

2.1 The preparation technology of macro-coating possessing the partial nano character

"The preparation of macro-coating possessing the partial nano character" actually points that when the coatings were prepared by the traditional surface engineering technologies, such as thermal spraying, electro-brush plating, lubrication additives, etc, some nano particles, for example the nano Al_2O_3 ceramic particles, metal Cu particles, and so on, were added into the materials, therefore the coatings possess certain nano components and characters, the properties of the coatings were improved obviously.

2.1.1 Nano thermal spraying technology

To the feed materials, namely the materials used to spray, their size in both nano thermal spraying and traditional thermal spraying are all within micrometers. Whereas the key difference is that the feed materials used in nano thermal spraying are the feed materials with micrometer reconstructed by mass of nano particles. Fig. 1 is the schematic diagram of the reconstruction of micro feed materials and nano thermal spraying[5]. The nano particles cannot be used directly in the thermal spraying due to their too small dimension and too light weight. When spraying, the nano particles will fly apart and burn. Fig. 2 is the Al_2O_3/TiO_2 composite micro feed particles after reconstruction to the nano Al_2O_3 and TiO_2 particles[6].

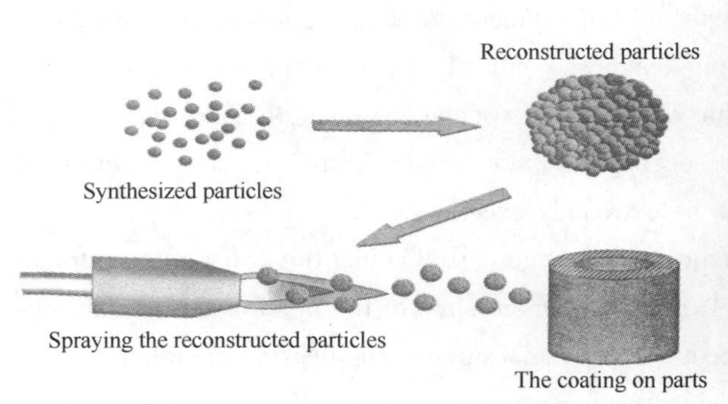

Fig. 1 Schematic diagram of reconstruction and nano thermal spraying

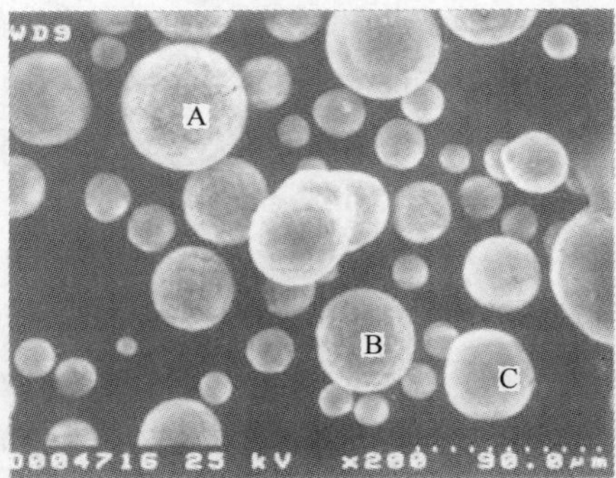

Fig. 2 Feed materials after reconstruction

The nano thermal spraying coatings have been employed by American Navy to repair the worn

components. In China, the laboratory studies have been completing and the great development will be achieved in the several years.

2.1.2 Nano electric-brush plating technology[7-9]

The nano electric-brush plating technology is the new electric-brush plating technology that the special nano ceramic particles are added into the plating bath, which mix evenly with matrix metal, and then the coatings with remarkable tribological properties are prepared.

Fig. 3a and b are the morphologies of the common nickel plating coating and nano plating coating added n-Al_2O_3 particles. The morphology of the later is more compact than that of the former, showing that the nano particles can refine evidently the microstructure of coating and increase numerously the grain boundary. Fig. 3c is the TEM morphology of nano plating coating. The phases pointed at by the arrows were n-Al_2O_3 particles, showing that the nano particles distributed diffusively in the coating and combined tightly with the matrix metal Ni.

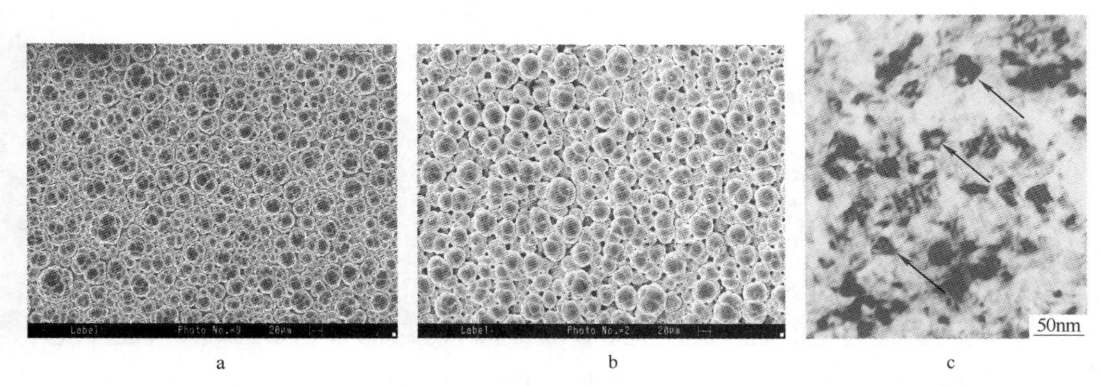

Fig. 3 Morphologies and microstructure of plating coating

a—Common nickel plating coating; b—Nano plating coating; c—TEM microstructure of b

Fig. 4 is the HRTEM morphologies of nano electric-brush plating coating, showing that the bonding between the nano Al_2O_3 particles and matrix metal Ni was compact. If the bonding between them was not the chemical bonding, there existed a great space in the interface in nanoscale. Combining with the results by XPS, it can be conclude that the bonding between nano particles and matrix metal is tight chemical one. The conclusion supports the theory foundation why the nano electri-brush plating coatings possess excellent wear-resistance and anti-fatigue.

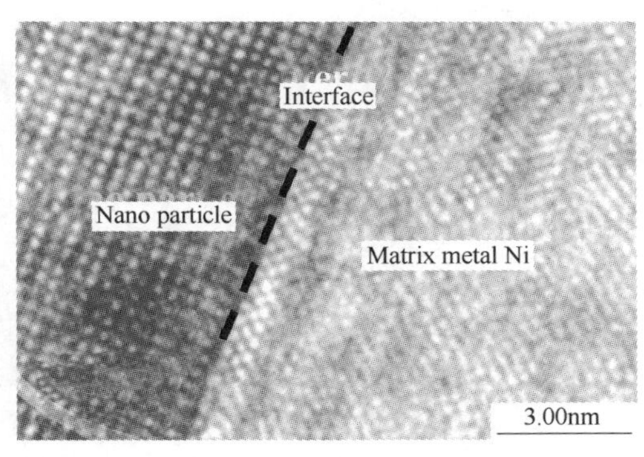

Fig. 4 HRTEM of the nano plating coating

The nano electric-brush plating technology has been utilized to repair the important components of vehicles and planes. By now the engine vanes of planes and the driving shafts of reducer of heavy-load vehicles have been repaired successfully.

2.1.3 Nano self-repairing anti-friction additives technology

The issues, such as anti-friction, wear-resistance and self-repairing, are the key ones to the precision frictional-parts. The additives technology is an important method to prolong the service life of counter-parts. The nano self-repairing anti-friction additives technology is a new in situ self-repairing technology. Once the additives containing nano particles, such as copper, were added into the lubrication oil, they were dispersed to contact surfaces of counterparts with the lubrication oil. Under the proper condition of temperature, press and friction force, the severe plastic deformation emerged. The nano particles could deposit on the frictional surfaces and interacted with them. When the temperature on friction surfaces increased to certain degree, the strength of nano particles decreased, and then formed the eutectic crystal with those grains on surfaces to produce a layer of repairing anti-friction films.

Our laboratory has developed the micro-level self-repairing anti-friction additives M3, with a good anti-friction effect. Based on it, the nano-level self-repairing anti-friction additives M6 has been developed successfully after the nano copper was added.

Fig. 5 is the wear variation of main journal with time after operation 300 hours in four engines under four kinds of lubrication conditions, namely M6, M3, MJ from abroad and common oil 50CC. It can be seen that the self-repairing effect of M6 was best. The wear-resistance of the additives was in the order of M6-MJ-M3-50CC[10].

M6 was used to repair successfully the cylinder-piston of automotive engines. The repaired piston was basically up to "zero wear". A compact, glossy golden yellow film could be seen on the piston surface after 300 hours test. Fig. 6 is the color elemental analysis result of worn piston surface. The red part is copper, showing the film comprising copper was surely formed.

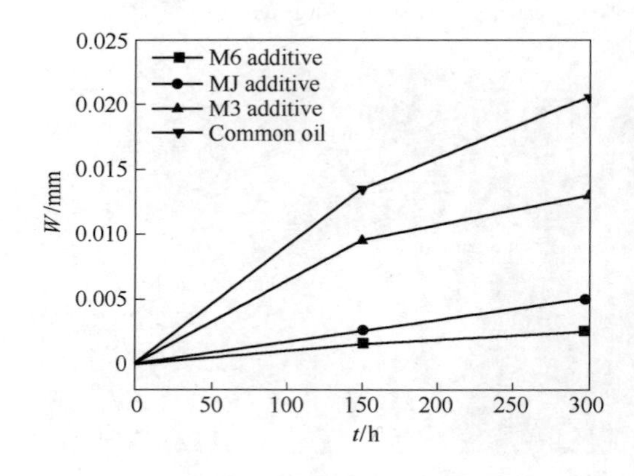

Fig. 5 Wear curves of main journal

Fig. 6 Worn piston morphology

2.2 Metal surface nanocrystallization

The metal surface nanocrystallization, simply speaking, it is a variation course of metal microstructure that under the condition of continuous impact, in metal surface, the crystal structure become gradually nano-crystal from coarse crystalline[11].

Fig. 7 is the cross-section morphology of pure iron after nanocrystallization for 60min[12], showing below surface 100μm, the microstructure changed greatly, a gradient deformation area emerged. According to the crystalline dimension, the deformation area can be divided into four parts: nano-crystal layer (0–15μm), sub-micro crystal layer (15–40μm), micro-crystal layer (40–60μm) and plastic deformation transition layer (60–110μm).

Fig. 8 is the TEM morphology of pure iron after nanocrystallization[13], showing the random equiaxial nano-crystals with dimension of 10–20nm were produced on the surface of pure iron.

Fig. 7 Cross-section morphology of pure iron after nanocrystallization for 60min

Fig. 8 TEM morphology of pure iron after nanocrystallization for 60min

The metal surface nanocrystallization changes the microstructure of materials surface. The high volume fraction formed in the interface of nanocrystals provides the ideal passage for the elements diffusion and makes it easy that the course of chemical heat treatment on the materials surface.

For example, when nitriding treatment was carried out on the pure iron, its temperature was up 500℃ and the time was over 20 hours. Whereas after surface nanocrystallization, the treatment temperature was only 300℃ and the time was only 9 hours[14]. When chromizing treatment, the diffusion performance of the chromium in pure iron after surface nanocrystallization was 9 magnitudes higher than that before nanocrystallization[15].

Due to its simpleness, low cost and universal applicability to all metals, the metal surface nanocrystallization technology can achieve the integrative design in the materials structure and performance, endowing the traditional metals with excellent performance.

3 Conclusions

Nano surface engineering is the combination among traditional surface engineering and advanced nano materials, nano technology, both reflects the inherent advantages of surface engineering, and expresses the elevated effect of nano technology, taking the wonderful remanufacturing effect with "1 + 1 > 2". Nano surface engineering technology will make great progress with the development of surface engineering and nano technology, and take the more important role in the field of remanufacturing and maintenance in the new century.

Acknowledgements

This research was granted by National Natural Science Foundation of China (50235030), "863" Project(2003AA331130) and "973" Project(G1999065009).

References

[1] Xu B S. Surface Engineering and Maintenance. Beijing: Mechanical Industry Press, 1996.

[2] Xu L T, Xu B S, Zhou M L. Study on the electric – brush plating composite coatings containing nano particles. China Surface Engineering. 1999, 12(3): 7 – 10.

[3] Shin K K, Dove D B. Ti/TiN, Hf/HF – N and W/W – N Multiplayer Films with High Mechanical Hardness. Applied Physics Letters. 1992, 61: 654 – 659.

[4] Xu B S, Ou Z W, Ma S N, etc. Nano surface engineering. Chinese Mechanical Engineering. 2000, 11(6): 707 – 712.

[5] Xu B S. Nano Surface Engineering. Beijing: Mechanical Industry Press, 2004.

[6] Liang X B, Deng Z C, Xu Y, et al. Preparation of the thermal spraying powder material and its coatings. Materials Engineering. 2001.

[7] Jiang B. Properties and strengthening mechanism of nano – particle reinforced nickel matrix composite coatings prepared by electric – brush plating: [Dissertation for Master Degree]. Chongqing: Chongqing University, 2003.

[8] Tu W Y, Xu B S, Jiang B, et al. Microstructure and co – deposition mechanism of n – SiO_2/Ni composite coating prepared by electro – brush plating. Chinese Journal of Materials Research. 2003, 10(5): 530 – 536.

[9] Xu B S, Liang X B, Dong S Y, et al. Progress of nano – surface engineering. International Journal of Materials & Product Technology. 2003, 18(4): 338 – 343.

[10] Shi P J. Preparation on antifriction and repairing lubricating additive and study on the engine bench test: [Dissertation for Master Degree]. Beijing: Department of Materials Science and Engineering, Academy of Armored Forces Engineering, 2003.

[11] Lu K, Lu J. Materials Science and Engineering: A. 2000, 286(1): 91 – 95.

[12] Tao N R, Wang Z B, Tong W P, et al. Acta Materialia. 2002, 50(18): 4603 – 4616.

[13] Wang Z B, Yong X P, Tao N R, et al. The influence on the friction and wear properties of low carbon steel after surface nanocrystallization. Acta Materialia Sinica. 2002, 37(12): 1251 – 1255.

[14] Tong W P, Tao N R, Wang Z B, et al. Nitriding iron at lower temperature. Science. 2003, 229: 686 – 688.

[15] Wang Z B, Tao N R, Tong W P, et al. Diffusion of chromium in nanocrystalline iron produced by surface mechanical attrition treatment. Acta Materialia. 2003, 51: 4319 – 4329.

Development of Surface Engineering in China*

Abstract Professor Tom Bell dedicated himself to the establishment of the discipline of surface engineering and contributed greatly to the development of surface engineering in China. This paper briefly overviews the four important stages in the development of surface engineering in China, i. e. single surface engineering, duplex surface engineering, nanosurface engineering and automation.
Key words surface engineering, nanotechnology, automatic technology

1 Professor T. Bell and Chinese Surface Engineering

Professor Tom Bell is recognised as the founder of the discipline of surface engineering and the most outstanding surface engineering scientist in the world. In the 1980s, he took the lead in putting forward the concept of surface engineering. In 1983, he established the first surface engineering institute in the world, the Wolfson Institute for Surface Engineering (WISE), in Birmingham, UK; in 1985, he launched *Surface Engineering*[1], the first international academic journal on surface engineering and at his suggestion, the International Federation for Heat Treatment changed its name to the International Federation for Heat Treatment and Surface Engineering (IFHTSE) to reflect the increased importance of surface engineering. For the next three decades, he continued to engage in the research, development and application of surface engineering.

Professor Bell made a major contribution in promoting the development of surface engineering in China and was a regular visitor involved in China-UK academic cooperation in surface engineering research. Realising this potential development, under his chairmanship, IFHTSE invited China to host the Third International Congress on Heat Treatment of Materials. As a result of his appointment as a heat treatment expert by the United Nations Industrial Development Organisation(UNIDO), a prestigious centre for heat treatment was set up in Shanghai.

In recent years, an Anglo – Sino governmental collaborative S&T project on nanocomposite coating and composite surface engineering used in advanced car components, enabled cooperation between the University of Birmingham and the National Key Laboratory for Remanufacturing[2]. As a result of his contribution to surface engineering in China, he was made one of the eight honorary members of the Chinese Mechanical Engineering Society (CMES) in 1986, and was further honoured as a foreign academician of the Chinese Academy of Engineering (CAE) in 2002.

2 Driver and Development of Chinese Surface Engineering

Failure of most engineering components occurs at the surface, corrosion begins from the sur-

* Reprinted from *Surface Engineering*, 2010, 26(1 – 2): 123 – 125.

face, wear occurs on the surface and fatigue cracks propagate inwards from the surface. In manufacturing industry, 70% of failure is caused by corrosion and wear, and one-third of energy is consumed by surface friction and wear during manufacture and in service.

For example, according to the Corrosion Investigation Report issued by CAE in 2003, the loss caused by corrosion was about ￥600 billion in 2002, about 5% of the year's GDP. From the Tribology Investigation Report issued by CAE in 2007, the loss caused by friction and wear was about ￥950 billion in 2006, about 4.5% of the year's GDP. The loss is enormous and therefore, developing surface engineering has significant advantages in building up resource saving and environment friendly treatments, thus saving energy and reducing pollution.

Surface engineering is a system of engineering to obtain the desired surface properties through surface coating, surface modification or duplex surface treatments on a materials surface to alter its morphology, chemical composition, microstructure and stress condition[3]. The major advantage of surface engineering is that the surface layer, being superior to the substrate, can be prepared on the components, to endow excellent properties such as high oxidation resistance, high corrosion resistance, wear reduction and good fatigue properties. Therefore, surface engineering is an important means of solving surface wear, corrosion and fatigue problems.

In 1988, *China Surface Engineering* was launched by CMES, which marked the formal establishment of Chinese surface engineering. The development of Chinese surface engineering has witnessed the following four stages:

(1) single surface engineering: single surface technology (such as thermal spraying, electro-brush plating, vacuum deposition, etc.) is utilised individually to protect the surface, but the more rigorous demands on coatings cannot be satisfactorily met.

(2) duplex surface engineering: combining two or more kinds of surface engineering technologies to obtain the "1 + 1 > 2" optimum strengthening effect. For example, the combination between thermal spraying and electron beam surface remelting to further enhance the properties of sprayed materials. Duplex surface engineering has become the "accelerator" to improve surface performance.

(3) nanosurface engineering: through incorporating nanoparticles within surface coatings and promoting the innovative combination of advanced nanomaterials and traditional surface engineering technologies to further greatly enhance the surface performance. Nanosurface engineering technologies with Chinese intellectual rights have been developed in China.

(4) automation in surface engineering: in order to meet the demands of the remanufacturing industry, surface engineering processes must be converted from manual to automatic operation. With the introduction of automation technology to surface engineering and the improved performance of the surface coating, the quality of remanufactured components has been greatly increased.

3 Duplex Surface Engineering

3.1 Combination surface technologies[4,5]

Several surface technologies could be utilised to improve the performance of coatings. For example, turbine vanes of aeroengines have been treated by thermal spraying and electron beam surface remelting. The fretting wear resistance of the electron beam remelted CoCrW coating, which was first prepared by plasma spraying, is 13 times higher than that of the original plasma sprayed CoCrW coating.

To fix the steel cables used in a wide span suspension bridge, thousands of rollers have been installed under the heavy loaded motherboard by some countries to realise the longitudinal shift. In China, when building the Xiling Yangzi River Bridge, the antifriction multimetal layers on the saddle and motherboard have been prepared by the electrobrush plating technology. The composite usage between the multi metal layers and the friction reducing additives not only reduces the friction coefficient, but also greatly alleviates the longitudinal thrust force from the pendant wire saddle, which is several tens of thousands of tons in weight.

3.2 Combination of surface materials[6,7]

Compared with single coating materials, composite coatings with several materials give much better performance due to the synergistic effect of these materials. For example, traditional antislip coatings are largely based on polymer coating systems. The coating degraded easily due to its short life span and poor bonding strength. This issue has been addressed through high velocity arc spraying using Al base Al_2O_3 powder cored wires. The anticorrosion Al coating wasprepared on the substrate first to protect the steel plate from corrosion by sea water, then the Al base Al_2O_3 composite antislip coating was sprayed, and finally the hole sealing was carried out. The application results demonstrated that the duplex treated composite coating has super anticorrosion and antislip performance.

Another example of the application of the combination of surface materials is high velocity arc spraying of rare earth (RE) containing Zn-Al-Mg-RE self-sealing anticorrosion coatings to prevent corrosion in ocean conditions. The self-sealing was achieved by the in situ formation of compact oxides in the course of corrosion, which sealed the pores in the coating, thus blocking the path to the substrate and improving the corrosion resistance. The key to the self-sealing of the sprayed coating is the addition of Mg and RE. It is known that Mg is so active that oxidation is unavoidable and the magnesium oxide is unstable. The alloying of the coating with a small amount of RE can increase not only the uniformity and density of the coating, but also the stability of the oxides.

4 Nanosurface Engineering

Nanotechnology is one of the three high technologies in the twenty-first century. Nanosurface engineering is a system of engineering which combines the nanomaterials and/or nanotechnology with surface engineering technologies to further enhance the performance of surface engineered materials[8].

4.1 Nanoparticle composite brush plating technology[9,10]

The nanoparticle composite electrobrush plating is a new brush plating technology, in which special nanoparticles are added into the bath and are evenly dispersed to enhance the wear resistance and fatigue properties of the brush plated coatings.

Two major technical barriers, i. e. aggregation of nanoparticles in the bath and codeposition of non-conducting inorganic nanoparticles and metals, need to be addressed for the use of the nanoelectrobrush plating technology. First, an innovative high power mechanical-chemical method has been developed to form homogeneous electrolysis with steady suspension of the nanoparticles. Second, through optimally controlling some key processing parameters, the codeposition of non-conducting inorganic nanoparticles with metals has been realised.

The nanoparticle composite coatings are characterised by very fine microstructure with evenly distributed nanoparticles and possess significantly improved hardness, wear resistance, fatigue properties and thermal stability. This technology has been employed to repair the gas booster vanes of aeroengines. This is a flexible and cost effective surface engineering technology, which has produced huge societal and economic benefits and impacts.

4.2 Nanoantifriction self-repairing additive technology[11]

A novel antifriction self-repairing additive technology has been developed in China by adding nanoparticles into lubrication oils as additives. Under certain combinations of temperature, pressure and friction force, severe plastic deformation occurs to articulating surfaces. The nanoparticles can embed in and react with the contact surfaces to produce a layer of self-repairing antifriction solid films.

For instance, the proprietary nanoself-repairing additive M6 has been bench tested for 300h for its durability. Good engine rig results have been obtained in terms of increased engine output power by 6%, increased torque by 2% and reduced oil consumption by 6%. The M6 additive has also been trialled in three buses and the results demonstrated that after road testing for 15000km, the engine's output power increased by 2%-5%, fuel consumption reduced by 3% –6% and the exhaust gas decreased by 30%-50%.

5 Automation in Surface Engineering

Remanufacturing has been regarded as one of the most effective technologies to reduce the consumption of energy and raw materials, to reduce waste disposal and to reduce green house gases, thus contributing to sustainable development. There is an increasing demand for automation in the remanufacturing industry to ensure the reproducibility of quality, to reduce costs and to improve working conditions.

Many surface engineering technologies have been successfully applied in the modern remanufacturing industry. However, some surface engineering processes, such as brush plating, were initially developed for manual operation. To this end, some automation technologies in surface engineering have been developed by the author's laboratory to substantially improve the performance of coatings and the quality of remanufactured components by robots or automated machinery[12].

5.1 Automatic high velocity arc spraying technology[13]

This technique uses the robotic arm or auto-operating machine to fix the high velocity arc spraying gun, to plan the moving path of the gun based on the control software, to adjust the spraying parameters in real time and to control the gun operation under the predetermined programs. This technology has been successfully applied to remanufacture automobile engine blocks and crankcases as well as other similar parts to improve efficiency. For instance, it took about 1.5h to remanufacture a crankshaft by manual spraying, but the time decreased to 20 min when automatic spraying was employed, thus increasing the spraying efficiency by a factor of 4.5. The remanufacturing of crankshafts using automatic spraying can avoid using the carbonitriding process which would otherwise be required to be carried out at a temperature of 400℃ for about 8h.

5.2 Automatic nanoparticle electrobrush plating technology[14]

After solving key issues, such as the continuous feeding of the electrolyte using solution circulation, monitoring plating parameters (voltage, current density and the temperature of the solution) and plating procedures (switching process and monitoring the thickness of the coating), a special machine for automatic nanoparticle electrobrush plating, aiming at batch remanufacturing for engine connecting rods, has been developed.

This purpose designed automatic brush plating machine can remanufacture six engine connecting rods at one time. The brush plating time for each piece reduced from 60 to 5 min and the production efficiency was increased by more than 10 times. The energy and material consumption of the remanufactured connecting rod are only 50% and 10% of the new one, and the cost was reduced by 90%.

5.3 Semiautomatic microplasma arc overlaying technology[15]

The microplasma arc has a high current density and a low heat input. This can effectively resolve the problem of the sensitivity of components to heat input and potential undue distortion of thin walled components. The bonding between the layer and the substrate is metallurgical, and could resist the impact load or cyclic loading.

The technology has been utilised to remanufacture the sealing cone of used engine exhaust valves. The deformation of the remanufactured valves is very low, and the hardness can meet the desired criteria. A new sealing cone is worth about £ 7, whereas the cost of remanufacturing a sealing cone is only about £ 1.

6 Summary

Professor Tom Bell is acknowledged as the founder of the discipline of surface engineering because he was instrumental in taking the lead in putting forward the concept of surface engineering. He made invaluable contributions in promoting the development of surface engineering in

China.

The development of surface engineering in China has gone through four technological stages, namely, single surface engineering, duplex surface engineering, nanosurface engineering and automation in surface engineering. These developments represent key developments in advanced manufacturing and high tech maintenance techniques.

Acknowledgements

The work described in this paper was financially supported by NSFC (No. 50735006), 863 Project (No. 2007AA04Z408) and 973 Project (No. 2007CB607601).

References

[1] Editorial: Director of the Britain Surface Engineering – Professor T. Bell visits Beijing, *Chin. Surf. Eng.*, 1989, 4, 1 – 2.

[2] B. S. Xu, S. C. Liu and H. D. Wang: Developing remanufacturing, constructing cycle economy and building saving – oriented society, *J. Cent. South Univ. Technol.*, 2005, 12, (S2), 1 – 6.

[3] B. S. Xu: Surface engineering and maintenance; 1996, Beijing, Mechanical Industrial Press.

[4] B. S. Xu: Magic surface engineering; 2000, Beijing, Tsinghua University Press.

[5] B. S. Xu, H. D. Wang and X. B. Liang: The good maintenance technologies based on nano – surface engineering, in Proc. Int. Conf. on Intelligent maintenance systems, 457 – 466; 2003, Changsha, National University of Defense Technology Press.

[6] B. S. Xu, X. B. Liang, S. Y. Dong and Y. Xu: Progress of nanosurface engineering, *Int. J. Mater. Prod. Technol.*, 2003, 18, (4), 338 – 343.

[7] B. S. Xu, H. D. Wang, S. Y. Dong and B. Jiang: Fretting wear resistance of Ni – base electro – brush plating coating reinforced by nano – alumina grains, *Mater. Lett.*, 2006, 60, (5), 710 – 713.

[8] B. S. Xu: Nano surface engineering; 2004, Beijing, Chemical Industrial Press.

[9] B. S. Xu, H. D. Wang, S. Y. Dong, B. Jiang and W. Y. Tu: Electrodepositing nickel silica nano – composites coatings, *Electrochem. Commun.*, 2005, 7, (6), 572 – 575.

[10] W. Y. Tu, B. S. Xu and S. Y. Dong: Chemical and electrocatalytical interaction: influence of non – electroactive ceramic nanoparticles on nickel electrodeposition and composite coating, *J. Mater. Sci.*, 2008, 43, (3), 2259 – 2265.

[11] P. J. Shi, B. S. Xu, Y. Xu, Y. L. Qiao and Q. Liu: Tribological behaviour of oil – soluble organo – molybdenum compound as lubricating additives, *Trans. Nonferr. Met. Soc. China*, 2004, 14, (S2), 386 – 390.

[12] B. S. Xu: Remanufacturing engineering and automation in surface engineering technology, *Key Eng. Mater.*, 2008, 373 – 374, 1 – 11.

[13] J. Y. Bai, Y. X. Chen, J. B. Cheng, X. B. Liang and B. S. Xu: An automatic high velocity arc spraying system, *Key Eng. Mater.*, 2008, 373 – 374, 89 – 92.

[14] B. Wu, B. S. Xu, B. Zhang and Y. H. Lü: Preparation and properties of Ni/nano – Al_2O_3 composite coatings by automatic brush plating, *Surf. Coat. Technol.*, 2007, 201, (16 – 17), 6933 – 6939.

[15] Y. H. Lü, B. S. Xu, Y. H. Xiang, D. Xia and C. L. Liu: Plasma transferred arc powder surfacing technology of thrust face, *Key Eng. Mater.*, 2008, 373 – 374, 43 – 46.

再制造工程设计及质量控制

Characterisation of Stress Concentration of Ferromagnetic Materials by Metal Magnetic Memory Testing*

Abstract The relationship between the degree of stress concentration and the $H_p(y)$ signals of metal magnetic memory was investigated by tension-tension fatigue tests. Sheet specimens of 45CrNiMoVA steel were machined and two types of notches were cut into them to obtain two stress concentration factors. During the testing process, $H_p(y)$ signals were collected from three measured lines on the specimen surface. In the absence of a precut notch, the $H_p(y)$ signal curve of a measured line at the central position was linear. However, the $H_p(y)$ signal curves of measured lines near a precut notch or across a notch showed a distribution of abnormal magnetic peaks. The magnetic gradient K of the abnormal magnetic peaks was found to increase as the stress concentration factor increased. This feature could potentially be used to characterise the degree of stress concentration of ferromagnetic materials.

Key words ferromagnetic material, metal magnetic memory, stress concentration, magnetic abnormal peak

1 Introduction

Ferromagnetic materials have a number of mechanical properties that make them useful in different engineering fields. However, a variety of defects, such as gas holes, inclusions, cracks, etc. inevitably appear in ferromagnetic materials during the refining ormanufacturing processes, which leads to microscopic or macroscopic discontinuities[1-3]. Most ferromagneticengineered components also contain geometrical discontinuities, such as holes, shoulders, and flanges. When components with these discontinuities are loaded, stress concentration zones are generated in the discontinuous areas[4]. A stress concentration zone is the most dangerous position in any ferromagnetic component, as the local stress becomes far greater than the nominal stress[5]. Consequently, evaluation of the degree of stress concentration by means of nondestructive testing methods becomes critical for assessment of in-service equipment safety. This type of assessment is a recurring theme in engineering research[6,7].

Metal magnetic memory testing (MMMT) was originally developed by Russian researchers[8], which was firstly presented at the 50th International Welding Conference in 1997[9]. MMMT is primarily based on magnetomechanical effects[10]. Ferromagnetic materials have magnetic domain structures and a self-magnetisation property. When an external load is applied to this type of material, the residual magnetic induction and the spontaneous magnetisation increase at stress

* Copartner: Dong Lihong, Dong Shiyun, Chen Qunzhi. Reprinted from *Nondestructive Testing and Evaluation*, 2010, 25 (2): 145 - 151.

concentration zones. This generates fixed nodes of magnetic domains, which appear on the surface in the form of a leakage field. The leakage field can be preserved after the applied load is removed, and therefore it can indicate the position of a stress concentration zone. The horizontal component $H_p(x)$ has a maximum value of field strength while the normal component $H_p(y)$ crosses the zero point[11]. Since a $H_p(y)$ signal exists a switch from the positive sign to the negative sign, it is easier to use a $H_p(y)$ signal than a $H_p(x)$ signal to detect the position of stress concentration zone.

MMMT is a passive nondestructive method that uses these self-leakage magnetic signals as a means of quality control for ferromagnetic materials. There is no requirement for an external magnetic field during measurement and the detected ferromagnetic components do not require demagnetisation after measurement. In this way, this technique differs from other magnetic nondestructive testing methods, such as the magnetic Barkhausen noise method and the magneto-acoustic emission method[12-15].

MMMT has been found to have a number of outstanding merits, and is receiving substantial attention from industry and academia in the form of both practical and theoretical investigations. For example, Luming et al. have studied the relationships between the Earth's magnetic field, residual stress and magnetic memory signals[16,17]. Wilson et al. have used a three-axis magneto-resistive magnetic field sensor to measure the residual magnetic fields, both parallel to an applied stress and the material surface (B_x) and perpendicular to the material surface (B_z). They have shown that the residual magnetic field variation and pattern can be used to assess residual stress, applied stress and defects[18].

However, due to the relatively short time span since its development, the physical mechanism underlying MMMT remains unclear, and the correct means for evaluating the degree of stress concentration using MMMT signals is still in question[19,20]. In addition, it has proven difficult to establish a general standard among different engineering fields. In the current paper, tension-tension fatigue specimens with two stress concentration factors, $K_t = 3$ and 5, were chosen to investigate the quantitative relationship between magnetic memory $H_p(y)$ signals and the degree of stress concentration.

2 Experiment

2.1 Specimen preparation

The composition and mechanical properties of the 45CrNiMoVA steel test material are given in Table 1 and Table 2. The shapes of tension-tension fatigue specimens are shown in Fig. 1. Different keen-edged precut notches were cut into the specimens to generate the different stress concentration factors.

Table 1 Chemical composition of tested material (wt%)

Material	C	Si	Mn	P	S	Cr	Ni	Mo	V	W
45CrNiMoVA	0.42 ~ 0.49	0.17 ~ 0.37	0.50 ~ 0.80	≤0.030	≤0.030	0.80 ~ 1.10	1.30 ~ 1.80	0.20 ~ 0.30	0.10 ~ 0.20	—

Table 2　Mechanical properties of tested material

Material	σ_s/MPa	σ_b/MPa	δ_5/%	ψ/%	α_k/J · cm^{-2}
45CrNiMoVA	600	920	7	35	39

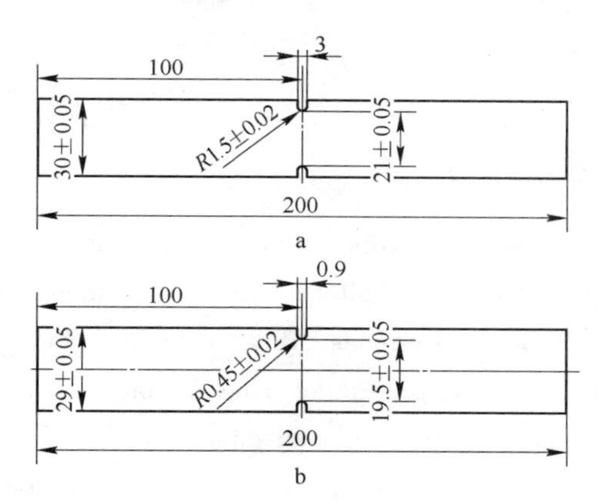

Fig. 1　Dimensions (in mm) of the notched sheet specimens with different stress concentration factor K_t

a—K_t = 3; b—K_t = 5

Before testing, all specimens were heated to 850℃ and kept for 30 min in a WZC-30 vacuum heat treatment furnace with a vacuum degree of 8×10^{-1} Pa, then cooled in the furnace to room temperature.

2.2　Experimental instrument

An MTS810 servo hydraulic system with dynamic error of ±1% was used and tension-tension fatigue tests of constant amplitude (sinusoidal waveform) were performed. The stress ratio R was zero, frequency f was 10 Hz. The maximum stress σ_{max} was 500 MPa.

During the fatigue test, after the specimen had been loaded to a predetermined cycle number, it was carefully taken off the machine and placed on a non-magnetic material platform in a south-north direction. The measured lines were arranged on the surface of the specimen, as shown in Fig. 2. Line 1 was located at the centre of the specimen, line 2 (line 2′) was on the front edge of the precut notch root, and line 3 (line 3′) was across the notch with an interval 2mm away from line 2 (line 2′).

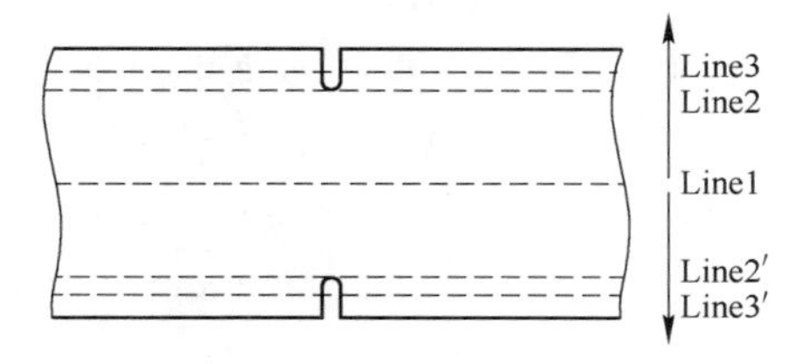

Fig. 2　The arrangement of all measured lines on the surface of the specimen

The magnetic signals on the surface of the specimen (i. e. $H_p(y)$ signals) were measured by an EMS – 2003 metal magnetic memory apparatus. The measuring instrument has a pencil-type probe with a 1 A/m sensitivity based on a Hall sensor. When measuring $H_p(y)$ signals, the probe was fixed onto a 3D electrical scanning platform made of non-magnetic materials and was placed vertical to the surface of the specimen, with a lift-off value of 1 mm. During the test, the scanning shelf moved the probe along the measured lines in the same manner, from one end to the other, to collect the magnetic memory signals for each line. $H_p(y)$ data were then input into a computer for further analysis.

3 Results and Discussion

In this research, the specimens were discontinuous regarding the position of the precut notch. The notches had a high local stress after loading and different stress concentration factors were obtained by altering the notch sharpness. Prior to loading, the specimens were demagnetised by a vacuum heat treatment to give them a clean initial magnetic state. This made the $H_p(y)$ signal variations on the measured lines easy to observe during the fatigue test. When an external tensile load was applied to a specimen, the magnetic properties of the precut notch changed according to magnetic mechanical effects, leading to a difference in $H_p(y)$ signal distribution between the zone with the precut notch and a zone without a precut notch.

As shown in Fig. 3, the initial value of $H_p(y)$ signals on three measured lines of two types of specimens are very small, close to zero. Thus, the initial magnetic curves of the specimens approximate a straight line, which indicates that the initial stray fields of the specimens were very weak. After loading, however, the distribution of $H_p(y)$ signals on the three measured lines changed immediately. The measured line 1 was positioned at the centre of the specimen. After even only 1 cycle, its magnetic curve rotated counterclockwise around the centre of the line and still presented good linearity. Divided by the centre point, the left signals of the magnetic curve were negative, while the right signals were positive. Consequently, the specimen was similar to a strip magnet and presented magnetic polarity under the effect of fatigue tensile stress. The slope coefficient of the magnetic curve increased concomitantly with the number of fatigue cycles.

The magnetic curves of line 2 and line 3 also showed a magnetic ordering distribution, in which left signals were negative and right signals were positive. The distributions of the $H_p(y)$ signals reconfirmed that the specimens became weak magnets after loading. However, the magnetic curves showed abnormal magnetic peaks that sharply crossed the zero line around a precut notch. In addition, the magnetic gradient K between these abnormal positive-negative peaks was very high. In fact, the precut notch could be treated as a macroscopic defect that penetrated the specimen. The specimen material was discontinuous at the position of the notch, but was continuous at the positions without notches. Accordingly, the segments of the magnetic curve away from notches were linear, while the part of the magnetic curve around the notches showed abnormal magnetic peaks. This distribution of magnetic curves was unchanged by increases in the number of fatigue cycles.

For example, for line 2, which existed near the tip of the notch, $H_p(y)$ signals were influenced by the notch since the measuring probe partly covered the notch during measurement. In con-

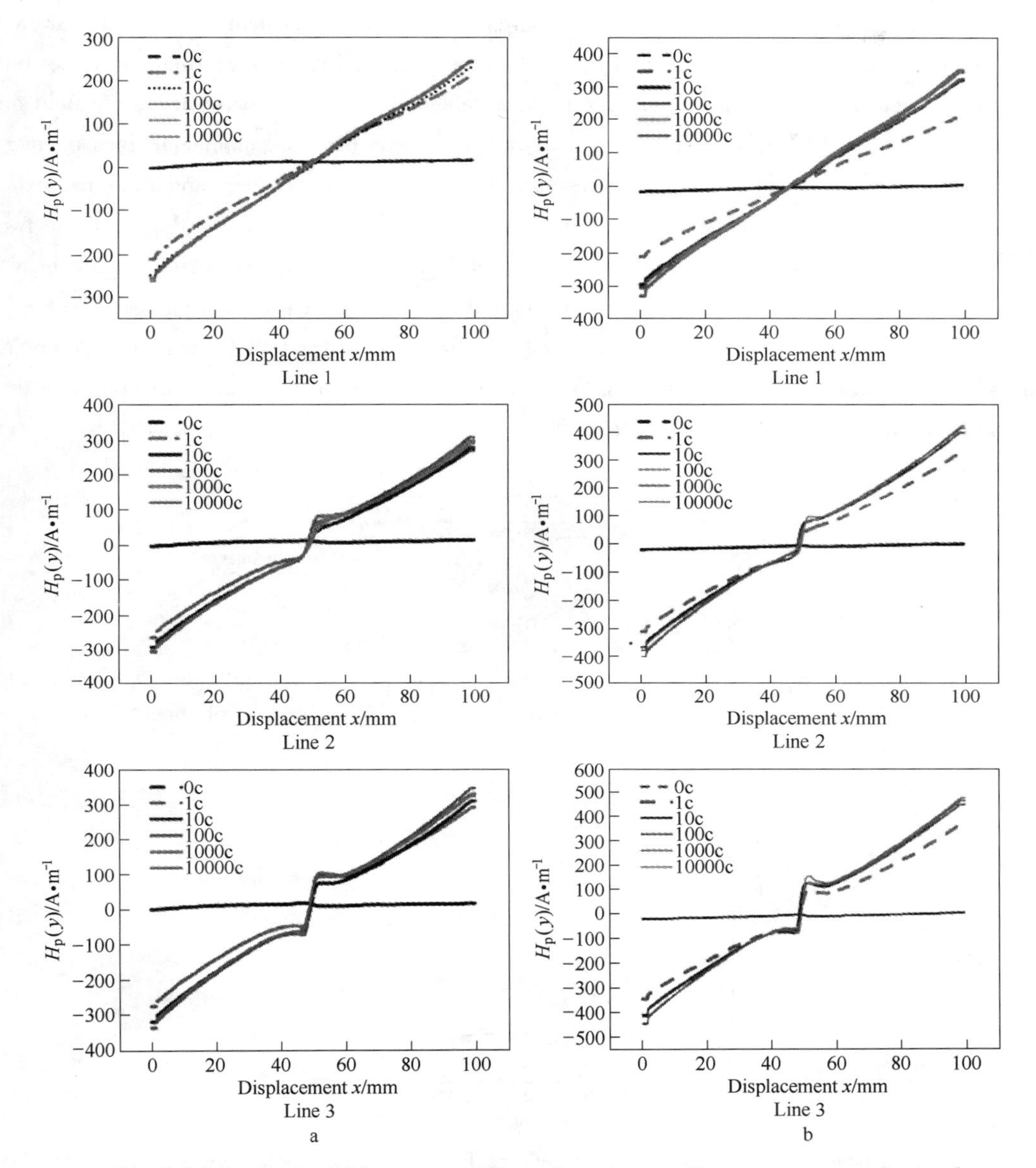

Fig. 3 The $H_p(y)$ distribution of the measured lines of specimens with two stress concentration factors

a—$K_t = 3$; b—$K_t = 5$

trast, line 3 fully crossed the notch, which resulted in an abnormal magnetic peak on this line that was stronger than that of line 2, since the measuring probe walked the whole discontinuous length. These experimental results indicated that it was the existence of a precut notch that influenced the distribution characteristics of the $H_p(y)$ signals on the surface of the ferromagnetic specimen. There was an abrupt change of permeability at the position of a precut notch since it had an air-gap, which caused a sharp change in the magnetic property of the specimen. The permeability decreased sharply at the position of the notch, and under the magnetisation effect of the Earth's magnetic field, opposite magnetic charges accumulated on either side of the notch. This induced the generation of the observed abnormal magnetic peak[12]. The appearance of an abnormal magnetic peak can be attributed to the permeability change from the measured area that lacked a defect to that which included a defect. On the other hand, the stray field on the surface of the ferromagnetic specimen near the notch was distorted since the lines of magnetic force bent around the notch. This resulted in the appearance of an abnormal magnetic peak in the $H_p(y)$ signal curve of the measured line 2 (as shown in Fig. 4). Moreover, the two types of specimens presented similar distributions of $H_p(y)$ signal curves during fatigue tests, and the abnormal peak values on the same measured line increased with increases in the stress concentration factor of specimen. Therefore, it is necessary to extract the magnetic gradient K with special software for further analysis. Fig. 5 shows the curve of the relationship between the magnetic gradient K and the fatigue cycles N.

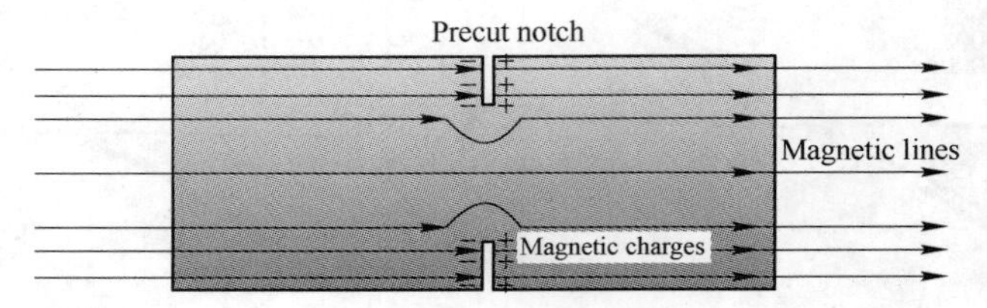

Fig. 4 Formation of an abnormal magnetic peak on a notch under the influence of the Earth's magnetic field

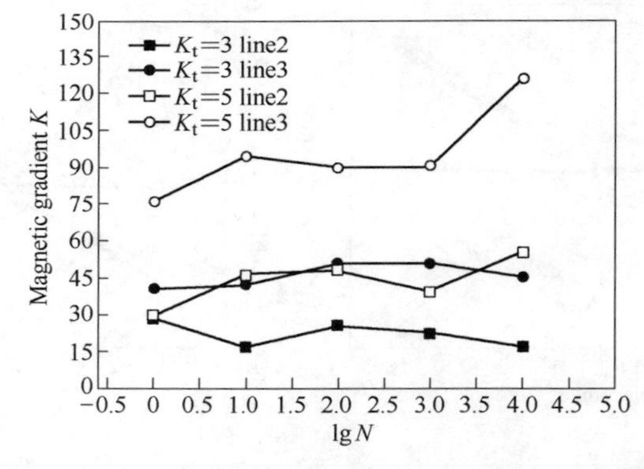

Fig. 5 Relationship curves between the magnetic gradient K of a magnetic abnormal peak on the notch position and the logarithm of the number of fatigue cycles N

As shown in Fig. 5, the magnetic gradient K increased when the stress concentration factor increased from 3 to 5 for the same measured line and the same fatigue cycles In contrast, for the different measured lines, the K was also different. For example, the magnetic gradient K at line 3 is higher than that at line 2. These results confirmed that the magnetic gradient K is an important parameter for describing the degree of stress concentration. At present, since only two-stress concentration factors, obtained by different sharpness of the precut notch (K_t = 3 and 5), were chosen for investigation, it is difficult to establish an empirical formula to quantitatively describe the degree of stress concentration. However, based on this observation, it can be concluded that the magnetic gradient K can be utilised to characterise the degree of stress concentration. Furthermore, the degree of stress concentration also appears to increase as the sharpness of the defect increases. For example, if a crack were to exist in a ferromagnetic component, its tip would be far sharper than that of a precut notch. Therefore, a crack would be expected to emit stronger abnormal magnetic signals after loading than would a precut notch. It could be inferred that the magnetic gradient K of an abnormal magnetic peak can be used to distinguish the degree of stress concentration in a ferromagnetic material.

4 Conclusion

The test results presented here confirmed that an abnormal magnetic peak appeared in the $H_p(y)$ signal curve at the position corresponding to a precut notch. The magnetic gradient K increased as the factor of stress concentration increased. However, the current study focused primarily on the changes in $H_p(y)$ signals that occurred in response to two stress concentration factors induced by a precut notch. Further theoretical investigation will be aimed at improving the understanding of the physical mechanism of MMMT by performance of fatigue tests using specimens with more stress concentration factors.

Acknowledgements

The authors would like to thank a main project of National Natural Science Foundation of China (Grant Nos 50735006), an ordinary project of National Natural Science Foundation of China (Grant Nos 50505052) and 973 Project (2007CB607601).

References

[1] S. Kustov, M. Corró, and E. Cesari, *Stress - induced magnetization in polycrystalline* Ni - Fe - Ga *ferromagnetic shape memory alloy*, Phys. Lett. 91 (2007), pp. 141907 - 1 - 141907 - 3.

[2] R. Clark, *Rail flaw detection: overview and needs for future developments*, NDT & E Int. 37 (2004), pp. 111 - 118.

[3] S. Yi, C. Eripret, and G. Rovsselier, *Influence of defect shape on damage evolution and fracture behavior of ductile materials*, Eng. Frac. Mech. 51 (1995), pp. 337 - 347.

[4] X. Z. Xiao, *Stress concentration*, China Machine Press, Beijing, 1986.

[5] Chyanbin Hwu and Y. C. Liang, *Evaluation of stress concentration factors and stress intensity factors from remote boundary data*, Int. J. Solids Struc. 37 (2000), pp. 57 - 72.

[6] J. L. S. Maurer, *Characterization of accumulated fatigue damage* in Ti - 6Al - 4V *plate material using transmission electron microscopy and nonlinear acoustics*, Ph. D. diss., University of Dayton, 2001.

[7] J. Spanner and G. Selby, *Sizing stress corrosion cracking in natural gas pipelines using phased array ultrasound*, NDE Eng. 22 (2002), pp. 68 - 71.

[8] A. A. Doubov, *A study of metal properties using the method of magnetic memory*, Metal Sci. Heat Treat. 39 (1997), pp. 401 - 402.

[9] A. A. Doubov, *Screening of weld quality using the magnetic metal memory effect*, Weld. World. 41 (1998), pp. 196 - 198.

[10] A. A. Doubov, *A technique for monitoring the bends of boiler and steam - line tubes using the magnetic memory of metal*, Therm. Eng. 48 (2001), pp. 289 - 295.

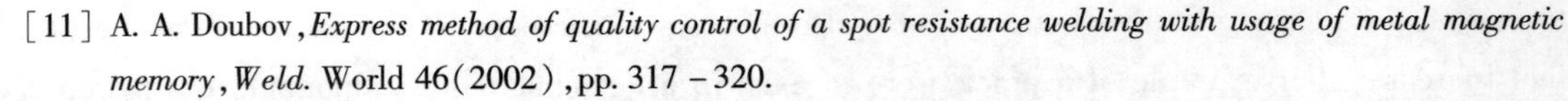

[11] A. A. Doubov, *Express method of quality control of a spot resistance welding with usage of metal magnetic memory*, *Weld.* World 46(2002), pp. 317 – 320.

[12] J. Pala, J. Bydžovský, and P. Švec, *Influence of magnetizing frequency and construction of pick – up coil on Barkhausen noise*, J. Electric. Eng. 55(2004), pp. 38 – 40.

[13] C. G. Stefanita, D. L. Atherton, and L. Clapham, *Plastic versus elastic deformation effects on magnetic Barkhausen noise in steel*, Acta Mater. 48(2000), pp. 3545 – 3551.

[14] M. Lindgren and T. Lepisto, *Effect of cyclic deformation on Barkhausen noise in a mild steel*, NDT&E Int. 36(2003), pp. 401 – 409.

[15] C. C. Yu, C. D. Qin, and D. H. L. Ng, *Nondestructive inspection of a fractured nickel bar by Barkhausen and magnetoacoustic emissions*, J. Appl. Phys. 79(1996), pp. 4750 – 4752.

[16] L. M. Li, S. L. Huang, X. F. Wang, S. S. Li, and K. R. Shi. *Stress induced magnetic field abnormality*, Trans. Nonferrous Met. Soc. China 13(2003), pp. 6 – 9.

[17] L. M. Li, S. L. Huang, X. F. Wang, S. S. Li, and K. R. Shi. *Magnetic field abnormality caused by welding residual stress*, J. Magnet. Magnet. Mater. 261(2003), pp. 387 – 391.

[18] J. Wilson, G. Y. Tian, and S. Barrans, *Residual magnetic field sensing for stress measurement*, Sensors Actuators A: Phys. 135(2007), pp. 381 – 387.

[19] L. H. Dong, B. S. Xu, S. Y. Dong, M. H. Ye, Q. Z. Chen, D. Wang, and D. W. Yin. *Monitoring fatigue crack propagation of ferromagnetic materials with spontaneous abnormal magnetic signals*, Int. J. Fatigue. 30 (2008), pp. 1599 – 1605.

[20] L. H. Dong, B. S. Xu, S. Y. Dong, M. H. Ye, Q. Z. Chen, D. Wang, and D. W. Yin. *Investigation of metal magnetic memory signals from the surface of low – carbon steel and low – carbon alloyed steel*, J. Cent. South Univ. Technol. 14(2007), pp. 24 – 27.

Metal Magnetic Memory Effect Caused by Static Tension Load in a Case-hardened Steel*

Abstract For investigating the magnetic abnormality influenced by stress in ferromagnetic materials, static tension tests on a case-hardened steel were carried out. Different loads, which covered tensile elastic loads up to plastic deformation and break, were applied. Meanwhile, the normal component of magnetic flux leakage, $H_p(y)$, was measured by metal magnetic memory testing. The results indicate that $H_p(y)$ values change with the tensile loads and positions. There exists a relationship between k, which is the inclination of the linear amplitude-locus magnetic flux leakage curve, and static tension load. A simple model is derived. Additionally, the mechanism of the magnetic memory effect can be explained by the theory of the interaction between dislocations and domains. The research provides the potential possibility of quantitative inspection for metal magnetic memory testing.

Key words metal magnetic memory effect, static tension load, case-hardened steel, magnetic flux leakage, dislocation

1 Introduction

It is known that the main source of aging of components is by fatigue rupture owing to the concentration of mechanical stresses, where the process of cracking, corrosion and creepage most intensively develop[1-3]. Applied load can result in the changes of the domain structure and therefore the internal magnetic field of ferromagnetic materials due to the magneto-mechanical effect, which is of growing interest of researchers[4-7], so it is possible to evaluate stress concentration, which is the ratio of the local stress under external load to the nominal stress, by means of detecting magnetic behavior.

In 1997, a new non-destructive testing technology, named metal magnetic memory testing, was elaborated[8,9]. Under the effect of the geomagnetic field and mechanical load, abnormal magnetic signals are generated in the stress concentration zones where the tangential component of magnetic flux leakage, $H_p(x)$, appears with a maximum value, as well as the normal component, $H_p(y)$, changes polarity and has a zero value. The magnetic state is still retained even if the load is removed. Therefore, the stress concentration zones can be detected by measuring the $H_p(y)$ values and their variable gradient without extra magnetic field, or even the residual lifetime of the components made of ferromagnetic materials could be predicted, which has been intensively recognized by researchers after the metal magnetic memory testing has been proposed[10-16]. But the physical mechanism of metal magnetic memory testing is still unclear so

* Copartner: C. L. Shi, S. Y. Dong, P. He. Reprinted from *Journal of Magnetism and Magnetic Materials*, 2010, 322: 413 – 416.

far because of the lack of systematic testing data and more profound theoretical support, thus how to use metal magnetic memory testing as a quantitative measurement method is a not yet solved problem[17,18].

In this paper, static tensile tests of flat-plate specimen made of 18CrNi4A steel, a kind of case-hardened steel, were accomplished. The development of the normal component $H_p(y)$ of the magnetic leakage field under different tensile loads were studied, the relation between tensile load and the inclination of the amplitude-locus magnetic flux leakage curve was also analyzed, as well as the interaction of dislocations and magnetic domains was used to explain the metal magnetic memory effect.

2 Experiment

2.1 Specimen

The specimen is made of 18CrNi4A steel having good mechanical properties by means of quenching and low-temperature tempering. The material is widely used to manufacture the heavy-duty gears and shafts in the field of aerospace and airplane industry. Its chemical composition and mechanical properties are shown in Table 1 and Table 2. Fig. 1 presents the geometry of specimen which was demagnetized using inductive demagnetization before tension test. Three scanning lines with a length of 70 mm were selected parallel with measuring positions for every 7mm, indicated as a, b and c.

Table 1 Chemical composition (wt%)

Steel	C	Mn	S	P	Si	Ni	Cr
18CrNi4A	0.15 – 0.20	0.30 – 0.60	≤0.010	≤0.015	≤0.35	3.75 – 4.25	0.80 – 1.10

Table 2 Mechanical properties

Steel	Tensile strength, σ_b/MPa	Yielding strength, σ_s/MPa	Extending rate, δ/%	Impact value, α_K/kJ · m^{-2}
18CrNi4A	1450	980	≥8	≥600

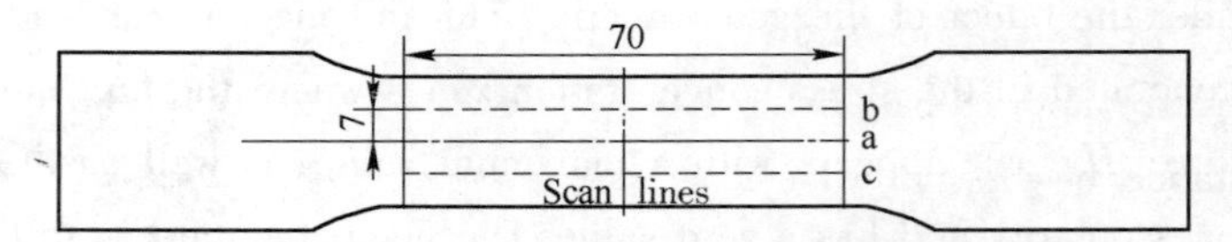

Fig. 1 Geometry of specimen (in mm) and scanning lines

2.2 Experimental instruments

The static tension tests were carried out on MTS810 hydraulic servo machine, whose static load error is ±0.5%. The specimen was loaded with low speed, approximately 0.5 kN/s, until a preset load value, then it was taken from the tensile machine and laid on the measurement platform in south to north direction. The $H_p(y)$ values along scanning lines were detected by the

EMS2003 metal magnetic memory device. The probe was gripped on a nonferromagnetic 3D electric scanning platform with a lift-off value of 0. 5 mm. After detection, the specimen was loaded again to a higher preset load value, the above procedure was repeated until the specimen failed. The fracture surface of the specimen was investigated by a Quant 200 scanning electronic microscope (SEM).

3 Results and Analysis

3. 1 Magnetic signals before failure

In the experiment, the variations of magnetic signals measured along the three lines (lines a-c in Fig. 1) were similar. Thus here only the results from the line a are shown. The amplitude-locus magnetic flux leakage curves in line a under different tensile loads are presented in Fig. 2.

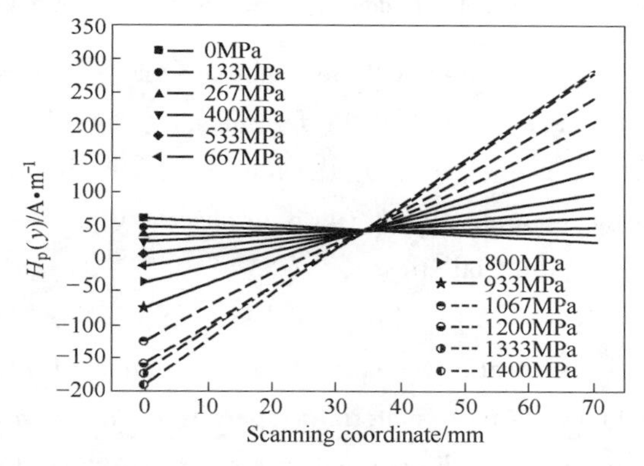

Fig. 2 Amplitude – locus magnetic flux leakage curves under different tensile loads in line a before failure

As shown in Fig. 2, the initial magnetic values are in the range of +25 to +60A/m, indicating that the specimen has only low residual magnetism after inductive demagnetization. The yield strength of 18CrNi4A steel is 980MPa and the tensile strength is 1450MPa. Accordingly, the solid lines in Fig. 2 present the $H_p(y)$ values in the elastic regime, and the dashed lines show the values in the plastic regime. All the measured curves are approximately linear, crossing at the distance of 35 mm where the value is not zero because of the residual magnetism. Moreover, there exists an obvious relationship between the value k, which is the inclination of the linear amplitude-locus magnetic flux leakage curve, and the static tensile load σ, shown in Fig. 3.

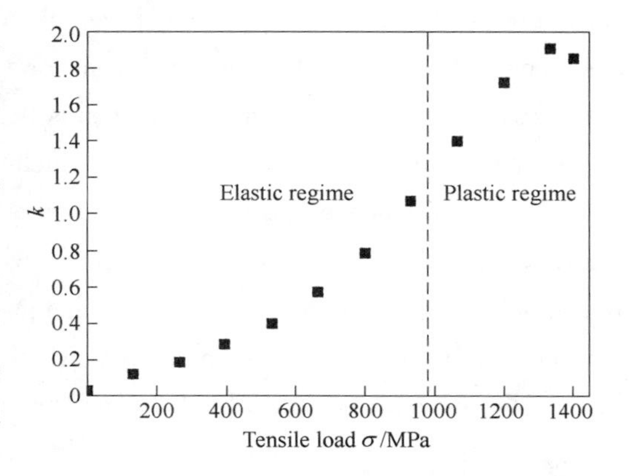

Fig. 3 Relationship between the k and applied tensile load σ

Fig. 3 shows that the inclination, k, continuously increases with the rise of the applied load at the elastic deformation stage, and the inclination increase is more and more pronounced. It is known that mechanical stress can change the magnetic behavior because of the magnetic domain movement, and it is easier for magnetization along the direction of uniaxial tensile stress[19]. Fig. 4 shows the domains' reorientation influenced by stress. When the specimen is loaded, the domains turn along the axial direction of the applied load based on the piezomagnetic effect, including movement of the domain walls, increasing of domains with favored magnetic direction and annihilation of the others and domains' rotation.

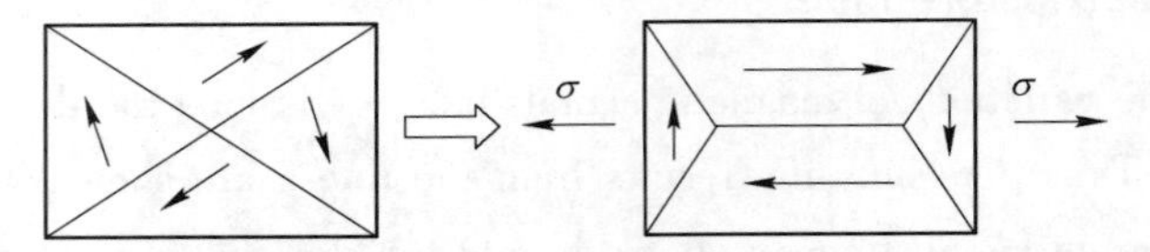

Fig. 4 Diagrammatic sketch of domains' reorientation influenced by stress

The magnetic domains are reoriented with the increase of applied stress, inducing also a change of the macroscopic magnetic field of the ferromagnet. Thus the magnetism of specimen becomes stronger and stronger, represented by the increase of the inclination in the elastic regime. According to previous research[20], Eq. (1) is to observe when the ferromagnet is affected by a weak magnetic field and uniaxial stress:

$$\mu = \mu_T(1 + bH/\mu_T)[a_0 + a_1 |\sigma|^m \exp(n|\sigma|)] \tag{1}$$

where μ_T is initial magnetic permeability relating to T; T is temperature; σ is stress; H is magnetic intensity; b is aconstant relating to material properties; a_0, a_1, m, n are the coefficients depending on the direction and value of the applied stress. Eq. (1) shows that the relation between magnetic permeability and stress is nonlinear, including the power function and exponential function. Accordingly, the magnetic permeability more and more rapidly increases and the ferromagnet is liable to be magnetized when the stress goes up, agreeing with the change of k induced by tensile load in the elastic regime.

As shown in Fig. 3, k initially goes up with the increase of stress at the plastic deformation stage, and a maximum value, which is equal to 1.91, appears. But then k decreases until the specimen fails. As we all know, the dislocation movement induced by residual deformation is very active in the plastic regime, and it pins the magnetic domain. But there are alloy elements in 18CrNi4A steel, such as Cr and Ni, which can easily increase the temper resistance and weaken the influence of dislocation on pinning the magnetic domain. At the beginning of the plastic stage, the dislocation density is small, which has little effect on pinning the domain, and the inclination still increases. With the increase of stress, the residual deformation increases leading to the rise of dislocation density, which powerfully pins the magnetic domain and decreases the magnetism. Accordingly, a maximum value comes out after the increase of inclination, and then the inclination decreases in the plastic regime.

According to the interaction of dislocation and magnetic domain, the variation of magnetic signals at the elastic and plastic deformation stage can be well explained. As shown in Fig. 3, there is a definite relationship between the stress and the inclination, but how much sensitive the effect is must be very important for quantitative inspection by metal magnetic memory testing. How good the correlation is can be characterized by the related coefficient, r as follows:

$$r = \left(\sum_{i=1}^{n} x_i y_i - n\,\overline{xy}\right) / \left[\left(\sum_{i=1}^{n} x_i^2 - n\,\overline{x}^2\right)\left(\sum_{i=1}^{n} y_i^2 - n\,\overline{y}^2\right)\right]^{1/2} \tag{2}$$

where x is stress σ; y is the inclination k. Based on the results in this research, r equals 0.974 approaching to 1, which shows that k is intensively related with σ. Then the relation between k and σ's modeled by an polynomial approach, and the related curve is shown in Fig. 5.

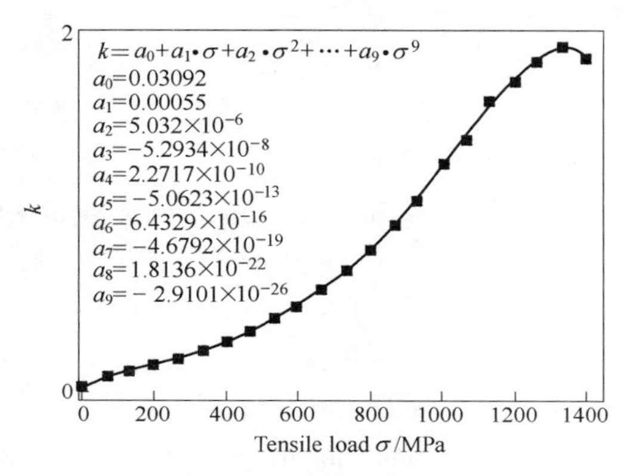

Fig. 5 Polynomial approach and modeled curve of k and σ

According to the related curve, the tensile stress acting on the 18CrNi4A steel specimen can be calculated by the inclination, k, detected by metal magnetic memory testing. It should be mentioned that the effects will be the same when another steel grade is investigated. Therefore, it is easy for metal magnetic memory testing to detect the stress distribution in a component made of steel when the suitable scanning device is used, even to predict the lifetime so far the behavior of the material is previously studied.

3.2 Magnetic signals after failure

When the applied stress reached 1450MPa, the specimen was fractured along the direction of 45°, where necking phenomenon occurred as shown in Fig. 6. Many dimples were observed in the fractured surface shown in Fig. 7, indicating that ductile fracture happened.

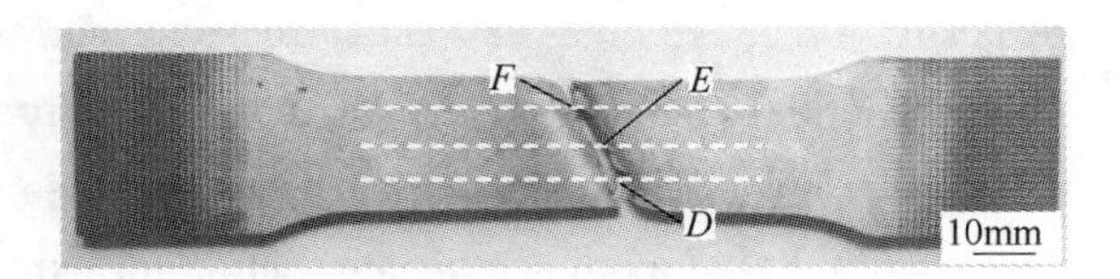

Fig. 6 Photograph of the tensile failure

The fractured specimen is inspected by metal magnetic memory testing and the results are presented in Fig. 8. The magnetic signals intensively change polarity and have a zero value in each measured line. According to Ref. [12], the phenomenon can be explained in terms of the interaction energy in ferromagnetic material.

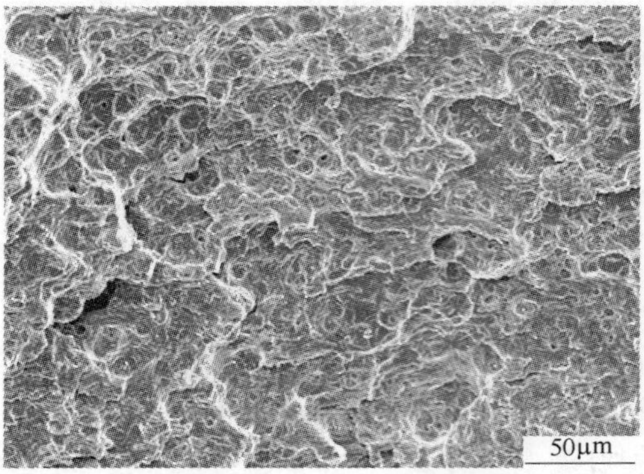

Fig. 7 SEM micrograph of the fractured surface

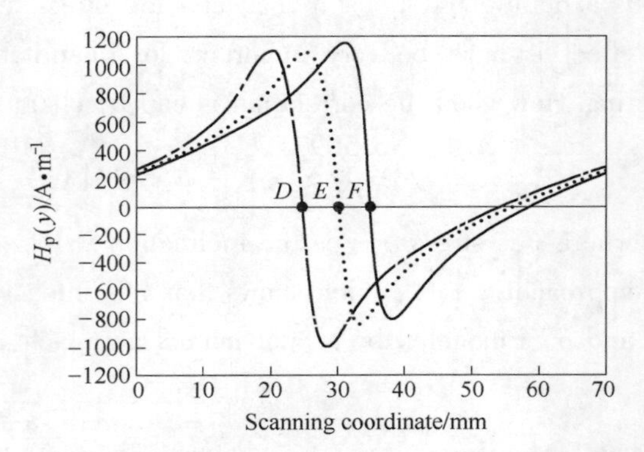

Fig. 8 Amplitude-locus magnetic flux leakage curves in different measured lines after failure

The internal energy, F, can be descried by Eq. (3) when the temperature is far below the Curie point:

$$F = F_k + F_\sigma + F_d \tag{3}$$

where F_k is magnetocrystalline anisotropy energy; F_σ is stress energy; F_d is demagnetization energy, Stress energy F_σ is far greater than F_k and F_d under the applied axial tensile stress. When the specimen is fractured, F_σ releases immediately. In order to recover the system to the balanced state, F_d increases dramatically. Therefore, the magnetic signals change positive – negative sign in the fractured zone. It should be noted that the positions of zero value in the measured lines correspond with the fractured points (D, E and F) in the experiment. Thus the accurate position of failure could be predicted by metal magnetic memory testing, even the position of microcracking could be inspected if the method has enough resolution.

4 Conclusions

From the investigation of magnetic signals induced by tensile loads in a case – hardened steel, it can be concluded that all the amplitude-locus magnetic flux leakage curves before failure are approximately linear. There obviously exist a relationship between the inclination, k, and the static tension load, σ, which is well described mathematically by statistics. Accordingly, the stress distribution could be detected by metal magnetic memory testing. After failure, the $H_p(y)$ values intensively change positive-negative sign, and the zero value corresponds with the position of fracture. Thus metal magnetic memory testing could be used to predict the development of cracking in ferromagnetic materials. The here presented work provides systematic testing data, and the theoretical analysis would give more useful information for explaining the metal magnetic memory effect. However, many factors, such as the influence of different types of load, machining processes, heat treatment conditions and external magnetic fields, intensively affect the metal magnetic memory signals, preventing metal magnetic memory testing from being widely used in practical projects. Therefore, how to exclude disturbing factors is the most important work in the future.

Acknowledgements

This work was carried with the financial help of the National Natural Science Foundation of China (Grant No. 50975287) and the National Key Technology R&D Program (Grant No. 2006BAF02A19 - 01). The authors also thank Dr. Gerd Dobmann for editing work.

References

[1] Tchankov, S. Ohta, A. Suzuki, Int. J. Fatigue 21(1999)941.
[2] B. S. Xu, W. Zhang, S. C. Liu, Remanufacturing technology for the 21st century, in: 15th European Maintenance Conference, Gothenburg, March 2000, pp. 335 - 339.
[3] D. Arola, C. L. Williams, Int. J. Fatigue 24(2002)923.
[4] A. E. Markaki, T. W. Clyne, Acta Mater. 53(2005)877.
[5] W. Y. Gong, Z. M. Wu, H. Lin, X. L. Yang, Z. J. Zhao, J. Magn. Magn. Mater. 320(2008)1553.
[6] B. K. Kardashev, K. V. Ouytsel, R. D. Batist, J. Alloys Compd. 310(2000)169.
[7] P. S. Papamantellos, J. R. Carvajal, K. H. J. Buschow, J. Magn. Magn. Mater. 250(2002)225.
[8] A. A. Doubov, Metal Sci. Heat Treat. 39(1997)401.
[9] A. A. Doubov, Therm. Eng. 48(2001)289.
[10] L. M. Li, S. L. Huang, X. F. Wang, J. Magn. Magn. Mater. 261(2003)385.
[11] D. W. Yin, B. S. Xu, S. Y. Dong, L. H. Dong, J. Cent. South Univ. Technol. 12(2005)107.
[12] L. H. Dong, B. S. Xu, S. Y. Dong, NDT&E Int. 41(2008)184.
[13] G. Dobmann, L. G. Debarberis, J. F. Coste, Nucl. Eng. Des. 206(2001)363.
[14] F. Liorzou, D. L. Atrerton, J. Magn. Magn. Mater. 195(1999)174.
[15] T. Yamasaki, S. Yamamoto, M. Hirao, NDT&E Int. 29(1996)263.
[16] V. E. Iordache, E. Hug, N. Buiron, Mater. Sci. Eng. A359(2003)62.
[17] E. Yang, L. M. Li, X. Chen, J. Magn. Magn. Mater. 312(2007)72.
[18] L. H. Dong, B. S. Xu, S. Y. Dong, D. Wang, J. Univ. Sci. Technol. B15(2008)580.
[19] M. A. Jiboory, D. G. Lord, Y. J. Bi, J. Appl. Phys. 73(1993)6168.
[20] K. C. Zhu, T. N. Liu, H. Q. Tang, Chinese J. Low Temp. Phys. 25(2003)121.

Investigation of Rolling Contact Fatigue Lives of Fe-Cr Alloy Coatings under Different Loading Conditions*

Abstract The aim of the present study is to address the rolling contact fatigue (RCF) life and behavior of thermal sprayed coatings. The Fe-Cr alloy coatings and Ni-Al alloys coatings were prepared by a plasma spray device as the surface coating and undercoating on steel substrates, respectively. RCF tests were conducted using the ball-on-disc tester under four kinds of loading conditions. The failure mechanisms were investigated on the basis of worn surface observations of failed coated specimens. The RCF lives of Fe-Cr coatings were evaluated by means of Weibull distribution plots. The results show the contact stress obviously influences the RCF life and failure behavior of the Fe-Cr coatings. The lives decreased as increasing the contact stress. At the same time, the main failure modes of the coatings changed from abrasion and spalling to delamination. The failure mechanisms of coatings are associated with the micro-defects, the bond strength of the coatings and the orthogonal shear stress (OSS) within the coatings. The *S-N* curve, based on RCF life data by using statistical approach, is helpful for the design of hard coatings.

Key words rolling contact fatigue, Fe-Cr coatings, contact stress, fatigue life, weibull plots

1 Introduction

Rolling contact fatigue (RCF) is one of typical surface damage modes in many types of industrial applications, which is responsible for the failure of rolling element bearings, gears, camshafts, and limits the service lives of these components, then reduces durability and reliability of mechanical products. RCF may be defined as cracking or pitting/delamination limited to the near-surface layer of bodies in the rolling/sliding contact[1,2]. There is an increased demand for improved life, reliability and load bearing capacity of surface materials and future applications call for them used in more hostile environments. The modification of surface by depositing protective materials is an effective approach to improve the RCF resistance of components and achieve the application in more severe tribological condition.

Thermal spraying is one kind of surface engineering technologies which is utilized in almost all industrial fields. Continuous advancements and enhanced understanding of thermal spraying technology have facilitated a synergetic approach towards a sustainable growth of its industrial applications[3,4]. Thermal spraying coatings deposited by techniques such as plasma spraying (PS), high velocity oxy-fuel (HVOF) spraying, arc wire spraying etc., can provide a cost-effec-

---

* Copartner: Piao Zhongyu, Wang Haidou, Pu Chunhuan. Reprinted from *Surface & Coating Technology*, 2010, 204 (9–10): 1405–1411.

tive solution for tribological applications in pure rolling or rolling/sliding contact[5-8]. Owing to the feature of the thermal spraying process, the microstructures of the coatings are often characterized by the lamellar structures with the existence of various pores, so the failure behavior and mechanism of the coatings and the bulk materials are obviously different. Although there have been some researches and applications of the thermal sprayed coatings under rolling/sliding contact, only fatigue resistant performances and failure modes of the coatings constituted by different materials were investigated and some studies of fatigue life were conducted under small sample space[9,10]. The RCF life based on large sample space and failure mechanism of the sprayed coatings under different tribological conditions are not thoroughly understood.

The aim of this study is to investigate the RCF lives of the plasma sprayed coatings under different loading conditions, for the contact pressure is an important factor controlling the RCF behavior. The Fe-Cr alloy coatings were deposited on steel substrate by using PS. RCF experiments were conducted by a ball-on-disk tester. The failure mechanisms of coatings were also investigated on the basis of the surface observations of the failed coating specimens using the scanning electron microscope (SEM). The RCF lives under different loading conditions were characterized by Weibull plots. The *S-N* curves under different failure probabilities were obtained by statistical methodology.

2 Experimental Procedures

2.1 Preparation of coatings

A commercially available Fe-Cr-B-Si self-fluxing alloy powder with the nominal composition Cr – 13.6, B – 1.6, Si – 1.1, C – 0.16, Fe-balance (wt%) was used as sprayed material for its higher micro-hardness and better wear-resistance at the room temperature. Boron and silicon are often added to promote the oxidation resistance of coatings, and chromium is added to increase the wear resistance of coatings by the formation of hard phases. The commercial Ni – Al alloy powder with the nominal composition Ni – 90, Al – 10 (wt%) was used as undercoating material. The bond strength between coating and substrate will be increased by the heat-producing reaction between melting aluminum and nickel as arriving at substrate. The AISI 1045 steel with ring-type geometry was used as substrate. The external diameter, internal diameter and thickness of the substrate were machined to 60mm, 30mm and 25mm, respectively. In order to reduce contaminations and form a clean and rough surface for enhancing the bond strength between the coating and substrate, the plane surface of substrate was cleaned in acetone solution and sandblasted by using corundum powder before spraying process.

A high-efficiency PS system with the novel PS gun was employed to prepare the Fe-Cr and Ni-Al alloy coatings, respectively. This PS device enabled supersonic spraying and improved greatly the coatings quality by obtaining lower porosity and high bond strength, but at low consumption[11]. Argon gas was used as the primary gas, hydrogen gas and nitrogen gas were used as secondary gases. The PS parameters were listed in Table 1.

Table 1 Plasma Spraying Parameters

Spaying material	FeCrBSi	Ni/Al
Argon gas flow/$m^3 \cdot h^{-1}$	3.4	3.4
Hydrogen gas flow/$m^3 \cdot h^{-1}$	0.3	0.3
Nitrogen gas flow/$m^3 \cdot h^{-1}$	0.6	0.6
Spraying current/A	380	320
Spraying voltage/V	150	140
Spraying distance/mm	110	150
Powder feed rate/$g \cdot min^{-1}$	30	30

After the PS process, the thickness of coatings ranged from 400 μm to 500 μm, and the surface was severely rough. Therefore the specimens were ground and polished to 0.5 μm at the coating surface. Finally, the Fe-Cr surface coating with the thickness of 150 μm and Ni-Al undercoating with the thickness of 50 μm were produced, respectively.

2.2 RCF tests

A ball-on-disc contact tester was used to evaluate the RCF lives of the Fe-Cr alloy coatings, as shown in Fig. 1. The specimen was fixed in the cup assembly by a clamp with gear shape, and the bearing with 11 uniformly distributed balls was fixed as the coupled contact element. These coupled bearing balls were fabricated by AISI52100 steel with surface roughness of 0.012 μm and radius of 5.5 mm. The rotation speed of the coupled bearing was controlled by a high-velocity motor through a driving belt, and the rotation speed could be monitored by speed sensor. Additionally, the vibration and torque sensor were used to monitor the variation of differ-

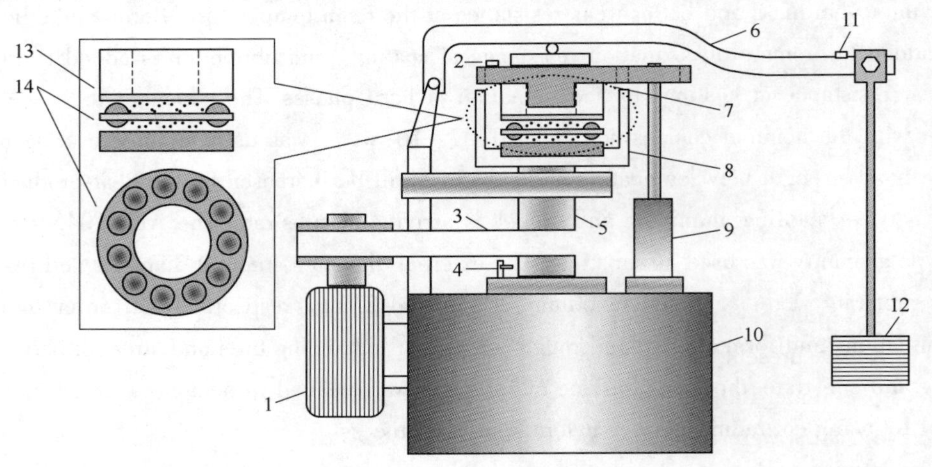

Fig. 1 Schematic of the ball-and-disc contact tester

1—Driving motor; 2—Temperature sensor; 3—Belt drive; 4—Speed sensor;
5—Driven spindle; 6—Loading lever; 7—Gear as a collet; 8—Cup assembly;
9—Torque sensor; 10—Stand; 11—Vibration sensor; 12—Weight;
13—Coated specimen; 14—Bearing with 11 balls

ent response parameters during the testing process. The signals obtained from these sensors were relayed to an amplifier and then imported a computer where the signals were converted to digital readings by dedicated software. The digital readings always varied along with the testing time. When the signals of sensors exceeded pre-installed threshold value owing to appearances of the coating failures, the test was stopped, and the fatigue life of specimen was calculated. All RCF tests were conducted under immersed lubrication condition.

For all RCF experiments, the rotation speed was kept to be a constant value of 1500r/min. In order to evaluate the RCF lives of the coatings under different loading condition, four grades of contact loads were applied in this study. They were 50N, 100N, 200N and 300N, respectively. The applied loads were transferred to the contact region by the lever with an amplification ratio of 8. Since the RCF life data exhibit extensive scatter, ten rolling contact tests under each loading condition were performed to obtain the statistical results.

2.3 Characterization of coatings

The microstructures of the coatings and the worn surface morphologies of the failed specimens after RCF tests were observed by using a Philips Quant 200 scanning election microscope (SEM). The elastic moduli of Fe-Cr and Ni-Al coatings were evaluated by using the Micro-Materials Nano-test 600 nano-indentation tester with a pyramidal indenter at room temperature. The test load was 15 mN and the dwell time was 10s.

3 Results

3.1 Contact stress calculation and analysis

When the thermally sprayed coatings applied in rolling contact, the key failure modes can be classified into different categories. The initiation and propagation of fatigue cracks are responsible for some failure modes. The maximum orthogonal shear stresses (OSS) beneath the surface are the drive forces of the initiations and propagations of fatigue cracks[12,13]. Therefore, quantitative analyses of these stresses are important for the design of the hard coatings in most tribological applications. However, both the theoretical and experimental approaches fail to predict successfully the distribution of the shear stress within layered media, which can be attributed to the fact that the thicknesses of the coatings are generally lower than that of the substrates[14-16]. In this case, the finite element method (FEM) is an effective approach to solve the contact stress problems. In the present study, the magnitudes and distributions of the OSS within the coatings and substrates under rolling contact were predicted by FEM. Firstly, the maximum contact stress and contact radius were calculated by the Hertzian equation[17]:

$$P_0 = \frac{3F}{2\pi a^2} \tag{1}$$

$$a = \left[\frac{3}{4}R_0\left(\frac{1-\nu_b^2}{E_b} + \frac{1-\nu_c^2}{E_c}\right)F\right]^{\frac{1}{3}} \tag{2}$$

where, P_0 is the maximum contact stress; a is the radius of the contact region (assumed round);

F is the applied load at the vertex of each coupled ball; R_0 is the radius of the coupled bearing ball; E_b is the elastic modulus of the coupled bearing ball, which is 220GPa; E_c is the elastic modulus of the Fe-Cr alloy coating, which was evaluated by the nano-indentation method, which is E_c = 187GPa; ν_b and ν_c are the Poisson's ratio of the bearing and coating, respectively. Here, the Poisson's ratios of all materials are taken as 0.3.

Hence, the magnitudes of maximum contact stress and radius of contact region under four loading condition can be calculated, respectively, they are 1.7114, 2.1123, 2.3874, 2.6841 GPa, a = 133.2, 154.9, 185.9, 208.9 μm. When the P_0 and a were given, the normal stress distribution in the contact region can be determined by the Johnson equation[18]:

$$p(x) = P_0\sqrt{1 - \left(\frac{x}{a}\right)^2}\ (0 \leqslant x \leqslant a) \tag{3}$$

where x denotes the distance from the contact center.

The finite element (FE) model was created using the commercial finite element analysis code ANSYS. This model was used to calculate the stresses inside the coatings. In order to reduce the data processing time, an axial symmetric problem was chosen. A Cartesian coordinate system $O-xy$ was introduced. More refined meshes were employed near the contact regions to improve the accuracy of calculation. The elastic moduli and Poisson's ratios of the Fe-Cr coating, Ni-Al undercoating and steel substrate were obtained by aforementioned process. Four grades of P_0 were applied to the contact region to calculate the OSS within the coatings configuration, respectively. Moreover, for all calculations, only the pure rolling contact cases were considered, i. e., the friction coefficient between the coatings and the coupling balls was assumed to zero.

Fig. 2 shows the distribution of the OSS under different loading conditions. It can be seen that the maximum OSS presented within the coatings under all loading conditions. Owing to the difference of the material mechanical property between the coating and substrate, there are obvious stress discontinuities at the coating/substrate interface.

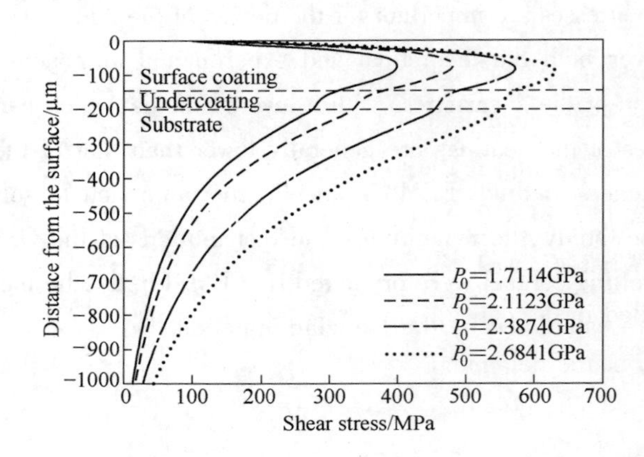

Fig. 2 Distributions of the orthogonal shear stresses within the coatings and substrate configurations under four kinds of loading conditions

3.2 Coating microstructure

Fig. 3 shows the cross-sectional microstructure of the coated system by SEM. The Fe-Cr alloy surface coating shown in Fig. 3, has almost no oxidation, lamellar structures and intra-lamella cracks. Only some pores can be observed. Under the cyclic loading, these micro-defects would propagate and result in the appearance of the coating failure. So these pores would have an important influence on the RCF performance of coatings. The Ni-Al alloy undercoating shown in Fig. 3, has few visible cracks in the coating/substrate interface. This indicates that the bond strength has been improved by the heat-producing reaction between melting aluminum and nickel.

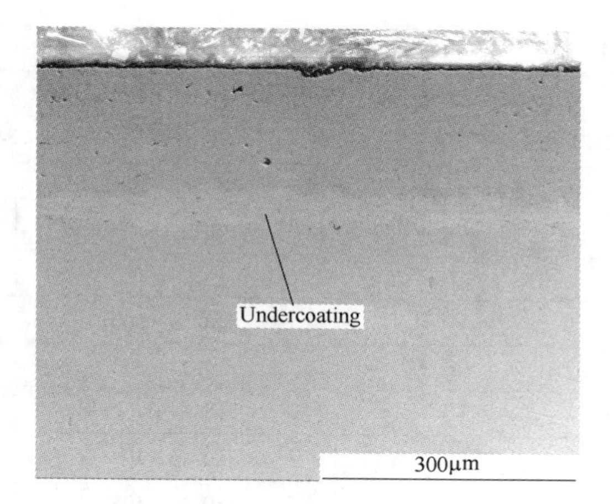

Fig. 3 Cross-sectional microstructure of the coated system

3.3 Worn surface morphologies observation

The RCF experiments results are listed in Table 2. Based on the observations of worn surface morphologies using SEM, four kinds of failure modes can be classified, respectively. They are abrasion, spalling, adhesive delamination and cohesive delamination. Where, the adhesive delamination is defined as the delamination at the interface between the coating and substrate, whereas the cohesive delamination is defined as the delamination within the coating. It can be seen that when P_0 was 1.7114 GPa, seven specimens failed in the abrasion and other ones failed in the spalling. When P_0 was 2.1123 GPa, seven specimens failed in the abrasion and spalling, and others failed in the cohesive or adhesive delamination. When P_0 was 2.3874 GPa, seven specimens failed in the delamination. When P_0 was 2.6841 GPa, all specimens subjected the delamination. It is obvious that the coating specimens failed in the adhesive delamination have extremely low RCF lives, which indicates that the adhesive delamination resulting in the coating failures in a short time is a destructive failure mode. So it can be seen that the contact stress plays an important role on the main failure modes of the PS coatings.

Table 2 RCF experimental results under four kinds of loading conditions

P_0 = 1.7114 GPa		P_0 = 2.1123 GPa	
Fatigue lives	Failure modes	Fatigue lives	Failure modes
0.91×10^6	SP①	0.61×10^6	AD③
1.36×10^6	SP	0.71×10^6	CD④
1.54×10^6	AB②	0.83×10^6	CD
1.71×10^6	AB	0.89×10^6	SP
1.95×10^6	SP	0.95×10^6	SP
2.13×10^6	AB	0.96×10^6	SP
2.35×10^6	AB	1.03×10^6	AB
2.54×10^6	AB	1.18×10^6	SP
3.21×10^6	AB	1.19×10^6	SP
4.05×10^6	AB	1.45×10^6	AB
P_0 = 2.3874 GPa		P_0 = 2.6841 GPa	
Fatigue lives	Failure modes	Fatigue lives	Failure modes
0.39×10^6	AD	0.28×10^6	AD
0.44×10^6	AD	0.29×10^6	AD
0.57×10^6	AD	0.36×10^6	AD
0.69×10^6	CD	0.37×10^6	AD
0.74×10^6	AD	0.44×10^6	AD
0.88×10^6	AD	0.46×10^6	AD
0.91×10^6	SP	0.48×10^6	AD
0.97×10^6	CD	0.51×10^6	AD
1.05×10^6	SP	0.61×10^6	CD
1.15×10^6	AB	0.70×10^6	AD

①SP—spalling; ②AB—abrasion; ③AD—adhesive delamination; ④CD—cohesive delamination.

Fig. 4 shows the abrasion morphology of the failed specimen with a fatigue life of 3.21×10^6 cycles when P_0 was 1.7114 GPa. There are many obvious micro-pits within the wear track. Although single pits have small dimensions, the occurrence of a high density of pits will influence the normal contact between the bearing balls and the coated surface under the high rotational speed, and result in the intensive vibration and noise.

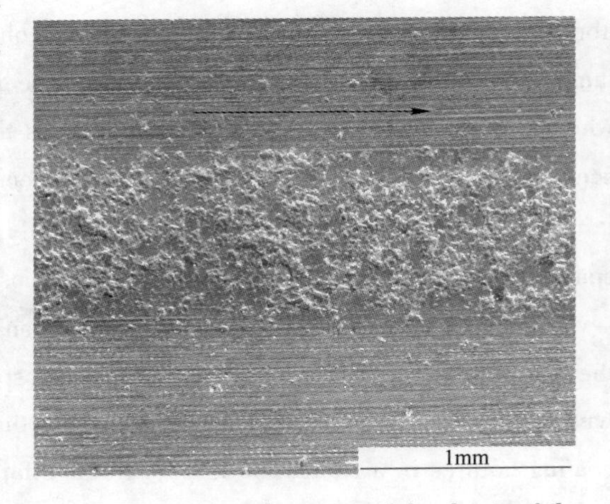

Fig. 4 Worn surface observation of the abrasion failure

Fig. 5 shows the spalling morphology of the failed specimen with a fatigue life of 1.19 ×

10^6 cycles when P_0 was 2.1123 GPa. The spalling exists within the wear track with larger dimension than the abrasion. And the spalling exhibits the irregular shape rather than ellipse or round as the spalling found in the failed bearing race.

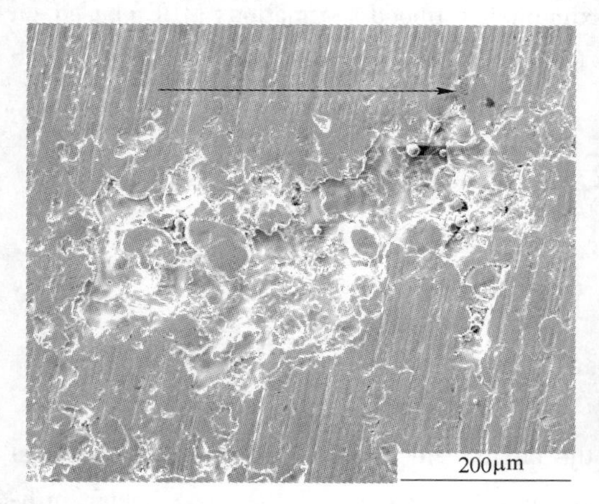

Fig. 5 Worn surface observation of the spalling failure

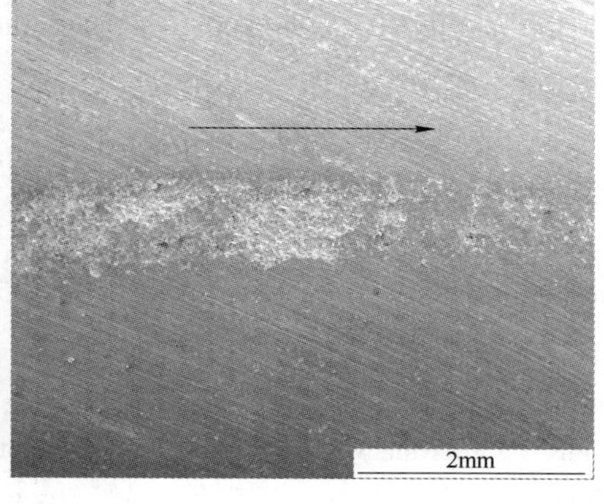

Fig. 6 Worn surface observation of the cohesive delamination failure

Fig. 6 shows the cohesive delamination morphology of the failed specimen with a fatigue life of 0.97×10^6 cycles when P_0 was 2.3874 GPa. The delaminated coating with a large surface area exhibits an approximate elliptical configuration. The long axis is almost parallel to the rolling direction and the minor axis is perpendicular to the rolling direction. The cohesive delamination is much deeper than spalling with the obvious layered edge.

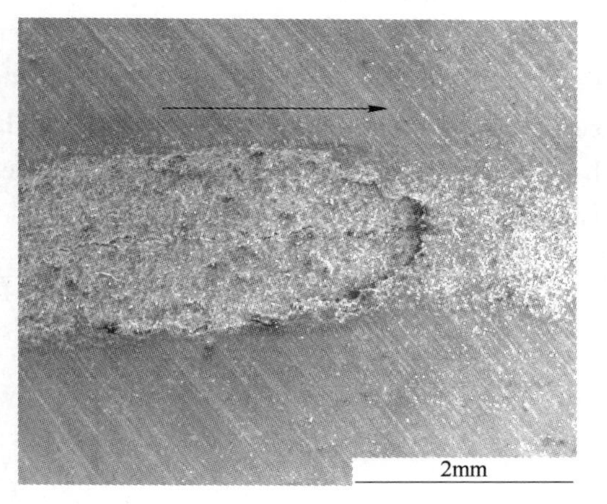

Fig. 7 Worn surface observation of the adhesive delamination failure

Fig. 7 shows the adhesive delamination morphology of the failed coated specimen with a fatigue life of 0.46×10^6 cycles when P_0 was 2.6841 GPa. The dimension of the delaminated coating also with an approximate elliptical shape is distinctly wider than the wear track. Partial substrate obviously exposes owing to the coating peeling off.

3.4 Weibull plots of RCF life data

Weibull distribution is one of the most widely utilized life distributions in the engineering reliability design. It is a versatile distribution that can take on the characteristics of other types of distribution, such as Gaussian distribution, exponential distribution, based on the value of the shape parameter (Weibull plot slope), β[19]. Two-parameter Weibull distribution was employed to characterize the fatigue life date of the coatings, which is given as:

$$F(N) = 1 - \exp\left[-\left(\frac{N}{N_a}\right)^{\beta}\right] \tag{4}$$

where two parameters of Weibull distribution i. e. Weibull plot slope β and characteristic lifetime N_a, were evaluated using the theory of Maximum Likelihood Estimation (MLE) based on the obtained life data with 90% confidence, the MLE relations are given as:

$$\frac{\sum_{i=1}^{n} N_i^{\beta} \ln N_i}{\sum_{i=1}^{n} N_i^{\beta}} - \frac{1}{n}\sum_{i=1}^{n} \ln N_i - \frac{1}{\beta} = 0 \tag{5}$$

$$N_a = \left(\frac{1}{n}\sum_{i=1}^{n} N_i^{\beta}\right)^{\frac{1}{\beta}} \tag{6}$$

where N_i is the fatigue life data of the coating and n is the number of the specimens tested to failure.

Fig. 8 shows the Weibull distribution plots of the fatigue life data of the specimens tested at different contact stresses. This figure illustrates the failure percentage of the specimens (ordinate) vs. the number of stress cycles to failure (abscissa). It can be seen that the fatigue lives of the specimens obviously decreased as the increasing of P_0. Generally, when the higher load was conducted, the Weibull plot slope became larger, which is a measure of the dispersion of the fatigue life. When the value of the slope goes up, the spread and scatter of life data becomes smaller[20]. That is when the coatings bear higher loading, the Weibull plot of fatigue life data becomes steeper, and the time to failure becomes more predicable. On the other side, under lower loading condition, the service life of the specimen is comparatively dispersal.

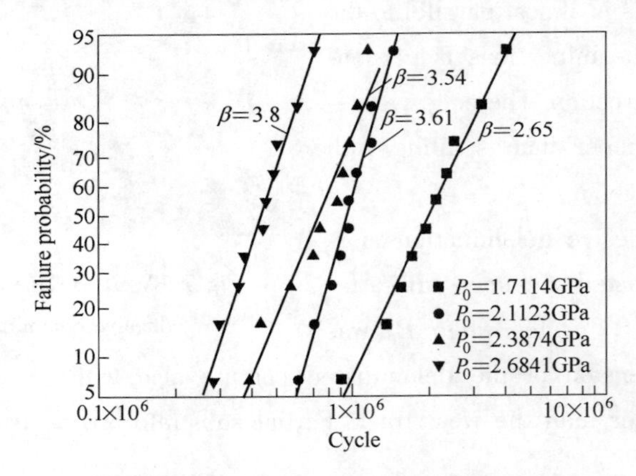

Fig. 8 Weibull comparison of the rolling contact fatigue lives under four kinds of loading conditions

3.5 *S-N* curve of Fe-Cr alloy coatings

The *S-N* curve is an effective approach to evaluate the fatigue performance of materials under various loading conditions. The fatigue process can be intuitionisticly understood by establishing the *S-N* curve[21-23]. The relation between the contact stress and the fatigue life can be described by means of a simple parametric relationship of the form[24]:

$$N = CS^{-m}$$

where N is the RCF life; S is the contact stress; C and m are the parameters which can be obtained by following relations:

$$-\frac{1}{m} = \frac{\sum_{i=1}^{n} X_i Y_i - \frac{1}{n}\sum_{i=1}^{n} X_i \sum_{i=1}^{n} Y_i}{\sum_{i=1}^{n} X_i^2 - \frac{1}{n}\left(\sum_{i=1}^{n} X_i\right)^2} \tag{7}$$

$$\frac{1}{m}\ln C = \frac{1}{n}\left(\sum_{i=1}^{n} Y_i + \frac{1}{m}\sum_{i=1}^{n} X_i\right) \tag{8}$$

where $X_i = \ln N_i$; $Y_i = \ln S_i$ and i is the number of stress grades. The MATLAB software was employed to calculate the aforementioned parameters and describe the S-N curve.

Fig. 9 shows the S-N curves of the Fe-Cr alloy coatings with four types of failure probabilities (P), i. e. $P = 10\%$, 50%, 63.2% and 90%.

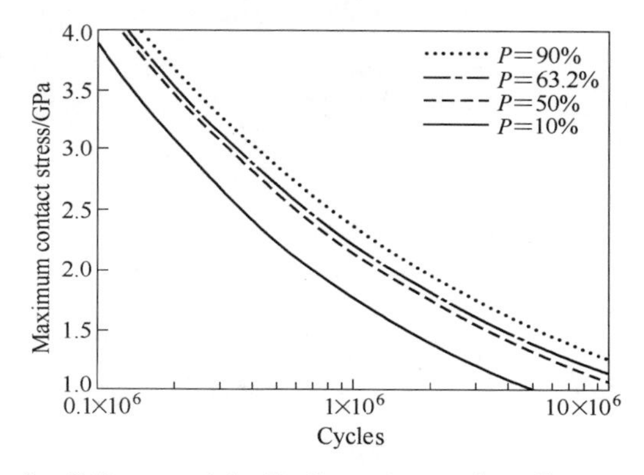

Fig. 9 S-N curve of the Fe-Cr coatings under rolling contact

4 Discussions

4.1 Failure mechanism

The main objectives of this section are to identify the tribological mechanisms leading to abrasion, spalling and delamination on the basis of the worn morphologies observation of the specimens. The previous studies have summarized four kinds of failure modes of the thermal sprayed coatings under the rolling contact[25,26]. All failure modes are identified in the present study.

For the abrasion failure, the asperity contact is considered to be the main reason. Although the RCF tests were conducted under the oil lubrication, there were also some micro-fractures or plastic deformations on the coating surfaces, which resulted in the micro-pitting of the surface. The abrasion failure process can be accelerated by the additional mechanism of three body abrasion, which is due to the action of the wear debris generated during the RCF tests[27].

For the spalling failure, there are some debates for the mechanism of the spalling of the coatings[28]. The asperity contact, the tensile stress at the edge and the lubricant entrapment mechanism are possible to result in this failure mode. Some phenomena in this study imply that the micro-defects on the surface and subsurface of the coatings, such as pores, unmelted or partially melted particles, under asperity contact and alternating contact stress may be the origins of spalling. Fig. 10 shows a micro-pitting after the unmelted particle peeling off. There are some obvious secondary cracks indicated by the arrows. These secondary cracks meet other crack or micro-defects, when they propagate from the surface to subsurface. Finally result in the spalling failure.

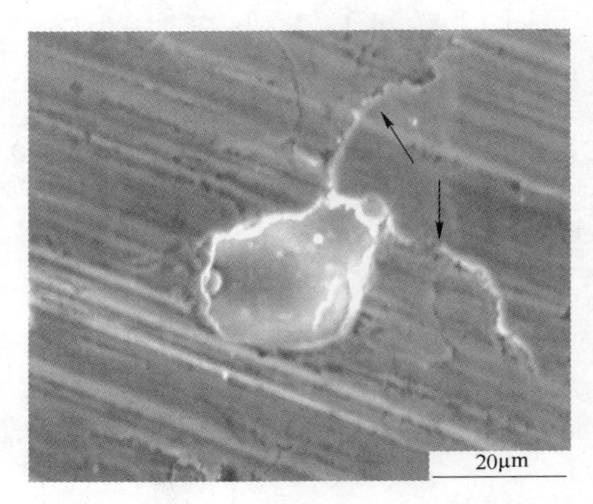

Fig. 10 Worn surface observation of the coating after unmelted particle peeling off

For the cohesive delamination failure, the influence of the OSS may be the failure mechanism. The stress concentrations owing to the pores and micro-cracks at the depth of the maximum OSS are driven to be the origins of fracture. These cracks initiate and propagate under the altering stress, after joining the adjacent cracks and arriving at the surface of the specimen, the final cohesive delamination formed owing to tensile radial stress at the contact edge.

For the adhesive delamination failure, the failure mechanism also has much to do with the influence of the OSS. However it seems that the stress concentrations on the coating/substrate interface play more important role on this failure mode than at the depth of the maximum OSS. Although the maximum OSS located within the coating, the interface altering shear stress is also sufficient to induce the micro-defects on the interface to be the fatigue cracks. Moreover the interface bears more intensive OSS as the contact stress increasing, which finally results in the formation of the adhesive delamination for the lower bond strength between the plasma sprayed coatings and the substrates. That is why the adhesive delamination is the key failure mode under the heavy loading condition.

4.2 Influence of contact stress

Experimental results show that the contact stress obviously influences the RCF behaviors of the Fe-Cr alloy coatings. When the contact stress was relatively low, the abrasion and spalling with higher RCF lives were the key failure modes. Whereas the cohesive and adhesive delamination with lower RCF lives became the main modes under the high contact stress. These phenomena indicate that there are some relationships between the contact stress and the RCF performance of the specimen. Fig. 2 illustrates that with the contact stress increasing, although the maximum OSS all locate within the coatings, the magnitudes of the maximum OSS are obviously increased. At the same time, the location moves towards the coating/substrate interface and the interface stress is also greatly promoted. So some viewpoints can be obtained on basis of the summary of experimental results and the worn surface observations of the failed specimens. When the loading condition is relatively low, the orthogonal shear stress is not enough to drive the fatigue fracture within the coatings, so the specimens are failed in the abrasion and spalling due to the asperity contact. As the contact stress increasing, the coatings are prone to subject the delamination failures for the high maximum and the interface OSS.

At the same time, the RCF life of the specimen is also obviously influenced by the stress condition. The lives of the specimens were high but exhibit more scatter under lower loading condition. But the experimental phenomena were different under higher loading condition. The reasons are also due to the failure mechanisms. The damage accumulating time of the abrasion and spalling failure are long, whereas the catastrophic delamination failure owing to the high shear stress often occurs in the early period for the poor bond strength of the PS coatings.

The prospective RCF life of the specimen under the given loading condition can be easily obtained with the requisite failure probability using the *S-N* curve. Therefore the *S-N* curve can offer some insights for the coating reliability designers.

5 Conclusions

The RCF behavior and life of the Fe-Cr alloy coatings prepared by the supersonic PS process were experimentally investigated under four kinds of contact stresses at the room temperature. The worn surfaces were examined by SEM to address the failure mechanism. The life data of the specimens were characterized using Weibull plot. The *S-N* curve of the coatings was established. On the basis of the above discussions, it can be concluded that:

(1) The contact stress plays an important role on the RCF failure modes of the plasma sprayed Fe-Cr alloy coatings. The specimens are failed in the abrasion and spalling under the lower loading condition, whereas the coatings are prone to subject delamination under the higher loading condition.

(2) The failure mechanisms are investigated on the basis of the worn surface of the failed coatings observation. Induced by asperity contact, the micro-defects on the surface and subsurface result in abrasion and spalling failure. Induced by the altering shear stress, the micro-defects at the depth of maximum OSS and the coating/substrate interface result in the cohesive and adhesive delamination, respectively.

(3) The RCF lives of the coatings are also influenced by the loading condition. RCF lives obviously decrease as the contact stress increasing. However the scatter of the RCF life data is generally lower under the high contact stress.

Acknowledgements

The authors are grateful for the support sponsored by 863 Project (2007AA04Z408), NSFC (50735006), the Foundation(9140C850103080C8510) and S&T Project(2008BAK42B03)

Reference

[1] R. Nieminen, P. Vuoristo, K. Niemi, T. Mantyla, G. Barbezat, Wear 212(1997)66.
[2] S. Stewart, R. Ahemd, Surf. Coat. Technol. 172(2003)204.
[3] M. Fujii, A. Yoshida, Tribol. Int. 39(2006)856.
[4] Y. Gao, X. L Xu, Z. J. Yan, G. Xin, Surf. Coat. Technol. 154(2002)189.
[5] R. Nieminen, P. Vuoristo, K. Niemi, T. Mäntylä, G. Barbezat, Wear 212(1997)66.
[6] S. Stewart, R. Ahmed, T. Itsukaichi, Wear 257(2004)962.
[7] A. Nakajima, T. Mawatari, M. Yoshida, K. Tani, A. Nakahira, Wear 241(2000)166.
[8] D. Stewart, P. Shipway, D. McCartney, Surf. Coat. Technol. 25(1998)13.
[9] S. Stewart, R. Ahmed, Wear 253(2002)1132.
[10] R. Ahmed, M. Hadfield, Tribol. Int. 30(1997)129.
[11] S. Zhu, B. S. Xu, J. K. Yao, Mater. Sci. Forum 475 - 479(2005)3981.
[12] M. Fujii, J. B. Ma, A. Yoshida, S. Shigemura, K. Tan, Tribol. Int. 39(2006)1447.
[13] A. Yoshida, M. Fujii, Tribol. Int. 35(2002)837.
[14] C. Morrow, M. Lovell, Wear 236(1999)360.
[15] X. C. Zhang, B. S. Xu, F. Z. Xuan, S. D. Tu, H. D. Wang, Y. S. Wu, Appl. Surf. Sci. 254(2008)3734.
[16] X. C. Zhang, B. S. Xu, F. Z. Xuan, S. D. Tu, H. D. Wang, Y. S. Wu, Wear. 265(2008)1875.
[17] H. Hertz, J. Reine Angew. Math. 92(1882)156.
[18] X. C. Zhang, B. S. Xu, F. Z. Xuan, S. D. Tu, H. D. Wang, Y. S. Wu, Int. J. Fatigue. 31(2009)906.
[19] C. K. Lin, C. C. Berndt, J. Mater. Sci. 30(1995)111.
[20] K. Hitesh, D. T. Gerardi, L. Rosado, Wear 185(1995)111.
[21] P. Ramaurty Raju, B. Satyanarayana. K. Ramji, S. Babu, Fati. Frac. Eng. Mater. Struc. 32(2008)119.
[22] J. H. Ahn, C. W. Sim, Y. J. Jeong, S. H. Kim, J. of Constru. Steel Research. 65(2009)373.
[23] L. Tom., R. Naman, Int. J. Fatigue 31(2009)70.
[24] E. S. Puchi - Cabrera, M. H. Staia, C. Tovar, E. A. Ochoa - Pérez, Int. J. Fatigue 30(2008)2140.
[25] R. Ahmed, M. Hadfield, Wear 230(1999)39.
[26] R. Ahmed, Wear 253(2002)473.
[27] K. Holmberg, A. Matthews, H. Ronkainen, Tribol. Int. 31(1998)107.
[28] S. Stewart, R. Ahmed, T. Ituskaichi, Surf. Coat. Technol. 190(2005)171.

Determination of Hardness of Plasma-sprayed FeCrBSi Coating on Steel Substrate by Nanoindentation*

Abstract In this paper, we develop a method for determining the hardness of FeCrBSi coating on a 1045 steel substrate. The nano-indents of the FeCrBSi coating exhibit obvious pile-up. The proposed method models the projected contact area as an equilateral triangle bounded by arcs to correct for the effect of pile-up on the contact area. The new method is described in detail and leads to improvement in obtaining coating hardness compared with the widely adopted Oliver-Pharr method.

Key words nanoindentation, hardness, pile-up

1 Introduction

Nanoindentation testing, as a convenient method, has widely been used to characterize the hardness of materials[1-7]. The Oliver-Pharr method for determining the hardness using a calibrated indenter tip area function is widely adopted[8-9]. Its attractiveness stems largely from the fact that mechanical properties can be obtained directly from indentation load-displacement measurements without the need to image the hardness impression. The basic assumption of the method is that the contact periphery sinks in, which limits the applicability of the method because it does not account for the pile-up of material.

Bolshakov et al. have examined the influence of pile-up on the contact area by using the finite element method[10,11]. When the pile-up is small, the contact areas given by the Oliver-Pharr method match very well with the true contact areas obtained from the finite element analyses. When pile-up is significant, the Oliver-Pharr method underestimates the contact area by as much as 50%. Therefore, it is important to develop a new method to extract the contact area accurately.

In the present paper, nanoindentation experiments were used to determine the hardness of FeCrBSi coating with significant pile-up. A new corrected method was proposed to calculate the real contact area, which leads to improvement in obtaining coating hardness compared with the Oliver-Pharr method.

2 Experimental

2.1 Deposition

The 1045 steel sample of 25 mm × 15 mm × 8 mm was coated with a FeCrBSi alloy powder with

* Copartner: Zhu Lina, Wang Haidou, Wang Chengbiao. Reprinted from *Materials Science & Engineering*, 2010, 528(1): 425 – 428.

an average diameter of approximately 40 μm, as shown in Fig. 1. The chemical composition of the FeCrBSi alloy powder is shown in Table 1. The 1045 steel was grit blasted using Al_2O_3 particles prior to air plasma spraying (APS). In order to improve the bonding strength between coating and substrate, Ni/Al alloy powder was coated as undercoat before the FeCrBSi coating was sprayed. The parameters used in the plasma spraying process are listed in Table 2. A 300 μm thick FeCrBSi coating was prepared.

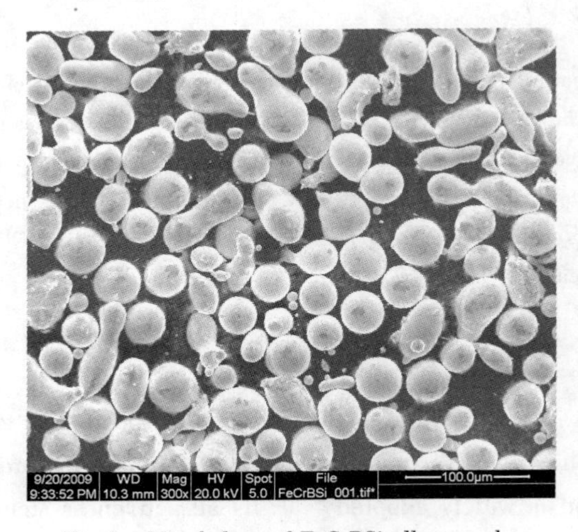

Fig. 1 Morphology of FeCrBSi alloy powder

Table 1 Chemical composition of the FeCrBSi alloy powder

Elements	Cr	B	Si	C	Fe
Composition/%	13.6	1.6	1.1	0.16	Bal.

Table 2 Plasma spray parameters

Parameters	Ni/Al	FeCrBSi
Primary gas, $Ar/m^3 \cdot h^{-1}$	3.6	3.6
Secondary gas, $H_2/m^3 \cdot h^{-1}$	0.22	0.25
Powder feed rate/$g \cdot min^{-1}$	35	40
Spraying current/A	340	380
Spraying voltage/V	140	135
Spraying distance/mm	140	120

2.2 Characterization

The phase constituents of the FeCrBSi coating were identified by X-ray diffraction (XRD). The melting characteristic of the coating was observed by scanning electron microscopy (SEM). The microstructures were also examined from cross-section morphology of the coating by SEM.

2.3 Nanoindentation tests

Prior to indentation, the FeCrBSi coating was mechanically ground and polished with 1.5 μm diamond polishing paste. Sharp indentation experiments were performed using a TriboIndenter® (Hysitron Corporation, USA) equipped with a diamond Berkovich indenter which was also used as an AFM tip and the indented surface was imaged after indentation. All images were collected with a resolution of 256 × 256 pixels, and taken in a 5 μm × 5 μm scan area with a scan rate of 0.5 Hz. The following loading sequence was applied: (1) loading to maximum load; (2) hold for 5 s at maximum load; (3) complete unloading. The maximum load of 3mN, 4.5mN, 6mN and 9 mN were applied, respectively and a 3 × 3 array indents were performed at each maximum load. The hardness of the FeCrBSi coating at each maximum load was averaged over the nine measurements.

3 Results and Discussion

3.1 Characterization

Fig. 2 shows the surface morphology of the FeCrBSi coating. It can be seen that the coating was well-molten. The cross-section morphology of the coating is shown in Fig. 3. The coating with some micro-pores shows obvious lamellar and dense microstructure. Fig. 4 shows the XRD pattern of the FeCrBSi coating. It is found that the diffraction peaks of the coating are mainly α-Fe. The coating shows high crystallinity.

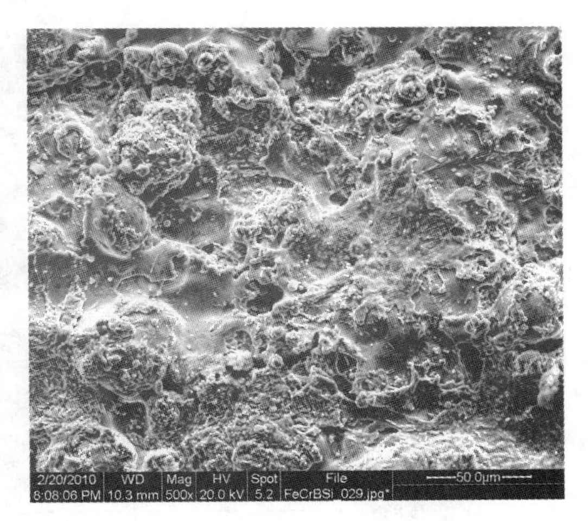

Fig. 2 The surface morphology of the FeCrBSi coating

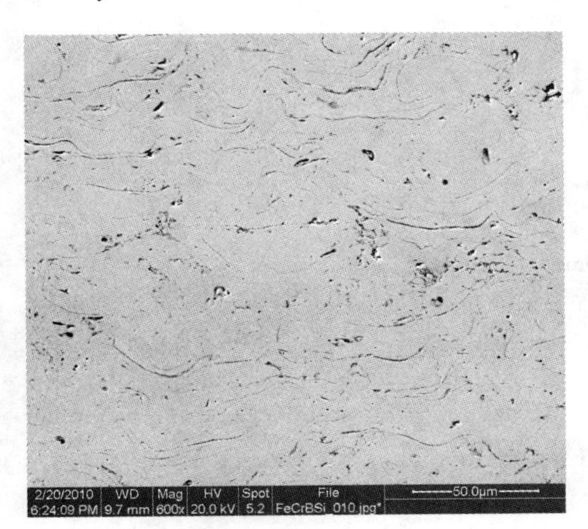

Fig. 3 The cross-section morphology of the FeCrBSi coating

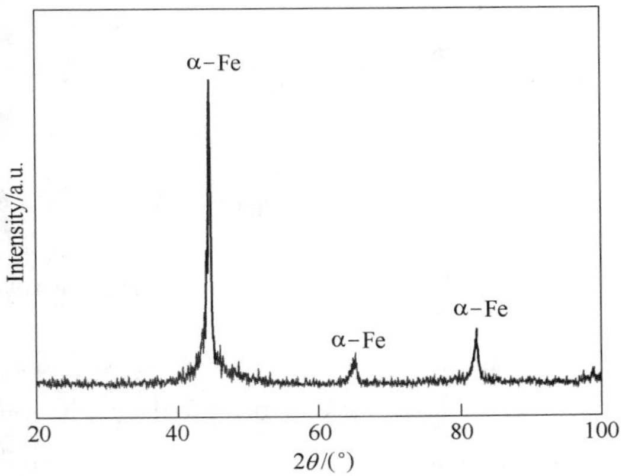

Fig. 4 The XRD pattern of the FeCrBSi coating

3.2 Hardness

The widely known Oliver and Pharr hardness denoted H_{OP}, which uses a calibrated indenter tip area function, is accurate only for indentations that do not exhibit significant pile-up or sink-in. The Oliver and Pharr method will overestimate or underestimate the hardness when pile-up or sink-in occurs.

Fig. 5 shows a typical AFM image and of cross-section profile nanoindent for the FeCrBSi coating at a fixed load of 9 mN. The light area along the sides of the triangular indent in Fig. 5a is the pile-up material and the height of this piled-up material can be easily determined from the cross-section of the indent in Fig. 5b. We now propose a new method to calculate a corrected contact area and compute a new corrected hardness denoted $H_{corrected}$.

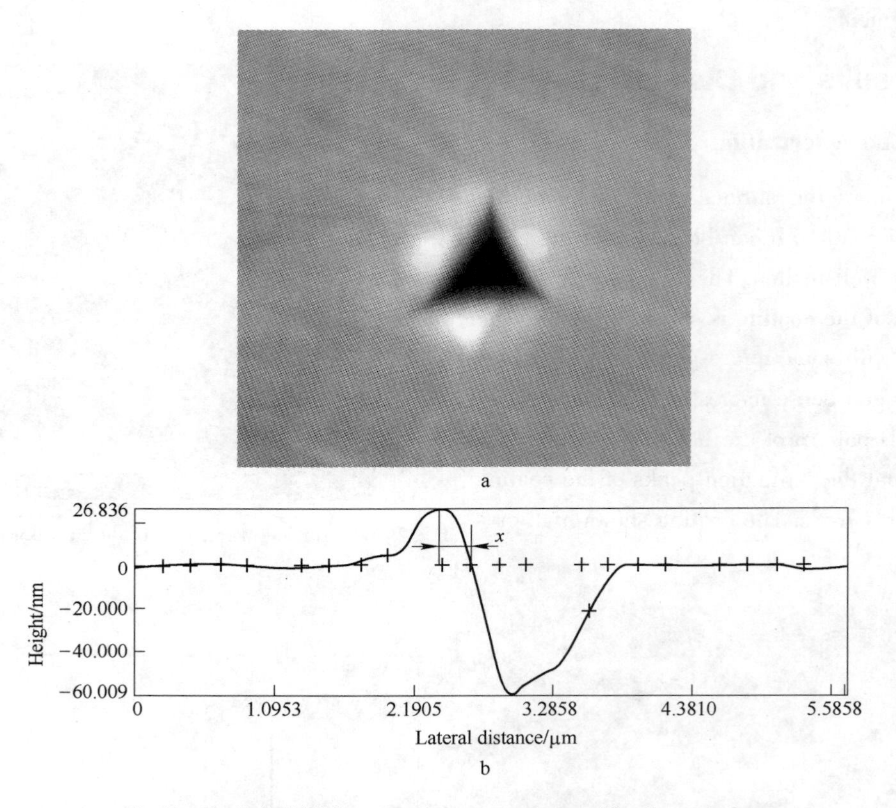

Fig. 5 Typical AFM image and cross-section profile of nanoindent for the FeCrBSi coating at a fixed load of 9 mN

a—AFM image; b—Cross-section profile

It is clear that the material piles-up only along the flat faces of the indent and not at the corners. Also the periphery of the piled-up material forms arcs along the triangular edges. Therefore, the projected contact area can be modeled as an equilateral triangle bounded by arcs as shown in Fig. 6[12]. The area of the three arcs A_1 can be calculated using geometrical relationships:

$$A_1 = 3\left(\frac{\theta}{360}\pi R^2 - \frac{1}{2}l \cdot \frac{\frac{l}{2}}{\tan\frac{\theta}{2}}\right) = 3\left[\frac{\theta}{360}\pi\left(\frac{\frac{l}{2}}{\sin\frac{\theta}{2}}\right)^2 - \frac{l}{2}l \cdot \frac{\frac{l}{2}}{\tan\frac{\theta}{2}}\right] \tag{1}$$

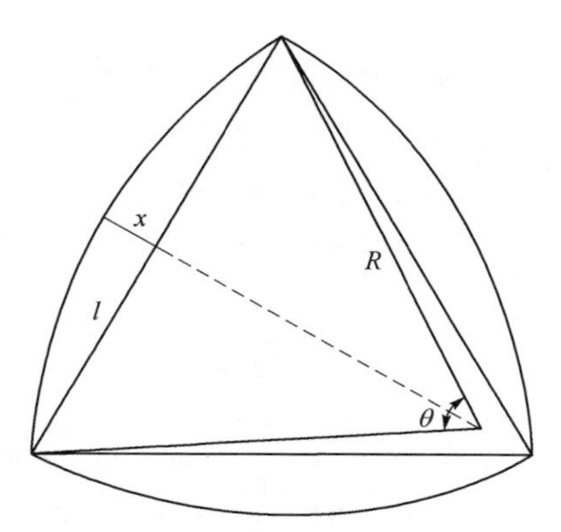

Fig. 6 Schematic representation of the projected contact area due to pile – up

The area of the equilateral triangle A_2 can be calculated as:

$$A_2 = \frac{1}{2}l \cdot \frac{\sqrt{3}}{2}l = \frac{\sqrt{3}}{4}l^2 \tag{2}$$

Then the real projected contact area A is:

$$\begin{aligned} A = A_1 + A_2 &= 3\left[\frac{\theta}{360}\pi\left(\frac{\frac{l}{2}}{\sin\frac{\theta}{2}}\right)^2 - \frac{1}{2}l \cdot \frac{\frac{l}{2}}{\tan\frac{\theta}{2}}\right] + \frac{\sqrt{3}}{4}l^2 \\ &= \frac{l^2}{4}\left(\frac{\theta\pi}{120\sin^2\frac{\theta}{2}} - 3\cot\frac{\theta}{2} + \sqrt{3}\right) \end{aligned} \tag{3}$$

The geometry of the Berkovich indenter is that the triangle, l, is equal to 7.53 h_c:

$$l = 7.53\ h_c \tag{4}$$

where h_c is the contact depth. h_c defined in the Oliver – Pharr method is always smaller than h_{max} where h_{max} is the maximum depth ($h_c < h_{max}$). However, for materials that pile-up, $h_c > h_{max}$. Hence, the real contact depth should be $h_c = h_{max} + h_{pile\text{-}up}$ for pile-up materials[13]. As the pile-up heights around nanoindent are different, the average height h_p^{ave} around the three sides of nanoindent is defined as the pile – up height, namely, $h_c = h_{max} + h_p^{ave}$.

Re-writing Eq. (3), we get :

$$A = 14.175\left(\frac{\theta\pi}{120\sin^2\frac{\theta}{3}} - 3\cot\frac{\theta}{2} + \sqrt{3}\right)(h_{max} + h_p^{ave})^2 \tag{5}$$

where θ can be determined using the geometrical relationships. θ will vary depending on the material properties of the coating and substrate[12].

$$x = R - R\cos\frac{\theta}{2} = \frac{\frac{l}{2}}{\sin\frac{\theta}{2}}\left(1 - \cos\frac{\theta}{2}\right) = 3.765\,\frac{1 - \cos\frac{\theta}{2}}{\sin\frac{\theta}{2}}(h_{max} + h_p^{ave}) \tag{6}$$

where x can be determined from Fig. 5b.

Equating Eq. (5) and Eq. (6), we can calculate the real projected contact area A.

From Fig. 5b and the load-depth curve of the FeCrBSi coating, the values of x, h_{max} and h_p^{ave} were obtained. According to Eq. (6), $\theta = 53.7°$ was calculated. Then θ was substituted in Eq. (5) to calculate the real contact area. The hardness H of the FeCrBSi coating can be determined by Eq. (7).

$$H = \frac{P_{max}}{A} \tag{7}$$

Using the Eq. (7), the hardness of the FeCrBSi at each maximum load were calculated by the above corrected method and Oliver-Pharr method, as shown in Fig. 7. The hardness calculated by the two methods exhibit no significant load effect. In this experiment, all the indentation depths are less than the one-tenth of the FeCrBSi coating thickness, and thus the substrate has no effect on the determined coating hardness. $H_{corrected}$ and H_{OP} vary between 4.59 GPa and 6.46 GPa, and between 8.07 GPa and 9.71 GPa, respectively. It is obvious that the corrected values are lower than the values obtained by the Oliver-Pharr method. The differences of the hardness obtained by the two methods are between 27% –49%. For the material with significant pile-up, the Oliver-Pharr method can overestimate the hardness by as much as 50% [11]. It is clear that the new corrected method leads to improvement in obtaining coating hardness compared with the Oliver-Pharr method.

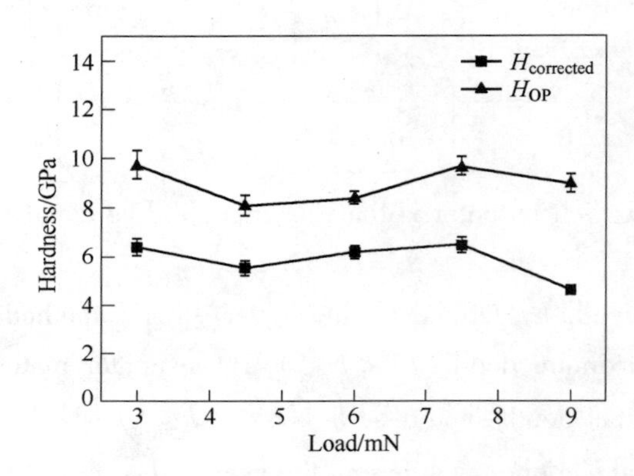

Fig. 7 Hardness calculated by the corrected method and Oliver-Pharr method at each maximum load

In the present study, the pile-up height of the residual indentation was assumed as the pile-up height formed during loading. However, after the fully removing of the indenter, the pile-up height underwent changes. As no method can be used to extract the pile-up height formed during loading at present, it is important to develop a much better technique to obtain the real contact depth for pile-up materials. Moreover, the top points on the residual surface profile are assumed as the contact boundary in this paper, but further study should be made in order to confirm its validity. Although the proposed method is based on some assumption, the measured hardness should be close to the real value compared to the commonly used Oliver-Pharr method.

4 Conclusions

A new method for extracting FeCrBSi coating hardness from AFM image has been described. In this method, the projected contact area is modeled as an equilateral triangle bounded by arcs. The real contact depth is redefined as the sum of maximum depth and pile-up height. The hardness calculated by the new method and widely adopted Oliver-Pharr method exhibit no significant load effect. $H_{corrected}$ and H_{OP} vary between 4.59 GPa and 6.46 GPa, and between 8.07 GPa and 9.71 GPa, respectively. The differences of the hardnesses obtained by the two methods are between 27%-49%. The new method leads to improvement in obtaining the hardness for pile-up materials. However, further study should be made in order to confirm its validity.

Acknowledgements

This paper was financially supported by Advanced Maintenance Research Project (9140A27-0304090C8501), NSFC (50875053, 50735006), Equipment Research Project and Fundamental Research Funds for the Central Universities (2009PY08).

References

[1] Seung Min Han, Ranjana Saha, William D. Nix. Determining hardness of thin films in elastically mismatched film-on-substrate systems using nanoindentation. Acta Materialia, 2006, 54: 1571 - 1581.

[2] Victoria M. Masterson, Xiaoping Cao. Evaluating particle hardness of pharmaceutical solids using AFM nanoindentation. International Journal of Pharmaceutics, 2008, 362: 163 - 171.

[3] Abhijit Kar, Sanjay Chaudhuri, Pratik K. Sen, Ajoy Kumar Ray. Evaluation of hardness of the interfacial reaction products at the alumina-stainless steel brazed interface by modeling of nanoindentation results. Scripta Materialia, 2007, 57: 881 - 884.

[4] H. Huang, Y. Q. Wu, S. L. Wang, Y. H. He, J. Zou, B. Y. Huang, C. T. Liu. Mechanical properties of single crystal tungsten microwhiskers characterized by nanoindentation. Materials Science and Engineering A, 2009, 523: 193 - 198.

[5] J. J. Roa, E. Gilioli, F. Bissoli, F. Pattini, S. Rampino, X. G. Capdevila, M. Segarra. Study of the mechanical properties of CeO_2 layers with the nanoindentation technique. Thin Solid Films, 2009, 518: 227 - 232.

[6] A. Concustell, N. Mattern, H. Wendrock, U. Kuehn, A. Gebert, J. Eckert, A. L. Greer, J. Sort, M. D. Baró. Mechanical properties of a two-phase amorphous Ni-Nb-Y alloy studied by nanoindentation. Scripta Materialia, 2007, 56: 85 - 88.

[7] C. L. Chen, A. Richter, R. C. Thomson. Mechanical properties of intermetallic phases in multi-component Al-Si alloys using nanoindentation. Intermetallics, 2009, 17: 634 - 641.

[8] Oliver W C, Pharr G M. An improved technique for determining hardness and elastic modulus using load and displacement sensing indentation experiments. Journal of Materials Research, 1992, 7 (6): 1564 - 1583.

[9] Oliver W C, Pharr G M. Measurement of hardness and elastic modulus by instrumented indentation: Ad-

vances in understanding and refinements to methodology. Journal of Materials Research, 2004, 19(1): 3 - 20.

[10] Bolshakov A, Pharr G M. Influences of pile-up on the measurement of mechanical properties by load and depth sensing indentation techniques. Journal of Materials Research, 1998, 13(4): 1049 - 1058.

[11] Bolshakov A, Oliver W C, Pharr G M. Influences of stress on the measurement of mechanical properties using nanoindentation finite element simulation. Journal of Materials Research, 1996, 11(3): 752 - 759.

[12] R. Saha, W. D. Nix. Soft films on hard substrates – nanoindentation of tungsten films on sapphire substrates. Materials Science and Engineering A, 2001, 319 - 321: 898 - 901.

[13] Xi Chen, Jin Yan, Anette M. Karlsson. On the determination of residual stress and mechanical properties by indentation. Materials Science and Engineering A, 2006, 416: 139 - 149.

加载条件下铁磁材料疲劳裂纹扩展自发射磁信号行为研究*

摘　要　研究了45CrNiMoVA铁磁钢中心裂纹试样加载条件下疲劳裂纹扩展过程中表面自发射磁信号的变化规律。结果表明：承受疲劳载荷前，各检测线呈现初始磁异变峰特征，切口中心检测线具有最大异变峰－峰值；加载后至疲劳裂纹萌生前，切口位置磁性反转，显示一次异变峰特征，远离切口检测位置磁信号呈现线性；疲劳裂纹扩展过程中，各检测线依次呈现二次异变峰；二次异变峰－峰值和正负峰间斜率可以动态表征疲劳裂纹扩展寿命。基于磁荷概念和等效磁场理论分析了加载条件下疲劳裂纹扩展自发射磁信号行为。

关键词　金属磁记忆　疲劳裂纹　加载　磁异变峰　磁荷

再制造毛坯剩余寿命评估是鉴别再制造生产对象有无再制造价值的关键技术[1,2]。再制造毛坯是经过长期服役的废旧零部件，在服役过程中，零部件可能产生不同程度的损伤。再制造前，依靠先进的无损检测技术预测废旧件剩余寿命是保证再制造产品质量的重要途径。

金属磁记忆技术是一项尚处于成长期的先进无损检测方法，它利用铁磁材料在地磁场环境下和服役过程中自发产生的弱磁信号来发现其应力集中部位，在早期损伤诊断领域极具潜力[3~6]，是再制造毛坯剩余寿命评估的重要技术手段之一。然而，金属磁记忆信号幅值微弱，易受多种因素影响，信号分布特征复杂，目前尚缺乏系统的基础实验研究，金属磁记忆技术应用于再制造毛坯剩余寿命评估的方式仍需探索[7,8]。

前期针对金属磁记忆技术的研究[9]发现，铁磁材料疲劳裂纹扩展过程中产生强烈的自发射磁信号，卸载状态下能检测到保留一次异变峰的磁记忆信号特征，一次异变峰－峰值与疲劳裂纹长度具有相关性。为深入研究疲劳裂纹扩展过程中产生的自发射磁信号的变化规律，探索金属磁记忆技术表征疲劳裂纹扩展寿命的方法，本文进一步研究了加载条件下含中心裂纹（CCT）试样表面不同位置随裂纹扩展产生的自发射磁信号行为，为建立再制造毛坯疲劳剩余寿命评估模型提供基础数据。

1　实验方法

实验材料选用45CrNiMoVA钢，其化学成分（质量分数,%）为：C 0.42～0.49，Si 0.17～0.37，Mn 0.50～0.80，P≤0.030，S≤0.030，Cr 0.80～1.10，Ni 1.30～1.80，Mo 0.20～0.30，V 0.10～0.20，Fe 余量。其力学性能为：屈服强度1323MPa，抗拉强度1470MPa，延伸率7%，断面收缩率35%，冲击韧性39J/cm^2。按照GB/T6398—2000加工含中心穿透裂纹的CCT试样。试样厚6mm，在中心采用线切割技术

* 本文合作者：董丽虹、董世运、薛楠。原发表于《金属学报》，2011，47(3)，257～262。国家自然科学基金项目50735006和50975283资助。

加工出人工预制切口，试样形状尺寸如图 1a 所示，中心切口总长度 $2a_0 = 11.35$mm。在试样表面布置检测线，切口正中心的检测线为中心线，向右每间隔 3mm 分别为 A1 ~ A5 线，向左每间隔 3mm 分别为 B1 ~ B5 线。所有检测线的长度为 50mm。实验前对试样进行真空热处理退磁。

采用 MTS810 型液压伺服试验机进行拉 - 拉疲劳实验，施加正弦波恒幅疲劳载荷，最大载荷 $\sigma_{max} = 260$MPa，应力比 $R = 0$，频率 $f = 10$Hz。表面杂散磁信号的检测采用 EMS - 2003 磁记忆检测仪，它包含有滤波、放大、A/D 转换等模块。检测时仪器探头安装在非磁性的电控三维扫描架上，垂直于试样表面，提离值 1mm。在实验过程中，试样经预定的循环次数后停机，保留最大静拉力，控制扫描架带动探头以相同的方式沿各条检测线从一端向另一端运动，获取各检测线表面的杂散磁场法向分量 $H_p(y)$，信号经处理后送入计算机分析。在线检测系统如图 1b 所示，中心裂纹长度由读数显微镜读取。

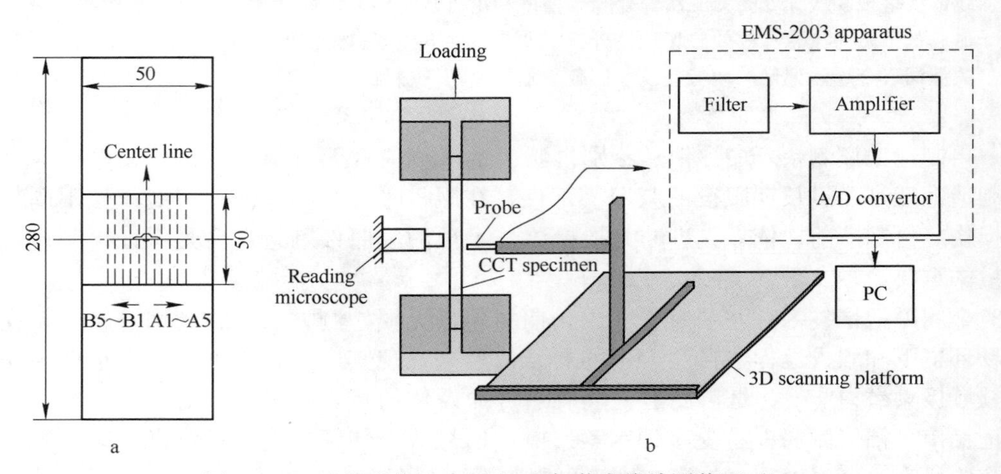

图 1　CCT 试样形状尺寸（a）和加载在线检测装置（b）

Fig. 1　Dimensions (unit: mm) and measured lines distribution of center - cracked tension (CCT) specimen (a) and experimental apparatus under loading condition (b)

2　实验结果

2.1　疲劳裂纹扩展过程中自发射磁信号变化规律

在拉 - 拉疲劳实验过程中，随疲劳循环次数的增多，疲劳裂纹将从预制中心切口两侧尖端萌生，沿着与检测线垂直的方向分别向试样边缘扩展，最终经 44147 cyc 疲劳循环后试样破断。拉 - 拉疲劳过程中，中心裂纹总长度 $2a$ 与疲劳循环次数之间的关系如图 2 所示。可见，初始阶段当疲劳循环次数较少时，裂纹增长缓慢；当疲劳循环次数大于 1.6×10^4 cyc 后，裂纹稳

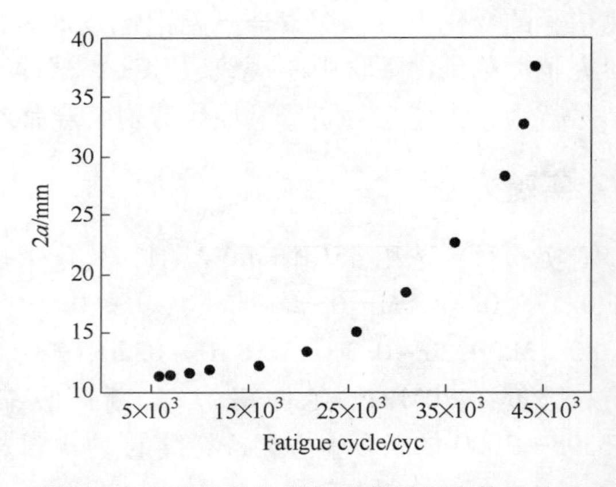

图 2　中心裂纹总长度与疲劳循环次数的关系

Fig. 2　Relationship between the total central crack length $2a$ and fatigue cycles

定扩展；循环次数超过 4.3×10^4 cyc 后裂纹快速扩展，直至断裂。

拉－拉实验过程中试样表面中心线两侧的两组检测线变化规律相似，图 3 给出了其中一组 A1～A5 线的检测结果。检测试样表面初始磁状态时，试样装夹在疲劳试验机上下夹头之间，此时试样两端仅受到试验机夹头的夹紧力，未受轴向疲劳载荷作用。观察初始的磁信号分布规律，发现各检测线对应预制切口部位存在磁异变峰，其特征为左侧波峰，右侧波谷。中心检测线跨过的切口长度最大，该检测线的初始异变峰最为显著，具有最大的峰-峰值；A1 线跨过的切口长度距离减小，其峰-峰值降低；A2 线位于切口尖端前缘，初始异变峰-峰值继续降低；A3～A5 线虽然位于材料无缺陷部位，但受到预制切口的影响，亦显示初始异变峰，其峰-峰值随着与切口距离增大依次减小，

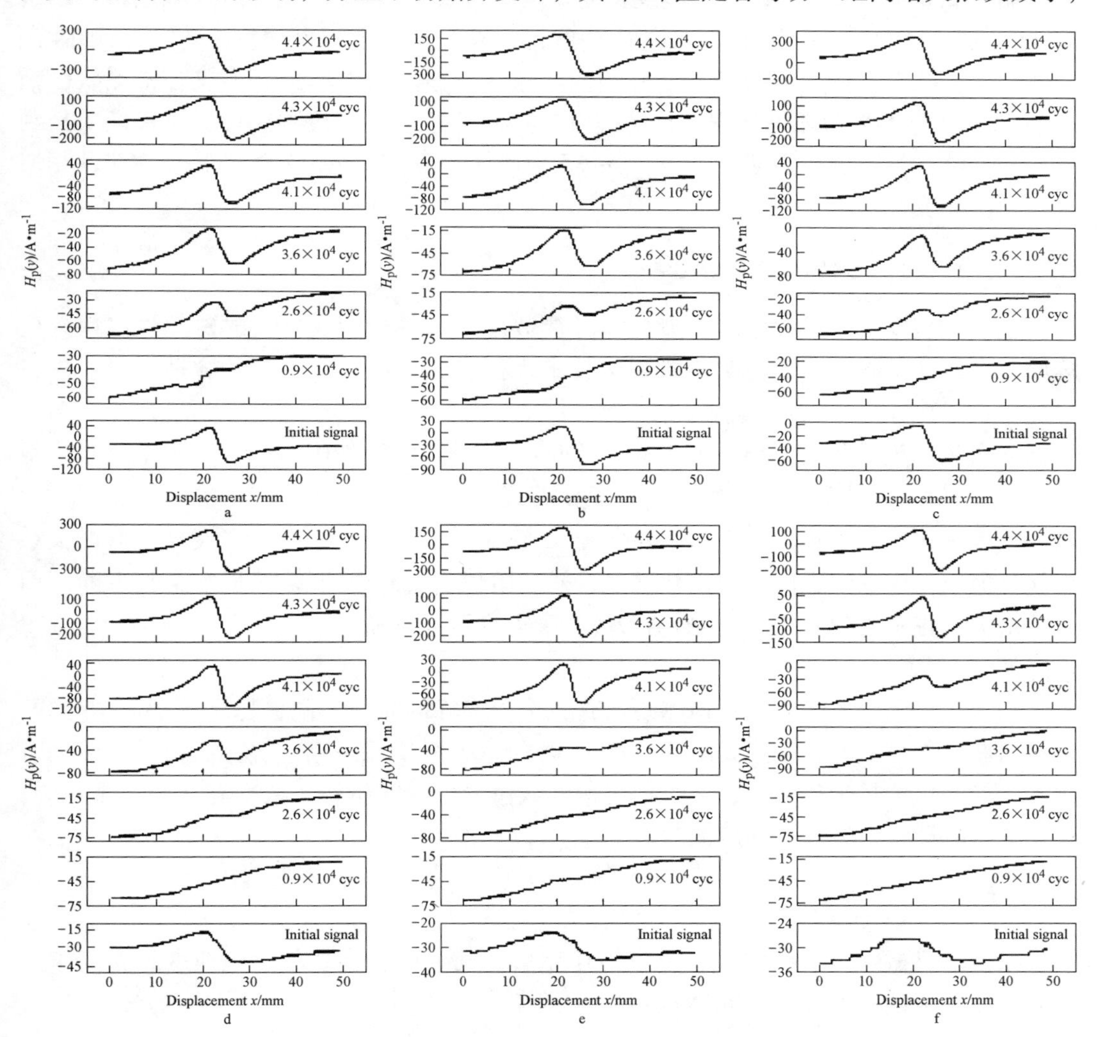

图 3　疲劳实验过程中中心线（a）及 A1～A5 线（b～f）在不同循环次数下杂散磁场法向分量 $H_p(y)$ 信号特征

Fig. 3　Normal component of stray field $H_p(y)$ signal distributions during the tension－tension fatigue test at different cycles measured on center line（a），A1 line～A5 line（b～f）

初始异变峰与卸载检测时分布特征相同。

施加疲劳载荷后，疲劳裂纹萌生前，发现各检测线初始异变峰－峰值迅速降低，随循环次数增加，逐渐失去初始异变峰的特征，穿过切口和靠近切口的检测线磁信号曲线呈现微弱的反转一次异变峰特征，左侧波谷，右侧波峰；远离切口的检测线磁信号则呈现良好线性。循环 6.2×10^3 cyc 后观察到疲劳裂纹萌生；循环至 9×10^3 cyc 时，中心裂纹长度 $2a=11.667$mm，此时萌生的疲劳裂纹还未扩展到 A2 检测线。分析检测结果发现，中心线——A3 线的磁曲线中波峰波谷特征略有增强，A4 和 A5 检测线磁信号仍然保持线性。

循环 2.6×10^4 cyc 后，中心裂纹长度 $2a=14.965$mm，此时疲劳裂纹刚刚扩展超过 A2 线，而中心线、A1 线和 A2 线的磁曲线都捕获到明显的二次异变峰，其左侧正峰、右侧负峰，分布特征与初始磁异变峰相同。随疲劳循环次数增加，不仅各检测线已捕获的二次异变峰-峰值不断增强，疲劳裂纹扩展新到达部位的检测线位置也逐渐采集到二次异变峰，位于切口尖端前方的 A3 ~ A5 检测线的自发射磁信号分布特征由线性相继呈现二次异变峰形貌。

2.2　自发射磁信号特征参量变化规律

在疲劳裂纹扩展过程中，随疲劳循环次数增加，各检测线自发射磁信号二次异变峰的分布特征不断变化。为明确磁记忆技术表征疲劳裂纹扩展寿命的合适参量，进一步提取各二次异变峰的峰-峰值 $\Delta H_{p2}(y)$ 及二次异变峰正负峰间的斜率 K_{p2}，作出两者与循环次数的关系曲线，如图 4 所示。可见，二次异变峰-峰值 $\Delta H_{p2}(y)$ 和正负峰间斜率 K_{p2} 是一对孪生特征参量，均能够表征疲劳裂纹扩展寿命。2 个特征参量随循环次数增加，呈现相反的变化趋势。二次异变峰-峰值 $\Delta H_{p2}(y)$ 与疲劳循环次数 N 之间的关系曲线类似于 a-N 曲线，随循环次数增加，初期阶段 $\Delta H_{p2}(y)$ 值缓慢增加，临近断裂前显著增大。正负峰间斜率 K_{p2} 在疲劳裂纹扩展寿命的初期阶段缓慢降低，断裂前斜率急剧降低。

图 4 还显示，不同位置的检测线捕获二次异变峰的时间不同，越靠近切口中心线，越早捕获二次异变峰，采集到二次异变峰-峰值 $\Delta H_{p2}(y)$ 和斜率 K_{p2} 特征参量的数据点越多；检测位置越远离切口，裂纹扩展达到该位置的时间越晚，二次异变峰出现越晚，采集到的特征参量数据点相应减少。

3　分析讨论

根据磁机械效应，铁磁材料受应力作用后，其表面磁性能将发生改变，相当于外加应力在铁磁试样上产生一个等效磁场作用[10~12]，使试样磁性能改变。等效磁场 H_σ 可用下式表示：

$$H_\sigma=\frac{1}{\mu_0}\times\frac{\partial E_\sigma}{\partial M}=\frac{3}{2}\times\frac{\sigma}{\mu_0}\left(\frac{\partial\lambda}{\partial M}\right)\sigma(\cos^2\theta-\nu\sin^2\theta)\tag{1}$$

式中，μ_0 为真空磁导率；E_σ 为应力能；M 为磁化强度，σ 为外加应力；λ 为磁致伸缩系数；θ 为外加应力方向与磁场强度之间的夹角；ν 为泊松比。

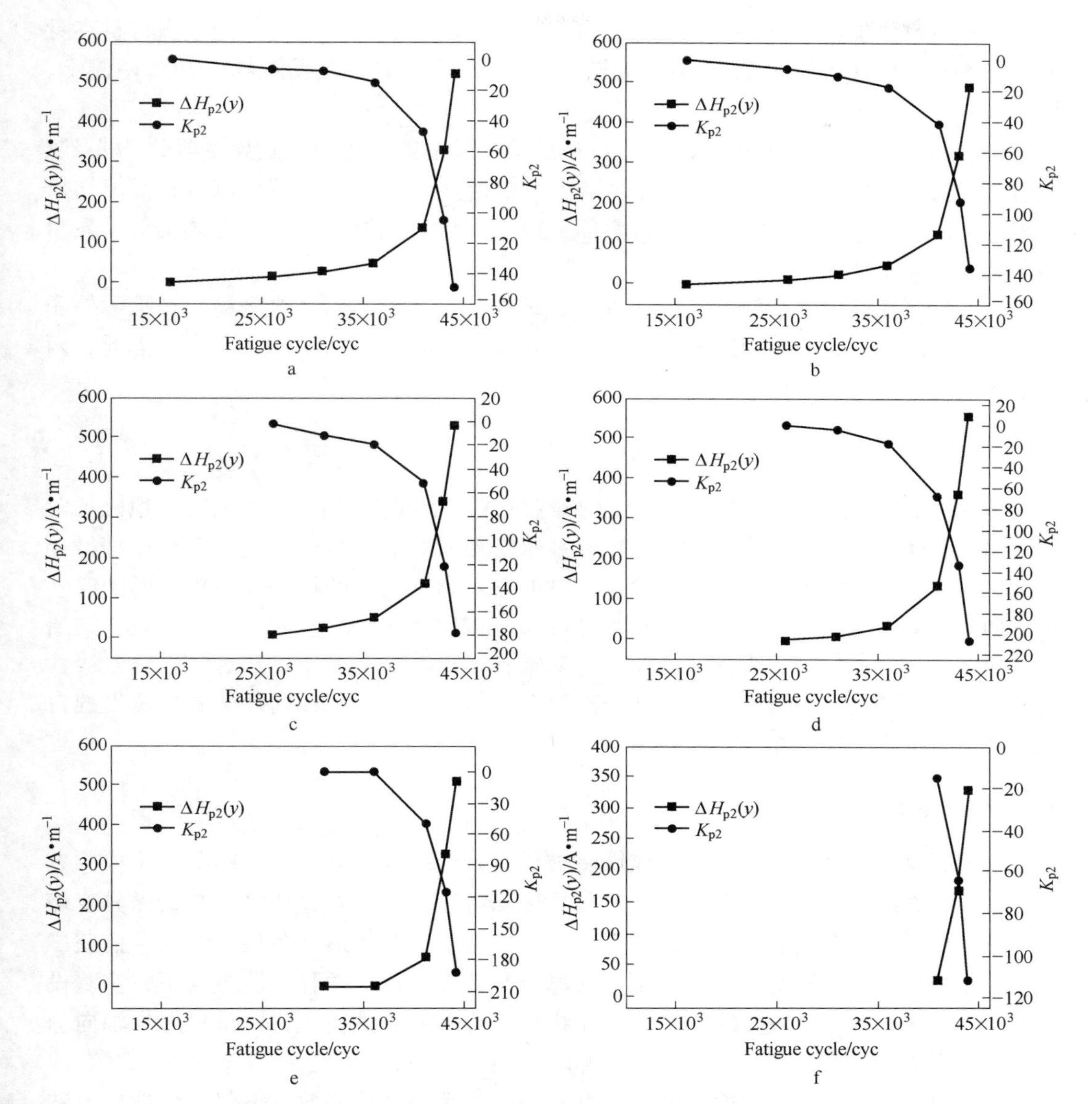

图 4 不同循环次数下中心线（a）及 A1 ~ A5（b ~ f）线二次异变峰 - 峰值 $\Delta H_{p2}(y)$ 和正负峰间斜率 K_{p2} 的变化

Fig. 4 Curves of the second abnormal peak - to - peak value $\Delta H_{p2}(y)$ and slope coefficient K_{p2} corresponding to the fatigue cycles during the tension - tension fatigue test measured on center line (a), A1 line ~ A5 line (b ~ f)

3.1 施加疲劳载荷前的初始磁状态

分析试样未加载前的初始磁状态，各检测线在对应预制切口位置呈现出磁异变峰特征，此时除试验机夹头夹持试样的压力外，试样检测区域未受外力作用。由于试验机的夹持力局限在试样两端，其对检测区域磁性的影响可以忽略不计，各检测位置无外加应力诱发生成的等效磁场。地磁场作为唯一的外磁场，对试样起到磁化作用。预制切口使得铁磁材料在该部位不连续，由于切口内的空气磁导率远低于铁磁材料本身的磁导率，导致在该位置地磁场磁力线的通路被截断，异性磁荷累积在切口两侧[9]，生成初始磁异变峰。

中心检测线跨越切口距离最长，正负磁荷累积最多，其初始磁异变峰最为强烈；随着与中心线距离的增加，磁荷累积量减少，初始磁异变峰的峰-峰值减弱；远离切口位置的检测线，布置在光滑连续材料表面，但受到切口尖端影响，磁力线绕行，显示微弱磁异变峰特征。

在线检测和离线检测两种实验条件下，试样放置方向不同，地磁场均是唯一外磁场，地磁场的矢量特性影响磁记忆信号幅值，但不影响信号的分布特征[13]，因此，两者的初始磁异变峰特征相同。

3.2 加载但未萌生疲劳裂纹的磁状态

施加疲劳载荷后，轴向拉应力产生的等效磁场与地磁场反向[14]，削弱了地磁场对试样的磁化作用，降低了切口产生的初始磁异变峰-峰值，使得试样磁性改变。随循环次数增加，等效磁场磁化作用占主导，切口上方及靠近切口的检测线呈现微弱的磁性相反的一次异变峰（左侧波峰，右侧波谷）；离开切口的检测位置材料光滑连续，初始磁异变峰完全消失，磁曲线显示线性，呈现磁有序状态；距离切口越远，磁信号线性越显著。萌生裂纹前，加载原位和卸载离位两种检测方式下，轴向疲劳载荷诱发的自发射磁信号变化规律相同[15]。

3.3 加载生成疲劳裂纹的磁状态

疲劳裂纹扩展过程中，裂纹导致铁磁材料原先连续部位开裂，在卸载离位检测方式下，疲劳裂纹扩展过程中仅呈现一次磁异变峰特征，一次磁异变峰与初始磁异变峰分布特征相反；而在线加载检测方式下，随着疲劳裂纹的扩展，各检测线先后捕获二次磁异变峰，二次磁异变峰与初始磁异变峰分布特征相同。两种检测方式下，实验设备和试样形式相同，最主要的区别在于在线加载检测时，试验机将对试样施加轴向静拉力，分析认为二次异变峰的出现与检测时存在拉应力直接相关。

静拉力导致检测时疲劳裂纹处于张开状态，弹性应变能释放，不仅以弹性应力波的形式发射出声发射信号，还将激发磁场能量，驱动切口及新形成的裂纹面两侧生成新的正负磁荷。当开裂部位裂纹宽度足够大时，积累的异性磁荷达到一定数量，则在该位置就捕捉到二次磁异变峰。疲劳裂纹向前扩展，到达新的检测位置，形成新的裂纹面，积累异性磁荷，在该检测位置生成新的二次异变峰；同时切口及已开裂部位的异性磁荷继续累积，表现为已形成的二次异变峰 - 峰值持续增大，远离切口位置新的二次异变峰不断形成，裂纹扩展过程中裂纹面磁荷累积如图 5 所示。二次异变峰 - 峰值和正负峰间斜率与疲劳裂纹扩展寿命直接相关。加载条件下铁磁材料疲劳裂纹扩展生成的自发射磁信号特征参量 $\Delta H_{p2}(y)$ 和 K_{p2} 可以动态表征铁磁材料疲劳裂纹扩展寿命，利用这一自发射的磁信号可以发展一种全新的适用于再制造毛坯质量监控的无损检测方法。

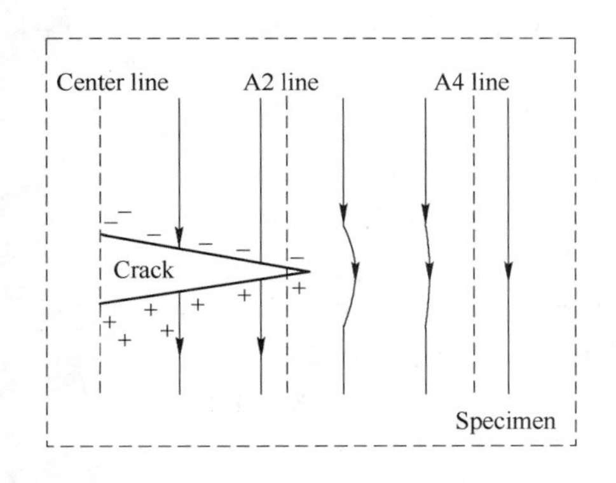

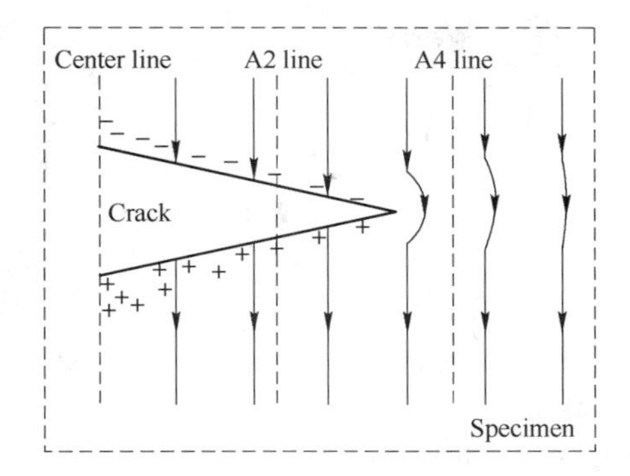

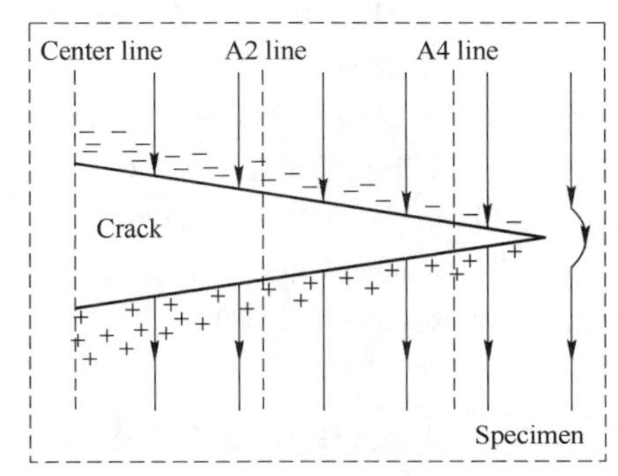

图 5　随疲劳裂纹扩展裂纹面磁荷累积示意图

Fig. 5　Magnetic charges accumulation on the crack surface during fatigue crack propagation

4　结论

（1）承受疲劳载荷前，中心裂纹试样表面各检测线呈现初始磁异变峰特征，切口中心初始磁异变峰最显著，离开切口距离增大，磁异变峰减弱；施加疲劳载荷后，轴向应力产生等效磁场，使试样磁性反转，显示微弱的一次异变峰特征，远离切口的检测线磁信号曲线呈现良好线性；疲劳裂纹萌生后，各检测线相继呈现二次异变峰特征。

（2）二次异变峰-峰值 $\Delta H_{p2}(y)$ 与正负峰间的斜率 K_{p2} 是疲劳裂纹扩展过程中可以动态表征疲劳裂纹扩展的自发射磁信号特征参量，随疲劳循环次数增加，两者变化趋势相反。

（3）初始磁异变峰和一次异变峰的形成与地磁场和应力产生等效磁场相互作用诱发切口累积异性磁荷相关；二次异变峰的形成与疲劳裂纹受静拉力作用处于张开状态，发射磁场能量，驱动切口及裂纹面异性磁荷重新排列累积增强有关。

参 考 文 献

[1] Xu B S, Liu S C. *Remanufacture and Recycling Economy*. Beijing: Science Press, 2007: 136.
徐滨士，刘世参．再制造与循环经济．北京：科学出版社，2007：136.

[2] Xu B S. *Theory and Technology on Remanufacturing Engineering*. Beijing: National Defence Industry Press, 2007: 124.

徐滨士．装备再制造工程的理论与技术．北京：国防工业出版社，2007：124.
[3] Doubov A A. *Weld World*, 1998, 41: 196.
[4] Doubov A A. *Therm Eng*, 2001, 48: 289.
[5] Doubov A A. *Weld World*, 2002, 46: 317.
[6] Li L M, Huang S L, Wang X F, Li S S, Shi K R. *Trans Nonferrous Met Soc*, 2003, 13: 6.
[7] Dong L H, Xu B S, Dong S Y, Chun Q Y, Wang D. *NDT&E Int*, 2008, 41: 184.
[8] Dong L H, Xu B S, Dong S Y, Qu M X, Wang D. *Surf Eng China*, 2010, 23: 106.
董丽虹，徐滨士，董世运，曲美霞，王丹．中国表面工程，2010，23：106.
[9] Dong L H, Xu B S, Dong S Y, Chen Q Z, Wang D. *Int J Fatigue*, 2008, 30: 1599.
[10] Lu L, Jiles D C. *IEEE Trans Magn*, 2003, 39: 3037.
[11] Lo C C H, Lee S J, Li L, Kerdus L C, Jiles D C. *IEEE Trans Magn*, 2002, 38: 2418.
[12] Chen Y, Jiles D C. *IEEE Trans Magn*, 2000, 36: 3244.
[13] Yu F Y, Zhang C X, Wu M. *Mech Des Manuf*, 2006, (5): 118.
于凤云，张川绪，吴淼．机械设计与制造，2006，(5)：118.
[14] Dong L H, Xu B S, Dong S Y, Chen Q Z. *NDT&E Int*, 2009, 42: 323.
[15] Dong L H. *PhD Thesis*, Academy of Armored Forces Engineering, Beijing, 2008.
董丽虹．装甲兵工程学院博士学位论文，北京，2008.

Self – emitting Magentic Signal Behavior of Ferromagnetic Material During Fatigue Crack Propagation under Loading Condition

Xu Binshi, Dong Lihong, Dong Shiyun, Xue Nan

(Science and Technology on Remanufacturing Laboratory,
Academy of Armored Forces Engineering, Beijing, 100072, China)

Abstract Variations of self-emitting magnetic signals during fatigue crack propagation of center-cracked tension (CCT) specimen of 45CrNiMoVA steel were studied. The results showed that the initial abnormal magnetic peak occurred in all the measured lines, and there was a maximum peak-to-peak value in the center measured line before loading. However, the first abnormal magnetic peak occurred on the position of the precut notch after loading, while magnetic signal curve presented linearity on the position without precut notch, at this time fatigue crack did not initiate. During fatigue crack propagation, the second abnormal magnetic peak occurred in the measured lines in turn. The peak-to-peak value of the second abnormal magnetic signals and the slope coefficient between peak-to-peak can be used to characterize fatigue crack propagation life dynamically. The behavior of self-emitting magnetic signals during fatigue crack propagation was analyzed on the basis of magnetic charge and effective magnetic field theory under loading condition.

Key words metal magnetic memory, fatigue crack, loading, abnormal magnetic peak, magnetic charge

再制造的热喷涂合金涂层的结构完整性与服役寿命预测研究*

摘　要　热喷涂技术是再制造工程的支撑技术，热喷涂涂层是再制造领域中常见的表面涂覆层，其初始质量和服役寿命为人关注。本文以等离子喷涂为例，研究了与工艺相关的涂层完整性和与服役条件相关的涂层寿命和失效机理。以不同 H_2 流量、功率和送粉量为条件，研究了工艺参数对涂层孔隙率和微观力学性能的影响。以接触疲劳过程为手段，研究了涂层寿命预测方法和寿命衰退机理。结果表明，工艺参数可以不同程度地影响涂层的结构完整性，通过优化设计可以大幅提高涂层质量；基于大样本空间建立的 $S-N$ 曲线可以直观预测涂层接触疲劳寿命，机理分析表明，点蚀、剥落和分层失效诱因不尽相同，分别由粗糙接触、近表面缺陷和剪切应力导致。

关键词　再制造　热喷涂　合金涂层　结构完整性　寿命预测

维修实践发现，装备的失效取决于最薄弱零件的失效，只要最薄弱零件的性能得以恢复或提升，装备的整体性能就能恢复或提升，装备的寿命就会延长。最薄弱零件或零件最薄弱处的失效，基本都是磨损和腐蚀失效。解决磨损和腐蚀问题，表面工程技术具有明显的优势。如果将大量的废旧装备集中起来，以拆解后的废旧零件为毛坯，利用表面工程技术对毛坯进行批量化修复，重新赋予废旧装备的服役能力，那么这一过程就是"再制造"[1]。当然，进入再制造流程的毛坯都经过严格的无损检测，材料内部严重损伤的废旧零部件予以报废，经过检测合格的毛坯进入再制造流程，进行表面材料恢复和功能升级处理。各种表面技术是再制造工程的主要技术支撑，其中热喷涂技术（高速电弧喷涂、超声速等离子喷涂等）是经常用于再制造工程实践的表面涂层制备技术[2]，应用广泛、效果良好。

1　热喷涂合金涂层的制备及结合机理

1.1　热喷涂涂层的制备

热喷涂涂层的制备主要分 3 个阶段[3]：（1）通过热源加热使喷涂材料成为液态；（2）通过气流或焰流使液态的材料细化成熔滴；（3）使熔滴高速冲击沉积到基体材料上形成涂层。上述过程是在极短的时间内进行的，特别是阶段（1）与（2）几乎是同时进行的。在粉末喷涂时，由于喷涂材料为粉末，就不存在阶段（2），而阶段（3）所占时间略长。在丝材火焰喷涂时，若熔滴的飞行速度为 50～80m/s，则熔滴从喷枪到达基材表面所需时间大约为 10^{-2}～10^{-3}s。对于高速电弧喷涂、超音速等离子喷涂等，由于其喷涂粒子的飞行速度已提高到接近音速或数倍于音速，故相应的飞行时间大大减

* 本文合作者：王海斗、朴钟宇、张显程。原发表于《金属学报》，2011，47(11)：1355～1361。国家重点基础研究发展计划项目 2011CB013405 和 2011CB013403 及国家自然科学基金项目 50735006 和 50975285 资助。

少，涂层与基体的结合强度明显提高。

利用数字高速摄像和脉冲激光光屏法观测高速电弧喷涂 Fe_3Al 粉芯丝材时，丝端（电极端部）接触短路后，电弧导致的熔滴形成与分离过程如图 1 所示[4]。测试表明，在宏观的电弧正常燃烧过程中存在着较频繁的瞬间断弧现象，电弧始终处于引弧 - 断弧 - 再引弧的循环之中；粒子（熔滴）的最初形态及尺寸与喷涂材料、电极极性、气流速度、电流大小等参数有关；粒子在飞行过程中受高速气流的作用而不断地雾化（如喷涂 Al 丝材时，平均粒子尺寸由距枪口 15mm 处的 42.1μm 下降到距枪口 315mm 处的 30.82μm）；高速电弧喷涂形成的粒子比普通电弧喷涂更为细小和均匀。

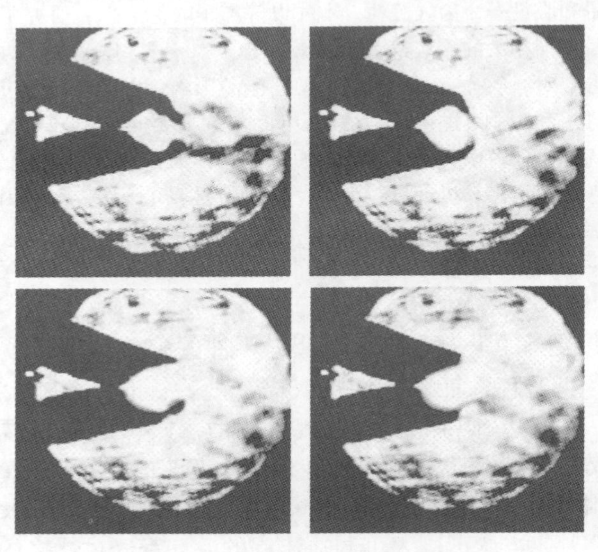

图 1　高速电弧喷涂时电极端部的 4 帧图像[4]

Fig. 1　Four frame images of electrode tip during high velocity electric arc spraying[4]

1.2　热喷涂涂层的结合机理

“晶内结合”和“晶间结合”是热喷涂涂层的两种结合方式。晶内结合是指涂层材料与基体材料共同参与在界面形成晶粒，或者相互间发生反应生成了金属间化合物；晶间结合是指涂层材料与基体材料之间不形成共同晶粒而只是相互接触。热喷涂涂层材料与基体的结合以晶间结合为主[5]，也就是说，涂层材料在基体表面上的结晶过程不是对基体晶格的外延，原因主要是基体与涂层材料的差异及基体温度较低。基体材料一般为黑色金属，而涂层材料以合金（或合金/陶瓷）为主，基体材料与涂层材料的晶格类型、晶格常数均不匹配，这使在基体上外延结晶比较困难。热喷涂过程中，基体的温度不高于 200 ~ 300℃，目的是减小基体的热变形，并确保涂层的化学成分不至于在喷涂过程中发生较大的偏离。基体温度低，一方面为熔滴的结晶提供了极大的过冷度，在基体表面上迅速发生形核和长大的结晶过程，另一方面使熔滴所具有的能量（包括热能和动能）还不足以克服原子间的势垒达到晶内结合，因此，涂层与基体之间有一条明显的晶粒界限，相互不能外延生长。尽管晶间结合的结合强度低于晶内结合强度，但已满足通常苛刻工况的要求。

为了进一步提高涂层的结合强度，可采取措施使基体与涂层材料之间发生扩散并生成化合物。比如，在正式喷涂之前，先在基体表面喷涂一层很薄的 Ni 包 Al 粉末涂层，其在 660 ~ 680℃ 时发生剧烈的放热反应，在熔滴到达基体表面之后，放热反应仍可持续一段时间，有充分的热量促成其与基体之间发生反应，在某些显微点上出现局部熔化，导致该点形成晶内结合，从而提高了涂层与基体的结合强度[6]。

2 表面喷涂层的结构完整性

多年实践表明，通过高速电弧喷涂和超音速等离子喷涂进行再制造后得到的表面涂层，其与基体的结合强度已满足服役要求。但是，表面涂层的综合服役性能及其服役寿命并不能仅用结合强度这一个指标来衡量，而需用涂层的结构完整性来衡量。

对涂层的结构完整性评价主要包括涂层结构中所含缺陷对满足预期功能要求及安全性和可靠性的影响程度。对再制造零件的表面涂层而言，其缺陷主要指涂层中的孔隙、微裂纹等结构缺陷，同时也涵盖了涂层微观力学性能等指标[7]。本课题组[8~10]以超音速等离子喷涂 NiCrBSi 涂层为研究对象，以涂层孔隙率和纳米硬度为评价标准，研究了气流量、喷涂功率和送粉量等参数对涂层结构完整性的影响规律。

2.1 涂层孔隙率的参数优化

采用基于灰度分析的分析软件对涂层的孔隙率在大样本空间下进行计算，采用两参数 Weibull 分布对具有一定分散性的数据进行统计学分析，分别得到不同 H_2 流量、喷涂功率和送粉量与涂层孔隙率的关系。图 2 显示了 95% 的置信水平下，H_2 流量对涂层孔隙率的特征值和名义值的影响[8]。可见，在 H_2 流量不同的条件下，特征值和名义值的上限和下限也不相同，但总体看来，涂层的孔隙率随着 H_2 流量的增加而降低。H_2 流量的增加显著提高主气的电离电压和等离子弧的热导率与热焓，从而提高喷涂颗粒的温度及颗粒的熔融程度，使颗粒更充分地铺展和搭接，所以显著地降低了孔隙等结构缺陷的产生[11,12]。

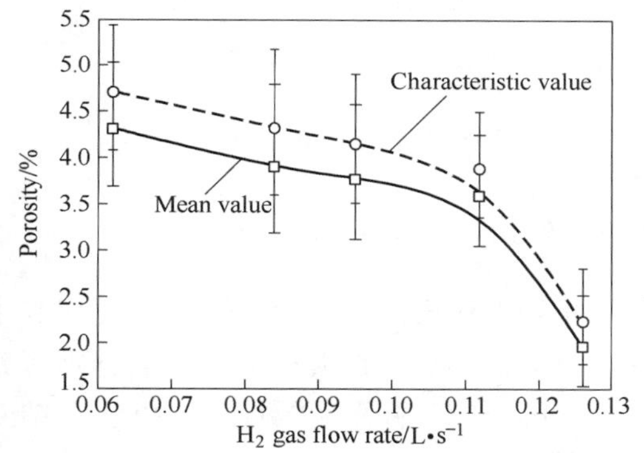

图 2 涂层孔隙率与 H_2 流量的关系[8]

Fig. 2 Variations of characteristic value and mean value of the coating porosity as functions of H_2 flow rate[8]

图 3 显示了在置信水平为 95% 的情况下，喷涂功率对涂层孔隙率大小的影响[8]。可见，当功率低于 50kW 时，孔隙率的特征值和名义值随着喷涂功率的增加而下降；当喷涂功率高于 50kW 时，孔隙率几乎不受功率大小的影响。这种现象产生的主要原因是低功率时，喷涂颗粒的熔化数量较低，而且很多颗粒处于半熔化状态，当这些半熔化熔滴或未熔颗粒沉积到基体表面时，不能完全铺展开，在冷却过程中，会产生局部的收缩，造成大量孔洞的产生；而功率较

高时，颗粒熔融充分能够很好地填充层状结构搭接边缘的微观缺陷，当喷涂功率达到50kW时，绝大部分喷涂颗粒已经被熔化，而且形成的熔滴流动性很好，此时如果继续增加喷涂功率，对涂层孔隙率的影响很小，但可能会导致涂层颗粒的过熔。

图4显示了在置信水平为95%的情况下，涂层孔隙率的特征值和名义值与送粉量的关系[8]。可见，孔隙率随送粉量的增加而提高，送粉量较低时，涂层微观结构致密，孔隙率较低，送粉量达到40g/min时，孔隙率较高。在等离子喷涂中，送粉量主要影响喷涂过程中喷涂颗粒与等离子焰流的相互作用[13]，从而对颗粒的温度和飞行速度产生影响。送粉量增加到一定程度，等离子焰流中的热量不足以使得所有的喷涂颗粒完全熔化，喷涂结束后，涂层内部保留着大量未熔和半熔状态的喷涂颗粒，严重影响涂层的致密度。

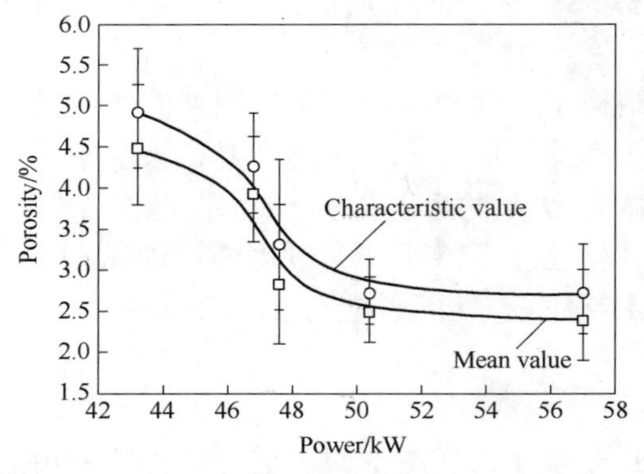

图3 涂层孔隙率与喷涂功率的关系[8]

Fig. 3 Variations of characteristic value and mean value of the coating porosity as functions of power[8]

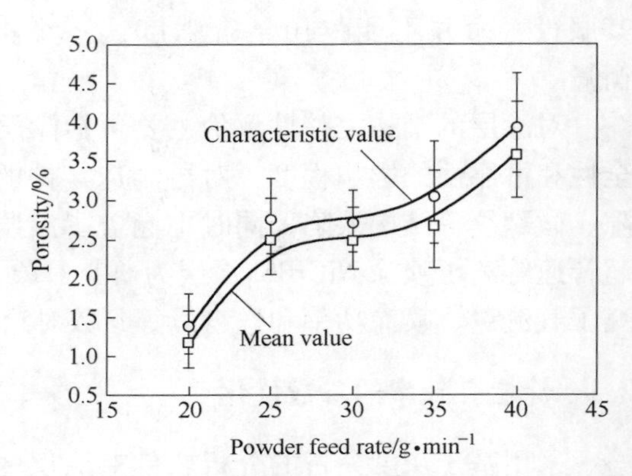

图4 涂层孔隙率与送粉量的关系[8]

Fig. 4 Variations of characteristic value and mean value of the coating porosity as a function of powder feed rate[8]

2.2 涂层的微观力学性能

再制造涂层的微观力学性能主要指其纳米硬度和弹性模量，它们都对涂层的服役质量有关键影响。在大样本空间下分别考察气流量、喷涂功率和送粉量对涂层微观力学性能的影响，并采用统计学的方法总结工艺参数对微观力学性能的关系。图5显示了在置信水平为95%的条件下，H_2流量对涂层硬度和Young's模量特征值和名义值的影响[9]。可见，随着H_2流量的增加，涂层硬度和Young's模量并非线性增加，而是增加到某一临界值后再下降。导致这种现象产生的主要原因有两个：首先，当H_2流量较低时，随着H_2流量的增加，涂层内部的孔隙率下降，但氧化物条带含量增加，涂层的微观力学性能增加，这与文献［14］和［15］给出的结果一致；其次，随着H_2流量的增加，在喷涂过程中，所有喷涂颗粒都被加热至熔融状态，沉积在基体表面的熔滴冷却速度降低，各熔滴的不均匀冷却可能导致涂层表面的微观力学性能测定值的分散程度较高，所以继续提高H_2流量，涂层的微观性能变化不大。考虑到实验误差可以推断，随着H_2流量的增加，涂层的硬度和Young's模量增加，当H_2流量增加到一个临界值后，涂层的硬度和Young's模量几乎不变甚至略有下降。

图 6 显示了在 95% 置信水平下，喷涂功率与涂层微观力学性能的关系[10]。可见，当喷涂功率较低时，涂层的硬度和 Young's 模量随着功率的增加而增加；当喷涂功率增大到一定程度，功率对硬度和 Young's 模量的影响不大。这是由于功率到达一定程度后，颗粒可以达到全熔，所以再提高功率微观硬度和 Young's 模量就没有明显增加了，Chwa 等[16]也得到了相似的结论。

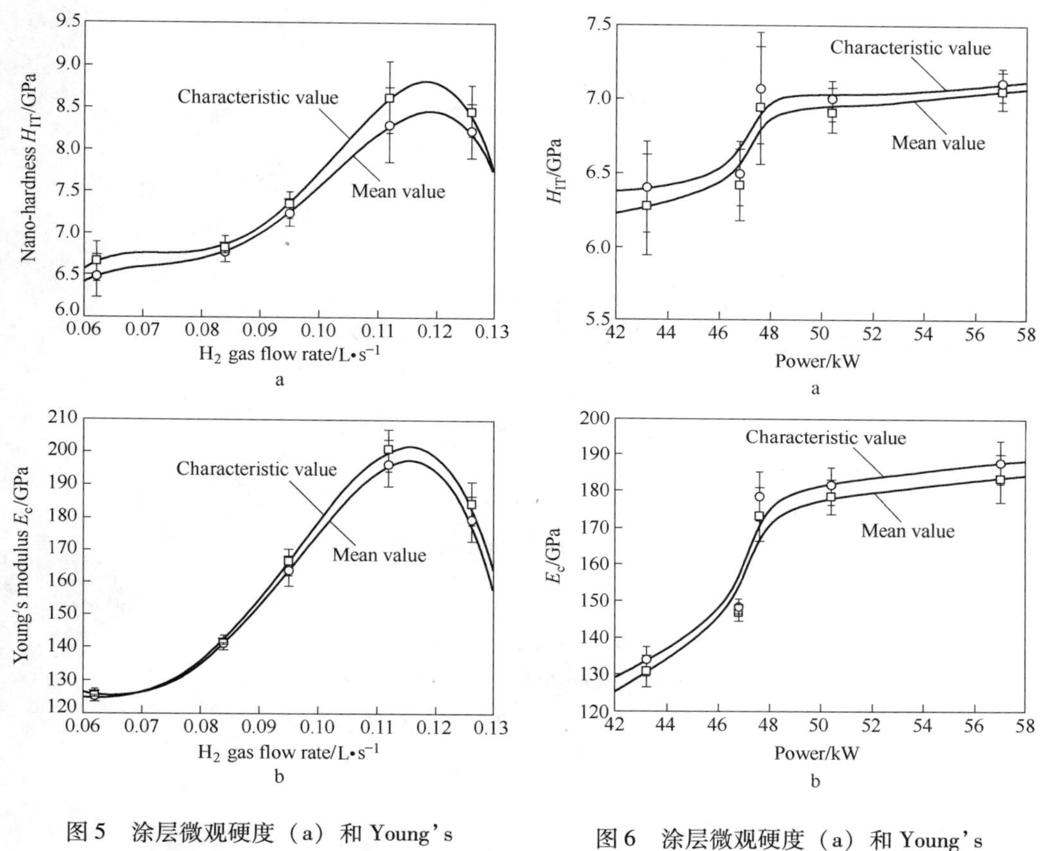

图 5 涂层微观硬度（a）和 Young's 模量（b）与 H_2 流量的关系[9]

Fig. 5 Variations of micro－hardness（a）and Young's modulus（b）of the coating with H_2 gas flow rate[9]

图 6 涂层微观硬度（a）和 Young's 模量（b）与喷涂功率的关系[10]

Fig. 6 Variations of micro－hardness（a）and Young's modulus（b）of the coating with spraying power[10]

图 7 显示了在 95% 置信水平下，送粉量与涂层微观力学性能的关系[11]。可见，随着送粉量的增加，涂层的硬度和 Young's 模量先降低后增加；当送粉量为 20g/min 时，涂层内部的微观缺陷含量很少，涂层组织致密均匀，所以涂层有很好的微观力学性能；当送粉量增加到 30g/min 时，虽然喷涂颗粒大部分被熔化，但是在熔滴冷却过程中，会产生一定的微观缺陷。由图 4 可知，随着送粉量的增加，涂层内部的微观缺陷增加，这些微观缺陷会导致涂层微观力学性能降低。如果送粉量继续增加，涂层内部还会存在着大量的未熔颗粒和半熔颗粒，由于硬度测定过程中，压头位置随机选取，这些颗粒会明显提高硬度测试值。

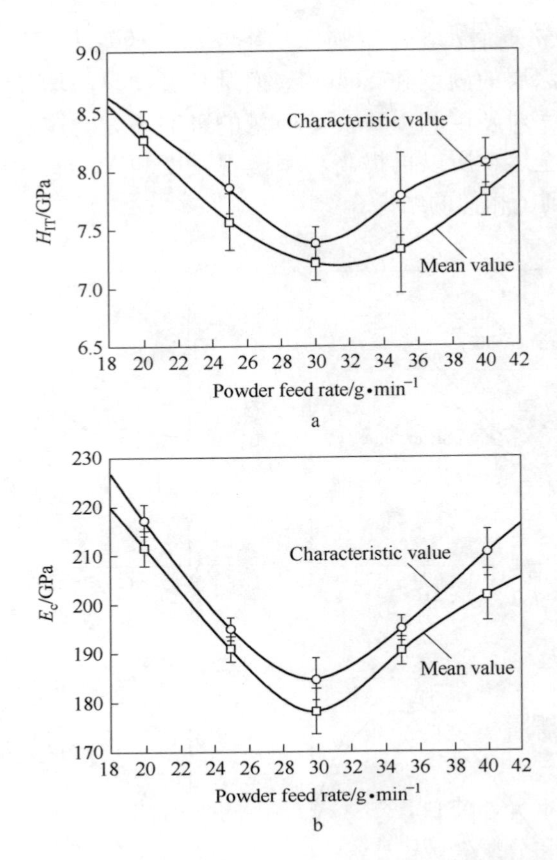

图 7　涂层微观硬度（a）和 Young's 模量（b）随送粉量的变化[11]

Fig. 7　Variations of micro - hardness（a） and Young's modulus（b） of the coating with the powder feed rate[11]

3　表面喷涂层的服役寿命预测

3.1　寿命预测方法

经过结构完整性控制的再制造零件，始终都要进入最终服役的流程，研究其服役过程中寿命衰退和功能丧失的内在规律，对建立寿命预测模型和丰富再制造基础理论都至关重要。在再制造零件寿命预测研究中，引入 Weibull 分布、最小二乘法等数学方法，建立 Weibull 失效概率曲线和 *S-N* 曲线的方式，预测涂层在一定应力范围内的寿命。

以超音速等离子喷涂 CrC-NiCr 金属陶瓷涂层的接触疲劳寿命预测为例[17,18]，在 4 种载荷条件下进行涂层的接触疲劳实验，不同接触应力作用下，涂层疲劳寿命的 Weibull 分布曲线如图 8 所示[7]，纵坐标为失效概率，横坐标为循环周次。结合最小二乘法建立了涂层的应力寿命（*S-N*）曲线，如图 9 所示[7]，图中寿命参数 L_{10}，L_{50} 和 L_{90} 分别为失效概率为 10%，50% 和 90% 时的循环周次，通过该曲线可以直观地预测涂层在一定应力范围的接触疲劳寿命。利用图 8 和图 9 所示的 Weibull 分布曲线和 *S-N* 曲线进行寿命预测时兼顾了大样本空间和随机性，且以实验数据为基础，真实可信同时又直观方便，是一种行之有效的寿命预测方法。

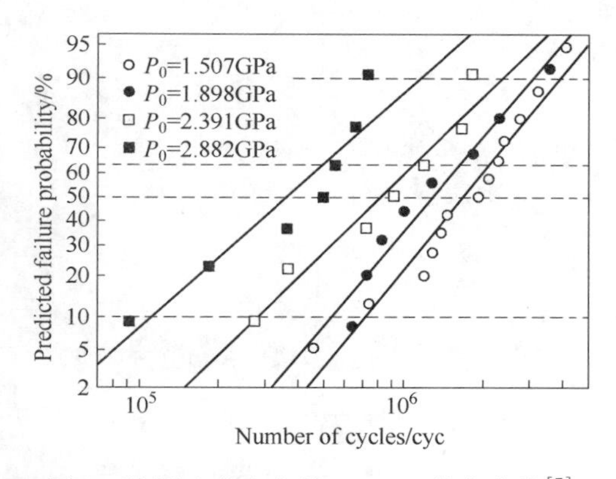

图 8　涂层疲劳寿命的 Weibull 分布曲线[7]

Fig. 8　Weibull distributions of the coating rolling contact fatigue lives[7]

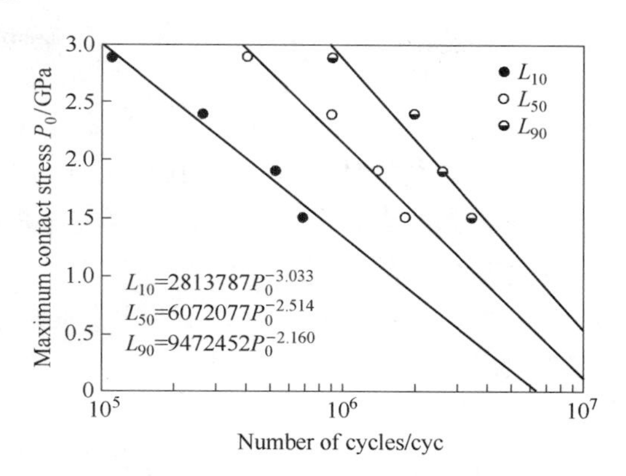

图 9　CrC-NiCr 涂层的 *S-N* 曲线[7]

Fig. 9　*S-N* curves for the CrC-NiCr coatings[7]

3.2　寿命演变规律

再制造涂层的寿命演变是功能丧失、材料去除的基本规律。仍以超音速等离子喷涂 CrC-NiCr 金属陶瓷涂层为例，通过对失效后涂层的断口进行分析，建立涂层的接触疲劳失效机制。通过对火效断口进行观察发现，典型的涂层接触疲劳失效形式主要有点蚀失效、剥落失效和分层失效[19]。

图 10 示出了典型的点蚀形貌[7]。从整体形貌来看，涂层表面的点蚀坑呈现密而浅的分布态势。点蚀的产生主要是由于涂层与对摩轴承球相互接触时，涂层表面粗糙不平，引起局部的塑性变形和接触区域的微观滑移以及随后的磨屑进入接触区域。实际上，任何零部件的表面都存在一定的粗糙度。如果在涂层接触疲劳实验过程中，润滑油的油膜厚度与涂层表面平均粗糙度比值较低，涂层表面的微凸体就会与轴承球相互接触。在接触区域形成的黏着磨损和微观滑移，会在微凸体的边缘产生很高的剪切应力。这种较高的微观剪切应力足够导致微凸体与涂层的分离，形成磨屑。磨屑的产生机制可能不尽相同，如微观剪切和犁沟等。一旦磨屑在接触区域产生，这些磨屑对涂层接触疲劳失效的作用不尽相同，主要依赖于磨屑大小、涂层厚度和表面粗糙度的关系、涂层硬度等。在实验早期，这些磨屑可能从微凸体尖端附近堆积；随着时间的延长，这些磨屑可能在接触区域堆积成层，形成一个新的摩擦副。在这种情况下，涂层最终可能由于三体磨损而加速失效[20]。

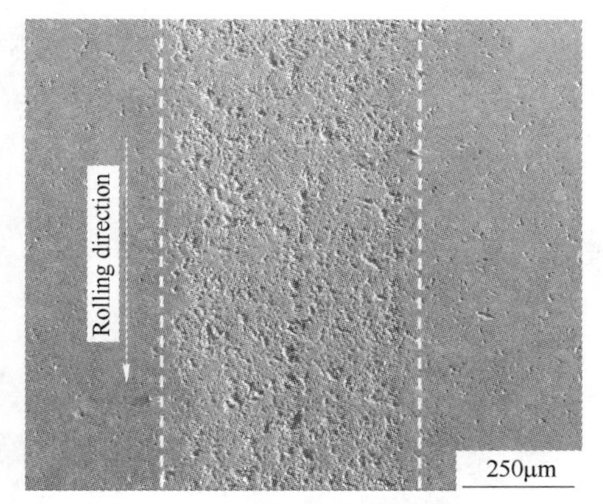

图 10　表面磨损导致的典型点蚀形貌[7]

Fig. 10　Morphology of pitting corrosion caused by surface abrasion[7]

图 11 示出了典型的剥落坑形貌[7]。可见，剥落坑宏观形貌呈现圆形或椭圆形，而且尺寸一般小于接触宽度，剥落坑底部底面比较平整，距离表面的距离较小。关于剥落失效的形成机制，目前还存在着一定的争议，本文中认为这类失效不应该由宏观的剪切应力所导致，而由涂层内部的微观应力所导致。在循环载荷作用下，涂层内部微观缺陷的周围会存在较大的应力集中，促使微观裂纹的萌生和扩展。这种微观裂纹的扩展方向可能是随机的，一旦相邻裂纹相互连接，可能会向涂层表面继续扩展，最终导致涂层层状结构逐层开裂，形成阶梯分布。另外，在循环载荷作用下，涂层次表面形成较大的剪切力作用。这样的剪切力不足以使涂层成分层剥落，但可能会在局部位置引发较大的塑性变形。这样的塑性变形可能会导致涂层内部硬质相的剥离。所以，涂层内部的未熔颗粒与涂层的剥离也可能是涂层剥落坑形成的一个因素。

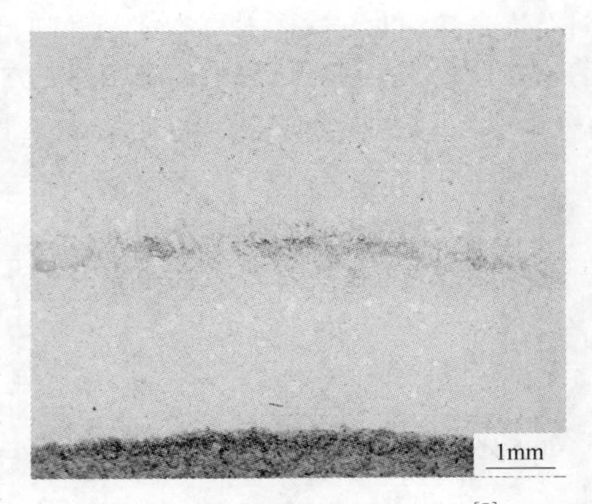

图 11　试样的表面剥落坑形貌[7]

Fig. 11　Morphology of surface spalling[7]

图 12 所示为典型的分层失效形貌[7]。总体分层有界面分层和层内分层两种形式：界面分层指涂层与界面开裂，涂层整体剥离；层内分层主要是指涂层内部开裂，但其形成了比磨痕宽的分层区域，同时也比剥落失效深很多。研究表明有几个原因影响着分层失效的形成。首先，涂层内部微观组织的不均匀性和涂层内部的微观缺陷大小可能影响分层失效的产生时间；其次，涂层和基体的界面状态可能影响分层失效产生的形貌。在涂层界面处，经常会堆积着由喷涂过程导致的污染物和微观缺陷，严重影响涂层和基体的结合强度。而且，由于涂层与基体材料失配，在界面区域会存在较大的残余应力不连续；再次，在接触应力作用下，涂层内部的最大剪切应力的位置可能影响着分层失效的深度。

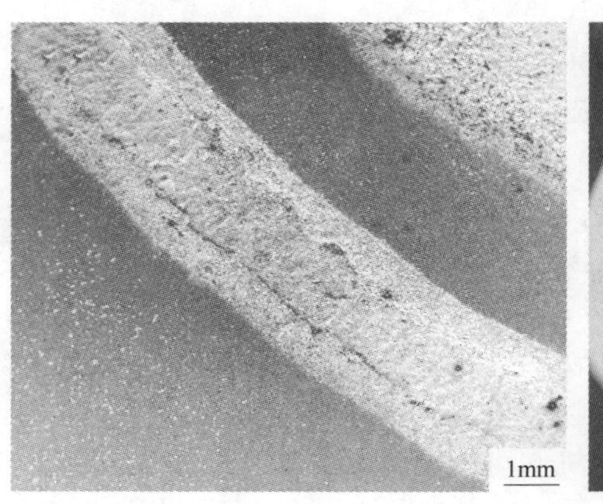

a

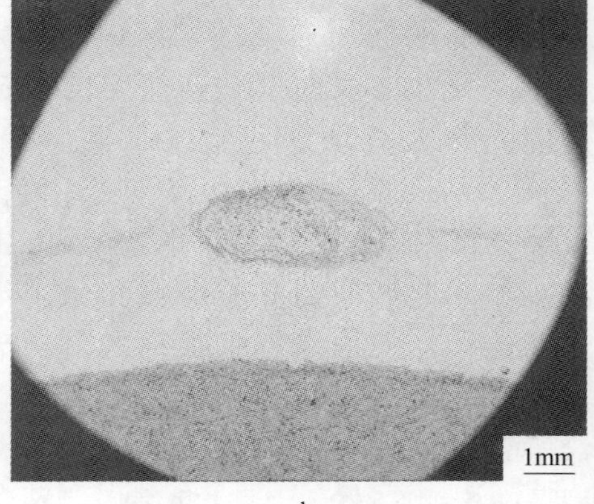

b

图 12　界面分层失效（a）和层内分层失效（b）形貌[7]

Fig. 12　Morphology of cohesive delamination（a）
and adhesive delamination（b）[7]

4 结论

（1）工艺参数显著影响涂层结构完整性，随着 H_2 流量的增加，孔隙率下降，硬度和弹性模量先升高后减小；随着功率的增加，孔隙率下降，硬度和弹性模量增大；随着送粉量的增加，孔隙率升高，硬度和弹性模量先减小后增大；喷涂颗粒的熔融状态和硬质相分布是影响孔隙率和微观力学性能的主要原因。

（2）疲劳寿命的 Weibull 分布曲线和 $S-N$ 曲线可在一定的应力范围内直观地预测接触疲劳寿命。

（3）点蚀、剥落和分层是典型的涂层接触疲劳失效形式，点蚀主要是由表面粗糙接触造成的，剥落是由于涂层近表层缺陷在接触应力诱发下引起的失效，分层主要是由于涂层内部的剪切应力引发失效。

参考文献

[1] Xu B S. *China Surf Eng*, 2010, 23: 1.
徐滨士. 中国表面工程, 2010, 23: 1.

[2] Xu B S. *Remanufacturing Engineering and Its Application*. Harbin: Harbin Institute of Technology Press, 2005: 1.
徐滨士. 再制造工程基础及其应用. 哈尔滨: 哈尔滨工业大学出版社, 2005: 1.

[3] Xu B S, Zhu S H. *Surface Engineering Theory and Technology*. 2nd Ed. Beijing: National Defense Industry Press, 2010: 1.
徐滨士, 朱绍华. 表面工程的理论与技术. 2 版. 北京: 国防工业出版社, 2010: 1.

[4] Zhu Z X, Liu Y, Xu B S, Ma S N. *Trans China Weld Inst*, 2005, 26: 1.
朱子新, 刘燕, 徐滨士, 马世宁. 焊接学报, 2005, 26: 1.

[5] Wu Z J. *Thermal Spray Technology and Applications*. Beijing: China Machine Press, 2005: 1.
吴子健. 热喷涂技术与应用. 北京: 机械工业出版社, 2005: 1.

[6] Piao Z Y, Xu B S, Wang H D, Pu C H. *Tribol Int*, 2010, 43: 252.

[7] Zhang X C. *PhD Thesis*, Shanghai Jiao Tong University, 2007.
张显程. 上海交通大学博士学位论文, 2007.

[8] Zhang X C, Xu B S, Xuan F Z, Wang H D, Wu Y X, Tu S D. *J Alloys Compd*, 2009, 467: 501.

[9] Zhang X C, Xu B S, Wu Y X, Xuan F Z, Tu S D. *Appl Surf Sci*, 2008, 254: 3879.

[10] Zhang X C, Xu B S, Tu S D, Xuan F Z, Wang H D, Wu Y X. *Appl Surf Sci*, 2008, 254: 6318.

[11] Pfender E. *Surf Coat Technol*, 1987, 34: 1.

[12] Bianchi L, Leger A C, Vardelle M, Vardelle A, Fauchais P. *Thin Solid Films*, 1997, 305: 35.

[13] Gawne D T, Liu B, Bao Y, Zhang T. *Surf Coat Technol*, 2005, 191: 242.

[14] Sampath S, Jiang X, Kulkarni A, Matejicek J, Gilmore D L, Neiser R A. *Mater Sci Eng*, 2003, A348: 54.

[15] Gnaeupel-Herold T, Prask H J, Barker J, Biancaniello F S, Jiggetts R D, Materjicek J. *Mater Sci Eng*, 2006, A421: 77.

[16] Chwa S O, Klein D, Toma F L, Bertrand G, Liao H L, Coddet C, Ohmori A. *Surf Coat Technol*, 2005, 194: 215.

[17] Zhang X C, Xu B S, Xuan F Z, Wang H D, Wu Y S, Tu S D. *Wear*, 2008, 265: 1875.

[18] Zhang X C, Xu B S, Xuan F Z, Tu S D, Wang H D, Wu Y X. *Appl Surf Sci*, 2008, 254: 3734.
[19] Zhang X C, Xu B S, Xuan F Z, Tu S D, Wang H D, Wu Y X. *Int J Fatigue*, 2009, 31: 906.
[20] Holmberg K, Matthews A, Ronkainen H. *Tribol Int*, 1998, 31: 107.

Investigation of Structural Integrity and Life Time Prediction of the Thermal Sprayed Alloy Coating for Remanufacturing

Xu Binshi, Wang Haidou, Piao Zhongyu, Zhang Xiancheng

(National Key Laboratory for Remanufacturing, Academy of Armored Forces Engineering, Beijing, 100072, China)

Abstract Thermal spray technique is one of the key techniques in remanufacture engineering. The thermal sprayed coatings are commonly used in remanufacturing applications, their initial performance and service lifetime are critical to the success of remanufacturing. In the present paper, structural integrity, lifetime and failure mechanism of plasma sprayed coatings were investigated. The influences of hydrogen gas flow, spraying powder and powder feed rate on porosity in coatings and their mechanical properties were described. The rolling contact fatigue (RCF) experiment was conducted to develop a method of life time prediction and to reveal the failure mechanism for plasma sprayed coatings. The results show that the structural integrity of coatings can be obviously influenced by spraying process and an optimal design of spraying process can remarkably promote the coating performance. For this purpose, the $S-N$ curve was established based on the large sample space to be used to easily predict coating lifetime. It is found that corrosive pitting, spalling and hierarchical failure are the main failure modes, those results from asperity contact, subsurface defect propagation and shear stress distribution, respectively.

Key words remanufacturing, thermal spray, alloy coating, structural integrity, life prediction

发动机旧连杆缺陷超声检测研究*

摘　要　连杆是传递发动机曲轴和活塞动力的重要零部件之一。对旧连杆进行再制造可以降低生产成本，提高企业竞争力。根据超声检测方法的优点，针对连杆材料及易损伤部位制作了相应的超声波探头，并采用小波分析方法对检测结果进行降噪处理。结果表明：连杆大头与杆身过渡处是易损伤部位，通过横波斜探头可实现该区域缺陷的检测；以 sym4 小波为母小波，对检测信号进行降噪处理，可在一定程度上抑制“噪声”，从而提高试验结果精度；建立降噪前后回波幅值与声程的关系式，并采用指数函数对其进行拟合，可初步实现缺陷定量的无损评价；根据试验验证结果可知，采用该方法对旧连杆能否进行再制造进行评价是可行的。

关键词　超声波　无损评价　旧连杆　小波分析

1　引言

连杆是传递曲轴和活塞之间动力的重要零部件之一，在服役过程中受拉伸和压缩交变载荷的作用，极易出现弯曲、折断、孔径磨损、孔径变形等破坏形式[1,2]。再制造工程[3]是旧产品高技术修复、改造的产业化，也是循环经济 4R 准则中最活跃的形式和首选途径。因此，大力发展再制造工程不仅可以节约资源[4]、保护环境、提高经济效益，还可以缓解就业压力[5]。研究表明，旧连杆缺陷、应力是影响连杆再制造后质量的关键因素[6]，因而，旧连杆缺陷的无损检测和评价方法也就成为影响再制造连杆质量的关键技术。

超声检测是五大常规无损检测方法之一，具有使用方便、安全等优点，因而，在缺陷、应力等检测领域得到广泛关注和应用[7]。相关研究表明，当超声波在材料中传播时，由于材料组织不均匀等，会出现干扰信号，即“噪声”，从而影响检测精度。小波分析作为一种较新的信号分析处理方法，不仅在时频两域具有表征信号局部特征的能力，而且还为有效解决降噪和特征信号丢失之间的矛盾，以及实现信号特征检测、信号重建等提供了良好的解决方法，因而在信号降噪处理中得到广泛应用。针对旧连杆超声质量无损评价，本研究在小波降噪基础上，对其质量进行无损评价，即对旧连杆能否进行再制造做出评判。

2　试验设备

以 JB/T 10659—2006《无损检测：锻钢材料超声检测连杆的检测标准》为依据，对汽车发动机旧连杆质量进行无损评价，并根据该标准选择了适于对汽车发动机旧连杆质量进行评价的检测设备。

---

* 本文合作者：董世运、刘彬、石常亮。原发表于《失效分析与预防》，2011，6(1)：19 ~ 22。国家自然科学基金项目（50975287）；“十一五”预研项目（513270102）资助。

所用检测设备为爱德森（厦门）电子有限公司和装甲兵工程学院共同研制的XZU－1超声波无损检测仪。根据发动机旧连杆的受力分析及考察断裂连杆件可知，连杆大头与杆身过渡处圆弧为连杆薄弱环节，即易损伤部位。为了实现易损伤部位旧连杆缺陷的检测及评价，针对连杆材料和检测部位，制作了相应的检测标块和超声波斜探头，标块宏观形貌及探头示意图分别见图1、图2。其中缺陷以JB/T 10659—2006《无损检测：锻钢材料超声检测连杆的检测标准》为依据，选择直径为2.0mm的通孔，其序号从左到右分别为1～5号；探头参数主要根据待检测分析连杆形状及材料设计得到，并委托汕头超声仪器研究所公司制作，其主要参数如下：中心频率为5MHz，曲率半径为44mm，入射角为45°。

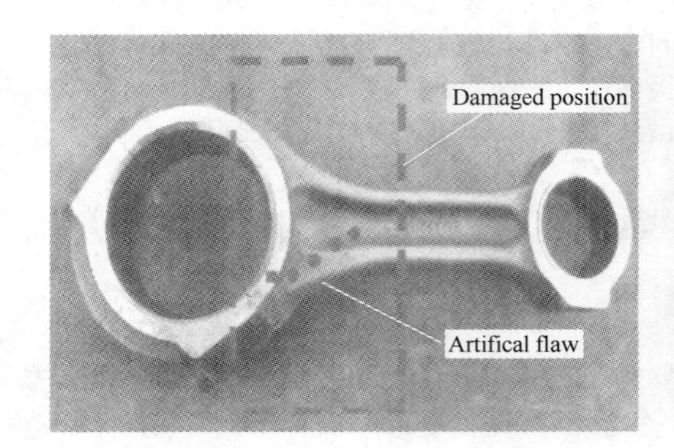

图1　旧连杆易损伤部位

Fig. 1　The damaged position of discarded connecting rod

图2　超声横波斜探头

Fig. 2　The ultrasonic probe of shear wave

3　试验结果与分析

为实现旧连杆缺陷大小的无损评价，本试验在“当量法”基础上采用超声横波斜探头，对图1中不同位置处的人工缺陷进行了检测分析，结果见图3。图3a～c分别为1号、3号和5号缺陷的检测结果。对比分析可知，当检测声程较小时，可同时检测到缺陷回波和棱边回波；随着检测声程的增加，缺陷回波信号幅值逐渐减小，并且棱边回波消失。这主要是因为随着检测声程的增加，连杆材料造成超声波的衰减程度逐渐增大，因而缺陷回波信号幅值逐渐减小，这与目前研究结果是相符的。

根据超声无损检测缺陷理论可知，超声回波信号的波形和回波峰值只与缺陷类型相关，与检测声程无关，即缺陷类型相同时，超声回波信号的波形相同并且峰值唯一。而根据实际连杆超声回波信号可知，其超声回波信号波形不同，并且峰值不唯一。分析认为：这主要是由于检测信号中存在大量的“噪声”，从而造成缺陷信号波形和峰值个数的不同。连杆超声回波信号中“噪声”的来源主要有以下3个方面：（1）连杆材料不均匀，超声波在连杆中传播时，会在阻抗不同物质的界面处发生反射、散射等现象；（2）工频噪声；（3）超声波探头。由于探头与连杆的耦合界面为凸圆弧面，因而，会在一定程度上引起超声波信号发散，其中以材料不均匀造成的噪声为主。

“噪声”一般是有害的，它不仅会影响检测结果的精度，甚至会造成结果误判[8,9]。

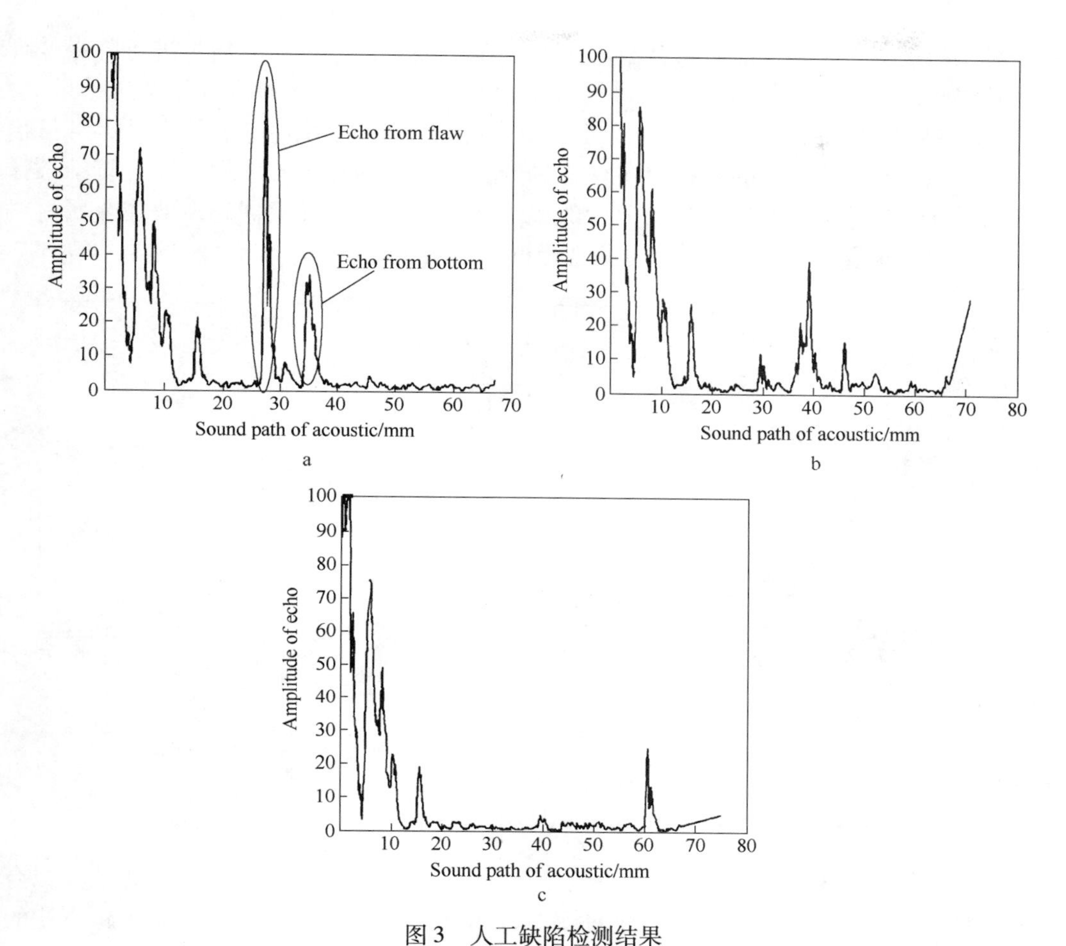

图 3　人工缺陷检测结果

Fig. 3　Testing results of artifical flaws

a—Testing result of No. 1 flaw；b—Testing result of No. 3 flaw；

c—Testing result of No. 5 flaw

因此，为保证检测结果的正确性及提高检测结果的精度，本试验采用小波分析的方法对连杆超声回波信号进行降噪处理。由于小波基和降噪方法不同时，小波降噪效果也不同，因而以 3 号缺陷超声回波信号为研究对象，结合功率谱分析，以 sym4 小波为母小波，采用软阈值消噪方法对其进行降噪处理，连杆超声回波信号降噪前后结果如图 4 所示。

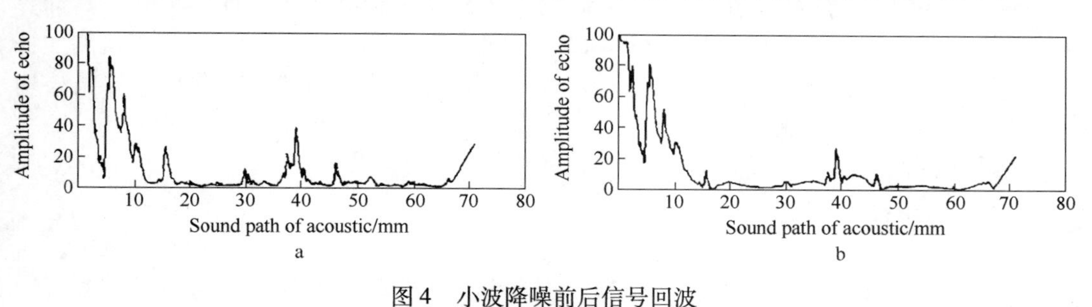

图 4　小波降噪前后信号回波

Fig. 4　The signal amplitude before and after denoising

a—Initial signal；b—Denoising signal

对比图4中小波降噪前后回波波形可知，对3号缺陷超声回波信号进行降噪处理后，其波形比较平滑，并且干扰信号的幅值会变小。说明在软阈值降噪方法基础上，以sym4小波为母小波降噪，可在一定程度上抑制连杆超声回波信号中的“噪声”信号。为了在评判缺陷大小的基础上实现连杆能否再制造的评价，依据上述分析结果计算不同尺寸缺陷超声回波信号降噪前后的幅值和检测声程，并根据超声波在金属材料中传播时的衰减规律 $P = P_0 e^{-\beta x}$[10] 对其进行拟合（图5）。

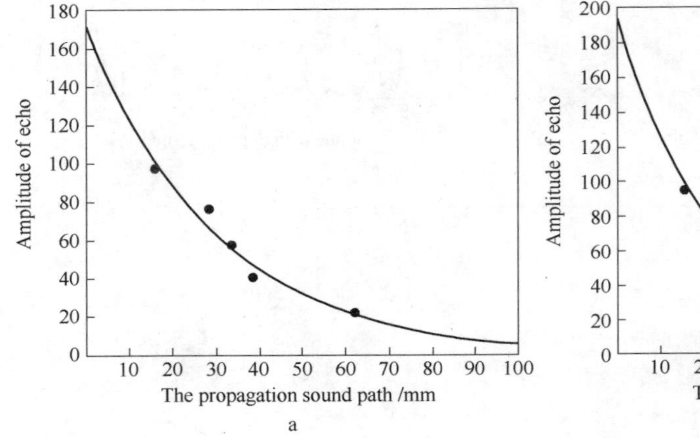

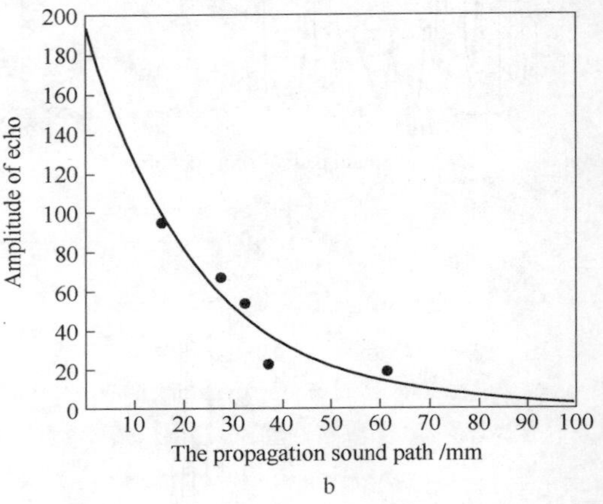

图5　拟合曲线

Fig. 5　Fitting curves

a—Fitting curve of initial data；b—Fitting curve of denoising data

图5a、b分别为超声回波信号降噪前后检测声程与信号幅值间关系曲线及其拟合结果。对比分析可知，随着检测声程的增加，回波信号幅值逐渐变小，这与超声波的衰减规律是相符的；而根据连杆超声回波信号的原始数据和降噪后数据可知，其拟合曲线分别为 $y = 173.1053\ e^{0.0339x}$ 和 $y = 170.4258\ e^{0.0396x}$。为了对其拟合效果进行评价，计算了其相关值，分别为0.85848和0.86647。根据对拟合程度的评价理论可知，拟合效果越好，其相关数值越大，这也说明了采用小波分析对连杆超声回波信号进行降噪处理可在一定程度上减小“噪声”对计算结果的影响，从而提高检测结果精度。

为了验证拟合曲线是否符合试验要求，对完好连杆中的人工缺陷（ϕ2.0mm通孔）进行检测。结果表明，当检测声程为17mm，缺陷当量尺寸为ϕ2.0mm通孔时，降噪前与降噪后超声波信号的回波信号幅值分别为102和93，而根据拟合曲线可知，其计算结果分别为97.2和86.7。根据JB/T 10659—2006《无损检测：锻钢材料超声检测连杆的检测标准》可知，当缺陷尺寸为ϕ2.0mm时，连杆报废，虽然降噪前后计算结果均表明连杆报废，但从降噪前后数据差可知，降噪后数值差要大于降噪前数值差，即采用小波分析可在一定程度上提高结果分析的可靠性。

4　结论

（1）所设计的超声检测探头——横波斜探头，可实现旧连杆大头与杆身过渡区域的检测；

（2）在软阈值消噪方法基础上，以sym4小波为母小波对旧连杆超声回波信号进行降噪处理，可在一定程度上抑制“噪声”信号，从而提高检测结果的精度；

（3）试验验证结果表明，采用本试验中的方法对旧连杆能否再制造进行评价是可行的。

参 考 文 献

[1] 周晓丽．汽车连杆热处理及缺陷分析［D］．吉林大学硕士学位论文，2004：1～4.

[2] 赵永利．发动机连杆的无损检测及疲劳寿命的研究［D］．辽宁工程技术大学硕士论文，2004：6.

[3] 徐滨士．维修工程的新方向——再制造工程在中国的发展（一）［J］．维修与管理，2009：17～19.

[4] 邢忠，姜爱良，谢建军．汽车发动机再制造效益分析及表面工程技术的应用［J］．中国表面工程，2004，17(4)：1～5.

[5] Ron G，Kevin G. Remanufacturing the Next Great Opportunity for Improving U. S Productivity [C/OL]. www. oem services. org，2003：2～3.

[6] Shi C L，Dong S Y，Xu B S，et al. Stress concentration degree affects spontaneous magnetic signals of ferromagnetic steel under dynamic tension load [J]. NDT& E International，2010，43(1)：8～12.

[7] 北京市技术交流站．超声波探伤原理及其应用［M］．北京：机械工业出版社，1982：1～25.

[8] Fedi M. Localized denoising filtering using the wavelet transform [J]. Pure and Applied Geophysics，2000，157(9)：1463～1469.

[9] 张家骏．超声检测技术某些新进展［J］．无损检测，1993(11)：324～327.

Ultrasonic Testing of Flaws in Discarded Connecting Rod of Engine

Dong Shiyun，Liu Bin，Xu Binshi，Shi Changliang

(National Key Laboratory for Remanufacturing，Academy of Armored Forces Engineering，Beijing，100072，China)

Abstract Connecting rod，usually used to transfer power，is a vital part of engine，so remanufacturing of discarded connecting rod can reduce production cost and improve company competitive power. In this paper，based on the merits of ultrasonic wave testing，by designing and preparing probe of shear wave，the region that is usually damaged firstly is tested，and the testing signals are treated and analyzed by the method of wavelet. The results show that the transition region from connecting rod big end to rod body，which was tested by probe of shear wave，was damaged firstly. When sym4 wavelet was regarded as mother wavelet，the noise signal could be suppressed by method of wavelet analysis，and the shape of ultrasonic wave became smoother than the shape of or－iginal wave. By calculating the fitting curve of data between sound path and amplitude of echo wave，the flaws and the size of flaws in discarded connecting rod were tested and evaluated. From the results，this method can be used to evaluate whether the discarded connecting rod can be remanufactured.

Key words ultrasonic wave，non－destructive evaluation，discarded connecting rod，wavelet

Health Condition Monitoring with Multiple Physical Signals in Tensile Test for Double-material Friction Welding*

Abstract The manifold physical signals including micro resistance, infrared thermal signal and acoustic emission signal in the tensile test for double-material friction welding normative samples were monitored and collected dynamically by TH2512 micro resistance measuring apparatus, flir infrared thermal camera and acoustic emission equipment which possesses 18 bit PCI-2 data acquisition board. Applied acoustic emission and thermal infrared NDT (non-destructive testing) means were used to verify the feasibility of using resistance method and to monitor dynamic damage of the samples. The research of the dynamic monitoring system was carried out with multi-information fusion including resistance, infrared and acoustic emission. The results show that the resistance signal, infrared signal and acoustic emission signal collected synchronously in the injury process of samples have a good mapping. Electrical, thermal and acoustic signals can more accurately capture initiation and development of micro-defects in the sample. Using dynamic micro-resistance method to monitor damage is possible. The method of multi-information fusion monitoring damage possesses higher reliability, which makes the establishing of health condition diagnosing and early warning platform with multiple physical information monitoring possible.

Key words micro resistance, pictorial infrared photography, acoustic emission, nondestructive testing, dynamic monitoring

1 Introduction

The life assessment of used equipment parts is an unresolved technical problem. In recent years, with the emergence of high precision micro resistance measurement instrument, the research on metal material injury with the micro resistance method increases evidently. The variation of micro resistance of metal components can sensitively and accurately reflect their microstructures[1-3]. The laws of resistance change during the testing of tension, high-temperature aging, high-cycle fatigue and creep damage etc. showed that the characterization of material damage based on micro-resistance method is possible, but the microscopic mechanism is still not very clear, lacking of adequate theory to support basic research. Moreover, currently, there was little study of the damage characterization using micro-resistance method to actual components in complex service environment, as well as the research of dynamic damage monitoring with micro-resistance method. The temperature change caused by defects arising and friction between

* Copartner: Zhang Yubo, Wang Haidou, Yang Daxiang, Zhu Lina. Reprinted from *Journal of Central South University of Technology*, 2012, 19(10): 2705 – 2711.

grains can be monitored during the whole process of metal ductile damage. The temperature field characteristics can be displayed directly in the form of false color to expand people's visual range from visible light to infrared light[4]. AE is a relatively mature nondestructive testing technique. In theory, it can detect tiny variation such as slip of crystal lattice because of its high sensitivity. This work applied acoustic emission and thermal infrared testing means, which have been studied relatively deeply, to verify the feasibility of using resistance method to monitor the samples' dynamic damage for the first time. The research of the dynamic monitoring platform was carried out with multi-information fusion including resistance, thermal infrared and acoustic emission, which can realize the synchronous output. Through analyzing the multiple physical signals, the critical failure point can be estimated and warned. Compared with the NDT means with only one signal monitoring, this platform possessed more superiority in the area of NDT. It is a new NDT method, which can enhance the reliability and accuracy of life evaluating for equipments. The research on micro mechanism of ductile damage for metal materials has significance to the production and popularization of remanufacture products.

2 Experimental

Circular standard sample (Fig. 1) made by actual engine exhaust possesses the structure of double – alloy friction welding. The material of the left part is 5Cr21Mn9Ni4N, and that of the right is 5Cr9Si3. Friction welding separates the two parts. Circular trough with diameter of 8 mm is used to fix the acoustic emission probe.

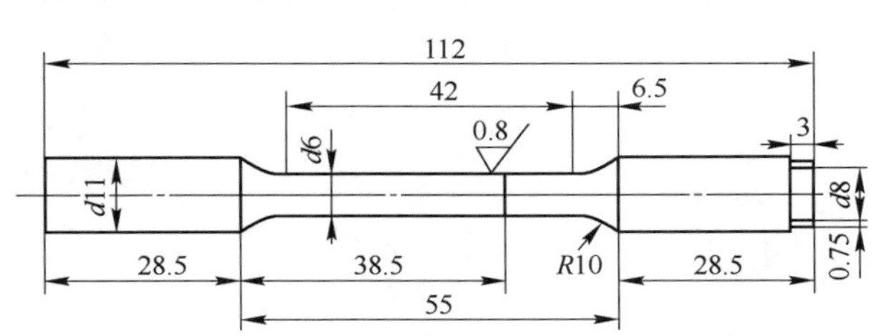

Fig. 1 Component drawing of standard sample (unit: mm)

Fig. 2 is the health monitoring platform which could export three different kinds of NDT data at the same time. Tensile test was performed on 5t electrohydraulic servo fatigue tester which can realize dynamic and static experiments. The control mode was set to "force" mode. Function generator type was defined as "Mono-Segment Wave". The loading rate is 0.1kN/s. The data acquisition software can collect the displacement and force. TH2512 micro resistance instrument was set to mΩ range. Resistance signal was set to high frequency acquisition, about 10 points per second. Considering the temperature increase caused by the environmental temperature and plastic deformation, we set the temperature interval of flir infrared thermography to 13.4 – 30℃. The resonant frequency of monitor probe with 18 bit acoustic emission signal was set to 140kHz. The preposed tone-up was 40dB. During the experiment, it was found that the signal amplitude of noise and mechanical friction was usually less than 50dB, so we set the threshold of signal as 50dB to collect the acoustic signal which can reflect the internal defects of specimens.

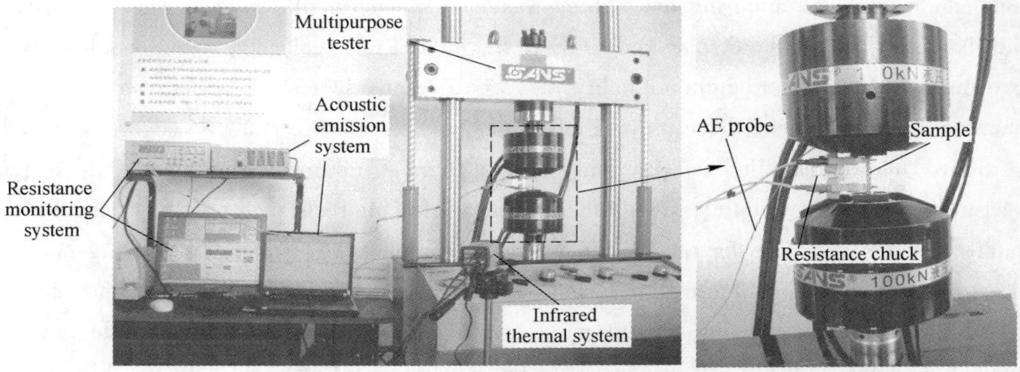

Fig. 2 Health monitoring platform which could export three different kinds of NDT data simutaneously

3 Results and Discussion

3.1 Analysis of mechanical property and fracture

Fig. 3 and Fig. 4 show the stress-strain curve and displacement-time curve, respectively. Fig. 3 shows that the elastic limit σ_e of the sample was about 21 kN, and it was stretched about 0.35 mm. Yield limit σ_s and yield extension stage are unclear, but when the load exceeded 21 kN, the plastic deformation rate began to increase obviously. In the strain hardening stage, the load was still increasing. Ultimately, the sample was failed to reach the tensile limit σ_b, and the rupture took place on the weld seams of the specimen. The specimen was extended 2.78 mm, and the corresponding maximum load was 30.57 kN.

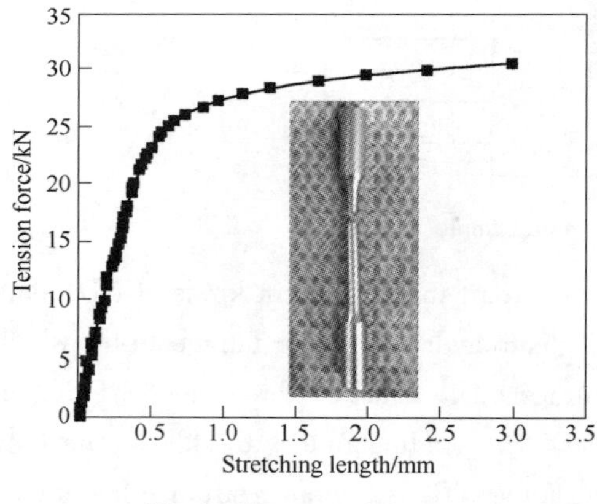

Fig. 3 Stress-strain curve

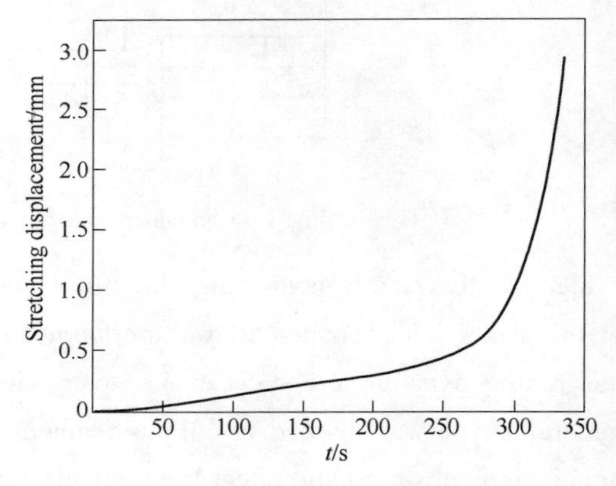

Fig. 4 Displacement-time curve

Fig. 5 shows the macroscopical and micro morphologies of fracture surface. For the metal materials with good ductility, ductile fracture often exhibits macro characteristics such as neck shrinking and shape of cup cone. The dimple pattern formed through micro cavity congregating can be observed in the micro structure. But the fracture surface didn't show the above - mentioned pattern characteristics (see Fig. 5a, b and d), and it presented quasi-cleavage fracture characteristic. Quasi-cleavage fracture is a kind of transition configuration between cleavage fracture and dimple rupture, which is a discontinuous rupture process. First, it generated cleavage micro cracks in different parts of the sample (Fig. 5c shows one rudimental crack with length about 1mm). Then the crack gradually grew, and finally the residual joint parts were torn by plastic way to form so-called laceration edges (see Fig. 5a and b). Quasi-cleavage fracture belongs to brittle fracture[5]. The reason that quasi-cleavage fracture rather than ductile fracture appeared is as follows: firstly, the alloy material is brittle due to high content of Cr. Secondly, due to the friction welding, the relatively more microscopic defects such as vacancy and inclusion, etc. were generated in the region of weld seam. Furthermore, the diffusion of alloy in this region was more inhomogeneous, which made the weld seam become the frailest part of the whole sample. Finally, brittle fracture occurred under tensile load.

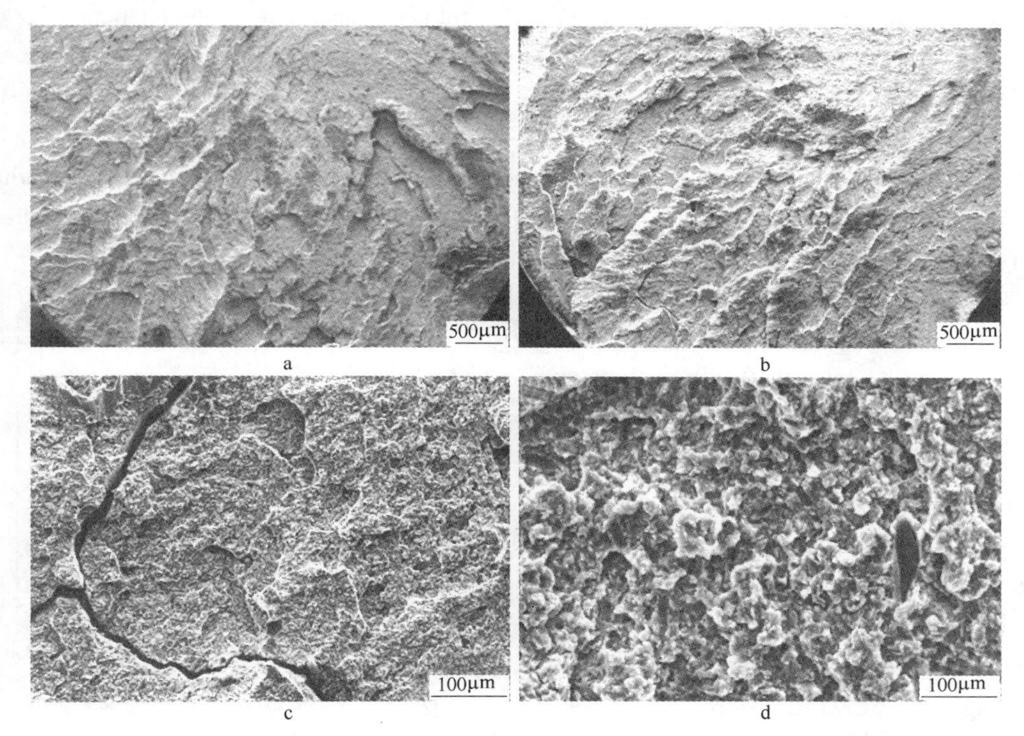

Fig. 5 Macroscopical and micro morphologies of fracture surface

a—Macroscopical morphology of weld seam fracture surface of NiCr20TiAl side;
b—Fracture macroscopical morphology of another side; c—Micro cracks remained on fracture;
d—Fracture microscopic morphology of NiCr20TiAl side

3.2 Analysis of micro resistance, infrared thermal signal and acoustic emission signal

3.2.1 Analysis of micro resistance

The variation of micro resistance of metal components is sensitive to the microstructure. During the tensile process, the variation of micro resistance will reflect the generation of macroscopic defects including dislocation, slipping, micropores and micro cracks. The change of micro resistance is caused by comprehensive action of both above damage factor and non-damage factors such as axial elongation, section contraction and temperature changes, etc. [6-8]. Later we will discuss the influence of all non-damage factors on the resistance results. The resistance of metal generally increases with temperature increasing. The resistance temperature coefficient C_{TCR} of most metals is about 0.004[9], e.g., Cr is 0.003 and Fe is 0.0065. In addition, the C_{TCR} will descend a little with deformation increasing. We chose C_{TCR} as 0.005:

$$\overline{C}_{TCR} = \frac{R_2 - R_1}{R_1(T_2 - T_1)} \tag{1}$$

The sample's original resistance is 0.45mΩ, when the temperature increases by 10℃, the corresponding increase of its resistance is about 0.0225mΩ. The part temperature field was monitored by the infrared thermovision instrument. It was found that the maximal increment of the temperature for this sample was within 10℃, so the influence amplitude of the temperature was less than 0.0225mΩ during the whole stretching process. However, this sample's resistance can reach above 10mΩ, which indicated that the temperature had a little influence on the resistance during the stretching process.

Fig. 6 shows the dynamic variation of resistance during the stretching process. Compared with the displacement-time curve, it was found that the resistance decreased rapidly and then tended to be stable in the elastic strain stage before 225s.

THOMAS originally found that the resistivity of alloys containing meta-elements such as Ni-Cr, Fe-Cr-Al, and Fe-Ni-Mo, etc. will descend obviously during cold working. The structure with k state is called inhomogeneous solid solution, which has stronger scattering effects on electrons. The cold working can seriously destroy the inhomogeneous solid solution, which clearly decreases the alloy resistivity[9]. The used sample in this work was produced by friction welding with 5Cr21Mn9Ni4N and 5Cr-9Si3, and abundant meta-elements existed in the welding seam.

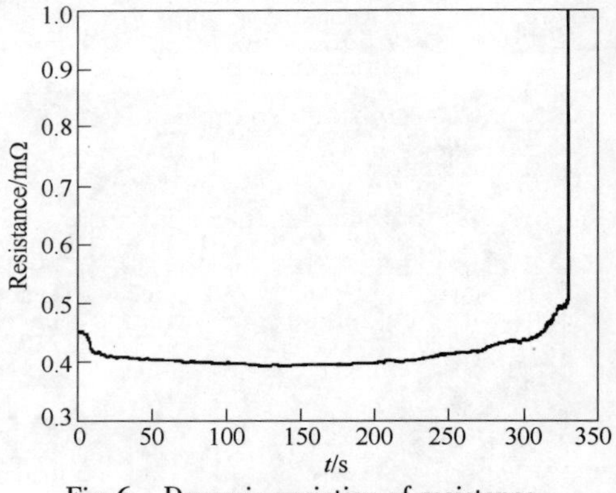

Fig. 6 Dynamic variation of resistance

The extension quantity and section shrinking can be neglected at the beginning stage of stretching process. In this stage, the payload can not reach the critical point to actuate dislocation sources. Acoustic emission signals approved that no dislocation was generated. Thus, the resistance decreased rapidly. As the stress continuously increased, favorable sliding system actuated, leading to the generation and propagation of dislocations. In addition, the influence of extension quantity and section shrinking on resistance increased. The combined actions of the three kinds of factors made the specimen resistance tend to be stable in the later stage of elastic deformation stage. When tensile load exceeded the elastic limit σ_e, strain rate was accelerated obviously. Accordingly, the resistance increased rapidly. Dislocation propagated and aggregated at crystal boundary and carbide inclusions to produce stress concentration. When the load increased to a certain degree, cleavage cracks were generated and expanded instantly. Ultimately, the sample fractured. The changing curve of resistance had good corresponding relationship with displacement-time curve in this stage. The non-damage factor had great effect on the resistance. Therefore, the resistance change was affected by both non-damage factor and damage factor.

3.2.2 Analysis of infrared thermal signal

In the process of metal plastic deformation, besides a majority of work was translated into heat, about 10% of deformation work was retained in the metal interior and stored in the form of residual stress and crystal lattice distortion. Most of the stored energy was consumed by the crystal lattice distortion caused by defects such as dislocation and vacancy[10]. This part of the stored energy can be reflected by resistance which is sensitive to the crystal structure. In tensile process, the factors that caused the heat change of interior material included thermal elasticity effect, thermal plasticity effect and thermal conduction effect[11]. The distribution of superficial temperature field can be monitored by flir infrared camera. Fig. 7 and Fig. 8 show the variation of temperature in weld seam and the representative infrared thermal images of elastic strain area, strain hardening area and fracture of specimen, respectively. The color difference of the infrared thermal images directly reflected the distribution of superficial temperature field in the diverse strain areas. Though the temperature has little effect on the resistance, the dynamic change curve of resistance has good agreement with the temperature change curve of weld seam. At the beginning of the elastic strain stage, the temperature of sample was mostly influenced by thermal elasticity effect. The temperature and elastic strain presented linear negative correlation. As the load increased, thermal elasticity effect can be neglected after the specimen went into plastic strain area. The thermal plasticity effect and thermal loss such as thermal conduction, thermal convection and thermal radiation made the temperature keep steady[12-15]. The consumption of plastic work will change the structure of material interior. The distributions of superficial temperature field and micro resistance are the external representation of inner energy accumulation effect. With the plastic deformation increasing, dislocation was generated and propagated quickly. The formation of slip and micro cracks as well as the friction between grains will be also accelerated. The released heat was increased and cumulated, so that the heat produced by thermal plasticity gradually exceeded the thermal loss, which increased the temperature rising rate of superficial temperature field. So the damage state of the specimen could be monitored and analyzed by flir infrared camera during the whole ductile damage process. Fig. 7 is similar to Fig. 6, but still different. The obvious inflecting point of resistance changing curve appeared about 45s earlier than that of the temperature changing curve, which confirmed that the micro resistance is more

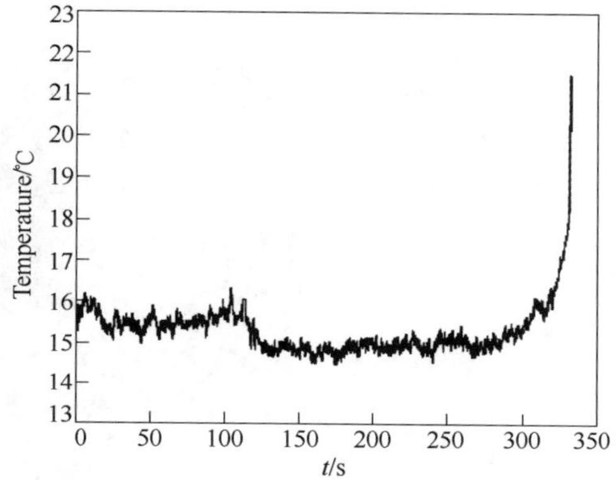

Fig. 7 Variation of temperature in weld seam

sensitive to structure.

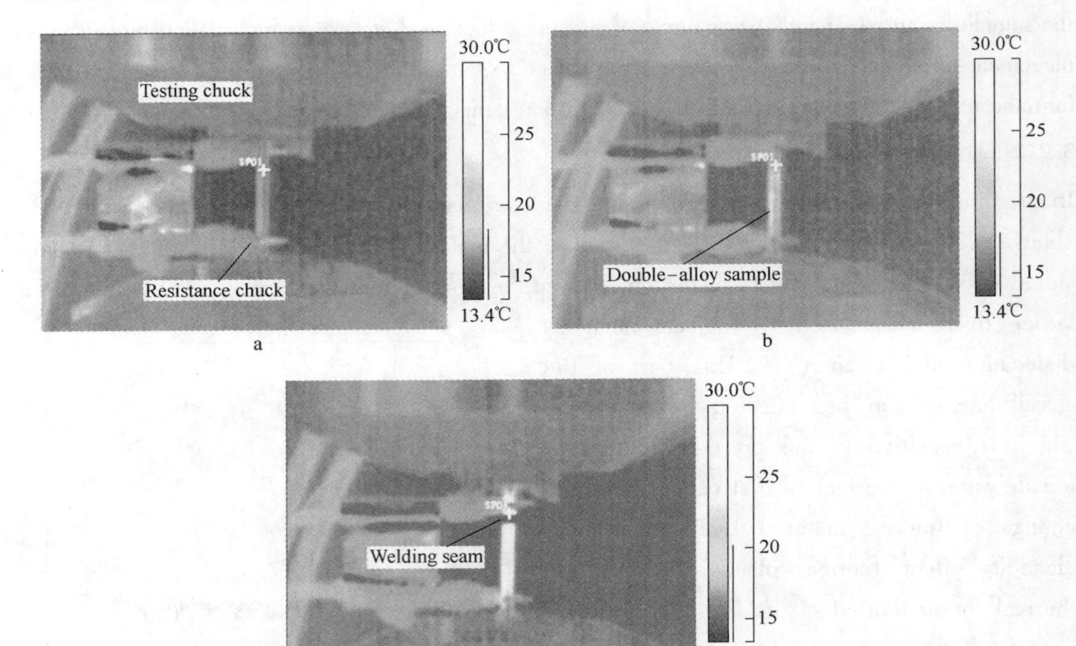

Fig. 8 Infrared thermal images of typical area

a—Infrared image of sample at elastic strain stage; b—Infrared image of sample at strain hardening stage; c—Infrared image of sample at instant of fracture

3. 2. 3 Analysis of acoustic emission signal

Metal plastic deformation was usually caused by dislocation motion or twinning deformation. Twinning along with high amplitude acoustic emission signal was often generated in the metals of Sn, Zn, and Ti, etc. Real-time monitoring results showed that the amplitude of acoustic emission signal seldom exceeded 70dB before fracture. Therefore, it can be concluded that the plastic deformation of the specimen was mainly caused by dislocation motion. In the ductile fracture process of metal materials, the acoustic emission signal sources include the moments of grain boundary sliding, dislocation source actuating, dislocation moving, sliding zone producing, micro crack forming, expanding and rupturing, etc. However, the acoustic emission signal was easily affected by complex noise signal. The threshold scope of acoustic emission was commonly selected from 55 to 65dB in the environment of strong noise[16,17]. In the early stage of high cycle fatigue experiment, the amplitude of noise signal was concentrated below 50dB, so we set the filter threshold as 50dB. At this stage, as no micro defect was generated, acoustic signals only involved noise and friction sound between chunk and specimen.

Fig. 9 and Fig. 10 show the distributions of AE amplitude signal and corresponding AE energy signal, respectively. The stage before 75s belongs to the real elastic deformation stage. At this stage, the outside work was almost stored in the material interior in the form of elastic energy[18], and no AE phenomenon occurred. After 75s, with the tensile load increasing, the amplitude of acoustic emission signal above 50 dB started to appear. In materials, the length of moving dislocation and moving distance of dislocation have a lower limit, below which acoustic emission signal cannot be detected[19-21]. The material with better toughness has few acoustic emission signal under less strain, and the signal energy is lower[22], which is coincident with the real damage process. Namely, with the load increasing, dislocation sources were generated, and dislocation rapidly propagated and moved. In the beginning, dislocation moved very smoothly, so acoustic emission energy and amplitude parameters were low. When plastic deformation was generated, dislocation movement reached the maximum, sliding deformation occurred and the stress was redistributed to produce frequent acoustic emission signals. In the later stage of material strain-hardening, the acoustic emission with the amplitude above 65dB occurred, which reflected the AE signal sources such as generation and propagation of micro cracks, the damage of interface between carbide inclusions and matrix, as well as the fracture of inclusion itself[19-21]. The tensile strain of the double-material welding specimen was less, and a lot of energy was released in the strengthening stage, so more acoustic emission signals were produced at this stage[23,24]. At the point of unstable fracture, strong acoustic emission signals were bursted, and the energy reached 2000 μV · s. Due to the fast signal speed, it was difficult to distinguish the details of fracture.

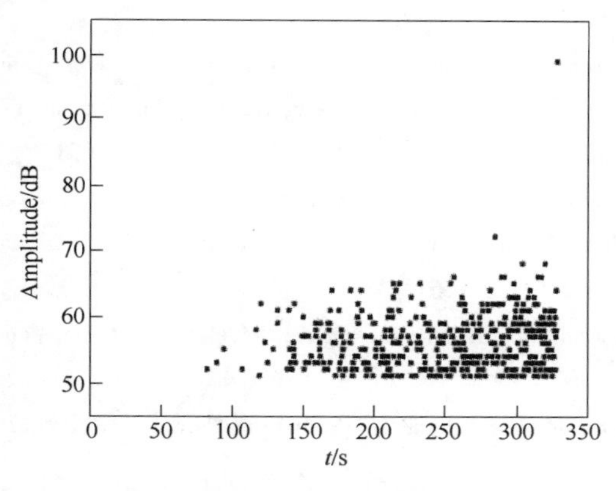

Fig. 9　Distribution of AE amplitude signal

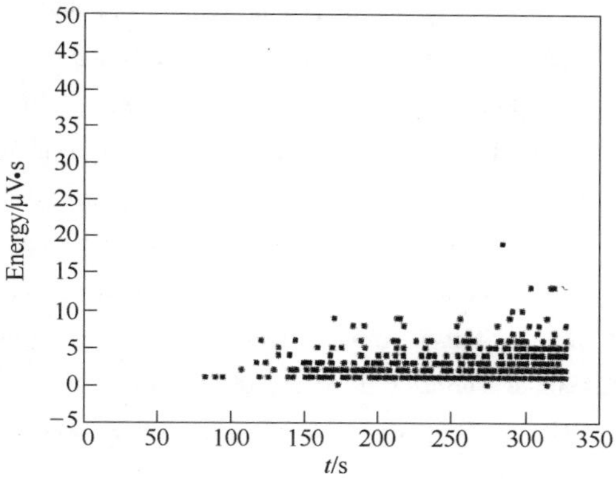

Fig. 10　Distribution of AE energy signal

4　Conclusions

(1) In the ductile damage process of metal materials, dynamic micro resistance signal, infrared thermal signal and acoustic emission signal have close inner mapping relationship. There is no interference among these signals. The changing curves of micro resistance and temperature of the weld seam have similar tendency. They both decrease firstly, and then tend to be stable. Finally, they increase rapidly till fracture. The different stages of curve reflect changes of microstructure. The resistance of the alloy sample containing meta-elements descends rapidly due to the damage of inhomogeneous solid solution.

(2) The inflection points of micro resistance and infrared thermal signal curves have good corresponding relationship with the yielding limit of stress-strain curve, so the yielding limit can

be caught through dynamic micro resistance and infrared thermal signals, and then early warning can be given. The micro resistance method to warn damage yielding is approved to be more sensitive under the same condition. In addition, electric signal need not to be translated, so the feedback is obtained conveniently to accomplish automation control.

(3) The transformation characteristics of micro defects during the whole toughness damaging process are monitored and analyzed in detail using the health monitoring platform which can export three different kinds of NDT data at the same time. The real-time acoustic emission signals monitored by this platform provide reliable evidence for the variation laws of micro resistance and infrared thermal signal. The three kinds of signals synergistically monitor the whole damaging process. Compared with the NDT method with only one signal monitoring, this platform possesses more superiority in the area of NDT. It is a new NDT method, which can enhance the reliability and accuracy of life evaluating for equipments. The research on micro mechanism of ductile damage for metal materials has significance to the production and popularization of remanufacture products.

References

[1] SEOK C S, KOO J M. Evaluation of material degradation of 1Cr – 1Mo – 0. 25V steel by ball indentation and resistivity[J]. Journal of Materials Science, 2006, 41: 1081 – 1087.

[2] STARKE P, WALTHER F, EIFLER D. Fatigue assessment and fatigue life calculation of quenched and tempered SAE 4140 steel based on stress – strain hysteresis, temperature and electrical resistance measurements[J]. Fatigue and Fracture of Engineering Materials and Structures, 2007, 30(11): 1044 – 1051.

[3] STARKE P, EIFLER D. Fatigue assessment and fatigue life calculation of metals on the basis of mechanical hysteresis, temperature, and resistance data: Extended version of the plenary lecture at the international conference low cycle fatigue[J]. Materials Testing, 2009, 51(5): 261 – 268.

[4] JIANG Shufang. Infrared thermal wave imaging for identifing different subsurface defects[D]. Beijing: The Capital Normal University, 2006(in Chinese).

[5] ZHONG Qunpeng, ZHAO Zihua. Fracture[M]. Beijing: Higher Education Publishing Company, 2005: 176 (in Chinese).

[6] SUN Binxiang, GUO Yimu. Prediction of high – cycle fatigue life based on resistance changes [J]. Journal of Mechanical Strength, 2002, 24(4): 81 – 85(in Chinese).

[7] SUN Binxiang, GUO Yimu. A high – cycle fatigue accumulation model based on electrical resistance for structural steels [J]. Fatigue and Fracture of Engineering Materials and Structures, 2007, 30 (11): 1052 – 1062.

[8] YOSHINOBU S, KEIKO O, AKIRA T, MASAHITO U. Detectability of bearing failure of composite bolted joints by electric resistance change method[J]. Key Engineering Materials, 2006, 321: 957 – 962.

[9] TIAN Shi. Material physics characteristic[M]. Beijing: Beijing University of Aeronautics and Aerospace Press, 2004: 41(in Chinese).

[10] LIU Zhien. Material science[M]. Beijing: Northwestern Polytechnic University Press, 2006: 2(in Chinese).

[11] BHALLA K S, ZENHDER A T, HAN X. Thermomechanics of slow stable crack growth: Closing the loop between experiments and computational modeling[J]. Engineering Fracture Mechanics, 2003, 70: 2439 –

2458.

[12] GUDURU P R. An investigation of dynamic failure events in steels using full field high - speed infrared thermography and high - speed photography[D]. Los Angeles: California Institute of Technology, 2001.

[13] RANC N, WAGNER D, PARIS P C. Study of thermal effects associated with crack propagation during very high - cycle fatigue tests[J]. Acta Materialia, 2008, 56(15): 4012 - 4021.

[14] CHARKALUK E, CONSTANTINESCU A. Estimation of the mesoscopic thermoplastic dissipation in high - cycle fatigue[J]. Comptes Rendus Mecanique, 2006, 334(6): 373 - 379.

[15] AMIRI M, KHONSAN M M. Rapid determination of fatigue failure based on temperature evolution: Fully reversed bending load[J]. International Journal of Fatigue, 2010, 32(2): 382 - 389.

[16] XIA Yongfa, LI Hailing. Application of acoustice mission(AE) technique in crack monitor during fatigue test of pump rod[J]. Material and Metallurgy Academic Journal, 2007, 6(1): 60 - 61.

[17] DRUMMOND G, WATSON J F, ACAMLEY P P. Acoustic emission from wire ropes during proof load and fatigue testing[J]. NDT & E International, 2007, 40(1): 94 - 101.

[18] SONG Weixi. Metallography[M]. Beijing: Metallurgy Industry University Publishing Company, 1980: 183 (in Chinese).

[19] LIU Guoguang, CHENG Qingchan, ZHOU Lihui, XUE Zhiyun, XU Guozhen. Acoustic emission monitoring of tensile test of A3 steel plate specimen[J]. Shanghai Metals, 2003, 25(3): 33 - 37(in Chinese).

[20] ROBERTS T M, TALEBZADEH M. Acoustic emission monitoring of fatigue crack propagation[J]. Journal of Constructional Steel Research, 2003, 59: 695 - 712.

[21] GRONDEL S, DELEBARRE C, ASSAAD J. Fatigue crack monitoring of riveted aluminium strap joints by Lamb wave analysis and acoustic emission measurement techniques[J]. NDT & E International, 2002, 35 (3): 137 - 146.

[22] ZHU Bo, WANG Chengguo, CAI Hua. Acoustic emission characteristic and relative research of material fracture toughness[J]. Physic Academic Journal, 2003, 52(8): 1960 - 1964(in Chinese).

[23] ENNACEUR C, LAKSIMI A, HERVE C. Monitoring crack growth in pressure vessel steels by the acoustic emission technique and the method of potential difference[J]. International Journal of Pressure Vessels and Piping, 2006, 83(3): 197 - 204.

[24] RAVISHANKAR S R, MURTHY C R L. Application of acoustic emission in drilling of composite laminates[J]. NDT & E International, 2000, 33: 429 - 435.

Effect of Residual Stress on the Nanoindentation Response of(100) Copper Single Crystal*

Abstract Experimental measurements were used to investigate the effect of residual stress on the nanoindentation of(100) copper single crystal. Equi - biaxial tensile and compressive stresses were applied to the copper single crystal using a special designed apparatus. It was found that residual stresses greatly affected peak load, curvature of the loading curve, elastically recovered depth, residual depth, indentation work, pile-up amount and contact area. The Suresh & Giannakopoulos and Lee & Kwon methods were used to calculate the residual stresses from load-depth data and morphology observation of nanoindents using atomic force microscopy. Comparison of the obtained results with stress values from strain gage showed that the residual stresses analyzed from the Suresh & Giannakopoulos model agreed well with the applied stresses.

Key words indentation, mechanical properties, atomic force microscopy

1 Introduction

The presence of residual stresses can affect mechanical properties and service lives of components in a positive or negativeway, depending on the nature and magnitude of residual stresses. Compressive residual stresses have been shown to increase the hardness, yield stress and fatigue resistance of components, whilst large tensile residual stress can cause deforming, cracking and dimensional accuracy reducing of components. Therefore the investigation of local residual stresses after production or during service is very important for assessing the reliability of components.

Load and depth-sensing indentation, also referred to as nanoindentation, has been widely used over the past three decades as an important tool for characterizing mechanical properties of materials. In the past decade, many researches have been done on how to measure residual stresses by nanoindentation[1-8]. Accurate measurement of residual stresses by nanoindentation methods requires a detailed understanding of the information contained in the indentation loading and unloading curves. Various nanoindentation parameters including indentation depth, indentation load, unloading behavior, pile-up deformation around nanoindents and contact area with and without residual stress can be compared to estimate the sign(tensile or compressive) and magnitude of residual stresses.

In this study, nanoindentation tests were carried out to investigate residual stresses on the nanoindentation response. In order to simplify the experimentation, nanoindentations were per-

* Copartner: Zhu Lina, Wang Haidou, Wang Chengbiao. Reprinted from *Materials Chemistry and Physics*, 2012, 136: 561 - 565.

formed on copper single crystal samples subjected to equi-biaxial tensile and compressive stresses which were applied by a special designed apparatus.

Based on the experimental results, a methodology was developed from which equi – biaxial residual stress can be extracted from indentation load-depth curves and atomic force microscopy observation of residual nanoindents from which the contact area can be accurately measured.

2 Experimental

An isotropic copper single crystal was used in this study in order to exclude the anisotropic deformation effects on the nanoindentation results. The 0.5 mm-thick (100) copper disc specimens with a diameter of 20 mm were made by the Hefei Kejing Materials Technology Corporation. The surface roughness of the specimens measured by an atomic force microscope (AFM) is (10.5 ± 2.5) nm.

The concentric bending apparatus in Fig. 1 was designed to generate equi-biaxial tensile and compressive stresses, respectively. In the present study, the apparatus similar to that but improved in the study of Yun-Hee Lee and Dongil Kwon[9], is composed of an upper die and a lower die. The two dies are joined by a screw. Equi – biaxial tensile and compressive stresses were applied on the convex surface and concave surface, respectively. The magnitude of the applied stress was measured by the strain gauge attached to the convex surface and concave surface of the specimen.

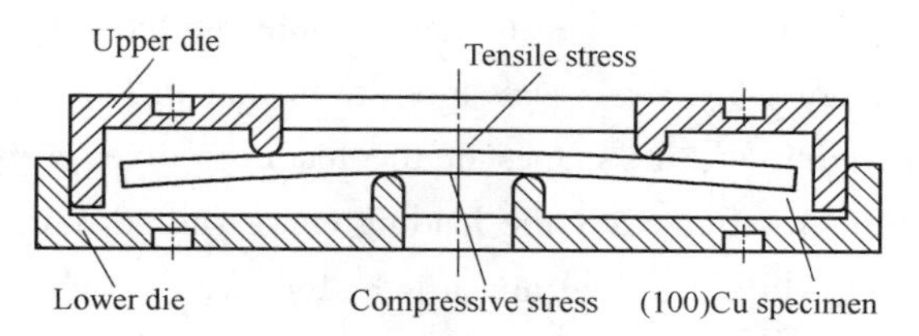

Fig. 1 Concentric bending apparatus designed to generate equi-biaxial tensile and compressive stresses

Nanoindentation tests on stress-free and stressed specimens were performed by employing TriboIndenter system (Hysitron, Inc.) equipped with a Berkovich diamond pyramid indenter which was also used as an atomic force microscopy (AFM) tip and the indented surfaces were imaged immediately after nanoindentation. The load and displacement resolutions are 0.1 μN and 0.1 nm, respectively. All nanoindentations were carried out to 700 nm. All images were taken in a 10 μm × 10 μm scan area with a scan rate of 0.5 Hz. Nanoindentations tests were made under the constant temperature of 20 ℃.

3 Results and Discussion

3.1 Effect on loading curve

The values of elastic modulus E and Poisson's ratio v of (100) copper single crystal are 108 GPa and 0.32, respectively. The strains measured from strain gauge were converted to applied stress by multiplying $E/(1-v)$. The strains of 4.309×10^{-4} and -8.653×10^{-4} correspond to the applied stresses of 68.4 MPa and −137.4 MPa. The variation of applied stresses during the nanoindentation tests was less than ±0.35 MPa.

Fig. 2 shows loading curves obtained for nanoindentations made to a maximum depth of 700 nm with different residual stress states of the (100) copper single crystal.

For stress-free, tensile, and compressive residual stress states, to penetrate to a depth of 700 nm, the required forces P_0, P_T, and P_C were 2.12 mN, 2 mN, and 2.21 mN, respectively. It is obvious that compared with the stress-free state, less force was required for tensile residual stress, hence the peak load P_{max} decreased, and an opposite effect was observed for compressive residual stress in that the peak load increased. Since the indentation stress acted perpendicularly to the specimen surface, the direction of contact shear stress beneath the indenter was identical with the tensile stress. Therefore, the tensile stress increased the magnitude of shear stress relative to the stress-free specimen. The increase of shear stress under tensile stress can enhance the indentation plasticity, and thus lower load was required than in the stress-free state for the same indentation depth[9]. Similarly, the effect of compressive stress was opposite to that in tensile stress state by decreasing the shear stress. In addition, the shapes of the loading curves were affected by residual stress. Compared with the stressfree state, the loading curve shifted down and exhibited a lower slope for the tensile stress, while it moved upwards and showed a higher slope for the compressive stress. The loading curve generally follows the relation described by Kick's Law:

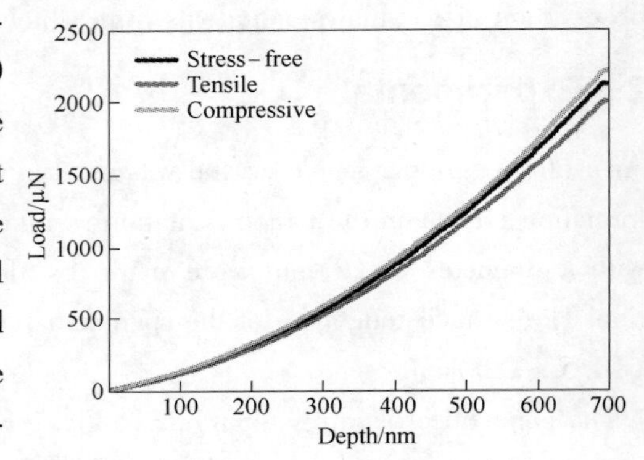

Fig. 2 Loading curves for nanoindentations made to a maximum depth of 700 nm with different residual stress states

$$P = Ch^2 \tag{1}$$

where P and h are the indentation load and indentation depth, respectively; C is the loading curvature. The existence of residual stresses strongly affected the loading curvature. The curvatures for stress-free, tensile, and compressive residual stress states are 4.33 GPa, 4.08 GPa, and 4.51 GPa.

3.2 Effect on unloading curve

Residual stresses also had influence on the unloading curves. Fig. 3 shows the ending part of unloading curves for nanoindentations made to a maximum depth of 700 nm with different residual stress states. A similar behavior was observed for the unloading curves. Compared with the stress-free state, the unloading curve shifted to the right for the tensile stress and left for the compressive stress. This resulted in a decrease in the elastically recovered depth h_e for the tensile stress and an increase for the compressive stress. Meanwhile, an opposite effect was obtained for the residual depth h_r and the ratio h_r/h_{max}. It is expected that since the unloading process of nanoindentation is a pure elasic process and the tensile stress in the material tended to pull the materials away from the indenter surface, which resulted in less elastic recovery, while the compressive stress gave an opposite effect, which induced more elastic recovery[10]. The unloading curve is usually well approximated by the power law relation:

$$P = \alpha(h - h_r)^m \tag{2}$$

where α and m are power law fitting constants.

The total indentation work W_t given by the area under the loading curve, and elastic work W_e given by the area under the unloading curve can be determined by integrating the loading curve and unloading curve, respectively:

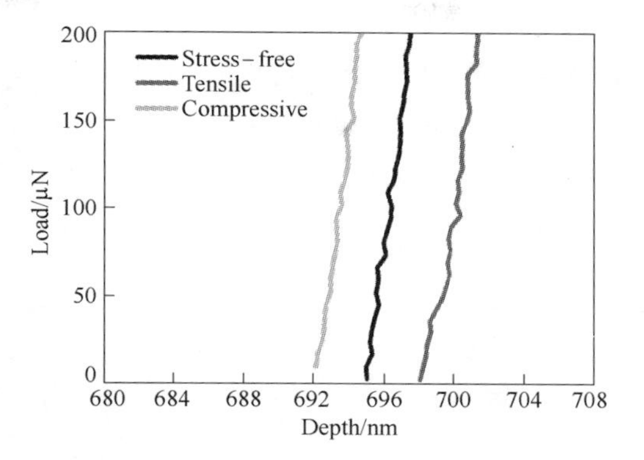

Fig. 3 The ending part of unloading curves for nanoindentations made to a maximum depth of 700 nm with different residual stress states

$$W_t = \int_0^{P_{max}} P(h)\,dh = \int_0^{P_{max}} Ch^2\,dh = \frac{Ch_{max}^3}{3} = \frac{P_{max}h_{max}}{3} \tag{3}$$

$$W_e = \int_{h_r}^{h_{max}} P(h)\,dh = \int_{h_r}^{h_{max}} \alpha(h - h_r)^m\,dh = \frac{\alpha(h_{max} - h_r)^{m+1}}{m+1} = \frac{P_{max}}{m+1}(h_{max} - h_r) \tag{4}$$

Thus, the plastic work W_p is the difference between W_t and W_e:

$$W_p = W_t - W_e = \frac{(m-2)h_{max} + 3h_r}{3(m+1)} \cdot P_{max} \tag{5}$$

The ratio W_p/W_t can be obtained from Eq. (6):

$$\frac{W_p}{W_t} = \frac{(m-2)h_{max} + 3h_r}{(m+1)h_{max}} = \frac{m-2}{m+1} + \frac{3h_r}{(m+1)h_{max}} \tag{6}$$

As the copper single crystal has low yield stress it behaves more plastically and has small elastically recovered depth, which results in the domination by plastic deformation. It can be found from Eq. (6) the ratio W_p/W_t is in direct proportion to the ratio h_r/h_{max}. Therefore, the indentation work was also affected by the presence of residual stresses. The ratio W_p/W_t decreased for tensile stress and increased for compressive stress.

3.3 Effect on pile-up deformation

It is well known that materials exhibiting low strain hardening tend to pile up around nanoindents due to the incompressibility of plastic deformation[11]. For copper single crystal, due to the low hardness and small strain hardening exponent, pile-up deformation occurred, as shown in Fig. 4. The amount of pile-up can be characterized by the height of the pile-up relative to the undeformed surface. Fig. 5a, b and c shows the cross-sectional profiles of nanoindents for stress-free, tensile stress and compressive stress, respectively. It is obvious that the pile-up height decreases for tensile stress and increases for compressive stress. Even for stress-free state pile-up occurred, which indicated that pile-up is not only related to the residual stress but is also attributed to the characteristic of the copper single crystal which has low hardness and small strain hardening exponent. For tensile residual stress, the materials were pulled away from the surface of indenter, which decreased the amount of pile-up; whilst compressive stress pushed the materials out to the surface of indenter which resulted in more pile-up.

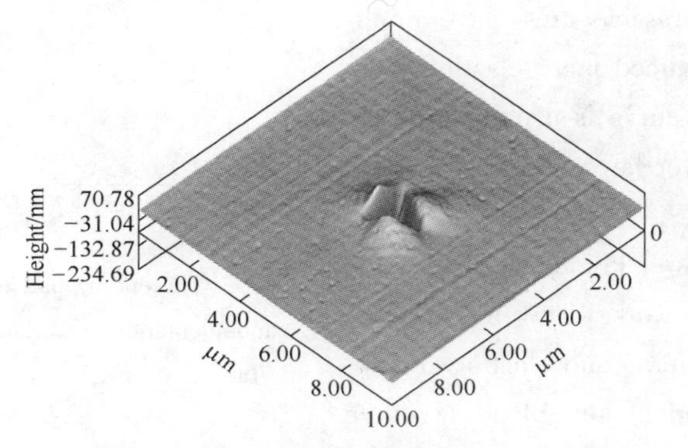

Fig. 4 Typical AFM morphology of nanoindent of (100) Cu single crystal

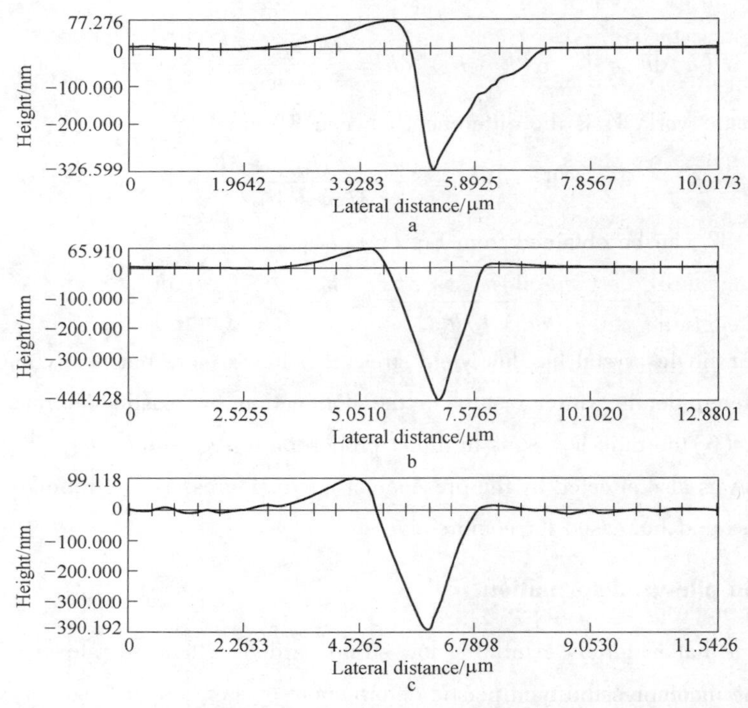

Fig. 5 Cross - sectional profile of nanoindents for different residual stress states
a—Stress - free; b—Tensile stress; c—Compressive stress

3.4 Effect on contact area

The contact area A_c can be measured from nanoindentation load-depth data based on the Oliver & Pharr method[12] from Eq. (7) and Eq. (8). However, the method does not account for the pile-up effects on the contact area. The study of Bolshakov and Pharr[13] has shown that if pile-up is neglected, the true contact area can be underestimated by as much as 60%.

$$h_c = h_{max} - \varepsilon \frac{P_{max}}{S} \tag{7}$$

$$A_c = \sum_{n=0}^{8} C_n (h_c)^{2-n} = C_0 h_c^2 + C_1 h_c + \cdots + C_8 h_c^{1/128} \tag{8}$$

In our previous study[14], we developed a method which models the projected contact area as an equilateral triangle bounded by arcs to correct for the effect of pile-up on the contact area. The real contact area can be calculated by Eq. (9) and Eq. (10).

$$A_c = 14.175\left(\frac{\theta\pi}{120\sin^2\frac{\theta}{2}} - 3\cot\frac{\theta}{2} + \sqrt{3}\right)(h_{max} + h_p)^2 \tag{9}$$

$$x = 3.765\frac{1 - \cos\frac{\theta}{2}}{\sin\frac{\theta}{2}}(h_{max} + h_p) \tag{10}$$

where θ is the angle subtended by the arc; h_p is the pile-up height; x is the height of the arc.

The contact areas for stress-free, tensile stress and compressive stress are referred to as A_c^0, A_c^T, and A_c^C, respectively. The ratios A_c^0/A_c^T and A_c^0/A_c^C at the fixed depth of 700 nm are 1.030 and 0.946, respectively. It is clear that the contact area decreases with tensile stress and increases with compressive stress.

3.5 Determination of residual stress from load-depth curves and residual nanoindents

Suresh and Giannakopoulos[1] proposed a method to extract equi-biaxial residual stresses from load-depth data based on the contact area difference between a stress-free material and the same material with residual stress. The residual stresses can be calculated by Eq. (11) and Eq. (12).

$$\sigma_r = H\left(\frac{A_c^0}{A_c^T} - 1\right) \text{ (for tensile residual stress)} \tag{11}$$

$$\sigma_r = \frac{H}{\sin\alpha}\left(1 - \frac{A_c^0}{A_c^C}\right) \text{ (for compressive residual stress)} \tag{12}$$

where α is the angle between indenter and the surface, which is 24.7° in the present study; H is the hardness.

Lee and Kwon[2] developed a modified stress-relaxation model to extract equi – biaxial residual stresses from load – depth data. The expressions are given by:

$$\sigma_r = \frac{3(P_0 - P_T)}{2A_c^T} \text{ (for tensile residual stress)} \tag{13}$$

$$\sigma_r = \frac{3(P_C - P_0)}{2A_c^C} \text{ (for compressive residual stress)} \tag{14}$$

Table 1 lists the results of residual stresses obtained from the Suresh & Giannakopoulos and

Lee & Kwon methods. It can be seen that the residual stresses obtained from Suresh and Giannakopoulos method showed a good agreement with the applied stresses from strain gage, compared with those from the Lee and Kwon method. This is due to that the Suresh and Giannakopoulos method can describe the well-known nonlinearity of the residual stress, but the stress-relaxation model proposed by Lee and Kwon considers the stress-relaxation process is linear. Moreover, indentation process is considered as an elasto-plastic problem, and residual stress cannot be relaxed alone while keeping the same indentation depth, because the stress-relaxation process will lead to a simultaneous change in the indentation load and depth[3].

The current models to measure residual stresses are mostly based on the comparison of nanoindentation parameters of stressed and stress-free samples. However, a stress-free sample is often not available, and some errors are inevitably made. This limits the application of these methods. In addition, the models only can measure plain residual stresses, and the residual stress along depth cannot be obtained. Therefore, in further work, it is necessary to develop new models which can determine the residual stresses in all the three directions and does not require a reference stress-free sample.

Table 1 Comparison of residual stresses obtained from the Suresh & Giannakopoulos and Lee & Kwon methods and the applied stresses (MPa)

Applied stresses from strain gage	Stresses from the Suresh and Giannakopoulos method	Stresses from the Lee and Kwon method
68.4	63.6 ± 1.4	95.4 ± 2.1
−137.4	−113.8 ± 8.2	−71.3 ± 10.7

4 Conclusions

Experimental tests were carried out to investigate the effects of residual stresses on the nanoindentation response of (100) copper single crystal. Equi-biaxial stresses (tensile and compressive) were applied to the copper single crystal using specially designed apparatus. It was found that the indentation parameters such as peak load, curvature of the loading curve, elastically recovered depth, residual depth, indentation work, pile-up amount and contact area were all affected by the existence of residual stresses. Compared with the stress-free state, tensile stress decreased the peak load, the curvature of loading curve, the elastically recovered depth, the indentation work, pile-up height and contact area, and increased the residual depth. Similarly, the effect of compressive stress on these parameters was opposite to that in tensile stress state.

The Suresh & Giannakopoulos and Lee & Kwon methods were used to calculate the residual stresses from load-depth data and AFM observation. It was found that the Suresh and Giannakopoulos method was more accurate than the Lee and Kwon method, when compared with the values obtained from strain gage. This was attributed to the difference of the two methods. The Suresh and Giannakopoulos method can describe the nonlinearity of the residual stress, but the Lee and Kwon method considers the stress-relaxation process is linear.

Acknowledgements

This paper was financially supported by NSF of Beijing(3120001), Distinguished Young Scholars of NSFC (51125023), 973 Project (2011CB013405), Equipment Research Project, and Fundamental Research Funds for the Central Universities.

References

[1] S. Suresh, A. E. Giannakopoulos, Acta Mater. 46(1998)5755 - 5767.
[2] Y. H. Lee, D. Kwon, Acta Mater. 52(2004)1555 - 1563.
[3] Q. Wang, K. Ozaki, H. Ishikawa, S. Nakano, H. Ogiso, Nucl. Instrum. Methods Phys. Res. B 242(2006) 88 - 92.
[4] B. X. Xu, B. Zhao, Z. F. Yue, J. Mater. Eng. Perform. 15(2006)299 - 305.
[5] J. G. Swadener, B. Taljat, G. M. Pharr, J. Mater. Res. 16(2001)2091 - 2102.
[6] Y. H. Lee, D. Kwon, J. Mater. Res. 17(2002)901 - 906.
[7] J. Dean, G. Aldrich - Smith, T. W. Clyne, Acta Mater. 59(2011)2749 - 2761.
[8] M. K. Khan, M. E. Fitzpatrick, S. V. Hainsworth, L. Edwards, Comput. Mater. Sci. 50(2011)2967 - 2976.
[9] Y. H. Lee, D. Kwon, Scripta Mater. 49(2003)459 - 465.
[10] Z. H. Xu, X. Li, Acta Mater. 53(2005)1913 - 1919.
[11] A. E. Giannakopoulos, S. Suresh, Scripta Mater. 40(1999)1191 - 1198.
[12] W. C. Oliver, G. M. Pharr, J. Mater. Res. 19(2004)3 - 20.
[13] A. Bolshakov, G. M. Pharr, J. Mater. Res. 13(1998)1049 - 1058.
[14] L. N. Zhu, B. S. Xu, H. D. Wang, C. B. Wang, Mater. Sci. Eng. A 528(2010)425 - 428.

Research on Tribological Behaviors of Composite Zn/ZnS Coating under Dry Condition*

Abstract A composite Zn/ZnS coating was prepared by a novel compound technology-combining high velocity arc spraying and low temperature ion sulfurizing in this paper. The surface and cross-section morphologies were observed by scanning electron microscopy (SEM). The X-ray diffraction (XRD) pattern for the Zn/ZnS coating implies that it mainly consists of Zn and ZnS. The nanohardness and elastic modulus were measured by a nano-indentation tester. The tribological behaviors were investigated on a ball-on-disk wear tester under dry condition. The results showed that the friction coefficient and worn depth of the composite Zn/ZnS coating were low and stable, indicating that it had excellent friction-reduction and anti-wear properties under dry condition.

Key words high velocity arc spraying method, low temperature ion sulfurizing, composite Zn/ZnS coating, tribological behavior

1 Introduction

Solid lubrication is using solid powders or films/coatings to protect the surfaces in relative motion from damage and reduce the friction and wear. It has been focused much attention because solid lubricant can be utilized in various special conditions such as high/low temperature, vacuum and heavy load, etc where common lubricating oil and grease cannot meet the requirements. The soft metals used commonly as the solid lubricant mainly include Pb, Zn, Sn, In, Au, Ag, etc. Zn has been applied widely because of its low cost. The electroplating or hot-dip methods are usually used to prepare the Zn and other soft metal layers[1-3]. ZnS is a kind of metal compound solid lubricant, and it possesses close-packed hexagonal lattice structure with low shearing strength. At present, a lot of studies have been done on ZnS nano-particles used as the additive in lubricating oil[4-7]. However, still few people have conducted the researches on the ZnS film/coating utilized as a solid lubricant.

In this paper, a composite Zn/ZnS coating was prepared by a novel compound technology-combining high velocity arc spraying and low temperature ion sulfurizing. A composite lubricating effect is expected[8-10].

2 Experimental

The substrate material was AISI 1045 steel, heat-treated by quenching and tempering, to a hardness of HRC 55. The surface roughness was $R_a = 3.2$ μm. Before deposition, the substrate was

* Copartner: Kang Jiajie, Wang Chengbiao, Wang Haidou, Liu Jiajun, Li Guolu. Reprinted from *Applied Surface Science*, 2012, 258(6): 1940 – 1943.

pretreated by grit-blasting; then the Zn coating was sprayed using a HAS – 01 spraying gun. The spraying parameters were: spraying voltage 35V, spraying current 160A, pressure of compressed air 0.7MPa, and spraying distance 150mm. A 300μm thick Zn coating was prepared finally.

The Zn coating was then treated by ion sulfurizing for 2h. During sulfurizing, the workpieces were connected to the cathode and the furnace wall was linked to the anode. The reaction gas was coming from the solid sulfur. A high potential direct current was applied on the cathode and anode to produce a plasma area in vacuum. The sulfur atoms were ionized and accelerated by the high-voltage field to impact at the surfaces of workpieces. The sulfur atoms diffused and penetrated into the Zn coating through the grain boundaries and defects and reacted with the iron atoms to form the ZnS coating.

The surace, cross-section, and wear scar morphologies and distribution of elements of composite Zn/ZnS coating were observed by scanning electron microscopy (SEM) equipped with energy dispersive spectroscopy (EDS). The phase structure was analyzed by X-ray diffraction (XRD). The nano-mechanical properties were characterized by a nano-indentation tester.

The friction and wear tests were carried out on a T-11 ball-on-disk tester under dry condition. The upper sample was 52,100 steel ball with hardness of HV770; the lower sample was the 1045 steel disc with composite Zn/ZnS coating on its surface. The sulfurized 1045 steel and original steel discs were also tested under the same conditions for comparison. Experimental parameters were: load 5N, sliding speed 0.2m/s, and experiment time 60min.

3 Results and Discussion

Fig. 1 shows the surface, cross-section morphologies and distribution of elements. The surface of the composite Zn/ZnS coating was relatively compact and smooth with a few irregular shape particles and flakes. The belt course with a thickness of 3 μm was the ZnS coating, as shown in Fig. 1b. The element distribution also proved the existing of ZnS coating.

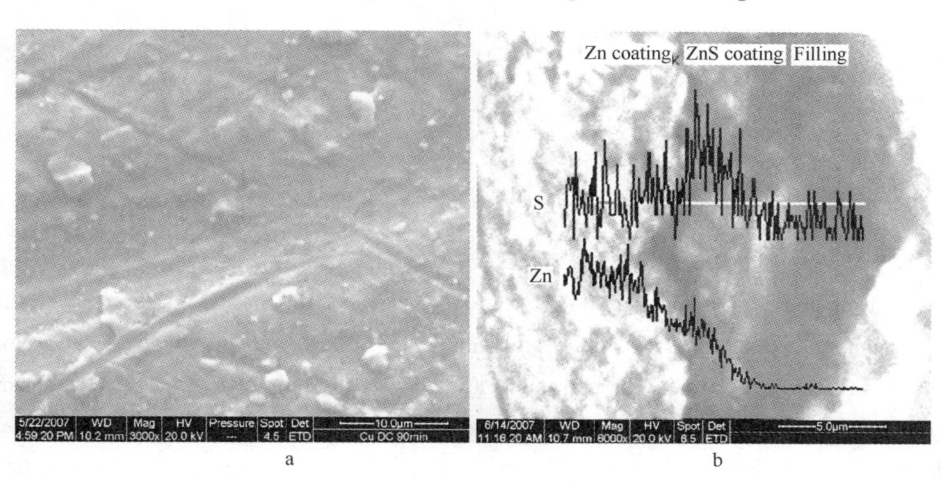

a b

Fig. 1 The surface, cross-section morphologies and element distribution of composite Zn/ZnS coating

a—Surface morphology; b—Cross – section morphology and element distribution

The phase structure of composite Zn/ZnS coating is shown in Fig. 2. The diffraction peaks were mainly ZnS and Zn, indicating that the solid lubricant ZnS film was undoubtedly formed as a result of the reaction between sulfur atoms and zinc atoms.

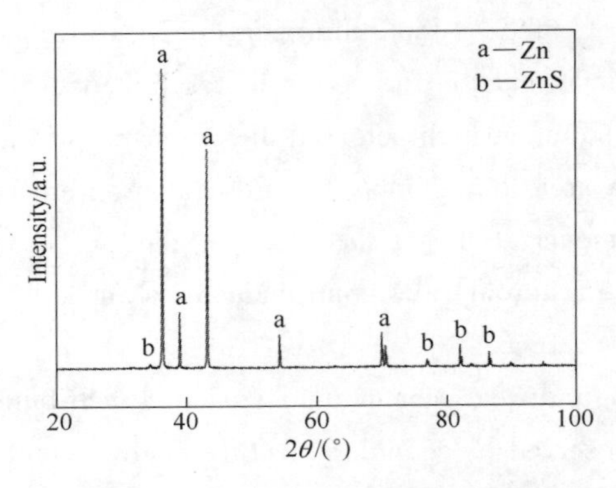

Fig. 2 XRD pattern of the composite Zn/ZnS coating

Fig. 3 shows the load-displacement curves of the Zn coating and composite Zn/ZnS coating obtained from nano-indetation tests. The five curves on the different positions of both Zn and composite Zn/ZnS coating almost matched together, indicating that the mechanical properties for the two coatings were both uniform and there were few defects on their surfaces. The nano-hardness and elastic modulus of the two coatings were 9. 35GPa and 179. 49GPa, 3. 86GPa and 123. 25GPa, respectively.

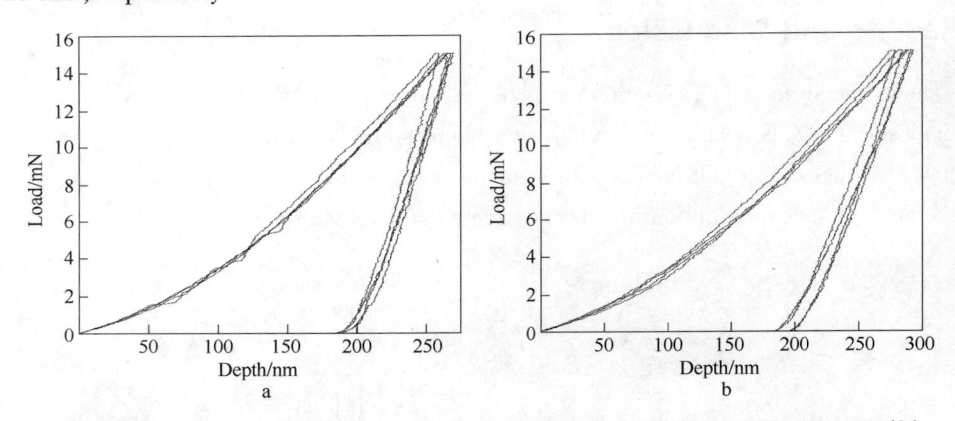

Fig. 3 Load – displacement curves of the Zn coating (a) and composite Zn/ZnS coating (b)

Fig. 4 shows the tribological curves of the 1045 steel, sulfurized 1045 steel, and composite Zn/ZnS coating under dry condition. The friction coefficients of the 1045 steel and sulfurized 1045 steel both increased sharply and kept at about 0. 8 and 0. 5, respectively. The friction coefficient of the Zn/ZnS coating was extremely low and stable, ranging from 0. 1 to 0. 15; during the whole test, it was much lower than that of the 1045 steel and sulfurized 1045 steel. The wear scar depth of the Zn/ZnS coating slowly increased and achieved a stable state finally; the wear scar depth of the 1045 steel increased linearly; the wear scar depth of the sulfurized 1045 steel increased fast at first, and then tended to be stable, but it was still much larger than that of the

Zn/ZnS coating.

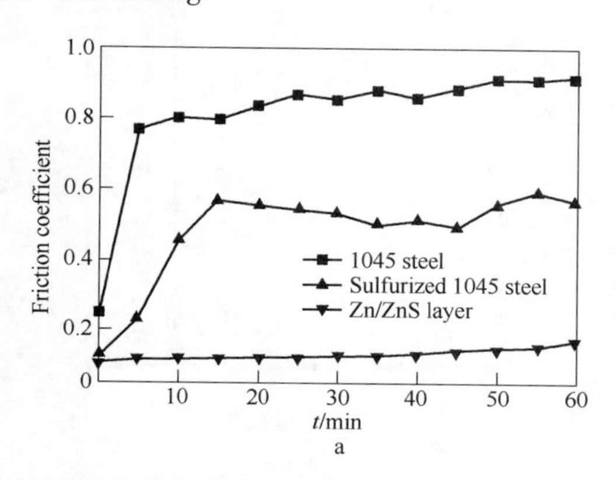

Fig. 4 Tribological curves of 1045 steel, sulfurized 1045 steel, and composite Zn/ZnS coating under dry condition
a—Variation of friction coefficient with time; b—Variation of wear scar depth with time

Fig. 5a, c and e shows the worn morphologies of the 1045 steel, sulfurized 1045 steel and composite Zn/ZnS coating after sliding of 60 min under dry condition. The compositions on the square areas for the 1045 steel, sulfurized 1045 steel and composite Zn/ZnS coating are shown in Fig. 5b, d and f. It can be seen from Fig. 5a that the wear of the 1045 steel was severe, and obvious wear scar occurred with a width of 500 μm. The wear of the sulfurized 1045 steel was milder compared with that of the 1045 steel and the former had a narrower trace, but the sulfurized 1045 steel of large area had flaked off and lost its lubricating property. The wear scar of the Zn/ZnS coating was the slightest with no obvious trace; only a little part of the coating was worn out. The element analysis in Fig. 5f indicates that there was still a certain amount of sulfur on the surface of Zn/ZnS coating at the end of test. This is a proof of existence of the ZnS solid lubricant after wear test. It can be concluded that the excellent tribological properties of composite Zn/ZnS coating was attributed to the combined effect of Zn and ZnS with close-packed hexagonal crystal structure and low shearing strength.

The composite Zn/ZnS coating possessed excellent tribological properties under dry condition; the main reasons were as follows:

(1) During the friction process, the composite Zn/ZnS coating is easily crushed and adhered to the surface of counterpart, which can effectively prevent the direct contact of metals and reduce the adhesion wear.

(2) When the ZnS coating is worn off, the Zn coating can continue to play the role of lubrication. In addition, the sprayed Zn coating possesses a porous structure, and its pores can accommodate the wear debris making the contact surface clean.

4 Conclusion

The composite Zn/ZnS coating with 3 μm thick upper coating ZnS and 300 μm thick lower coating Zn was prepared by a novel method-combining high velocity arc spraying and low temperature ion sulfurizing. The composite Zn/ ZnS coating had lower friction coefficient and wear

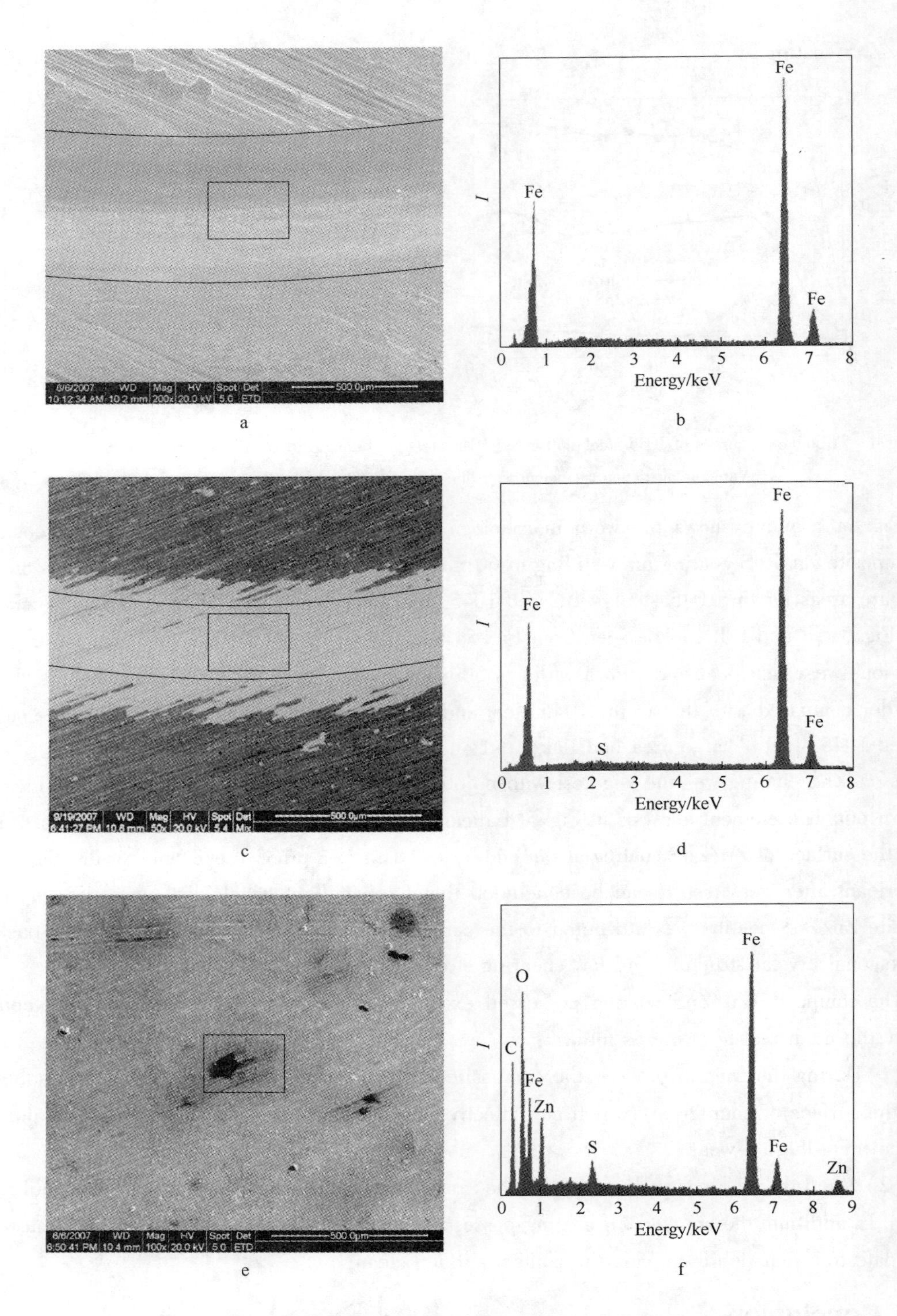

Fig. 5 Worn morphologies and compositions of the 1045 steel, sulfurized 1045 steel and composite Zn/ZnS coating under dry condition

a—Worn morphology of the 1045 steel; b—Composition of the 1045 steel;
c—Worn morphology of the sulfurized 1045 steel; d—Composition of the sulfurized 1045 steel;
e—Worn morphology of the composite Zn/ZnS coating; f—Composition of the composite Zn/ZnS coating

scar depth compared to 1045 steel substrate and suflurized 1045 steel. ZnS and Zn could Combine to take role of friction-reducing and wear-resisting. The composite Zn/ZnS coating possessed excellent tribological properties under dry condition.

Acknowledgements

The paper was financially supported by the Fundamental Research Funds for the Central Universities(2009PY07), NSFC(50975285), Advanced Maintenance Research Project (9140A270304090C8501), Equipment Maintenance Project and Equipment Research Project.

References

[1] T. Ben Nasr, N. Kamoun, C. Guasch. Physical properties of ZnS thin films prepared by chemical bath deposition. Applied Surface Science 254(2008)5039 – 5043.

[2] K. R. Murali, A. Clara Dhanemozhi, Rita John. Brush plated ZnS films and their properties. Journal of Alloys and Compounds 464(2008)383 – 386.

[3] Qi Liu, Mao Guobing, Ao Jianping. Chemical bath – deposited ZnS thin films: Preparation and characterization. Applied Surface Science 254(2008)5711 – 5714.

[4] Huaqiang Shi, Xiaodong Zhou, Xun Fu, Feng Gao, Zhengshui Hu. Preparation of organic fluids containing Cyanex 302 – modified ZnS particles with high loading concentration and their tribological properties. Colloids and Surfaces A: Physicochemical and Engineering. Aspects 317(2008)482 – 489.

[5] Guangbin Yang, Hongxia Ma, Zhishen Wu, Pingyu Zhang. Tribological behavior of ZnS – filled polyelectrolyte multilayers. Wear 262(2007)471 – 476.

[6] Weimin Liu, Shuang Chen. An investigation of the tribological behaviour of surface – modified ZnS nanoparticles in liquid paraffin. Wear 238(2000)120 – 124.

[7] Shuang Chen, Weimin Liu. Characterization and antiwear ability of non – coated ZnS nanoparticles and DDP – coated ZnS nanoparticles. Materials Research Bulletin 36(2001)137 – 143.

[8] Lina Zhu, Chengbiao Wang, Haidou Wang, Binshi Xu, Daming Zhuang, Jiajun Liu, Guolu Li, Tribological properties of WS_2 composite film prepared by a two – step method, Vacuum 85(2010)16 – 21.

[9] Jiajie Kang, Chengbiao Wang, Haidou Wang, Binshi Xu, Jiajun Liu, Guolu Li, Characterization and tribological properties of composite 3Cr13/FeS layer, Surface & Coatings Technology 203(2009)1927 – 1932.

[10] J. J. Kang, C. B. Wang, H. D. Wang, B. S. Xu, J. J. Liu, G. L. Li, Microstructure and tribological properties of composite FeCrBSi/FeS layer, Part J: Journal of Engineering Tribology 224(2010)807 – 813.

Investigation of Acoustic Emission Source of Fe-based Sprayed Coating under Rolling Contact*

Abstract Acoustic emission(AE) technique was used to monitor rolling contact fatigue(RCF) experiments of plasma sprayed coatings. AE signal response was investigated. Worn morphologies of the coating were observed. Results show that typical AE signal mode owns three stages, i. e. running – in, stable and jump stage. Observations of worn coating release there are three kinds of AE signal sources, i. e. the crack initiated from surface, cracks embedded within the coating and cracks on the coating/substrate interface. The combined effect of orthogonal shear stress(OSS) and radial stress (RS) is main reason of the initiation and propagation of cracks within the coatings.

Key words rolling contact fatigue, plasma sprayed coating, acoustic emission, crack source

1 Introduction

The rolling contact fatigue(RCF) is a typical mode of mechanical product whose surface endured altering loading[1,2]. RCF failure is common in industrial applications. For prolonging RCF lifetime, some surface engineering techniques are introduced to deal with the surfaces of mechanical products. Previous studies released that thermal spray technique is an effective method for resisting RCF. The material with excellent RCF – resistance can be deposited on component surface by thermal spray technique[3-5]. RCF lifetimes and failure mechanisms of thermal sprayed coatings were widely investigated based on statistical method and fracture analysis, respectively[6-8]. The rules of RCF lifetime decline of sprayed coatings were successfully characterized by Weibull distribution plot[9,10]. The failure mechanism of sprayed coating seems complex. The orthogonal shear stress(OSS) resulted from rolling contact is recognized the driver for RCF crack within the coating. Furthermore, the distribution of shear stress is also obtained based on finite element(FE) method. But the original morphologies of RCF crack sources driving by OSS seem difficult to be found by traditional fracture analysis approaches. Because fracture analyses are always conducted after the appearances of catastrophic coating peeling – off, the sub – critical failure state has been destroyed.

Process monitoring of RCF experiment of sprayed coating will release some information about crack sources. Acoustic emission(AE) technique is an effective approach for process monitoring. AE phenomenon is defined as transient elastic waves generated from a rapid release of strain energy caused by a deformation of the material under loading[11,12]. Fracture, elastic and

* Copartner: Piao Zhongyu, Wang Haidou, Wen Donghui. Reprinted from *International Journal of Fatigue*, 2013, 47(2): 184 – 188.

plastic deformation of the surface and subsurface are all the potential AE sources[13]. The technical feature of AE technique decides that it is appropriate to monitor RCF process of sprayed coating. AE technique has been employed to on – line monitor the appearances of RCF cracks of hard machining surface[14,15]. The result showed that the surface integrity of hard machining surface can be evaluated by analysis of AE signal. There are always no published reports about AE monitoring of RCF process of sprayed coating. So the aim of the present study is to investigate RCF crack sources of thermal sprayed coating by AE technique.

2 Experimental Procedures

2.1 Preparation of coatings

Commercially Fe-Cr-B-Si self-fluxing alloy(wt% : Cr – 13.6, B – 1.6, Si – 1.1, C – 0.16, Fe – balance) and Ni-Al alloy(wt% : Ni – 90, Al – 10) powders were used as surface coating and undercoating materials, respectively. AISI 1045 steel with ring – type geometry was used as substrate. The external diameter, internal diameter and thickness of the substrate were machined to 60mm, 30mm and 25mm, respectively. The substrate was cleaned by acetone solution in an ultrasonic bath and sandblasted by the corundum powders of mesh 48 size. The sandblasting process was conducted with blasting pressure of 0.6 MPa, blasting angle of 70°, standoff distance of 100 mm and blasting time of 15 – 25s. Finally, sandblasting velocity was about 100m/s.

High-efficiency plasma spraying (PS) system was employed to deposit the coating system, i. e. the surface coating and undercoating. During PS process, argon gas was used as the primary gas, hydrogen gas and nitrogen gas were used as the secondary gases. Table 1 summarizes the relevant details of plasma spraying parameters. After PS process, the coatings were ground to obtain the thickness of 200 μm on rubbing bed by abrasive disk. After surface machining, surface roughnesses of the coatings were all about 0.4 μm.

Table 1 Plasma spraying parameters

Spaying material	FeCrBSi	Ni/Al
Argon gas flow/$m^3 \cdot h^{-1}$	3.4	3.4
Hydrogen gas flow/$m^3 \cdot h^{-1}$	0.3	0.3
Nitrogen gas flow/$m^3 \cdot h^{-1}$	0.6	0.6
Spraying current/A	380	320
Spraying voltage/V	150	140
Spraying distance/mm	110	150
Powder feed rate/$g \cdot min^{-1}$	30	30

2.2 RCF tests

A ball-on-disk tester was used to conduct RCF experiments of the coatings. The schematic of the tester and the detail of RCF experimental process can be found in author earlier publication[16]. In the present study, the test applied load was kept at 100 N and rotation speed of basic shaft at 1500 r/min. RCF experiments were all conducted under immersed lubrication condition. The machine oil(SAE 46) was used as lubrication.

PCI - 2 numerical AE monitoring instrument was used to on-line monitor RCF experimental process of plasma sprayed coating. NANO-30 AE sensor was directly assembled on the substrate, while vacuum grease served as a coupling media for resisting the signal attenuation. AE sensor has a 140 kHz resonant frequency and connects to an 18 bit data acquisition board which was connected with a computer. Before the signals were relayed to computer, they were sent through a preamplifier at a gain 40 dB to strengthen the signal intensity. A signal-processing package, AEwin[17], was employed to on-line analysis AE signals. The details of AE instrument are listed in Table 2. The selections of AE instrument components such as sensor and preamplifier were decided vased on the features of RCF tests of sprayed coatings. The size of sensor ensures the sensor can be easily assembled in RCF tester. The operating frequence range of sensor ensures the accurate detections of RCF cracks. The setting value of preamplifier ensures the intensity of obtained signals.

Table 2 AE instrument parameters

AE instrument components	Main parameters
Sensor	Type: NANO - 30; size (DIA HT, mm × mm): 8 × 8; operating temperature (℃): -65 to 177; operating frequence range (kHz): 125 - 750; resonant frequency (kHz): 140
Transducer (data acquisition board)	Type: PCI - 2; channel number: 2; sampling rate (samples/s): 40M; system frequence range (kHz): 1 - 3M
Preamplifier	Amplification range (dB): 20, 40 and 60

2.3 Characterization of coatings

The worn surface morphologies of the coating after RCF tests were observed by scanning election microscope(SEM). The distributions of the stresses within the coatings were investigated by finite element(FE) analysis method.

3 Results and Discussions

3.1 AE signal response mode

Actually, the visual signals resulting from RCF process on the computer of AE monitor instrument are electrical signals. The analyses of electrical signals are conducted by special software, i. e. AEwin. And then various signal parameters are obtained, the different signal parameter can be used to response the different features of AE sources. Amplitude and energy of AE signal are usually considered to sensitively monitor the material fracture on the basis of the previous researches[18,19]. In this study, amplitude and energy signals were selected as the characteristic parameters to evaluate the initiations and propagations of RCF cracks within the coatings. The typical AE signal response mode was investigated in large sample space.

Fig. 1 shows the typical AE signal response mode. It is obvious that three stages present in whole AE signal response for RCF experiment. They are defined as running - in, stable and jump stage in this study, respectively. In running – in stage (numbered 1), there were some signal fluctuations. The asperity contact between Fe – base coating and coupling ball induced the micro – fractures on the coating surface. Some plastic deformations of the coating surface also appeared for the high vertical pressure. So micro – fracture and plastic deformation were the reasons of signal fluctuations in running – in stage. In stable stage(numbered 2), asperity contact almost disappeared for the micro – bulges on the coating surface had been ground off. At the same time, the stable wear trace had been formatted, plastic deformation also disappeared. So the signal exhibited more smooth than running – in stage. In jump stage(numbered 3), the signal exhibited shape rise for the initiation and propagation of RCF crack within the coating.

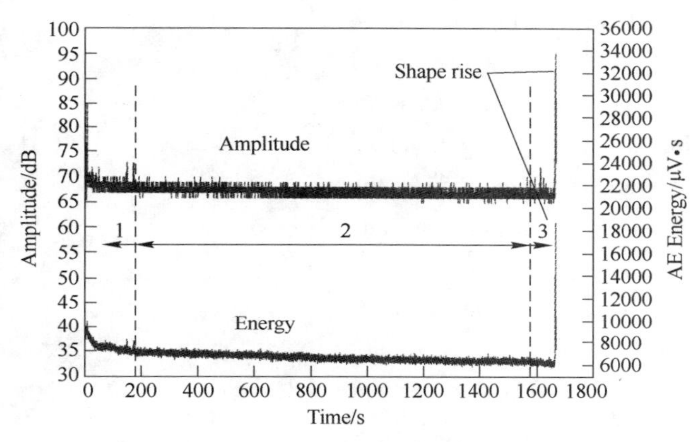

Fig. 1 Typical AE signal response mode of sprayed coating during RCF test

3.2 AE signal sources

When AE signal exhibited shape rise, RCF experiment of the coating was immediately stopped. The morphologies of sprayed coating surface were observed by SEM. Fig. 2a shows the morphologies of sprayed coating surface experienced RCF test after shape rise of AE signal. There is no obvious material loss on the coating surface, but some cracks. The early author

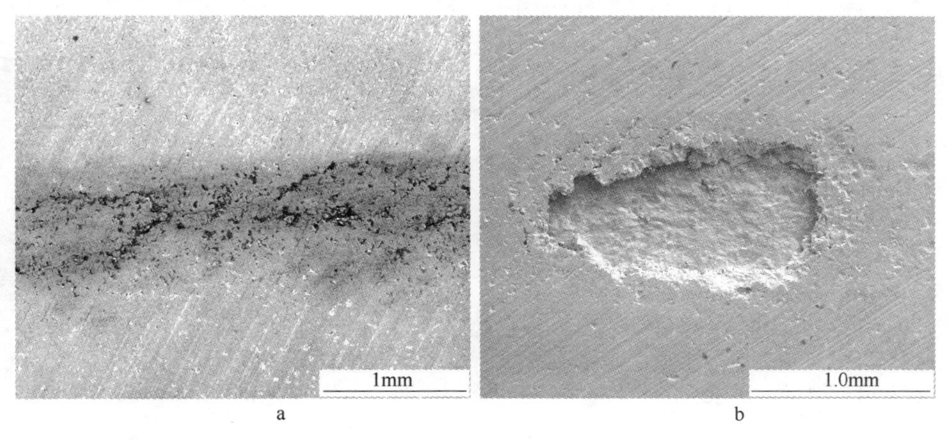

Fig. 2 Morphologies of the worn coatings

a—After shape rise of AE signal; b—After shape rise of vibration signal

publications show SEM images of the morphologies of sprayed coating surface after shape rise of traditional vibration signal, as shown in Fig. 2b[20]. There is obvious the coating peeling – off. The sensitivities of AE and vibration signal on RCF cracks have also been investigated in author early publication[20]. AE signal parameters are proofed to be much more sensitive to the initiations and propagations of the fatigue crack than traditional vibration signal. The high sensitivity of AE signal for the initiation and propagation of RCF crack is obviously positive for the investigation of RCF crack origins. The cross – sectional morphologies of the coatings experienced RCF tests after shape rise of AE signal were investigated by SEM to address RCF crack origins. Results showed that three kinds of RCF cracks were found within the coatings.

Fig. 3 shows the crack initiated from surface (as indicated by arrows). The crack propagates along the vertical direction. For the cracks within coating always initiate and propagate along the direction with the angle about 45° to the coating surface[21]. So the present crack initiate from coating surface under high vertical pressure during RCF test. Fig. 4 shows the cracks embedded within the coating (as indicated by arrows). The cracks obviously initiate and propagate near the unmelted particles. After these cracks propagate to the coating surface, the delamination of the coating will inevitably occur. Fig. 5 shows the cracks on the coating/substrate interface (as indicated by arrows). The cracks on the interface are just beneath the surface wear trace resulted from RCF tests. These cracks will induce the peeling – off of the whole coating. Generally, the aforementioned three kinds of RCF cracks all are AE signal origins. And they will induce the various failures of the coatings if RCF tests go on. So AE technique is proofed effective to detect the initiations and propagations of cracks and predict the coming failure by on – line monitoring RCF tests.

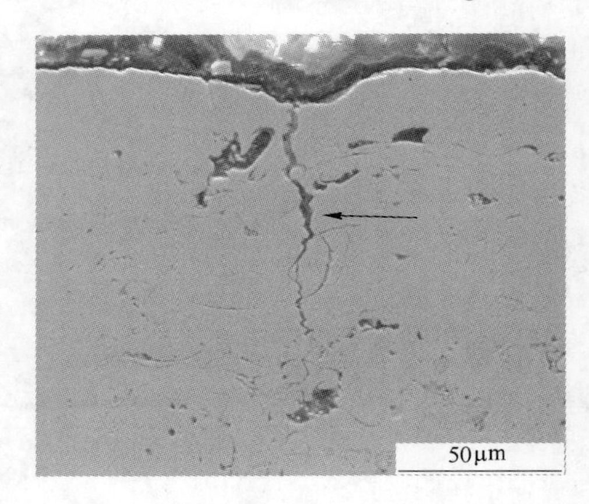

Fig. 3 Morphology of crack initiated from surface

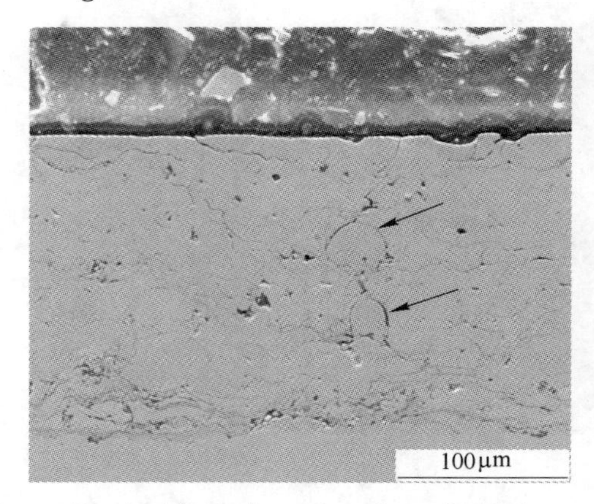

Fig. 4 Morphology of cracks embedded within the coating

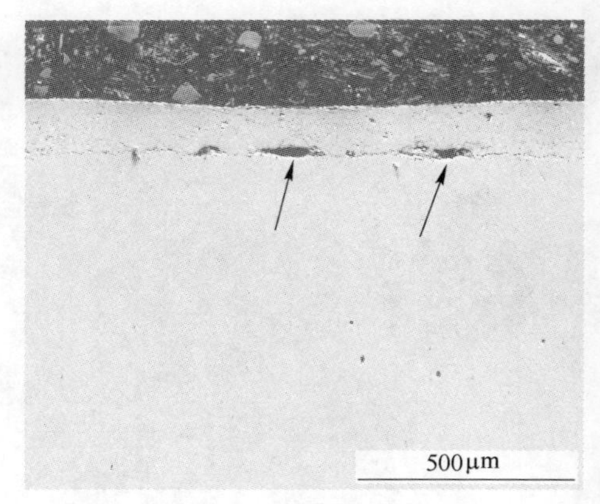

Fig. 5 Morphology of cracks on the coating/substrate interface

3.3 Fatigue mechanism

The stresses on the coating surface or within the coating resulted from rolling contact are the main reasons for the initiation and propagation of RCF cracks. The distribution of orthogonal shear stress (OSS) within the coating and radial stress (RS) on the coating surface were calculated by FE method. The commercial finite element analysis code ANSYS was used to investigate the distribution of the stresses. A elastic axial symmetric model was established. The details about FE model establishments and calculations can be found in author early work[22]. Fig. 6a shows the distribution of OSS within the coating. The maximum magnitude of OSS located within the coating. So the microstructures near the depth of maximum OSS were inclined to format crack origins, such as unmelted particles, oxides and pores, as shown in Fig. 4. Then the cracks always propagated alone the direction with the angle about 45° to the coating surface. At the same time, the coating/substrate interface also endured intensive OSS as shown in Fig. 6a. Because of the mechanical bond between the coating and substrate, the bond strength was weak. Consequently the microstructures on the interface were easy to be driven and propagated under OSS, as shown in Fig. 5. Fig. 6b shows the distribution of RS on the coating surface. It can be seen that the maximum tension RS located in the edge of wear trace. So firstly the cracks initiated on the coating surface where endured intensive RS, and then propagated towards the coating interior, as shown in Fig. 3. Generally, OSS drives the cracks within coating or on the interface; RS drives the cracks on the coating surface. The cracks driven by OSS propagate towards the coating surface and meet the cracks driven by RS near the coating surface. Finally, the combined crack system formats and a mass of coating material peel off. Fig. 7 shows the proof of aforementioned mechanism. It can be seen that the crack driven by RS numbered 1 (indicated by arrow) combines with the crack driven by OSS numbered 2 (indicated by arrow). So the combined effect of OSS and RS is the main reason of the delamination of sprayed coating.

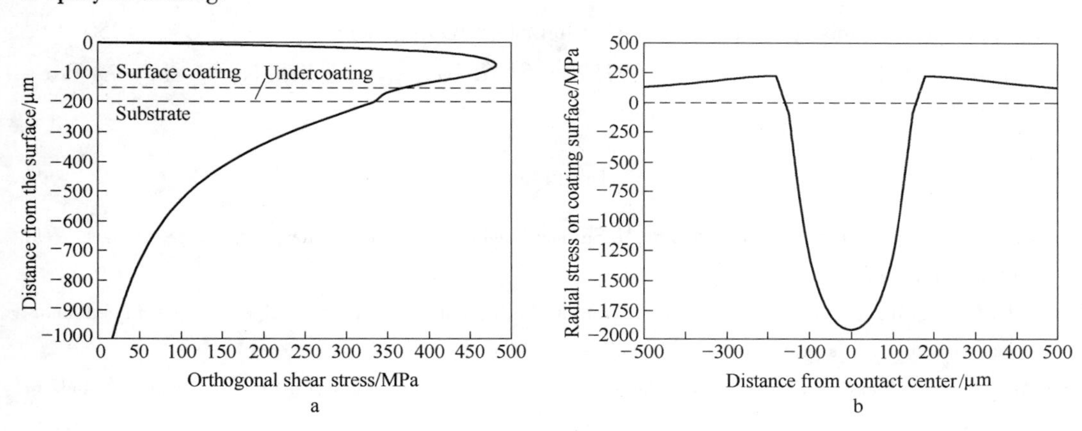

Fig. 6 Distribution of OSS and RS within or on the coating

a—Distribution of OSS; b—Distribution of RS

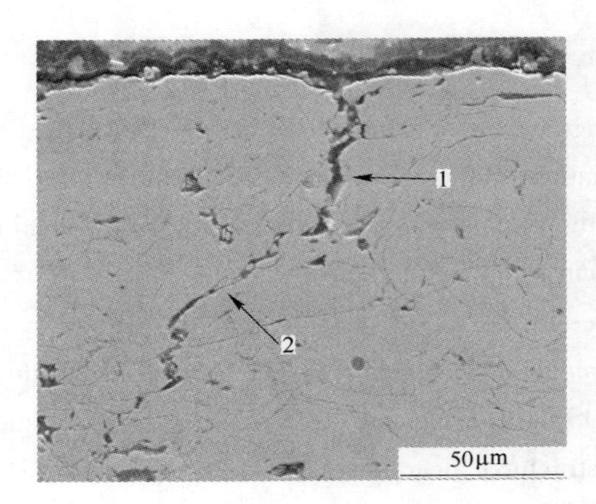

Fig. 7 Combination of crack initiated from the coating surface and crack embedded within the coating

4 Conclusions

(1) AE technique was introduced to monitor RCF tests of Fe – based coatings. The typical AE signal response mode was summarized. Three stage were appeared in RCF test, i. e. running-in, stable and jump stage.

(2) AE sources within the coating were investigated. Three kinds of cracks were found, i. e. , the crack initiated from surface, cracks embedded within the coating and cracks on the coating/substrate interface.

(3) The mechanism of RCF failure was investigated. The cracks within the coating and on the interface were driven by OSS. The cracks on the coating surface were driven by RS. The combined effect of OSS and RS is the main reason of the failures of the coatings.

Acknowledgements

This paper was financially supported by Distinguished Young Scholars of NSFC(51125023), 973 Project(2011CB013405), NSFC(50975285, 51075368), Young Scholars of Zhejiang Province(R1111149).

References

[1] Kapoor A, Franklin F J, Wong S K, Ishida M. Surface roughness and plastic flow in rail wheel contact. Wear 2002, 253: 257 – 264.

[2] Scharf T W, Singer I L. Role of the transfer film on the friction and wear of metal carbide reinforced amorphous carbon coatings during run – in. Tribol Lett 2009, 6(1): 43 – 53.

[3] Ahmed R, Hadfield M. Rolling contact fatigue performance of detonation gun coated elements. Tribol Int 1997, 30: 129 – 137.

[4] Stewart S, Ahmed R. Contact fatigue failure modes in hot isostatically pressed WC – 12% Co coatings. Surf Coat Technol 2003, 172: 204 – 216.

[5] Nieminen R, Vuoristo P, Niemi K, Mantyla T, Barbezat G. Rolling contact fatigue failure mechanisms in plasma and HOVF sprayed WC – Co coatings. Wear 1997, 212:66 – 77.

[6] Fujii M, Ma J B, Yoshida A, Shigemura S, Tani K. Influence of coating thickness on rolling contact fatigue of alumina ceramics thermally sprayed on steel roller. Tribol Int 2006, 39:1447 – 1453.

[7] Stewart S, Ahmed R, Ituskaichi T. Rolling contact fatigue of post – treated WC – NiCrBSi thermal spray coatings. Surf Coat Technol 2005, 190:171 – 189.

[8] Ahmed R, Hadfield M. Failure modes of plasma sprayed WC – 15% Co coated rolling elements. Wear 1999, 230:39 – 55.

[9] Zhang X C, Xu B S, Xuan F Z, Tu S D, Wang H D, Wu Y X. Fatigue resistance and failure mechanisms of plasma – sprayed CrC – NiCr cermet coatings in rolling contact. Int J Fatigue 2009, 31:906 – 915

[10] Piao Z Y, Xu B S, Wang H D, Pu C H. Investigation of rolling contact fatigue lives of Fe – Cr alloy coatings under different loading conditions. Surf Coat Technol 2010, 204:1405 – 1411.

[11] Warren AW, Guo Y B. Acoustic emission monitoring for rolling contact fatigue of superfinished ground surfaces. Int J Fatigue 2007, 29:603 – 614.

[12] Miguel J M, Guilemany J M, Mellor B G, Xu Y M. Acoustic emission study on WC – Co thermal sprayed coatings. Mater Sci Eng A 2003, 352:55 – 63.

[13] Sun J, Wood R J K, Wang L, Care I, Powrie H E G. Wear monitoring of bearing steel using electrostatic and acoustic emission techniques. Wear 2005:1482 – 1489.

[14] Guo Y B, Ammula S C. Real – time acoustic emission monitoring for surface damage in hard machining. Int J Mac Tools Manu 2005, 45:1622 – 1627.

[15] Guo Y B, Warren A W. The impact of surface integrity by hard turning vs. grinding on fatigue damage mechanisms in rolling contact. Surf Coat Technol 2008, 203:291 – 299.

[16] Piao Z Y, Xu B S, Wang H D, Pu C H. Influence of undercoating on rolling contact fatigue performance of Fe – based coating. Tribol Int 2010, 43:252 – 258.

[17] Lohr M, Spaltmann D, Binkowski S, Santner E, Woydt M. In situ Acoustic Emission for wear life detection of DLC coatings during slip – rolling friction. Wear 2006, 256:469 – 479.

[18] Schwach D W, Guo Y B. A fundamental study on the impact of surface integrity by hard turning on rolling contact fatigue. Int J Fatigue 2006; 28:1838 – 1844.

[19] Guo Y B, Schwach D W. An experimental investigation of white layer on rolling contact fatigue using acoustic emission technique. Int J Fatigue 2005, 27:1051 – 1061.

[20] Piao Z Y, Xu B S, Wang H D, Pu C H. A separation of experimental study on coatings failure signal responses under rolling contact. Tribol Int 2011, 44:1304 – 1308.

[21] Piao Z Y, Xu B S, Wang H D, Pu C H. Investigation of fatigue prediction of Fe – Cr alloy coatings under rolling contact based on acoustic emission technique. Appl Surf Sci 2011, 257:2581 – 2586.

[22] Piao Z Y, Xu B S, Wang H D, Pu C H. Effect of thickness and elastic modulus on stress condition of fatigue – resistant coating under rolling contact. J Cent South Univ Technol 2010, 17:899 – 905.

Investigation of a Novel Rolling Contact Fatigue/ Wear Competitive Life Test Machine Faced to Surface Coating*

Abstract A novel rolling contact fatigue(RCF)/wear competitive life test machine based on double - roll mechanism was designed to investigate the RCF/wear failure mode and predict the competitive life regularity of surface coatings. The test machine was designed for variable slip ratio to simulate the complex service conditions. The failure can be monitored using signals which are acquired from acoustic emission(AE) of the contacting rollers. In the experimental studies, the typical morphologies of RCF damage such as pits, surface abrasion, and delamination were observed. The effectiveness of the AE signals used as a RCF failure monitoring tool was analyzed. The *P* - *N* and *P* - *S* - *N* plots were established to predict the life regularity of surface coatings.

Key words rolling contact fatigue, slip - rolling, thermal spray coatings, acoustic emission

1 Introduction

Rolling contact fatigue(RCF) and wear are the most common failure mode of mechanical components and engineering structural parts such as shafts, gears, cams, and rollers, etc. RCF and wear can cause a huge economic loss and restrict the service property of equipments. Wear is mainly in the form of surface material removal caused by plow and adhesion under slip contact condition. RCF often occurs on the surface of friction pairs under rolling contact condition. It is a persistent damage process which involves the crack initiation, crack propagation and crack induced fracture due to the generated shear stress in the superficial layer under the action of alternating load. RCF and wear are both subject to the surface material failure with a common characteristic of contact, friction, and surface accumulative damage.

Thermal spraying as a kind of convenient and high efficient technology is often used to prepare surface coatings with high bonding strength and hardness for repairing the failure parts due to RCF and wear, which has attracted extensive attention of researchers in the field of tribology[1-3]. The wear - resistance and contact fatigue resistance properties are the key indicators to evaluate the quality and durability of the repairing coatings[4-6]. However, it should be noted that surface coatings have multicomponent composition and high irregular metastable structure, which endows surface coatings special characteristics - high free energy and multi - interface different from those of homogeneous materials. High free energy contributes to the improvement of the wear - resistance property, while multi - interface tends to induce the initiation of micro

* Copartner: Kang Jiajie, Wang Haidou, Wang Chengbiao. Reprinted from *Tribology International*, 2013, 66: 249 - 258.

– cracks and reduce the RCF life. Therefore, the life evolution laws of surface coatings are non-linear, which leads to the complexity of life prediction. Under the "slip – rolling" complex condition, the failure mode of surface coatings is determined by the competition between RCF and wear. Therefore, it has great scientific significance to research competitive life of surface coatings under "slip – rolling" mode which is close to real serving condition.

A novel RCF/wear competitive life test machine was designed to simulate the real contact condition i. e. "slip – rolling" of surface coating. Acoustic emission (AE) technology was selected to monitor the failure of the surface coating, especially RCF failure. The energy and amplitude of acoustic emission signals can monitor sensitively the stress wave which is generated from the fracture of materials due to fatigue crack propagation[7-11]. Moreover, acoustic emission signals can prewarn the failure occurrence ahead of the commonly used vibration signals. This contributes to reduction of human attention and human error, and accurate determination of failure point as well.

2 RCF/wear Competitive Life Test Machine

2.1 Design criteria

The novel RCF/wear competitive life test machine was designed to simulate the complex service condition and provide a reliable test platform. It should meet the requirement of the following criteria[12]:

(1) The friction pairs contact with each other as a line contact mode to simulate the service condition of gears, rollers, and cams.

(2) The test machine should have controllable drive device to run the friction pairs to achieve any desired speed and realize accelerated life tests with high efficiency.

(3) The relative motion state between the friction pairs can switch conveniently among pure slip, slip – rolling, and pure rolling.

(4) The test machine should have reliable loading facility to give a contact pressure up to 2.6 GPa.

(5) The test machine should have sensitive system to monitor failure process accurately, and realize automatic stop when determining failure point.

2.2 Modules of the test machine

The test machine was designed in the form of modules. Fig. 1 shows the schematic diagram of the test machine. The test machine consists of mechanical system and a measurement and control system. The mechanical system includes roller assembly module, drive module, loading module, and lubrication module. The measurement and control system is composed of measurement module and data acquisition and processing module.

Table 1 shows the technical parameters of the test machine. The significant features of the test machine are as follows:

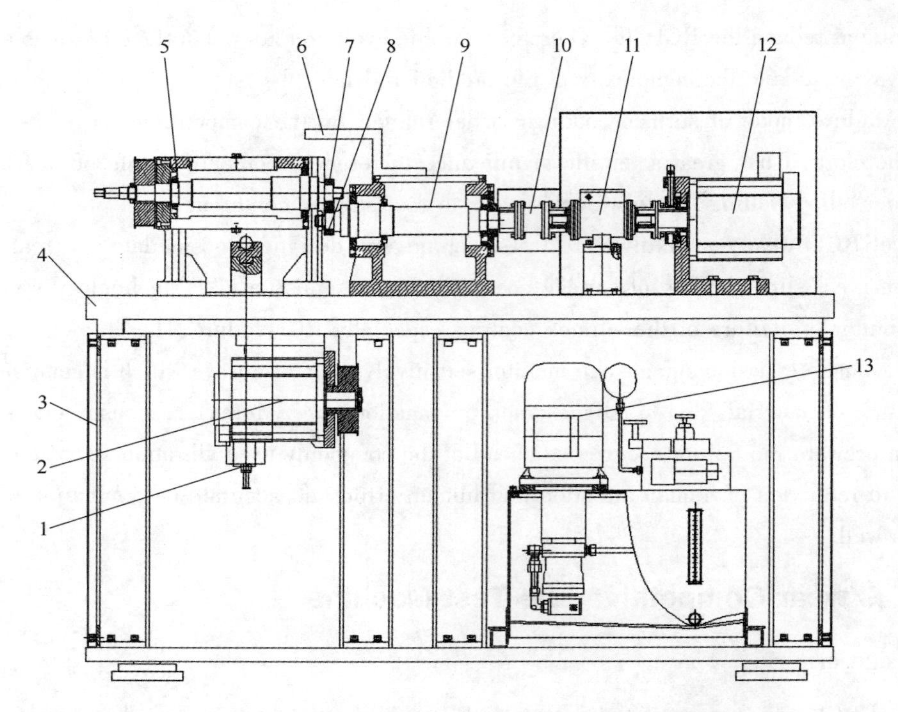

Fig. 1 Schematic diagram of RCF/wear competitive life test machine

1—Hydraulic piston;2—Servo - actuator driving standard roller;3—Chest;4—Working flat;
5—Main shaft of standard roller;6—Oil interceptor;7—Standard roller;8—Test roller;9—Main shaft of test roller;
10—Flexible coupling;11—Torque sensor;12—Servo - actuator driving test roller;13—Hydraulic station

Table 1 Technical parameters of the test machine

Technical parameters	Values
Loading range of test force/kN	1.2 - 30(relative error≤ ±0.5%)
Measurement range of friction torque/N · m	1 - 20(relative error≤ ±1%)
Rotational speed/r · min^{-1}	5 - 2000(infinitive variable)
Test time	1 s - 9999 min
Measurement range of test revolution	0 - 999999999
Measurement range of temperature/℃	-25 - 650
Slip ratio/%	0 - 100
Power of servo - actuator/kW	5
Torque output of servo - actuator/N · m	23
External dimensions/mm × mm × mm	1690 × 960 × 1210

(1) The test roller and standard roller are driven by respective servo - actuator to realize the accurate control of slip ratio(0-100%). Where slip ratio is the ratio of velocity difference between test roller and standard roller to the velocity of standard roller. The rolling - slip contact condition can change conveniently by adjust the slip ratio randomly.

(2) The convex part of the test roller was designed to make the machine specific for testing surface coatings.

(3) Hydraulic and lever loading device ensures the test machine can load continuously without interruption. The loading value can be controlled by computer.

(4) AE technique was introduced to make the test machine accurately judge the failure point. The typical AE waveforms and frequency spectra of different stages in RCF process can be extracted to judge the RCF failure.

(5) The test machine can realize automatic stop when the selected parameter (energy of AE signal, friction torque, and test time) is over the pre-set value.

2.2.1 Roller assembly module

Fig. 2 shows the schematic diagram of roller assembly module. The roller assembly module of the test machine mainly includes a test roller and a standard roller. The test roller and standard roller were assembled on respective main shaft, and both locked by a nut with the diameter of 35 mm. The main shaft of test roller was fixed on the working flat by a bearing block. The main shaft of standard roller was fixed on one side of the hydraulic and lever loading device. The parallelism between the two main shafts is relatively high to ensure the tight contact between the two rollers.

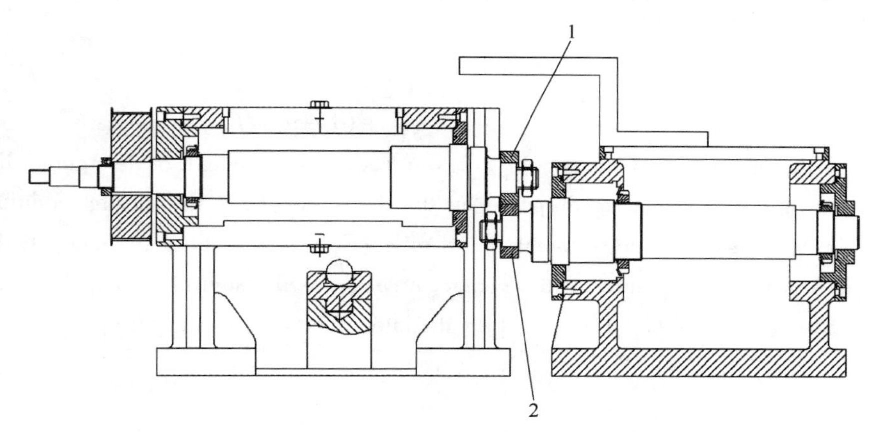

Fig. 2 Schematic diagram of roller assembly module
1—Standard roller; 2—Test roller

Fig. 3a and b shows the configuration of the test roller and standard roller, respectively. The tested coating should be prepared on the external circle surface. The design of convex part of the test roller makes it specific for surface coatings. The relative small area of coating on the convex part can weaken the negative effect of the elastic mismatch on the bonding strength often occurring in lager area coating. The edge of convex part of the test roller was beveled with the dimension of 0.5 mm to prevent the occurrence of stress concentration on the edge of the coating. The length of contact line between the rollers is 5 mm.

Supersonic plasma sprayed Al_2O_3/40% TiO_2 composite ceramic coating (AT40 coating) and quenched + low temperature tempered AISI 52100 steel were taken as an example to calculate the contact stress. The Poisson's ratio and elastic modulus of the coating and steel were 0.3, 173 GPa; and 0.3, 219 GPa; respectively. The maximum contact stress can reach 2.6 GPa through calculating by Hertz equation when the applied load was 30 kN as maximum. The Hertz

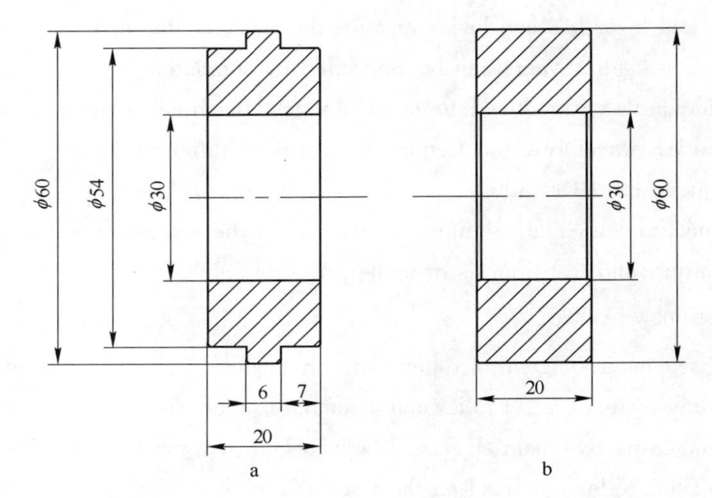

Fig. 3 Configuration of the rollers

a—Test roller; b—Standard roller

equation is shown below:

$$\sigma_{\max} = \sqrt{\frac{F(\Sigma\rho)}{\pi L[(1-\nu_1^2)/E_1 + (1-\nu_2^2)/E_2]}} \tag{1}$$

where $\sigma_{\max}$ is the maximum contact stress; F is the load; L is the length of the contact line; ν_1 is the Poisson's ratio of the coating prepared on the test roller; E_1 is the elastic modulus of the coating; υ_2 is the Poisson's ratio of the standard roller; E_2 is the elastic modulus of the standard roller; ρ is the principal curvature at the contact area; $\Sigma\rho$ is the sum of principal curvatures of test roller and standard roller, which can be calculated according to Eq. (2):

$$\Sigma\rho = \frac{1}{R_{11}} + \frac{1}{R_{12}} + \frac{1}{R_{21}} + \frac{1}{R_{22}} \tag{2}$$

where R_{11} and R_{21}, which are vertical to the rolling direction, are the principal curvatures of the test roller and standard roller. R_{12} and R_{22}, which are along the rolling direction, are the principal curvatures of the test roller and standard roller.

2.2.2 Drive module

Slip ratio is the key index which indicates the relative motion state between the test roller and standard roller. The relationship between slip ratio and relative motion state is shown in Table 2. Two same servo – actuators were used to drive the test roller and standard roller, realizing infinitive adjust and control of slip ratio. Note that "slip" shown in Table 2 means the significant slip between the rollers due to the different speeds, which is different from the microslip occurring in the contact area even in pure rolling condition. Fig. 4 shows the schematic diagram of test roller drive module. The main shaft of test roller was horizontally installed on the working flat. The main shaft and servo – actuator were connected in series by flexible coupling and torque sensor. Fig. 5 shows the schematic diagram of standard roller drive module. The servo – actuator and transmission shaft, as well as the transmission shaft and main shaft were connected by respective synchronal circle – arc tooth belt to realize synchronous revolution. The transmission shaft was installed on the fulcrum position of the lever loading device by a bearing block. The axial lead of the transmission shaft coincides with that of the lever fulcrum. The main shaft of standard roller was assembled on one side of the lever holder. It can rotate around the axial lead of the lever fulcrum at a certain angle. In this arrangement, it is convenient to install and disassemble the test roller.

Table 2 Relationship between slip ratio and the relative motion state

Slip ratio	Slip ratio = 0%	0% < Slip ratio < 100%	Slip ratio = 100%
Relative motion state	Pure rolling	Slip + rolling	Pure slip

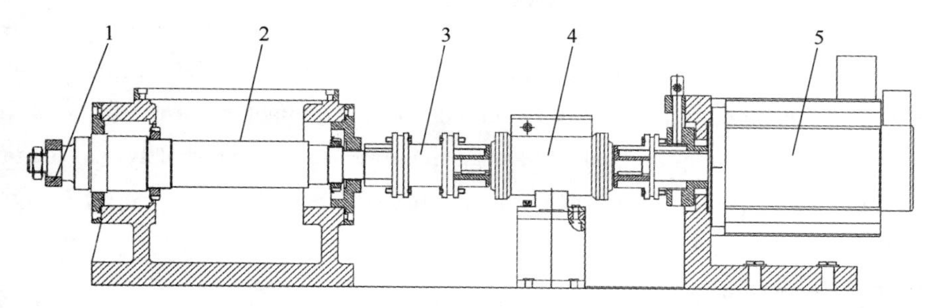

Fig. 4 Schematic diagram of test roller drive module
1—Test roller; 2—Main shaft of test roller; 3—Flexible coupling;
4—Torque sensor; 5—Servo – actuator driving test roller

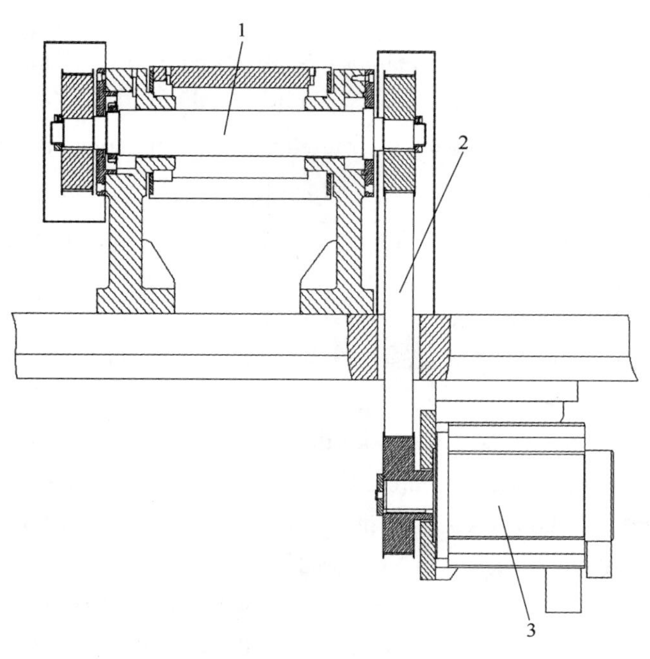

Fig. 5 Schematic diagram of standard roller drive module
1—Transmission shaft; 2—Synchronal circle – arc tooth belt;
3—Servo – actuator driving standard roller

2.2.3 Loading module

Traditional RCF test machine often combines lever with weights or springs to load. When loading by the lever combining weights, not only continuous loading cannot be realized, but also the accuracy of monitoring signals (e.g. vibration) is reduced due to the jump during weights changing. Although the combining of lever and spring avoids the occurrence of impact load, the error accumulation during the long test will decrease the accuracy of test data.

The schematic diagram of hydraulic and lever loading module is shown in Fig. 6. When loading, the crown bar jacks the loading terminal of the lever holder to apply load on the main shaft of standard roller through the lever. The hydraulic and lever loading system is able to load continuously and change the loading value smoothly by the control of computer. Therefore, the accelerated life test with the accelerating factor of load can be carried out. In addition, the stableness and accuracy of hydraulic loading increase the precision of the life test data. The loading facility can give a test force up to 30 kN, which ensures the contact stress reach the maximum of 2.6 GPa to meet the test requirement. When the measurement and control system judge the occurrence of the failure point, the test machine can unload immediately to keep the original failure morphology of specimen

2.2.4 Lubrication module

The oil was dripped to adequately lubricate the contact region between the test roller and standard roller. The oil dripping speed is controlled by a control valve at the inlet side. An oil box is placed below the rollers to collect the used oil, which is filtered by a filter in the outlet line of the oil box before flowing into the oil tank. The remnant particles in the filtered oil will deposit in the oil tank by gravity. In the end, the decontaminated oil will be circularly supplied to the inlet side by an oil pump. Furthermore, not only the oil container should be cleaned at set intervals, but also the lubricating oil should be regular replaced to ensure the lubricating oil is clean.

2.2.5 Measurement module

Measurement module includes friction torque sensor, rotational speed sensor, load sensor, AE sensor, and temperature sensor.

2.2.5.1 Friction torque and rotational speed sensor

A strain torque and rotational speed sensor gives the digital display of both friction torque and rotational speed of the rollers. The operating principle of the friction torque measurement is that the resistance values of bended strain gages will change when small deformation occurs on the strain shaft under the effect of torque force. The resistance change of the measuring bridge composed of strain gages can be transformed into voltage signal. By this method, the friction torque value under rotational condition can be measured accurately.

The operating principle of the rotational speed measurement is that an impulse signal with a certain periodic width can be obtained when the speed measuring code disk rotates. Therefore, the rotational speed can be calculated according to the teeth number of the disk and the frequency of the output signals.

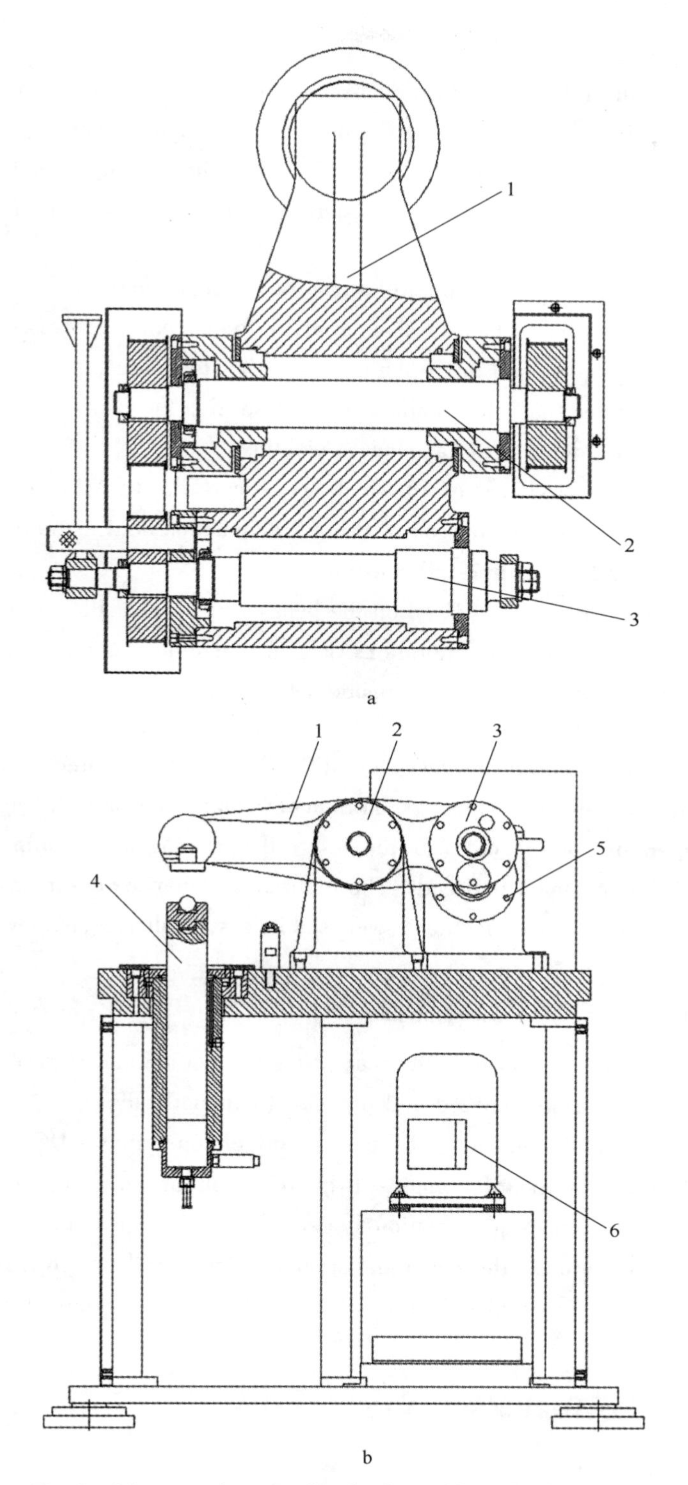

Fig. 6　Schematic diagram of hydraulic and lever loading module

a—Top view; b—Lateral view

1—Lever holder; 2—Transmission shaft; 3—Main shaft of standard roller; 4—Crown bar;
5—Main shaft of test roller; 6—Hydraulic station

2.2.5.2 Pressure sensor

The load applied on the rollers is measured by a strain pressure sensor. The generated strain caused by the pressure can be transformed into electrical signals which are linear with the strain. The strain pressure sensor exhibits excellent yaw stiffness, high accuracy, and reliable property. It should be preheated to ensure the stability and reliability of the output signals.

2.2.5.3 AE sensor

Acoustic emission (AE) is the propagation of stress waves at frequencies from 0.1 to 1 MHz. It is generated within or on the surface of a material by fundamental processes that define RCF and wear such as deformation and micro - fracture[13]. AE technique was chosen as the online monitoring method to judge the failure point of coatings during the RCF/wear test because of its high sensitivity to plastic deformation and micro yielding.

The PCI - 2 AE device produced by Physical Acoustics Corp. (PAC) was assembled on the test machine. The AE device is composed of PCI - 2 acquisition card, portable case, preamplifier, Nano - 30 sensor, and signal transmission line, etc.

By a specific jig, the AE sensor is fixed on the bearing block near the test roller as it is difficult to fix the AE sensor on the rotating roller. AE signals can also be well detected because AE signals hardly attenuate when transmitting inside solid body.

2.2.5.4 Temperature sensor

As the rollers are rotating during the test, it is difficult to measure directly the temperature of the contact region between the rollers. A platinum resistance sensor is fixed in the oil box to measure the temperature of the oil near the test roller, aiming at obtaining the temperature changing curve of the specimen indirectly. The platinum resistance sensor has several advantages, such as high sensitivity and accuracy, well stableness, small size, light weight, and simple structure, etc.

2.2.6 Data acquisition and processing module

Under the control of software, not only the real - time acquisition and storage of signals is realized, but also the data can be analyzed and processed automatically.

The data is received and transmitted by using a 64 channel PIO - D64 board which connects to the PCI bus. The board contains two 16 bit input ports and two 16 bit output ports. The signals of friction torque, rotational speed, load, AE, and temperature will be transmitted to the computer through the communication interfaces and the corresponding channels on the PIO - D64 board. The real - time test data can be shown dynamically in form of curves or sheets.

3 Quality for the Test Machine

3.1 Coating deposition

In order to evaluate the reliability and accuracy of the novel test machine, a series of RCF tests under different contact stresses were carried out. The tested coating and standard material were plasma sprayed AT40 coating and quenched + low temperature tempered 52100 steel, respectively.

The AT40 coating was deposited on the quenched + low temperature tempered 1045 steel roller by plasma spray technology. Ni/Al alloy powders were coated as bonding coating in order to improve the bonding strength between the AT40 coating and substrate. The parameters used in the plasma spraying process are listed in Table 3. A 400μm thick AT40 coating was successfully prepared. Fig. 7 shows the cross section morphology of the AT40 coating. The AT40 coating was relative dense with a few micro - defects.

Table 3 Plasma spraying parameters

Plasma spray parameters	Spraying materials	
	AT40	Ni/Al
Argon gas flow/$m^3 \cdot h^{-1}$	2. 8	3. 4
Hydrogen gas flow/$m^3 \cdot h^{-1}$	0. 4	0. 3
Nitrogen gas flow/$m^3 \cdot h^{-1}$	0. 6	0. 6
Spraying current/A	440	320
Spraying voltage/V	110	140
Spraying distance/mm	100	150
Powder feed rate/$g \cdot min^{-1}$	30	30

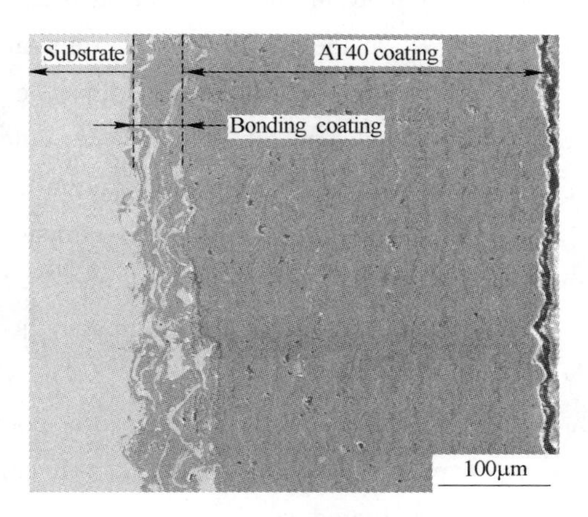

Fig. 7 Cross section morphology of the AT40 coating

3. 2 RCF test

The RCF tests of AT40 coating were carried out under four different contact stresses of 0. 75 GPa, 1 GPa, 1. 25 GPa, and 1. 5 GPa. Ten coated specimens were tested under each contact stress. The main failure modes of the AT40 coating were pit, surface abrasion, and delamination, as shown in Fig. 8. Pit was the main failure mode under the lowest contact stress (0. 75 GPa); while delamination dominated under the other higher contact stress, especially under 1. 25 and 1. 5 GPa.

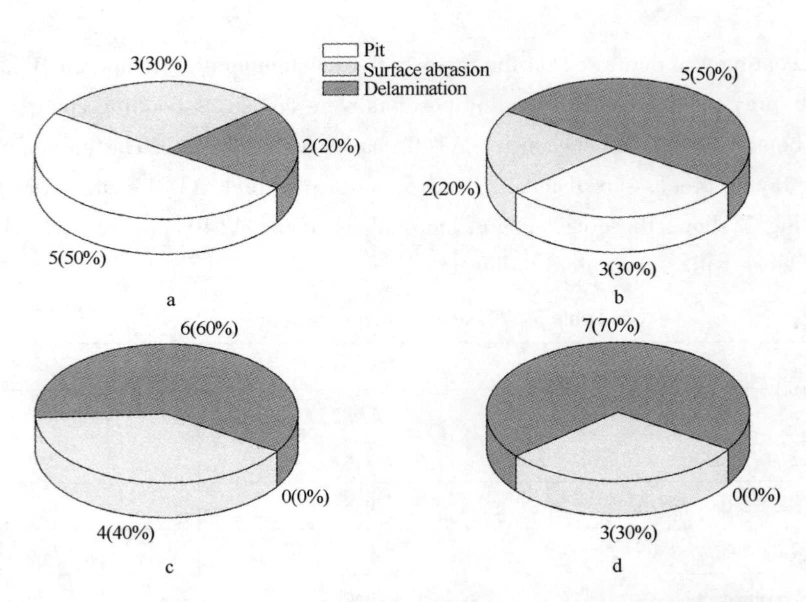

Fig. 8 Pie chart of failure modes of AT40 coating in the RCF test under four contact stresses

The typical morphologies of pit, surface abrasion, and delamination observed by optical microscope (OM) and scanning electron microscope (SEM) are shown in Fig. 9 – Fig. 11, respectively. It can be seen from Fig. 9 that the typical morphologies of pit is a superficial material removal with the diameter of about 1 mm. When the pits reach a certain number or pitting area accounts for a certain proportion of contact area, failure occurs. The surface abrasion failure (Fig. 10) was due to the grain abrasion between the coating and standard roller caused by the remaining abrasive grains[14]. When intermediate layer forms due to the accumulation of the abrasive grains, the abrasion mode is transformed from grain abrasion to three – body abrasion. This can accelerate the RCF failure process of the coating[15]. Fig. 11 shows the typical delamination morphologies of the failed coating. The delamination is almost propagated along the rolling direction and the delamination area is very large[16]. The width of the delamination area is about 3 mm, which is more than half of the length of the contact line. Because of the coating loss, partial substrate was exposed, as shown in Fig. 11a. It can be seen that there is a cliff edge at the boundary of the delaminated coating, as shown in Fig. 11b. A few micro – cracks exist near the edge of the delamination area. The maximum orthogonal shear stress (OSS) induced by the alternating stress which locates within the AT40 coating is considered to be the key failure mechanism of delamination. The OSS distribution within the AT40 coating was analyzed by finite element method (FEM), as shown in Fig. 12. As contact stresses increased, the maximum OSS became larger, and its location got closer to the interface between coating and substrate. Fig. 13 shows the propagating fatigue cracks near the interface. It can be concluded that the initiation and propagation of the fatigue cracks were attributed to not only the OSS which surpasses the coating shear strength, but also the micro – defects within the composite ceramic coating.

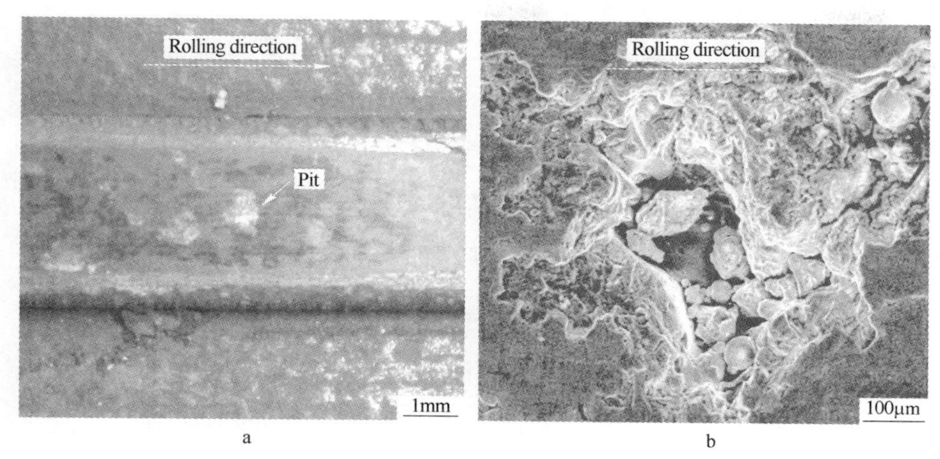

Fig. 9　Typical morphologies of pit failure of AT40 coating

a—OM image; b—SEM image

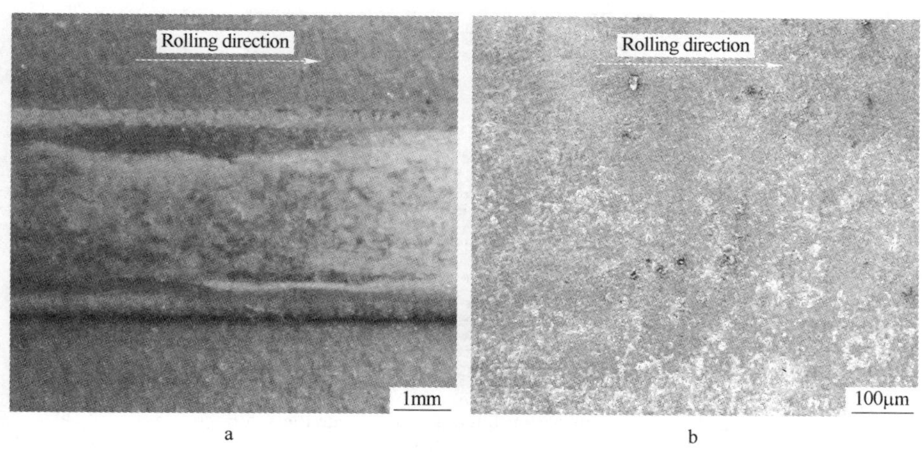

Fig. 10　Typical morphologies of surface abrasion failure of AT40 coating

a—OM image; b—SEM image

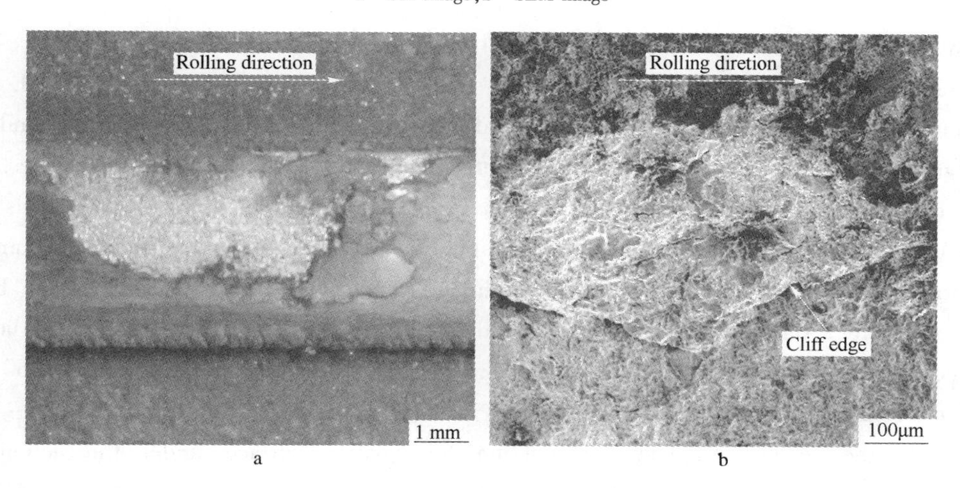

Fig. 11　Typical morphologies of delamination failure of AT40 coating

a—OM image; b—SEM image

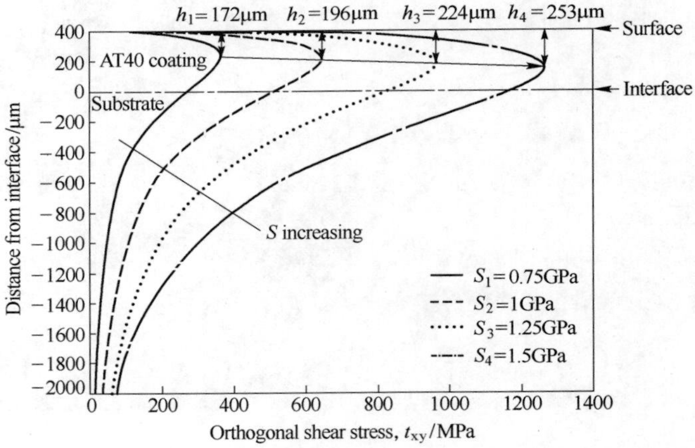

Fig. 12 The OSS distribution within the AT40 coating from FEM

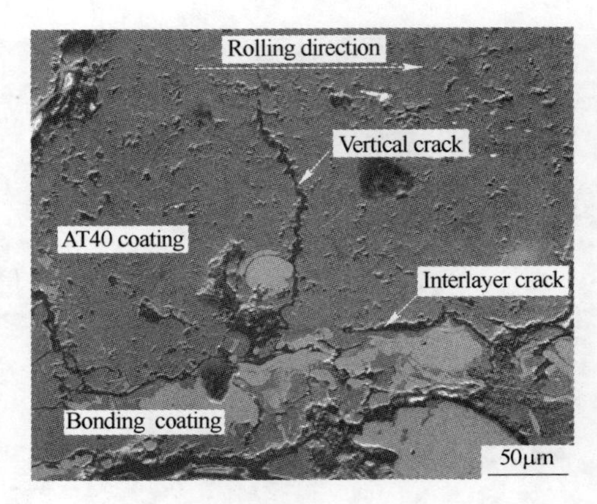

Fig. 13 The propagating fatigue cracks near the interface

3.3 AE signal analysis

The major drawback of traditional vibration analysis method is the low signal frequency(< 10 kHz) range, which causes that the signals are easily affected by machine noises. As AE can overcome this shortcoming, it's getting more and more popular in RCF monitoring[17]. Kinds of AE signal characteristic parameters such as energy, amplitude, count and RMS were investigated to describe the RCF process. In the current study, AE amplitude which is defined as the maximum AE signal excursion during an AE hit was chosen to characterize the RCF damage regularity of AT40 coatings. It is proved that the AE amplitude is very sensitive to the RCF damage. The RCF damage process can be divided into three stages in terms of the variation tendency of AE amplitude, as shown in Fig. 14. Only AE data over threshold value of 60 dB were collected. The AE amplitude was relatively high(≈72 dB) at the run - in stage(0 - 200 s) due to the removal of sharp salient on the coating surface under the action of rolling. Subsequently, at the stable stage(200 - 1300 s), AE amplitude kept a stable level(less than 68 dB) because of the decrease of surface roughness. As micro - fracture and plastic yield occurred on the surface of coatings, continuous AE signals with the maximum amplitude of 68 dB were present. In the end, the AE amplitude continually reached 75 dB, and presented a sharp rise with the value of 81 dB at 1380 s. Based on a number of experiments, it was found that the RCF failure of AT40 coatings all occurred when the AE amplitude was over 80 dB. Therefore, the critical value of AE amplitude was set as 80 dB to judge the RCF failure of AT40 coatings. The sharp rise of the AE amplitude was due to the brittle fractures of the coating, which were caused by the sudden joint between the primary cracks and adjacent micro - defects. It should be noted that the redetermination of critical value of AE amplitude must be made when the tested coating changes.

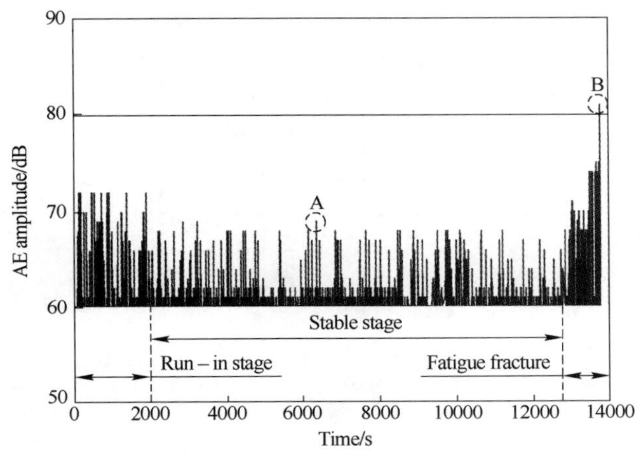

Fig. 14 Typical AE signal during the RCF damage process of AT40 coating specimen

The typical AE signals in the stable stage and fatigue fracture stage were marked as A (at 6500 s) and B (at 13800 s) in Fig. 14. Fig. 15 and Fig. 16 show the waveforms and frequency spectra of A and B, respectively. Both of the waveforms shown in Fig. 15a and Fig. 16a exhibit sharp rise and short decaying time, which was associated with the high brittleness and lamellar structures of the AT40 coating. The peak value of B reached about 800 mV, which was much

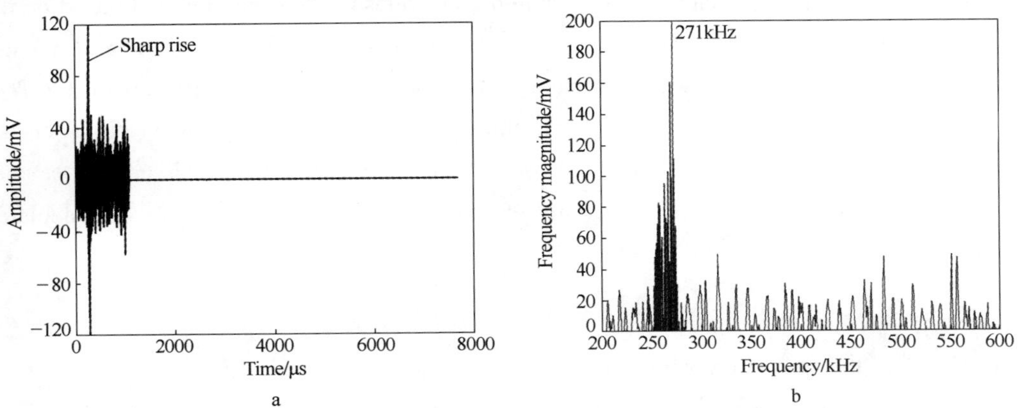

Fig. 15 AE waveform and frequency spectrum at stable stage during RCF process

a—Waveform; b—Frequency spectrum

higher than that of A. The frequency spectra distributions at different RCF stages were similar, and the corresponding frequencies of peak values were 271kHz and 265kHz, respectively. Nevertheless, the frequency magnitude of peak value at fatigue fracture stage was much higher than that at stable stage. The AE signals in the stable stage were released mainly by the initiation and propagation of the fatigue cracks due to the OSS within the coating. The typical burst AE waveform of B was mainly due to the transient fracture and materials removal, which were caused by the unstable growth and sudden closure of fatigue cracks. It is obvious that the AE amplitude was sensitive to the transient fracture during the RCF damage process. Meanwhile, the analysis of waveforms and frequency spectra provide a reliable way to judge the different stages of the RCF process.

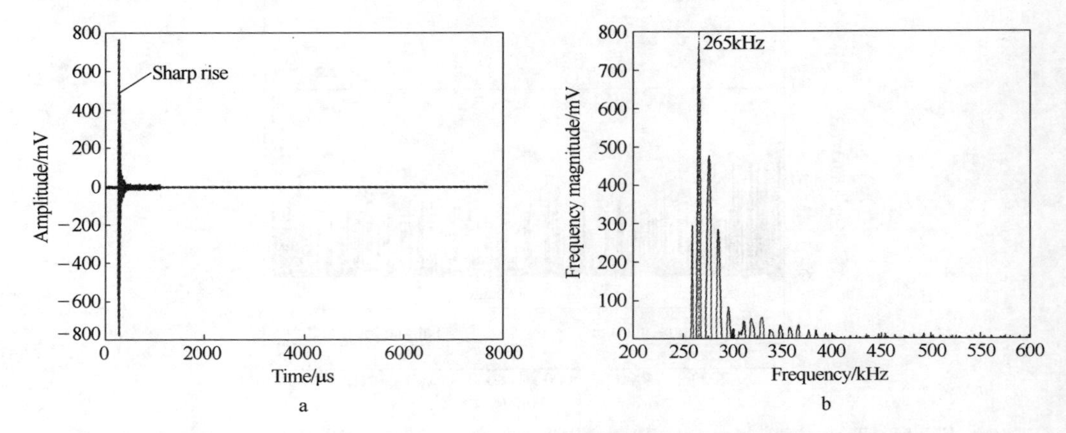

Fig. 16 AE waveform and frequency spectrum at fatigue fracture stage during RCF process
a—Waveform; b—Frequency spectrum

3.4 Life prediction plots

3.4.1 *P* – *N* plots

The P (Predicted failure probability) – N (Number of cycles) plots were obtained based on the Weibull distribution, which is one of the most widely used life distributions in reliability engineering, as shown in Fig. 17. Two – parameter (shape parameter β and characteristic life N_a) Weibull distribution was used to characterize the RCF life of coating specimen based on the obtained life data. β and N_a were evaluated by using the theory of maximum likelihood estimation (MLE) [18]. Then, the failure probability of specimen under any cycle can be calculated by Eq. (3):

$$P(N) = 1 - \exp\left[-\left(\frac{N}{N_a}\right)^{\beta}\right] \tag{3}$$

where $P(N)$ is the failure probability and N is the number of cycles.

The straight fitting lines under different contact stresses (0.75 GPa, 1 GPa, 1.25 GPa, and 1.5 GPa) were acquired by regressing the data which is obtained by repeating the RCF experiments more than 10 times. It is well known that the value of shape parameter β has an inverse –

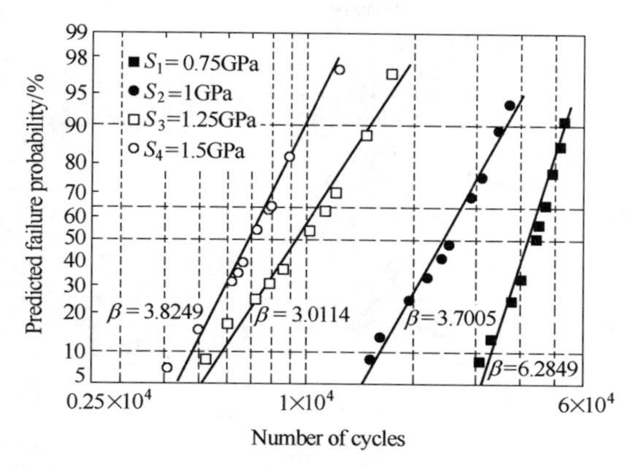

Fig. 17 $P-N$ plots of RCF life data of AT40 coatings

correlation with the scatter of life data. The number of cycles at any failure probability under four certain contact stresses can be predicted in terms of $P-N$ plots. It is obvious that the scatter of life data under the minimum contact stress of 0.75 GPa is much smaller than that under any other contact stress. At any cycle, the higher the contact stress, the higher the failure probability.

3.4.2 $P-S-N$ plots

In order to predict the RCF life under any contact stress, $P-S-N$ plots were introduced to treat life data. The parameters for $P-S-N$ plots can be calculated according to the Weibull distribution function (Eq. (3)). The relationship between contact stress and number of cycles is as follows:

$$N = CS^{-m} \tag{4}$$

where S is the contact stress, C and m are undetermined parameters which can be calculated by Eq. (5) and Eq. (6):

$$-\frac{1}{m} = \frac{\sum_{i=1}^{n} X_i Y_i - 1/n \sum_{i=1}^{n} X^i \sum_{i=1}^{n} Y_i}{\sum_{i=1}^{n} X_i^2 - 1/n(\sum_{i=1}^{n} X_i)} \tag{5}$$

$$\frac{1}{m}\ln C = \frac{1}{n}\left(\sum_{i=1}^{n} Y_i + \frac{1}{m}\sum_{i=1}^{n} X_i\right) \tag{6}$$

$P-S-N$ plots can be obtained once C and m are determined, as shown in Fig. 18. The number of cycles under any contact stresses at four certain failure probabilities (10%, 50%, 63.2%, 90%) can be directly predicted based on $P-S-N$ plots. Note that, the failure probability in $P-S-N$ plots can be arbitrary, not limited to the four failure probabilities. Namely, the RCF life of coating specimen can be characterized by $P-S-N$ plots under any contact stress at any failure probability.

4 Conclusions

A novel RCF/wear competitive life test machine was designed in the form of modules, which in-

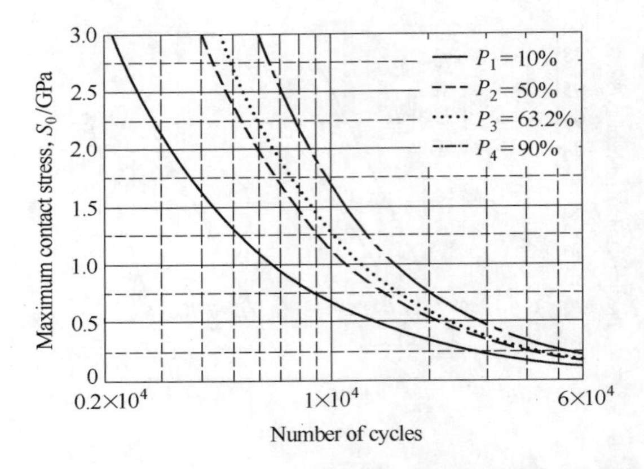

Fig. 18 $P-S-N$ plots of RCF life data of AT40 coatings

cludes roller assembly module, drive module, loading module, lubrication module, measurement module, and data acquisition and processing module. The prominent characteristic of the test machine is that the slip ratio can be changed conveniently to simulate the complex service conditions including pure slip, slip – rolling, and pure rolling. The test machine is able to load continuously, and it ensures that the contact stress can reach the maximum of 2. 6 GPa to meet the test requirement. The typical failure modes of the plasma sprayed AT40 composite ceramic coating were pits, surface abrasion, and delamination in the RCF tests under different contact stresses of 0. 75 GPa, 1 GPa, 1. 25 GPa, and 1. 5 GPa. The maximum orthogonal shear stress induced by the alternating stress is the key failure mechanism of delamination, which is considered to be the most serious RCF failure mode. The AE amplitude was sensitive to the transient fracture during the RCF damage process. Meanwhile, the analysis of waveforms and frequency spectra provide a reliable way to judge the different stages of the RCF process. The $P-N$ plots show that the RCF life of AT40 coating was significantly affected by the contact stress level. At any failure probability, the higher the contact stress, the lower the RCF life. The $P-S-N$ plots can predict the RCF life of coating specimen under any contact stress at any failure probability.

Acknowledgements

The paper was financially supported by Distinguished Young Scholars of NSFC (51125023), 973 Project(2011CB013405), NSFC(51275151), NSF of Beijing(3120001).

References

[1] Zhang X C, Xu B S, Xuan F Z, Tu S D, Wang H D, Wu Y X. Rolling contact fatigue behavior of plasma – sprayed CrC – NiCr cermet coatings. Wear 2008, 265: 1875 – 1883.

[2] Fujii M, Yoshida A, Ma J, Shigemura S, Tani K. Rolling contact fatigue of alumina ceramics sprayed on steel roller under pure rolling contact condition. Tribology Inter national 2006, 39: 856 – 862.

[3] Scharf T W, Singer I L. Role of the transfer film on the friction and wear of metal carbide reinforced amorphous carbon coatings during run – in. Tribology Letters 2009, 6: 43 – 53.

[4] Karamis M B, Yildizli K, Cakýrer H. Wear behaviour of Al – Mo – Ni composite coating at elevated temper-

ature. Wear 2005,258:744 - 751.
[5] Canadinc D, Sehitoglu H, Verzal K. Analysis of surface crack growth under rolling contact fatigue. International Journal of Fatigue 2008,30:1678 - 1689.
[6] Bouras S, Zerizer I, Gheldane F, Bouazza TM, Bouzabata B. Study of the resistance to crack propagation in alumina by acoustic emission. Ceramics International 2008,34:1857 - 1865.
[7] Rahman Z, Ohba H, Yoshioka T, Yamamoto T. Incipient damage detection and its propagation monitoring of rolling contact fatigue by acoustic emission. Tribology International 2009,42:807 - 815.
[8] Chang H, Han E H, Wang J Q, Ke W. Acoustic emission study of fatigue crack closure of physical short and long cracks for aluminum alloy LY12CZ. International Journal of Fatigue 2009,31:403 - 407.
[9] Harvey T, Wood R, Powrie H. Electrostatic wear monitoring of rolling element bearings. Wear 2007;263:1492 - 1501.
[10] Löhr M, Spaltmann D, Binkowski S, Santner E, Woydt M. In situ acoustic emission for wear life detection of DLC coatings during slip - rolling friction. Wear 2006,260:469 - 478.
[11] Warren A W, Guo YB. Acoustic emission monitoring for rolling contact fatigue of superfinished ground surfaces. International Journal of Fatigue 2007,29:603 - 614.
[12] Manoj V, Manohar Shenoy K, Gopinath K. Developmental studies on rolling contact fatigue test rig. Wear 2008,264:708 - 718.
[13] Douglas R M, Steel J A, Reuben R L. A study of the tribological behaviour of piston ring/cylinder liner interaction in diesel engines using acoustic emission. Tribology International 2006,39:1634 - 1642.
[14] Holmberg K, Matthews A, Ronkainen H. Coatings tribology - contact mechanisms and surface design. Tribology International 1998,31:107 - 120.
[15] Ahmed R, Hadfield M. Failure modes of plasma sprayed WC - 15% Co coated rolling elements. Wear 1999,230:39 - 55.
[16] Zhang X C, Xu B S, Xuan F Z, Tu S T, Wang H D, Wu Y X. Fatigue resistance of plasma - sprayed CrC - NiCr cermet coatings in rolling contact. Applied Surface Science 2008,254:3734 - 3744.
[17] Zhang Z Q, Li G L, Wang H D, Xu B S, Piao Z Y, Zhu L N. Investigation of rolling contact fatigue damage process of the coating by acoustics emission and vibration signals. Tribology International 2012,47:25 - 31.
[18] Piao Z Y, Xu B S, Wang H D, Pu C H. Influence of undercoating on rolling contact fatigue performance of Fe - based coating. Tribology International 2010,43:252 - 258.

纳米压痕法测量等离子喷涂铁基涂层表面的残余应力*

摘　要　利用无损的绿色检测技术——纳米压痕法测量等离子喷涂铁基涂层表面的残余应力。通过对比无应力和有应力涂层的载荷－位移曲线，得出涂层表面存在残余拉应力。由于无应力和有应力涂层的压痕周围均有明显的凸起变形，目前广泛使用的 Oliver 法提取残余应力的特征参量——真实接触面积的公式已不适用。利用之前建立的适用于压痕周围有凸起变形的材料的真实接触面积计算公式得到无应力和有应力涂层的真实接触面积，分析两者之间的差异，最终计算出涂层表面的应力值为 188MPa，与 X 射线法得到的 162MPa 较为符合。

关键词　纳米压痕　铁基涂层　残余应力　凸起变形

1　前言

涂层中存在的残余应力对涂层的界面韧性[1]、结合强度[2]、抗疲劳[3]等性能都有着显著的影响，同时也是导致涂层表面产生裂纹、涂层剥落失效以及被涂覆零件变形的一个重要因素。因此，为了提高表面涂层的使用寿命，预测其内部残余应力的大小以及分布规律具有重要的工程意义。

目前传统的残余应力检测技术（钻孔法、X 射线法、中子衍射法、超声法以及磁性法）都存在很多不足之处，其发展空间也受到种种条件的限制。有些技术对材料产生破坏，而有些技术则局限于表面应力或特殊的材料。近年来，基于纳米压痕技术的方法来测量残余应力已经引起了国内外研究学者的广泛关注[4~8]。纳米压痕技术使材料在非常小的范围内产生变形，不会破坏材料的结构完整性，且允许在微/纳米范围内进行残余应力的测量。在目前的测量工作中，Oliver 法[9]占据着主导地位，其重要贡献是给出了接触投影面积的标定方法[10]，而接触投影面积是表征残余应力的特征参量，因此精确计算出接触投影面积至关重要。但 Oliver 法在对压痕周边有凸起变形的材料进行面积计算时会产生较大的误差，之前已经建立了凸起变形材料的真实接触面积公式[11]。

本文利用纳米压痕技术测量了等离子喷涂 FeCrBSi 涂层表面的残余应力，利用之前建立的适用于凸起变形材料的真实接触面积公式得到了无应力和有应力涂层的真实接触面积之间的差异，最终计算出涂层表面的残余应力，并与传统的 X 射线法进行了比较。

---

* 本文合作者：王海斗、朱丽娜。原发表于《机械工程学报》，2013，49(7)：1~4。北京市自然科学基金重大（3120001）、国家杰出青年科学基金（51125023）和国家重点基础研究发展计划（“973”计划，2011CB013405）资助项目。

2 试验方法

2.1 涂层制备

喷涂基体选择 45 钢，经调质处理后的 45 钢拥有优良的综合力学性能（较高的硬度和较好的韧性）。喷涂前用丙酮对 45 钢进行超声波清洗，去除表面的污染物。喷涂材料选用 FeCrBSi 自熔剂合金粉末。喷涂前，对基体 45 钢在 150℃下进行预热，然后对其进行喷砂处理，砂料为棕刚玉，粒度为 1.2mm，以毛化表面，增加结合力。为了提高喷涂层与基体之间的结合强度，喷一层厚度约为 50μm 的 Ni/Al 过渡层。为获得较为致密的喷涂层，在喷涂时，等离子焰流的轴线与 45 钢表面之间的角度为 90°。等离子喷涂具体的工艺参数见文献［11］。最终制备出厚度为 200μm 的 FeCrBSi 涂层。

2.2 纳米压痕试验

用线切割将 FeCrBSi 涂层的表面切去约 0.5mm 厚的薄层，使应力得到充分释放，作为无应力的参考试样。采用 TriboIndenter®（Hysitron 公司，USA）纳米压痕仪对有应力和无应力涂层表面进行了固定深度 175nm 的压痕试验。每个载荷和深度下至少压 20 个点，各点间隔 10μm。进行纳米压痕试验之前，涂层表面经过抛光处理，抛光后的表面粗糙度为 10.3nm ± 1.2nm。

残余应力通过以下公式计算得出[12]：

残余拉应力为：

$$\sigma_1 = H\left(\frac{A_0}{A} - 1\right) \tag{1}$$

残余压应力为：

$$\sigma_2 = \frac{H}{\sin\alpha}\left(1 - \frac{A_0}{A}\right) \tag{2}$$

式中，H 为涂层的硬度；A_0 和 A 分别为无应力和有应力时的接触投影面积；α 为压头边界与材料表面的夹角，对于玻式压头，$\alpha = 24.7°$。

对于涂层硬度，采用式（3）进行计算：

$$H = \frac{P_{max}}{A} \tag{3}$$

式中，P_{max} 为最大压入载荷。

对于真实接触面积，基于原子力显微镜（Atomic force microscope，AFM）对压痕形貌的观察，采用之前建立的真实接触面积公式进行计算[11]：

$$A = 14.175\left(\frac{\theta\pi}{120\sin^2\frac{\theta}{2}} - 3\cot\frac{\theta}{2} + \sqrt{3}\right)(h_{max} + h_p^{ave})^2 \tag{4}$$

式中，θ 为压痕周围凸起材料的圆弧所对应的夹角；h_{max} 为最大压入深度；h_p^{ave} 为压痕三角形各边长凸起材料高度的平均值。

2.3 X 射线法测量涂层表面残余应力

采用 X－350A 型 X 射线应力仪对涂层表面三个不同位置进行了残余应力测量，扫描方式：$\theta-\theta$ 扫描，2θ 最小步距为0.01°。

3 试验结果与讨论

图1为无应力和有应力涂层的载荷－位移曲线。在固定相同的深度下，有应力的涂层表面所需要的最大压入载荷（7.5mN）明显低于无应力涂层表面（7.8mN），表明涂层的表面存在残余拉应力，这是因为拉应力对压入过程起到了促进作用。M. K. Khan[13]的有限元结果表明，残余压应力会增大弹性恢复深度，减小残余深度；反之，残余拉应力会减小弹性恢复深度，增大残余深度。从图1中可以看出，与无应力的涂层相比，有应力的涂层表面的弹性恢复深度较小，而残余深度较大，同样证明了涂层的表面存在残余拉应力。

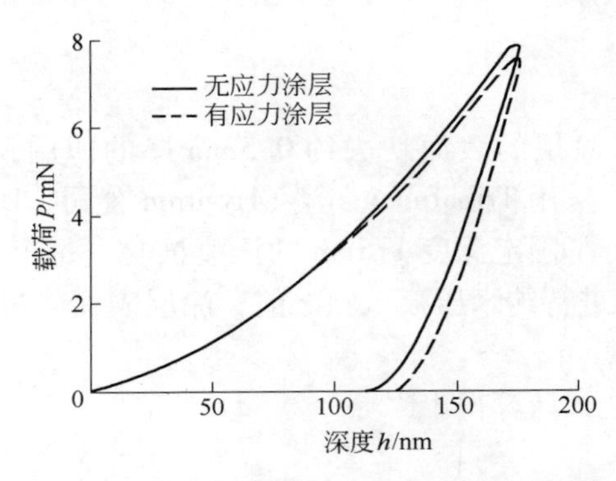

图1　无应力和有应力涂层的载荷-位移曲线

图2为FeCrBSi涂层的二维和三维压痕形貌。涂层的压痕周围产生明显的凸起变形。图3为无应力和有应力的FeCrBSi涂层在固定深度175nm下的压痕形貌轮廓图。无应力和有应力的FeCrBSi涂层的压痕周围均产生明显的凸起变形，且有应力涂层的凸起高度显著低于无应力涂层，这与它们表面的残余应力有关，含有残余拉应力的涂层倾向于将材料从压头的表面拉开，从而减小凸起变形量。

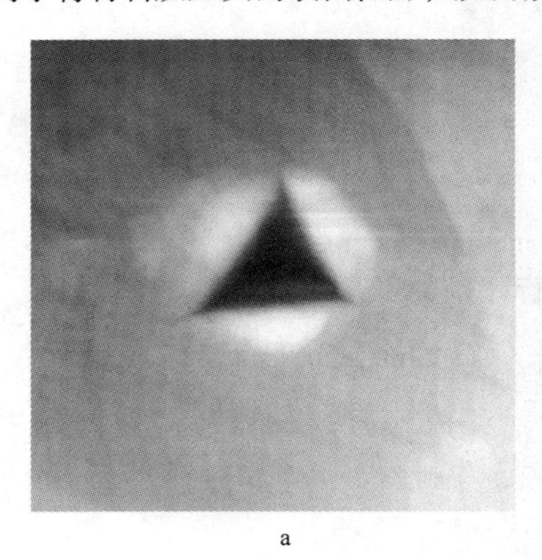

a

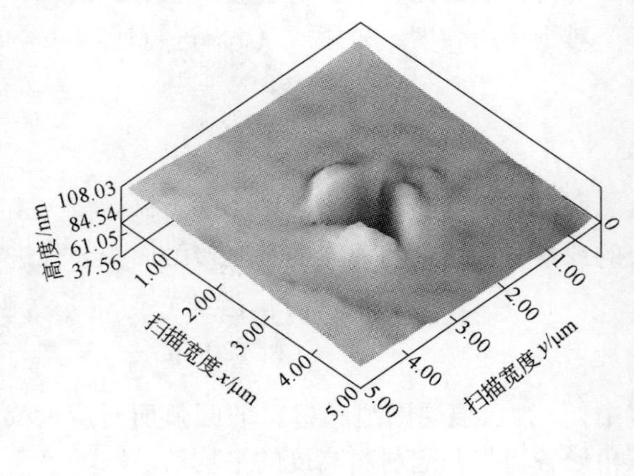

b

图2　FeCrBSi涂层的二维和三维压痕形貌

a—二维形貌；b—三维形貌

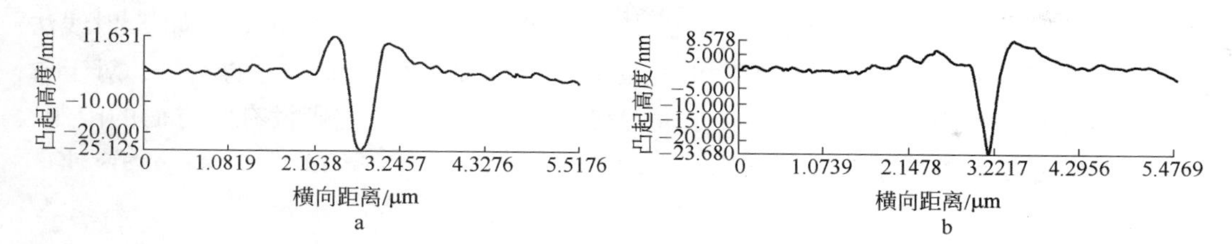

图 3　无应力和有应力涂层在固定深度 175nm 下的压痕形貌轮廓图

a—无应力涂层；b—有应力涂层

利用式（3），分别用 Oliver 法和之前建立的 AFM 法计算出无应力和有应力涂层的硬度，如图 4 所示。由于之前建立的 AFM 法考虑了凸起变形部分的面积，得到的硬度值明显小于 Oliver 法的计算结果。一般认为，随着应力的增大，残余压应力会使材料的硬度增大，拉应力会使材料的硬度减小，如图 4 中 Oliver 法得到的有应力涂层的硬度明显低于无应力涂层，而用之前建立的 AFM 法计算出的有应力涂层和无应力涂层的硬度几乎相等，那是因为 Oliver 法计算真实接触面积时没有考虑材料的凸起变形。实际上，对于压头周围有显著凸起变形的材料，其硬度与残余应力大小无关，并不随应力值的变化而发生改变。

利用式（1）计算 FeCrBSi 涂层表面的残余应力，各特征参量的计算结果如表 1 所示，其中硬度值为采用 AFM 法的结果，且取有应力和无应力时硬度的平均值。最终计算出 FeCrBSi 涂层表面的残余应力为 188MPa。图 5 为 X 射线法测量得到的涂层表面三个不同位置的应力值，分别为（161 ± 25）MPa、（184 ± 14）MPa 和（141 ± 22）MPa，涂层表面的残余应力分布较为均匀，平均值为 162MPa。

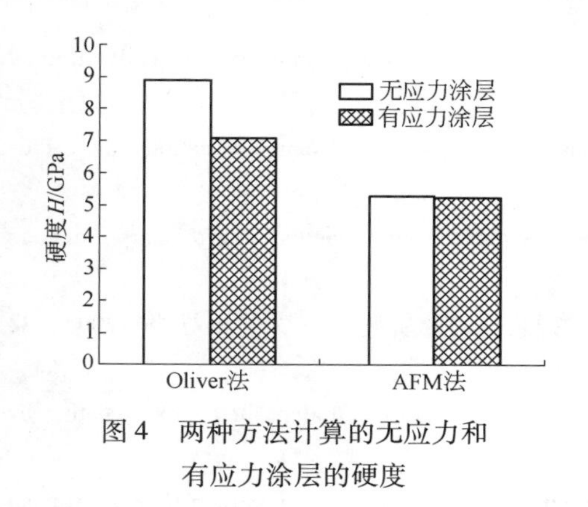

图 4　两种方法计算的无应力和有应力涂层的硬度

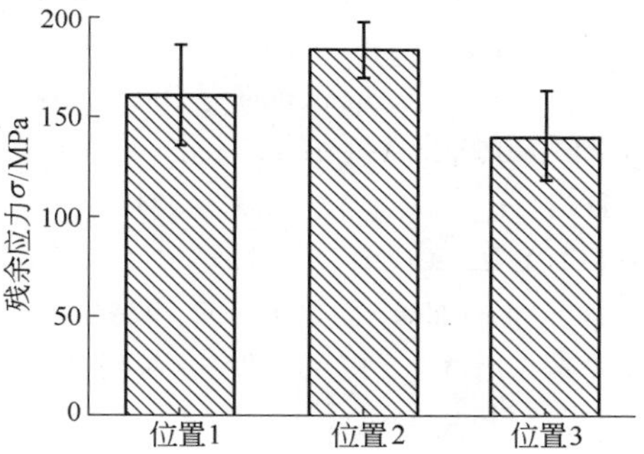

图 5　X 射线法测量 FeCrBSi 涂层表面的残余应力

表 1　残余应力公式的各特征参量

特 征 参 量	数　值
有应力涂层的接触面积 A/nm^2	1344665
无应力涂层的接触面积 A_0/nm^2	1389762
涂层的硬度 H/GPa	5.60

对比可知，纳米压痕法与传统的 X 射线法的测量结果较为一致。可见，纳米压痕

法能够精确地计算出涂层中的残余应力，并且能够克服X射线法的不足，如X射线法是基于晶体材料在有应力和无应力时的晶面间距的变化进行残余应力的测量，对于含有非晶成分的材料，测量结果会产生很大的误差，而纳米压痕法则没有此方面的限制，具有良好的应用前景。

4 结论

（1）FeCrBSi涂层表面存在残余拉应力。与无应力涂层相比，含有拉应力的涂层最大压入载荷和弹性恢复深度较小，而残余深度较大。

（2）无应力和有应力涂层的压痕周围均有明显的凸起变形，而含有残余拉应力的涂层倾向于将材料从压头的表面拉开，从而减小凸起变形量。

（3）纳米压痕法测量出FeCrBSi涂层表面的残余应力为188MPa，与X射线法得到的162MPa较为一致。纳米压痕法可以弥补X射线法无法测量非晶材料的不足，具有良好的应用前景。

参考文献

[1] LESAGE J, CHICOT D. Role of residual stresses on interface toughness of thermally sprayed coatings [J]. Thin Solid Films, 2002, 415: 143~150.

[2] YANG Y C. Influence of residual stress on bonding strength of the plasma-sprayed hydroxylapatite coating after the vacuum heat treatment [J]. Surf. Coat. Technol., 2007, 201: 7187~7193.

[3] 张秀林，宋德玉．宏观残余应力与金属构件疲劳强度的关系［J］．冶金设备，2000，(5)：54~55.

ZHANG Xiulin, SONG Deyu. Relationship between macroscopic residual stress and fatigue strength of metal components [J]. Metallurgical Equipment, 2000(5): 54~55.

[4] LEE Y H, KWON D. Measurement of residual-stress effect by nanoindentation on elastically strained (100) [J]. Scripta Mater., 2003, 49: 459~465.

[5] XU Z H, LI X. Estimation of residual stresses from elastic recovery of nanoindentation [J]. Philos. Mag., 2006, 86: 2835~2846.

[6] LEE Y H, KWON D. Estimation of biaxial surface stress by instrumented indentation with sharp indenters [J]. Acta Mater., 2004, 52: 1555~1563.

[7] 章莎，周益春．应用纳米压痕法测量电沉积镍镀层残余应力的研究［J］．材料导报，2008，22(2)：115~118.

ZHANG Sha, ZHOU Yichun. Measurement study of residual stress in electrodeposited nickel coating by instrumented nanoindentation [J]. Materials Review, 2008, 22(2): 115~118.

[8] CRAIG A T, MARK F W, WILSON K S C. Residual stress measurement in thin carbon films by Raman spectroscopy and nanoindentation [J]. Thin Solid Films, 2003, 429: 190~200.

[9] OLIVER W C, PHARR G M. An improved technique for determining hardness and elastic modulus using load and displacement sensing indentation experiments [J]. J. Mater. Res., 1992, 7: 1564~1583.

[10] 陈伟民，李敏，徐晓，等．纳米压痕仪接触投影面积标定方法的研究［J］．力学学报，2005，37(5)：645~652.

CHEN Weimin, LI Min, XU Xiao, et al. Comments on the calibration technique of the projected contact area of nanoindentation tester [J]. Chinese Journal of Theoretical and Applied Mechanics, 2005, 37(5): 645~652.

[11] ZHU L N, XU B S, WANG H D, et al. Determination of hardness of plasma – sprayed FeCrBSi coating on steel substrate by nanoindentation [J]. Mater. Sci. Eng., A, 2010, 528: 425 ~ 428.
[12] SURESH S, GIANNAKOPOULOS A E. A new method for estimating residual stresses by instrumented sharp indentation [J]. Acta Mater., 1998, 46: 5755 ~ 5767.
[13] KHAN M K, FITZPATRICK M E, HAINSWORTH S V, et al. Effect of residual stress on the nanoindentation response of aerospace aluminium alloys [J]. Comput. Mater. Sci., 2011, 50: 2967 ~ 2976.

Measurement of Residual Stress of Plasma Sprayed Fe – based Coating by Nanoindentation

Wang Haidou[1], Zhu Lina[1,2], Xu Binshi[1]

(1. National Key Lab for Remanufacturing,
Academy of Armored Forces Engineering, Beijing, 100072, China;
2. School of Engineering and Technology,
China University of Geosciences, Beijing, 100083, China)

Abstract The nanoindentation method, which is a non – destructive testing technique, is used to measure the residual stress of plasma sprayed Fe – based coating. Tensile stress was proved to be generated on the coating surface through comparing the load – depth curves of stress – free and stressed coatings. As significant pile – up deformation was observed around the indents of both stress – free and stressed coatings, the commonly used Oliver method is not available to calculate the true contact area, which is the characteristic parameter of residual stress. The early established formula for the calculation of true contact area of pile – up material is used to obtain the true contact areas of stress – free and stressed coatings. Based on the difference of the contact areas, the residual stress with the value of 188MPa was obtained, which agreed well with that (162 MPa) by XRD method.

Key words nanoindentation, Fe – based coating, residual stress, pile – up

中国特色的再制造零件质量保证技术体系现状及展望*

摘　要　再制造在实现资源能源节约的条件下生产出经济发展所需要的产品，是建设资源节约型、环境友好型社会的有效途径。针对中国特色的再制造模式，结合再制造产品及其生产环节，分析影响再制造零件质量的主要因素，指出为保证再制造零件质量应把控的几个环节，即再制造毛坯质量评价、再制造生产过程监测、再制造零件/涂层质量检验，并提出再制造零件质量保证技术体系，介绍我国近几年在再制造标准方面的主要成绩，并指出为适应我国再制造产业快速发展的需要，急需加强对再制造成型技术方法和再制造质量检测评价方法等方法标准的研究。阐述再制造毛坯和再制造涂层质量无损监/检测评价方法，并结合车辆发动机典型零件再制造实践，介绍无损检测评价技术在再制造质量保证中的应用。展望中国特色再制造零件质量保证技术的未来发展。

关键词　再制造零件　质量保证　技术体系

1　前言

资源、能源和环境作为当今社会发展的主题，多年来得到了世界各国的高度重视。自20世纪末，中国政府开始实施可持续发展战略，大力发展循环经济，在生产活动中大力倡导节能、节材和降低环境污染。在此背景下，中国特色的再制造工程应运而生，并已成为发展循环经济、促进社会可持续发展的有效途径。

再制造工程以机电产品全寿命周期理论为指导，以旧件实现性能跨越式提升为目标，以优质、高效、节能、节材、环保为准则，以先进技术和产业化生产为手段，对旧件进行修复和改造。再制造的重要特征是再制造产品的质量和性能要达到或超过新品，成本仅是新品的50%左右，节能60%左右，节材70%以上[1]。再制造在实现资源能源节约的条件下生产出经济发展所需要的产品，是建设资源节约型、环境友好型社会的有效途径。

近年来，我国政府和企业大力推动再制造产业化发展，已把再制造列为社会经济发展“十二五”规划中的“战略性新兴产业”，在再制造领域的科研投入不断加大，再制造成型技术和应用方面取得了显著成果[2,3]。可以说，再制造在中国有着良好的发展环境和机遇，已步入了发展的“快车道”。但同时，再制造零件质量已成为行业内共同关注的重点问题之一。业界关注重点已不仅仅局限于研发再制造成型技术，如今更加重视如何保证再制造零件质量，不断加强在再制造零件质量保证技术方面的工作，在注重环境保护、资源节约的背景下大力发展绿色再制造产业。

本文结合中国特色的再制造模式的技术特点和中国的再制造产业化发展，阐述中

* 本文合作者：董世运、史佩京。原发表于《机械工程学报》，2013，49(20)：84~90。国家自然科学基金(50735006)和国家重点基础研究发展计划（“973”计划，2011CB013405）资助项目。

国特色的再制造零件质量保证技术内容及其应用。

2 中国特色的再制造模式

欧美等工业发达国家在20世纪60年代已开始发展再制造产业，在传统制造业基础上逐渐发展并完善了以“换件法和尺寸修理法”为核心的再制造模式，即对于损伤程度较重的零件直接更换新件；对于轻度损伤的零件则利用车、磨、镗等机械加工手段，在改变零件尺寸的同时恢复零件的几何精度，再与“加大”尺寸的非标新品零件配副。英美的再制造企业，例如李斯特派特公司和康明斯公司，均采用这种再制造模式。针对该模式，他们在新品设计阶段，就设计预留好了以后再制造时的加工尺寸[4]。因此，他们在再制造过程中，一般采用制造业技术标准保证再制造产品质量。

中国自1999年正式提出再制造的概念以来，在探索过程中自主创新出了中国特色的“表面恢复和性能提升法”再制造模式[5]。该模式把先进的无损检测评价技术、表面工程技术和熔覆成型技术引入再制造，利用激光熔覆、纳米表面工程等先进表面工程技术，使失效零件尺寸恢复到原始设计尺寸，并恢复甚至提升功能部位性能。这种再制造模式既可以对轻度损伤零件进行再制造，也能实现表面重度损伤零件的再制造；并且，在恢复零件损伤部位几何尺寸的同时，可以通过选用性能更优异的再制造材料提升零件的服役性能[2]。十余年的实践证明，中国自主创新的再制造模式很好地弥补了国外再制造模式的不足，节能节材环保效果更加显著。

现阶段，我国企业在再制造生产中，有效融合了国内外再制造模式，实现了融合创新，推动中国再制造产业获得了快速发展。针对上述两种再制造模式，为保证再制造机械零件质量，国外再制造生产基于制造业加工技术和制造标准，中国特色的再制造模式基于维修和表面工程技术，其质量影响因素更加复杂，所需技术更加先进，无法直接采用制造标准。针对中国特色的再制造模式，就技术方面而言，需要评价再制造涂层和再制造前后零件的质量与性能，这是设计制造阶段所没有考虑的。因此，为保证再制造零件质量，急需建立相应的再制造零件质量保证技术体系。

3 再制造零件质量内涵及其影响因素

再制造产品不是新品，但也不属于二手产品，这已成为共识。再制造零件虽然不同于新品零件，但是其质量和性能却达到甚至超过新品零件，这是对再制造零件质量的根本要求。如何实现该要求，保证再制造零件的质量，这需要掌握影响再制造零件质量的主要因素。

再制造零件的质量是从其几何尺寸、材料组织及服役性能和寿命等多方面对再制造零件进行综合评价的，主要考虑再制造零件是否存在气孔和裂纹等缺陷，性能是否达到服役条件要求，寿命能否经历下一个服役周期等。

图1给出了机械零部件再制造工艺流程[1]。图1表明，回收的废旧产品首先经过拆解，对零件损伤情况进行初步判断，并将零部件分类，严重损坏而无法再制造的零件即直接报废，不再进入下一步工序；外观完好的零件，经清洗后才能进入下一步工序。由图1可以看出，从技术方面讲，再制造产品质量与再制造毛坯质量、再制造成

型过程、再制造涂层质量等密切相关。

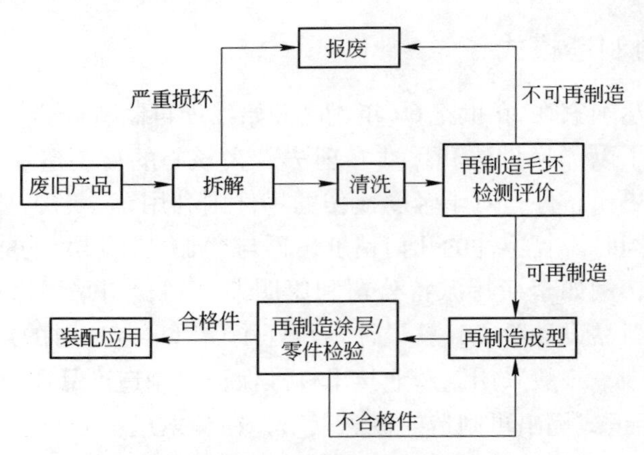

图1　机械零部件再制造工艺流程

再制造毛坯质量是保证再制造零件质量的前提。再制造毛坯质量主要取决于零件原始制造质量和服役过程。只有原始制造质量合格且经历一个服役周期后仍具有足够剩余寿命的零件，才是质量优良的再制造毛坯，才能通过先进再制造技术获得合格的再制造零件。因此，再制造毛坯质量检测评估在再制造生产中占据重要地位。

再制造成型过程是根据再制造毛坯的失效特点和服役性能的要求，采用纳米电刷镀、电弧喷涂、激光熔覆等先进的再制造成型技术手段，恢复零件尺寸和性能。在此过程中，在科学设计技术方案的前提下，影响再制造零件质量的主要因素是再制造成型技术工艺。理论上讲，采用合理的技术方案和优化的技术工艺参数，应当获得质量合格的再制造零件。但是，设备运行的稳定性、人员误操作以及外界环境干扰等不定因素，会造成工艺过程的扰动或混乱，进而影响再制造零件质量。因此，有必要对再制造成型过程工艺参数或成型质量特征参数进行监/检测和调控。

对再制造涂层/零件检测评价是对再制造零件的质量进行严格“把关”，确保所再制造出的零件质量和性能满足服役要求的重要技术环节，是提升用户对再制造产品信任度的直接有效手段。

综上所述，从再制造技术角度分析，影响再制造零件质量的因素主要来自于再制造毛坯和成型过程两个环节，主要包括再制造毛坯质量和再制造成型工艺。为确保交付用户的再制造产品是合格产品，在掌握先进再制造技术方案和工艺的前提下，应当严格把控“入口关”和“出口关”，即对再制造毛坯和再制造涂层/零件严格检测评价。为此，需要有实用可靠的质量保证技术。

4　再制造产品质量保证技术体系

近年来，随着中国再制造产业的快速发展，中国特色的再制造零件质量保证技术也得到了相应的快速发展。

4.1　再制造质量保证技术手段

再制造生产过程中，对再制造毛坯和再制造涂层/零件进行检测评价一般不能损伤零件，需要采用无损检测评价技术方法。结合再制造工艺流程，为确保再制造零件性能不低于新品，应建立与中国特色再制造模式相适应的再制造零件质量保证技术体系，见表1。

表1　再制造零件质量保证技术体系

质量控制环节	检测内容	检测评价技术	备　注
再制造毛坯无损检测评估	几何尺寸测量	卡尺测量、三维反求等	N
	缺陷检测	涡流、超声无损检测方法等	N
	残余应力测定	X射线残余应力测定等	N
	剩余寿命评估	金属磁记忆等方法	N
再制造成型过程在线监/检测	再制造工艺监测	红外热像、温度监测等方法	N
	产品质量在线监/检测	涡流、超声等无损监测	N
再制造涂层/零件检测评估	涂层缺陷检测	光学观察，涡流、超声等无损检测	N
	涂层结合强度测定	拉伸法、划痕法、压痕法	D
	涂层残余应力测定	X射线残余应力测定等	N
	涂层/零件服役寿命预测		N
再制造装备质量验证	服役性能综合指标	台架试验、实车考核试验等	D

注：D为破坏（有损）评价法；N为无损检测评价法。

在表1所提出的再制造质量保证技术体系中，除了再制造毛坯、再制造成型过程和再制造涂层/零件三个控制环节之外，还提出了再制造装备质量验证环节。在此环节，主要是通过台架试验或实车考核试验的途径，综合考核经前几个环节检测评估后认为合格的再制造零件及再制造装备的服役性能指标。但是，一般而言，经过此试验后，再制造零件会发生损伤甚至耗尽一个寿命周期，因此，这种考核验证并不是针对每个再制造零件实施的，而主要针对采用一定再制造技术方案所再制造出的一批产品，用于考核验证和优化再制造技术方案。上述再制造零件质量保证技术体系，主要建立在无损监/检测评价理论和技术方法基础上，尤其是前三个质量控制环节，从再制造生产实际出发，采用无损检测评价技术实现，满足再制造生产中对检测评价的快速、无损的要求。

首先，在实施再制造之前，采用涡流检测、超声检测等无损方法检测评价再制造毛坯的损伤情况以及裂纹等缺陷状态，采用金属磁记忆等无损方法预测其剩余寿命，判断它能不能再制造。再制造毛坯已经经历了一个服役周期的使用过程，在此过程中，零件可能发生了不同程度的疲劳损伤，在其表面和内部可能已产生裂纹等缺陷。如果再制造毛坯的损伤达到一定程度，或局部部位产生了危险缺陷，那么对其进行再制造，就难以保证其再制造质量。

其次，在再制造成型过程中，采用红外热像等方法对热喷涂、熔覆沉积成型等工艺过程的温度场和流场进行监/检测，对再制造成型过程中的工艺参数进行监/检测；采用涡流检测、超声检测等具有在线实时检测功能的快速无损检测技术，在线监/检测再制造成型零件。这样，可以实时掌握生产过程中再制造成型工艺稳定性和再制造成型零件状态，最大限度地避免再制造成型不合格零件进入后续工序，以免造成再制造成本和工时浪费。

最后，采用涡流、超声等方法无损检测评估再制造成型零件/涂层以及涂层结合区域的缺陷，采用金属磁记忆、超声等方法无损预测再制造零件/涂层的服役寿命。必要时，根据零件服役要求，测定评价涂层残余应力以及结合强度等其他性能。只有检测评价结果表明合格的再制造零件，才能装机使用。

另外，为了进一步考核再制造零件性能，可以根据需要，对再制造零件进行台架试验或实车考核。同时，这也是进一步验证再制造技术方案合理性的有效措施。但是，台架试验或实车考核试验周期长、经费投入大，一般而言，只在某型零件验证再制造技术方案时才实施。如果再制造零件通过了该类试验验证，那么就确定了该类零件的再制造方案，就可以制定其再制造规范或者标准，实施产业化生产。

4.2 再制造标准

如上所述，再制造零件质量依靠先进再制造技术、先进无损检测评价和实际考核技术予以保证。通过检测评估和实际考核，成熟的再制造成型技术方法和检测评价方法等应当形成再制造相关标准，为再制造生产提供依据，保证工业生产中再制造产品的质量。

中国再制造产业正快速发展，急需结合我国国情和再制造模式，在国家、行业和企业的不同层面上制定相应的再制造标准。近几年来，再制造标准得到了政府相关部门、公司企业以及科研单位和相关学术组织的重视，再制造标准工作也在快速发展，例如，我国成立了“全国绿色制造技术标准化技术委员会再制造分技术委员会”等从事再制造标准工作的专门学术组织，提出了再制造标准体系。国家部委资助立项了多项再制造标准研究项目，增强了在再制造标准研究方面的投入力度。目前，我国已经发布实施了10余项再制造标准（表2），已经报批或在研的再制造标准有近20项。这为保证我国再制造产业高质量快速发展提供了支撑。但是，由表2可以看出，目前我国实施的再制造标准主要是基础标准、产品标准以及再制造拆解和清洗标准等，仍然缺乏再制造成型技术方法和再制造涂层或产品缺陷检测、寿命评估/预测与质量检验方法等方面的方法标准，还远不能满足再制造生产和产业发展需要。因此，企业和科研单位应当加强合作，进一步加强再制造技术方法标准和质量检测评价方法标准等系列标准的制定，不断充实中国特色的再制造标准体系。

5 无损检测评价技术在再制造生产中的应用

5.1 无损检测评价方法的选用原则

无损检测评价是再制造质量保证的关键技术内容，但只有根据零件对象特征和检测内容需要，选用正确的无损检测方法，才能达到检测评价目的。目前已有200余种不同的无损检测方法，其中，最常用的是五大常规方法，即渗透检测、磁粉检测、涡流检测、超声检测和射线检测。此外还有多种非常规无损检测方法，如工业CT法、中子射线照相法、声发射法、巴克豪森噪声法和金属磁记忆法等。在再制造生产中，应根据各检测方法的特点，针对检测对象的形状结构、材质、可能的缺陷特征以及检测

表2　中国已经实施的再制造标准

序号	标准号	中文标准名称	归口单位	牵头单位	标准类别	实施日期
1	GB/T 27611—2011	再生利用品和再制造品通用要求及标识	全国产品回收利用基础与管理标准化技术委员会	中国标准化研究院	管理	2012-05-01
2	GB/T 28615—2012	绿色制造金属切削机床再制造技术导则	全国绿色制造技术标准化技术委员会	重庆大学	基础	2012-12-01
3	GB/T 28618—2012	机械产品再制造通用技术要求	全国绿色制造技术标准化技术委员会	再制造重点实验室	基础	2012-12-01
4	GB/T 28619—2012	再制造术语	全国绿色制造技术标准化技术委员会	再制造重点实验室	基础	2012-12-01
5	GB/T 28620—2012	再制造率的计算方法	全国绿色制造技术标准化技术委员会	再制造重点实验室	基础	2012-12-01
6	GB/T 28672—2012	汽车零部件再制造产品技术规范交流发电机	全国汽车标准化技术委员会	上海汽车工业总公司	产品	2013-01-01
7	GB/T 28673—2012	汽车零部件再制造产品技术规范起动机	全国汽车标准化技术委员会	东风汽车公司	产品	2013-01-01
8	GB/T 28674—2012	汽车零部件再制造产品技术规范转向器	全国汽车标准化技术委员会	上海汽车工业总公司	产品	2013-01-01
9	GB/T 28675—2012	汽车零部件再制造拆解	全国汽车标准化技术委员会	上海汽车工业总公司	方法	2013-01-01
10	GB/T 28676—2012	汽车零部件再制造分类	全国汽车标准化技术委员会	济南复强动力有限公司	方法	2013-01-01
11	GB/T 28677—2012	汽车零部件再制造清洗	全国汽车标准化技术委员会	中国人民解放军第六四五六工厂	方法	2013-01-01
12	GB/T 28678—2012	汽车零部件再制造出厂验收	全国汽车标准化技术委员会	潍柴动力再制造有限公司	方法	2013-01-01
13	GB/T 28679—2012	汽车零部件再制造装配	全国汽车标准化技术委员会	东风汽车公司	方法	2013-01-01

内容和目的等，选用适当的无损检测评估方法[6~8]。再制造生产中合理选择无损检测评价方法，一般应遵循技术性原则和经济性原则。

技术性原则是指应根据检测对象的特征、检测内容和目的，选择合适的无损检测评价方法。为此，应尽可能充分掌握检测对象信息，分析被检工件的材质、成型方法、加工过程、服役经历、再制造成型方法和成型过程等，对缺陷的可能类型、方位和性质进行预先分析，以便有针对性地选择恰当的无损检测评价方法。没有哪一种无损检测评价方法是万能的，各种无损检测评价方法都有其适用范围，一般而言，射线检测对体积型缺陷比较敏感，渗透检测则用于对表面开口缺陷的检测，而涡流检测对导电材料表面开口或近表面缺陷都具有很好的适用性，超声检测可以评价金属和非金属内部缺陷以及残余应力等，金属磁记忆检测方法可以检测评价铁磁性材料零件的应力集中和疲劳损伤。因此，对特定零件进行再制造时，应首先按照技术性原则，科学选择检测方法、设计检测评价方案。

经济性原则就是在能够实现检测评价目的的前提下，综合考虑因实施无损检测评价方法方案而增加的仪器设备投入、再制造生产成本增加以及因及时剔除不合格件和提高再制造产品质量带来的效益，也就是说应当综合考虑其“投入”与“产出”。一般而言，在满足再制造检测评价目的的条件下，应当优先选择检测仪器设备成本低、操作简单、检测评价过程无污染、对人体无危害的检测评价方法。

5.2 无损检测评价在汽车发动机再制造中的应用

在现阶段的汽车发动机再制造生产中，无损检测评估技术主要完成如下任务：

（1）发动机旧件缺陷的快速无损检测评价。例如：利用磁粉法、涡流法和超声波法等，实现零件表面和内部裂纹等缺陷的无损检测。

（2）旧件疲劳损伤程度或剩余寿命的快速无损评估。例如：采用金属磁记忆技术评估曲轴等特定零件的疲劳损伤程度，快速测定零件应力集中部位及应力集中程度，评判其再制造的可行性与可靠性。

（3）再制造零件涂层质量的快速无损检验与评价。例如：采用涡流方法快速测定金属镀层的厚度，采用超声方法检测涂层/基体结合界面缺陷，采用 X 射线衍射法快速测定涂层残余应力。

实践表明，在发动机零件再制造生产中，需要综合应用多种无损检测评价方法[8~10]。下面针对发动机曲轴、连杆、缸体、缸盖、气门等发动机再制造典型零件，进一步阐释其目前采用的主要无损检测评价方法，见表 3。

针对表 3 中的发动机缸体再制造实例，在原来的再制造生产中，采用目视观察和“打水压”的方法检测缸体缸筒壁是否存在明显可见缺陷，是否发生泄漏（存在穿透性裂纹），而对存在的非穿透性裂纹及其裂纹深度、局部应力集中等情况无法检测，导致再制造缸体存在安全隐患。装备再制造技术国防科技重点实验室与爱德森（厦门）电子有限公司联合，研制出了涡流/磁记忆综合检测评估方法及其工业化应用仪器设备，并在再制造生产中获得了实际应用。

表 3　发动机典型零件再制造生产中旧件无损检测方法

典型零件	检测评价的主要内容	采用的无损检测评价方法
曲轴	表面裂纹	磁粉法
	内部裂纹	超声检测法①
	疲劳损伤	金属磁记忆检测法①

续表3

典型零件	检测评价的主要内容	采用的无损检测评价方法
缸体	缸筒泄露（筒壁贯穿性裂纹）	“打水压”检测法
	筒壁内表面裂纹（非贯穿裂纹）等	涡流/磁记忆复合检测法①
缸盖	水/汽路穿透性裂纹	“打水压”检测法
	“鼻裂”非穿透性裂纹	涡流检测法①
连杆	表面裂纹	磁粉法
	内部裂纹	超声检测法①
	疲劳损伤	金属磁记忆检测法①
气门	气门杆摩擦焊焊缝裂纹	敲击法、超声检测法①
	气门盘过渡段裂纹	涡流检测法①

①为作者课题组增加应用的方法。

由表3可以看出，一个零件需要综合应用多种无损检测方法的同时，一种具体的无损检测方法也可以应用于检测不同的零件，也就是说，无损检测方法具有通用性，但是针对具体的零件需要研究其具体的检测工艺方法。例如，在原来的再制造生产中，采用荧光磁粉方法检测旧连杆和旧曲轴表面及近表面裂纹等缺陷，作者课题组又研发应用涡流方法检测发动机缸盖、缸体以及气门等零件，研究采用超声方法检测气门、曲轴、连杆等零件，研制的检测仪器设备和技术方法均在再制造生产中得到了应用。

6 展望

在注重资源能源效益和环境保护的当今社会经济中，再制造更显现出其巨大潜力。中国以规模宏大的机械行业为依托，中国特色的再制造技术及其产业规模已达到了国际先进水平。随着再制造产业领域拓展和规模扩大，中国再制造产业对再制造质量控制技术将提出更细致、更广泛的需求，再制造质量保证技术方法必将越来越丰富。本文提出的再制造质量保证技术体系是一个体系框架，其中需要监/检测的具体内容及其具体的监/检测技术方法将随着再制造产业化的发展而不断丰富。

再制造标准是再制造质量保证技术体系的重要内容，也是再制造生产规范化的重要依据。为适应再制造产业发展，中国再制造国家标准、行业标准和企业标准均需要更快发展。

在“十二五”社会经济发展规划所提出的“战略性新兴产业”中，节能环保产业被列为第一个产业领域，中国特色的再制造以其显著的技术和产业优势在节能环保产业发展中具有巨大发展潜力。随着再制造质量保证技术的不断发展和完善，中国再制造产业必将更加健康发展，为节能环保产业和循环经济发展发挥更大作用。

参考文献

[1] 徐滨士．装备再制造工程的理论与技术［M］．北京：国防工业出版社，2007.

XU Binshi. Theroy and technology of equipment remanufacture engineering [M]. Beijing: National Defense Industry Press, 2007.

[2] 徐滨士．中国再制造工程及其进展［J］．中国表面工程，2010，23(2)：1~6.

XU Binshi. Remanufacture engineering and its development in China [J]. China Surface Engineering, 2010, 23(2): 1~6.

[3] 徐滨士，董世运，朱胜，等. 再制造成形技术发展及展望 [J]. 机械工程学报，2012，48(15)：1~8.

XU Binshi, DONG Shiyun, ZHU Sheng, et al. Development in remanufacture forming science and technology [J]. Journal of Mechanical Engineering, 2012, 48(15): 1~8.

[4] 徐滨士. 再制造与循环经济 [M]. 北京：科学出版社，2007.

XU Binshi. Remanufacture and cycling economy [M]. Beijing: Science Press, 2007.

[5] XU Binshi, ZHU Sheng. Advanced remanufacturing technologies based on nano - surfaceengineering [C] //Proc. 3rd Int. Conf. on Advances in Production Eng, Guangzhou, 1999: 35~43.

[6] 张俊哲. 无损检测技术与应用 [M]. 北京：科学出版社，1993.

ZHANG Junzhe. Non-destructive testing technology and its applications [M]. Beijing: Science Press, 1993.

[7] 任吉林，林俊明. 电磁无损检测 [M]. 北京：科学出版社，2008.

REN Jilin, LIN Junming. Electo-magnetic non-destructive testing [M]. Beijing: Science Press, 2008.

[8] 王丹，徐滨士，董世运. 涂层残余应力实用检测技术的研究进展 [J]. 金属热处理，2006，31(5)：48~52.

WANG Dan, XU Binshi, DONG Shiyun. Progress of the residual stress detecting technology for coatings [J]. Heat Treatment of Metals, 2006, 31(5): 48~52.

[9] 刘彬. 再制造涂层质量的超声波/磁记忆综合无损评估研究 [D]. 哈尔滨：哈尔滨工业大学，2012.

LIU Bin. Non - destructive evaluation on quality of the remanufactured coatings by the ultrasonic and metal magnetic memory complex methods [D]. Harbin: Harbin Institute of Technology, 2012.

[10] 徐滨士，董丽虹，董世运. 再制造前废旧曲轴 R 角部位金属磁记忆信号采集装置：中国，201110003051.6 [P]. 2012-07-04.

XU Binshi, DONG Lihong, DONG Shiyun. A device to collect the metal magnetic memory signals from the R position of the old crank - shaft: China, 201110003051.6 [P]. 2012-07-04.

States and Prospects of China Characterised Quality Guarantee Technology System for Remanufactured Parts

Xu Binshi, Dong Shiyun, Shi Peijing

(National Key Laboratory for Remanufacturing, Academy of Armored Forces Engineering, Beijing, 100072, China)

Abstract Remanufacturing has become an effective way to develop the resource - saving and environment - friendly society, which can provide the needed products for economy development while saving resources and energy. The China characterises remanufacturing mode was stated. With consideration of the remanufactured products and the production procedures, the principle factors affecting the remanufactured product quality are analyzed. It is pointed out that the three procedures, such as core assessing, production process inspecting and remanufactured products or coatings assessing, should be strictly controlled to guarantee quality of the remanufactured products. It puts

forward the China characterized remanufactured product quality guarantee technology system, then it introduces the non – destructive testing and inspecting technology for the core and remanufactured coating quality control and their applications to remanufacturing practice of the typical vehicle engine components. It introduces the recent progress of Chinese remanufacturing standards, and suggests that it is urgent to study the standards on remanufacturing forming techniques and remanufacturing quality assessing techniques to meet requirements of the rapid – developing remanufacturing industry in China. It prospects future development of the China characterises remanufacturing quality guarantee technology.
Key words remanufactured parts, quality guarantee, technology system

再制造关键技术

The Remanufacturing Engineering and Automatic Surface Engineering Technology*

Abstract Entering into the 21st century, remanufacturing engineering has been developed rapidly in China, especially from 2005, lots of remanufacturing laws and regulations have been released. Remanufacturing engineering is the industrialization of high technology maintenance to the waste productions, and the advanced period of the maintenance engineering and surface engineering. The basic character of surface engineering is synthesis, intercross, compounding, and optimization. Surface engineering takes the "surface" as core. Nano surface engineering is the integration and creation between the nano materials and traditional surface engineering. To adapt the demand of remanufacturing industrialization, five kinds of automatic and intelligentized technologies, namely automatic nano electro-brush plating technology, automatic high velocity arc spraying technology, semi-automatic micro plasma arc welding technology, automatic laser cladding technology, and intelligentized self-repair technology, have been independently innovated.

Key words remanufacturing engineering, surface engineering, automatic technology, intelligentized technologies

1 Rapid Development of RM Engineering in China

Entering into the 21st century, the momentous strategy, namely developing circular economy and constructing saving-oriented society, is made by Chinese government. As one of the important components, remanufacturing engineering has been developed rapidly. Especially from 2005, lots of remanufacturing laws and regulations have been released.

In May 2005, the "Strategic Consultant Study on Constructing the Saving-Oriented Society", which the sub-project "4R Engineering" including remanufacturing was listed, was started by CAE. The General Report pointed that the remanufacturing ratio to waste mechanical products will be up to 50% in 2010 and to 80% in 2020. In June 2005, the No. 21 and No. 22 documents, released by China State Department, pointed out that the government will support greatly the remanufacturing to waste electromechanical products, and the "Green Remanufacturing Technology" was listed as one of the technologies, which will be highlighted to develop and popularize by the government.

In Nov 2005, the document "Notification on Organizing the First Time Circular Economy Demonstration", released by National Development & Reform Committee (NDRC), et al, publicized the 42 demonstration enterprise lists. Remanufacturing was listed as one of four highlight areas. The only engine remanufacturing enterprise, Jinan Fuqiang Power Co., Ltd., is one of the

---

* Reprinted from *Key Engineering Materials*, 2008, 373 – 374: 1 – 10.

remanufacturing area demonstration enterprises.

In April 2006, vice premier Zeng Pei yan commented in the Report on Development of Vehicle Components Remanufacturing and Relative Countermeasures: "Agreeing on the remanufacturing demonstration in the automobile components, exploring the experiments, developing technologies, meanwhile getting ready for revising the laws and regulations".

According to the direction from vice premier Zeng, the series actions about automobile components remanufacturing have been organized by NDRC. The Chinese Automobile Industry Society was designated to establish a special study group, which He Guangyuan, former minister of Mechanical Industry ministry, and academician Xu Binshi were advisers. By now, the automobile enterprises such as Yiqi, Erqi, Shangqi, pay more and more attention to the automobile components remanufacturing. In 2007, the first "Circular Economy Law" in China, instituted by State Department, has been submitted to the Legal System Office of State Department, and will be released in Oct, 2007. The law mentions definitely "Electromechanical Products Remanufacturing" and its policies.

2 Connotation of Remanufacturing Engineering

Remanufacturing engineering is a general designation of all techniques and engineering treatments to maintain and rebuild worn products, which taking the productive whole life period design and management as instruction, taking the great upgrade of the performance of waste productions as goal, taking the good quality, high efficiency, energy-saving, environment-protecting as rule, and taking the advanced techniques and industrializing process as measures[1].

Simply, remanufacturing engineering is the industrialization of high technology maintenance to the waste productions. The important character of remanufacturing engineering is that the quality of the remanufactured productions is as same as or superior to that of the new productions, and the cost is only a half of the new one, saving energy 60%, saving materials 70%, and decreasing the bad influence to environments. The remanufactured products are not the used ones but belong to the new products.

The whole expense of production, from preliminary study, design, manufacturing, using, maintenance, to abandoning, was regarded as whole life period expense. The first half life, which only possessed 20% - 30% expense, was paid great attention in the traditional notion. Whereas the after life possessed 70% - 80% expense was ignored.

3 Surface Engineering and Nano Surface Engineering

The advanced surface engineering and nano surface engineering technology is the key technology of remanufacturing engineering.

Surface engineering is a system engineering[2] to obtain desired surface properties through surface coating, surface modification or duplex surface treatments on a pretreated metallic or non-metallic surface to alter its morphology, chemical composition, microstructure and stress condition. The basic character of surface engineering is synthesis, intercross, compounding, and optimization. Surface engineering takes the "surface" as core. The major advantage of surface engineering is that the surface functional coating, being superior to the substrate materials, can be prepared on the components by a lot of surface methods, to endow the properties such as temperature-resistance, anti-corrosion, wear-resistance and anti-fatigue with the components. Comparing with the substrate materials, the coatings are thin, small in area, but hand on the main shoulders of components.

There are three developing stages in surface engineering, namely single surface engineering, composite surface engineering, and today's nano surface engineering[3].

The single surface engineering technologies, such as thermal spraying, electro - deposition, are beyond the rigor environment. The combined surface technologies that integrate two or more kinds of surface engineering technologies enable to obtain "$1+1>2$" optimum strengthening effects on material surface. The composite technology has been the "accelerator" to improve the surface performance.

The third stage is today's nano surface engineering. Nano surface engineering is a system engineering that combining the nano materials and nano technology with traditional surface engineering, through particular processing techniques or methods to alter farther morphologies, compositions, microstructures of surfaces and impart them new properties much more[4]. In 2000, Professor Xu Binshi presented firstly the concept Nano Surface Engineering in Chinese Mechanical Engineering.

Professor T. Bell, the international founder of surface engineering, foreign nationality fellow of CAE, academician of Britain Royal Academy of Engineering, praised greatly nano surface engineering. He determined to cooperative study the nano surface engineering with Chinese colleagues. The project named Nano Composite Coating and Composite Surface Engineering Used in Advanced Car Components has been rated as the Sino-Britain governmental collaborative project of science and technology in 2002.

4 Automatic Surface Engineering Technology

Lots of surface engineering technologies, aiming at the repair and remanufacturing of damage components, were developed in the recent years, such as Materials Preparation and Forming Integration High Velocity Arc Spraying Technology[5], Nano Electric Brush Plating Technology[6], Nano Selfrepairing Antifriction Additive Technology[7], Nano Thermal Spray Technology[8], Nano Solid Lubrication Technology[9], as well as Residual Life Evaluation Technology[10].

The remanufacturing course if a process and manufacture course with industrialization and batch. To adapt the demand of remanufacturing industrialization, the surface engineering technology must be from manual to automatic. Therefore in the period of Tenth-five and Eleventh - five years, some automatic and intelligentized surface engineering technologies have been developed by our lab, to improve more the performance of surface coatings and quality of remanufacturing. Automatic Nano Particle Composite Brush Plating Technology. The nano particle composite brush plating technology is one of the advanced remanufacturing technology. With it, nano-particle composite brush plating coating which has excellent performance can be prepared. The service temperature of the coating enhances from 200℃ to 400℃, and the contact fatigue performance improves from 105 cycles to 106 cycles. And it has been succeed to apply to remanufacturing of arming component. For example, the life of the imported engine compressor blade which was remanufactured with nano particle composite brush plating technology was over 300 hours bench test. But the manual brush plating technique has many disadvantages, such as low production efficiency and high working intensity. And it should be mentioned specially that the quality of remanufacturing components by manual brush plating technique is affected by the skill of operator easily. Accordingly automatic nano-particle composite brush plating technology has been developed by National Key Laboratory for Remanufacturing on base of manual brush plating technology.

4.1 The automatic nano particle composite brush plating technology

(ANPCBP) is one of the surface engineering technologies which can prepare excellent performance nano-particle composite coating. It achieves the process of brush plating automatically on base of virtual instrument technology and automatic control technology. The automatic process of brush plating has been achieved by solving the key problems of feeding solution continuously, using solution circularly, monitoring processing parameter of brush plating (brush plating voltage, current density and temperature of solution) and procedure of brush plating (switching process and monitoring thickness of coating) real-timely. A model machine has been developed (Fig. 1).

Fig. 1 The model machine of ANPCBP

The n-Al_2O_3/Ni composite coating prepared by the automatic brush plating is relatively dense and uniform and has a smaller crystalline microstructure than that of the coating prepared by manual plating (Fig. 2).

Table 1 shows the micro-hardness of composite coating prepared by automatic brush plating and manual plating. It can be seen that micro-hardness of manual plating coating is uneven distribution and its hardness value shows great fluctuation in different micro-area. Micro-hardness distribution of automatic brush plating coating is more uniform than that of manual plating coating while hardness value of the former doesn't show great increment compared with the latter.

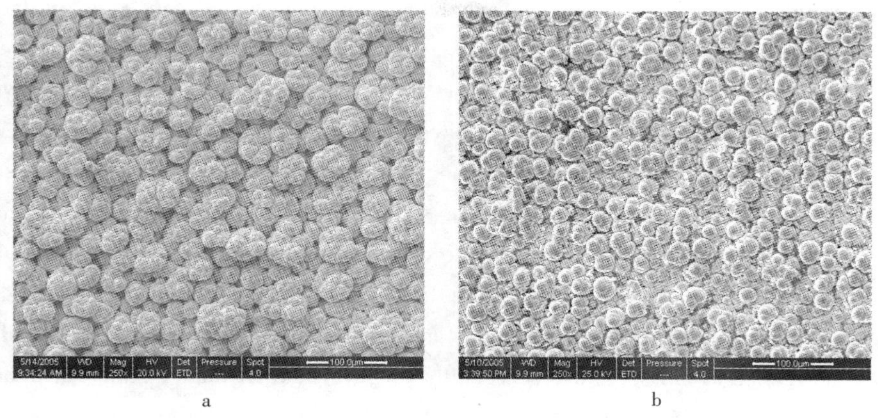

a b

Fig. 2 Surface morphologies of $n-Al_2O_3/Ni$ composite coating

a—Manual brush plating; b—Automatic brush plating

Table 1 Micro – hardness of nano – composite coating prepared by two methods

Coatings	Test load/g	Load time/s	Micro – hardness $HV_{0.1}$				
			1	2	3	4	5
Auto	100	10	655	660	646	649	648
Manual	100	10	666	641	670	681	639

A special machine of automatic nano-particle composite brush plating, which is aimed at the batch remanufacturing problem of engine's connecting rods, has been developed on base of model machine and technology of automatic brush nano-particle composite plating, in Fig. 3. The special machine solves many actual problems in the batch remanufacturing, such as positioning accuracy, remanufacturing quality, continuous working and adjustment of contact. It achieves the automatization in remanufacturing process. The special machine can remanufacture 4 to 6 engine's connecting rods one time and reduces the working time from 60min to 30min. This increases the production efficiency and reduces working intensity greatly. Consequently, it accelerates the development of demonstration plant for remanufacturing.

Fig. 3 Automatic nano – particle composite brush plating machine

Automatic High Velocity Arc Spraying Technology. The high velocity arc spraying technology is a type of coating preparing process that using the ultrasonic air flow to atomize the melted metal wire feed stocks and drive the atomized jet to the surface of substrate, thus the flying particles deposit on the substrate from layer to layer. As compared with traditional are spraying technology, the high velocity arc spraying technology has such advantages as higher stability, higher processing efficiency and the coating has more compact structure. The new method of using the Fe-Al/ Cr_3C_2 cored wire to prepare inter-metallic compound and ceramic composite coating by high velocity arc spraying technique, has been independent innovated by NKLR, and is applied in boiler pipes of power station resistant from high temperature corrosion and erosion – corrosion, to realize the integrated processing of preparing Fe-Al inter-metallic compound and forming coating. The serving period of both Fe-Al/Cr_3C_2 coating and TAFA 45CT coating is about 3 to 5 years, but the maintenance cost per year of the former coating is only 50 percent of the later, which implies the Fe-Al/ Cr_3C_2 coating has a wide applying prospect in the field. The 1Cr18Ni9Ti/Al "pseudo" alloy coating is prepared by the on high velocity arc spraying technology has been successfully applied in remanufacturing STYER automobile engine block and crankcase. It has been established that, due to the micro-poles in the coating has the ability of oil-storing, the wear resistance of the pseudoalloy coating is about 9% higher than that of the traditional 1Cr18Ni9Ti coating under the oil lubrication sliding condition. The decrease of hardness of the pseudo alloy coating is easy for the subsequent mechanic processing.

The newly developed automatic high velocity arc spraying technology combined the advantages of automatic technology based on industry robot and high-velocity technology based on new design of spraying gun, using the motion arm of robot or auto-operate machine to fix the high velocity arc spraying gun, planning the moving path of the gun based on the control software, adjusting the spraying parameters in real-time and controlling the gun worked under the predecided programs. The developed automatic spraying system contains four units, rectangular Cartesian coordinate auto-operate machine with four motion freedoms, high load-supporting positioner with two freedoms, center controlling unit and arc spraying equipments. The system has such features as: (1) the motion devices has six freedoms that enhanced the spraying stability; (2) the optimized controlling unit enables the wholly automatic action; (3) the real-time adjusting or controlling process increases the accuracy of spraying process; (4) the optimized configuration of spraying gun improves the quality of coating; (5) the development of inverse power supply improves the arc stability.

As compared with the manual arc spraying process, the automatic spraying process makes the thickness of deposited coating more uniform and improves the coating quality. On the other hand, the automatic spraying process let the workers far away from the high strength and polluted environment (high temperature, dense dust and strong noise), and increase the working efficiency. This automatic spraying technology is suitable for spraying several types of metal wires, such as cord wires, pseudo alloy wires and some special alloy composite wires. This technology has been accepted by the Jinan Fuqiang Power Co., Ltd., to remanufacture the automobile engine block and crankcase and other typical parts, which enhances the performance, decreases the cost, saves the materials and energy, and improves the efficiency. Illustrationally, the remanufactured crankshafts use the discarded crankshaft as workblanks, thus it saves a great deal of rough stocks as compared with that of producing a new crankshaft. In addition, the remanufactured crankshafts avoid the nitridation process in the condition of 400℃ high temperature for about 8 hours. It will take about 1. 5 hours to remanufacture an engine crankcase by manual arc spraying process, but by automatic arc spraying technology it will decrease to 20 minutes. Fig. 4 shows the arc spraying process of using the automatic high velocity arc spraying technology to remanufacture a discarded engine block, and the remanufactured domain (i. e., the crankshaft bearing shell hole) is shown in Fig. 5.

Fig. 4 Process which the AHVAS technology remanufactures a discarded STYER engine block

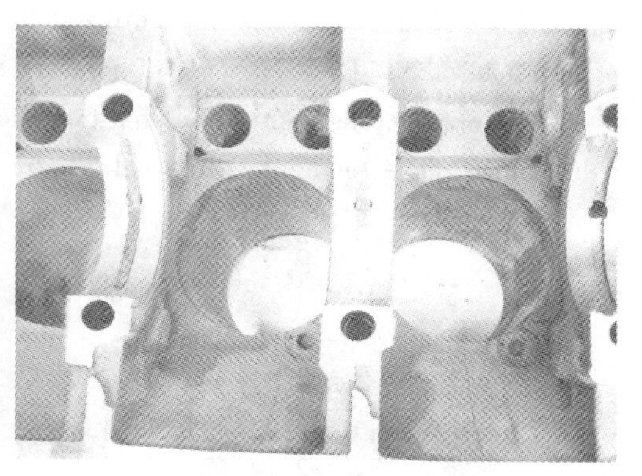

Fig. 5 The remanufactured crankshaft bearing shell hole (after machining process)

4. 2 Semi-automatic micro plasma arc welding technology

The micro plasma arc welding (MPAW) technology is the plasma welding technique with the current less than 30A. Semi-automatic MPAW utilizes micro-plasma arc to melt the metal power, clad another type of metals on the surface of remanufacturing or repairing components under the control of computer or single-chip computer. The micro-PAW system is developed by National Key Laboratory, which includes the micro-PAW power, the operation station, the positioner, powder-feeder system, and the gas-feeder system, as shown in Fig. 6.

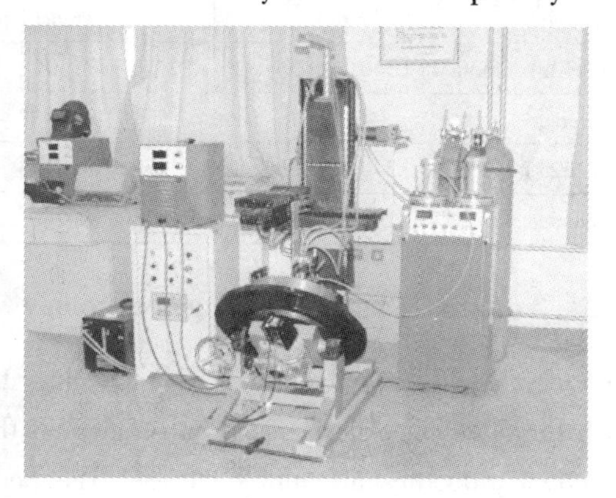

Fig. 6 Micro-PAW system

The bonding between the cladded layer prepared by MPAW and substrate is metallurgical bonding, and it avoids the problem that the spray coating tends to flake away under the impact load or alternate load. MPAW has a high current density and a low heat input, therefore, it can resolve effectively the problem which the components that are sensitive to the heat input can not be repaired by the other methods. And the deformation of the middle and small sized components caused by the heat input is resolved. High frequency inverter MPAW power with the invert frequency of 70kHz, which is higher than the usual invert frequency of 20kHz, is independently developed by National Key Lab for Remanufacturing. As a result, the volume of the equipment decreases, the responding-characteristic of the system increases, and the technological process of MPAW is more stable.

The characteristic of the semi-automatic micro plasma arc welding technology is so excellent that makes the remanufactured components have good performance, high efficiency, low cost, andenergy-saving, materials-saving, environment-protection. The technology has an extensive application foreground in the field of remanufacturing, such as camshafts, Al alloy cases, crankshafts, etc.

The MPAW Technology has been utilized to remanufacture the sealing cone of the used engine exhaust valves. The ST6 and Ni25 power were employed. The deformation of the remanufactured valves decreases, and their hardness renew to the standard. The micro-hardness is listed in Table 2. The remanufactured valve were showed in Fig. 7.

a b

Fig. 7 The remanufactured sealing cone of used engine exhaust valves

a—Remanufactured and no machining; b—Remanufactured and machining and fit

Table 2 The vickers hardness of the sealing cone of valve (HRC)

Samples	Substrate			Transition area				Cladding layer			Top of layer	
1(ST6)	330.48	340.57	367.68	492.57	471.04	479.2	450.89	537.83	525.75	517.75	671.58	643.89
Average	346.2			473.4				527.2			657.8	
2(Ni25)	330.48	340.57	367.68	403.44		407.19		412.09	448.39	427.92	546.12	535.38
Average	346.2			405.3				429.5			540.8	

4.3 Automatic laser cladding technology

The laser cladding technology is the technology that the laser is used to heat the damage surface to melt the cladding materials and substrate, and then the metallurgic layers with lower dilution and excellent properties are rapidly formed. The performance of remanufactured components, such as wear-resistance, anti-corrosion, heat-resistance and oxidation-resistance is improved greatly.

The automatic laser cladding technology (ALCT) means a technology that the damage surface is repaired and remanufactured automatically by the laser operated by the industrial robot or operating - machine which has been input the moving programs and controlled by computer. This technology has many advantages, for instance, less distortion of substrate, metallurgic layers, superior properties, and less machining process after cladding.

Fig. 8 Automatic laser cladding remanufacturing to the tooth surface of heavy load gear

ALCT has been successfully utilized to remanufacture the severe disabled tooth surface of heavy load gears by National Key Laboratory of Remanufacturing (seeing Fig. 8). The problems, such as layer cracks, over heat in part of substrate, keeping the precision of gear surface and improving the properties, have all been resolved. The performance of wear - resistance and anti contact fatigue of the remanufactured components is up to the demand similar with new one.

Due to its huge size, the CO_2 laser that is used to laser cladding is not suit for the field operation. Meanwhile, the CO_2 laser has the long wavelengths and low energy conversion rate. The absorption of cladding materials to laser energy is also lower. Therefore, the high - power fiber coupled solid laser and high power diode laser have been developed by our laboratory to realize the field rapid remanufacturing for the components, break through the technique limitation of laser cladding in field, and solve the key technique and theory problems, including new materials used for rapid formation remanufacturing, accuracy control and quality control to the remanufactured components formation (seeing Fig. 9 and Fig. 10).

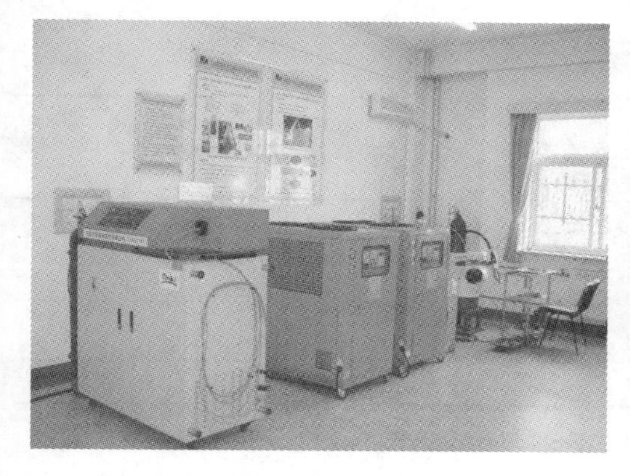

Fig. 9 Fiber coupled solid laser system with robot operating

Fig. 10 The high power diode laser system with manipulator operating

4.4 Intelligentized self-repairing technology

The intelligentized self-repairing technology is a kind of technology that bases on the intelligentized biology technology and micro-nano technology, through the friction chemistry between the metal substrates and self-repairing materials, a reaction thin film with good friction reduction and self-repairing performance is formed on the worn surface, a dynamical equilibrium is achieved between wear and dynamic repair, and the self-repairing on the worn surface is realized without disassembly and shut down of the machine.

The Cu nanoparticles, with an average size of 20 – 50nm, have been successfully prepared by liquid phase reducing method in National Key Laboratory for Remanufacturing. The high-energy mechanical decentralization method and surface modifying method were utilized to creatively solve the great problem that nanoparticles decentralize stably in the oil medium. Fig. 11 and Fig. 12 show the XRD spectra and TEM morphology of Cu nanoparticles, respectively.

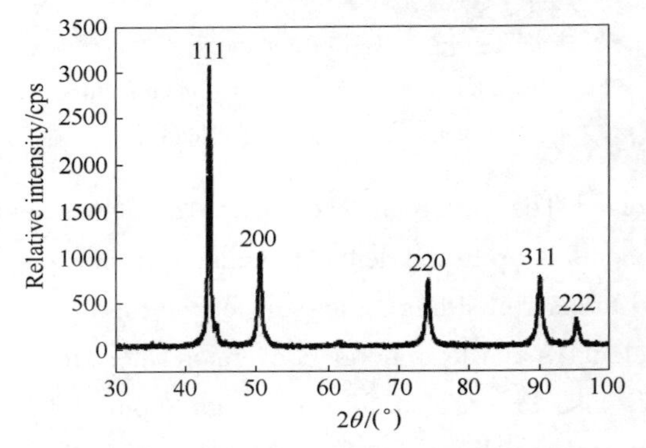

Fig. 11 XRD analysis of Cu nanoparticles

50nm

Fig. 12 TEM morphology of Cu nanoparticles

Table 3 Bench experiment results of Jeep engine with six cylinders

Total power		Maximal power/rotating speed(kW/rpm)	Maximal torque/rotating speed(N · m/rpm)	Minimal oil consumption ratio/rotating speed(g/(kW · h · rpm))
No. 1 engine	Before experiment	118. 22/4500	280. 04/2250	318. 82/3000
	After experiment	114. 72/4500	278. 60/2250	317. 26/3000
No. 2 engine	Before experiment	112. 52/4750	269. 45/2250	310. 97/2250
	After experiment	119. 36/4750	274. 82/2250	292. 36/2250

Note: 1rpm = 1r/min.

A kind of the in-situ nano friction reduction and self-repair additive was independently prepared by our laboratory. Its formula is novel including the self-repairing elements such as Cu nanoparticles and rare earth compound, and other functional additives with the function of cleanness, dispersion, oxidation resistance and corrosion resistance. The additive is granted the national patent for invention in 2004.

The mechanism which the additive reduces friction, decrease wear and self-repairing was studied and explained. A protective Cu self-repairing film is formed on the worn surface after friction. The film possesses the excellent mechanical performance and friction reduction property, good bonding with substrate, and can in-situ repair the early damage on the worn surface and improve the lubrication.

The bench experiment for Jeep engine lasting 300 hours was carried out. The common oil was used in No. 1 engine, and the oil containing the additive was used in No. 2 engine. The endurance experiment results showed that the No. 1 engine's maximum power and torque decreased, whereas the maximum powder of No. 2 engine increased by 6.08%, the maximum torque increased by 2% and the fuel consumption reduced by 5.98%, respectively. The above results indicate the good economical efficiency and power ability. Fig. 13 shows the element analysis result of worn cylinder liner surface after 300 hours endurance experiment. It can be seen that a red protective film that composed of copper is formed on the worn cylinder liner surface. The film improves the lubricating condition of piston ring/cylinder frictional pairs and prolongs the service life of the engine.

Fig. 13 Element analysis result of worn cylinder liner surface

The application test on the heavy-load vehicles with duration of 1.5 years was performed. The Results indicate that the lubricating conditions of the vehicles are improved obviously, wear of materials is reduced, dynamic performance is improved and machine oil consumption is reduced.

5 Conclusions

(1) Remanufacturing engineering is the industrialization of high technology repair and reformation to the waste productions. The important character of remanufacturing engineering is that the quality of the remanufactured productions is as same as or superior to that of the new productions, and the cost is only a half of the new one, saving energy 60%, saving materials 70%, and decreasing the bad influence to environments. The remanufactured products are not the used ones but belong to the new products.

(2) There are three developing stages in surface engineering, namely single surface engineering, composite surface engineering, and today's nano surface engineering. The advanced surface engineering and nano surface engineering technology is the key technology of remanufacturing engineering.

(3) Five kinds of automatic and technologies, namely automatic nano electro – brush plating technology, automatic high velocity arc spraying technology, semi-automatic micro plasma arc welding technology, automatic laser cladding technology, and intelligentized self-repair technolo-

gy, have been independently innovated. These technologies have been applied in the manufacturing demonstration enterprise.

Acknowledgements

The paper was supported by NSFC (50735006), 973 Project (2007CB607601), 863 Project (2007AA04Z408), Consultative project of CAE (12/2002A) and Sino - Britain collaborative project (2002/209 M3).

References

[1] Xu B S: Consultative report of China Academy of Engineering Vol. 12(2000), 12(in Chinese).

[2] Xu B S: Theory and Technology of Surface Engineering(National Defence Industry Press, China 1999).

[3] Xu Bin Shi : Nano Surface Engineering(Chemical Industry Press, China 2004).

[4] Xu B S, Liu S C. China Materials Engineering Canon. (Chemistry Industry Press, China 2006).

[5] Xu B S, Zhu Z X and Zhang W: Wear Vol. 257(2004), p. 1089.

[6] Xu B S, Wang H D and Dong S Y: Electrochemistry Communications Vol. 7(2005), p. 572.

[7] Xu B S, Wang H D and Liang X B: Proceedings of International Conference on Intelligent Maintenance Systems. 2003: 457 - 466.

[8] Zhu S, Xu B S: Key Engineering Materials Vol. 280(2005), p. 1203.

[9] Xu B S, Liu S C and Liang X B: Chinese Journal of Mechanical Engineering Vol. 39(2003), p. 21(in Chinese).

[10] Zhang X C, Xu B S and Wang H D: Journal of Applied Physics Vol. 101(2007) p. 083530-6.

Erosion Properties of Fe Based Amorphous/ Nanocrystalline Coatings Prepared by Wire Arc Spraying Process*

Abstract Fe-Cr-B-Si-Mn-Nb-Y and Fe-B-Si-Nb-Cr amorphous/nanocrystalline coatings were fabricated by wire arc spray process. The microstructure and phase compositions of the coatings were characterised by means of SEM, EDAX and XRD. The formation mechanism of amorphous phase and nanocrystalline phase was discussed. The elevated temperature erosion behaviour of the coatings was also investigated using an air solid particle erosion rig. The microstructures of the coatings consist of amorphous phase and α-(Fe, Cr) nanocrystalline phase. Both coatings are fully dense with low porosity. The coatings exhibit better erosion resistance at lower impact angles. With increasing erosion temperature, the erosion rates of the coatings decrease. The Fe-B-Si-Nb-Cr coating has better elevated temperature erosion resistance than that of the Fe-Cr-B-Si-Mn-Nb-Y coating.

Key words amorphous and nanocrystalline phases, coatings, microstructure, elevated temperature erosion

1 Introduction

Amorphous and nanocrystalline alloys are becoming attractive for their unique properties, such as high strength, excellent corrosion and wear resistance and good magnetic properties[1]. Amorphous alloys, which are free from lattice defects and grain boundaries, are important candidate materials for corrosion resistant applications[2-4]. And nanomaterials (< 100 nm grain sizes) show many superior properties in the sense of mechanical and chemical to conventional crystalline structure[5,6]. Thus, as a promising advanced material, their application to various engineering components has been widely studied[7-9].

There has been an increasing interest in producing Fe based amorphous alloys because of their attractive combinations of engineering properties: high strength and hardness, improved wear resistance, superior corrosion resistance and lower materials cost[10,11]. In industrial fields, however, the application of amorphous alloys as engineering materials is restricted because of the limited size, low toughness at room temperature and high costs[12]. In order to widen the industrial application fields, thermal spraying as a potential candidate is used to produce amorphous alloy coatings for large area industrial applications. Recently, plasma spraying[12,13] and high velocity oxy fuel spraying[13,14] techniques have been used to prepare amorphous coatings. But the preparation cost is high and the materials are amorphous alloys. Compared with the thermal spray methods mentioned above, wire arc spray is considered as a simple, low cost and

* Copartner: X. B. Liang, J. B. Cheng, J. Y. Bai. Reprinted from *Surface Engineering*, 2010, 26(3): 209 - 215.

high efficiency coating process. The wire arc spray process is highly beneficial to the development of amorphous phase and nanocrystalline phase due to rapid solidification of the spray droplets[15]. During wire arc spraying, individual splat was estimated to cool at a rate of about 10^5 K/s[16], which is suitable for forming amorphous phase. The glass precursor, when heated to its crystallisation temperature, readily transforms into nanostructure[17]. And there has recently been a significant effort toward the development of amorphous and nanocrystalline coatings used in industrial applications. Branagan et al. [17] have deposited amorphous/nanocrystalline composite coatings by wire arc spraying SHS7170 cored wires. The composite coatings present excellent mechanical properties and elevated temperature erosion resistance. Georgieva et al. [15] have developed TAFA 140MXC cored wires to deposit nanocrystalline coatings using wire arc spray process. The coating exhibit high hardness and excellent wear resistance. This low cost manufacturing process could potentially solve many wear and corrosion problem by fabricating high performance coatings.

To explore the possibility of preparing amorphous/nanocrystalline coatings by thermal spraying conventional feedstock wires, two exploratory studies on the multicomponent Fe base amorphous and nanocrystalline coatings production by wire arc spray process were carried out in the present work. Both novel Fe base amorphous/nanocrystalline coatings produced are believed to have a considerable potential for advanced engineering applications due to their advantageous properties and low cost. In this study, the microstructures of the coatings were characterised. The porosity and microhardness were examined. The erosion resistance of the coatings was evaluated with a laboratory elevated temperature erosion tester.

2 Experimental

The non – sprayed surface of AISI 1045 steel substrate (25mm × 16mm × 5mm) was treated by powder calorising (powder: 15% Al, 84% Al_2O_3, 0.5% NH_4Cl and 0.5% KHF_2; process condition: 900℃, 4h) to reduce the effect of oxidation at high temperature on the erosion test. Then the non-calorised surface of the substrate was degreased by acetone, dried in air, and then grit blasted. A selfdesigned HAS-2 wire arc gun system was employed for coatings preparation. The iron base cored wires which consist of conventional alloy powders, such as ferroboron and ferrosilicon, were used as feedstock. They contain specific atomic ratios of elements to maximise glass forming ability. Both cored wires are proprietary, which chemical compositions contain: <15 wt% Cr, <5 wt% B, <2 wt% Si, <7 wt% Nb, <3 wt% Mg, <5 wt% Y, Fe balance and <5 wt% Cr, <7 wt% B, < 2 wt% Si, <9 wt% Nb and Fe balance respectively. The wire arc spraying parameters are as follows: spraying voltage, 36 V; wire feed rate, 2.7 m/min; compressed air pressure, 700 kPa; the standoff distance, 200 mm; the torch scan speed, 0.03 m/s; the interpass distance, 10 mm; the overall number of scans, 4. The surfaces of the coatings were not calorised.

After spraying, samples were cut along cross-section direction from the coated plates by linear cutting machine and were mounted for metallographic examination. Then they were ground by silicon carbide abrasive paper (400 to 1500 grit), and polished by grain size 0.5μm Al_2O_3

powder. After that the samples were cleaned in an ultrasonic cleaner with acetone and dried. Finally, the microstructures of the coatings were characterised by using SEM with EDXA (Philips Quant 200). The phase structures of the coating were analysed by XRD(D8-Advance with Cu K_α radiation). The image analysis method was employed for the measurement of porosity. The SEM images with a magnification of ×1000 and with at least 20 view fields from different positions and cross-sections of the coatings were used for the porosity measurement. The microhardness profile along the depth direction of the coating was evaluated by the Vickers hardness tester with a testing load of 1.96 N and a dwelling of 15 s. For each sample, 12 averages Vickers hardness values are performed along the depth direction and each average hardness value is calculated by five Vickers indentations.

An elevated temperature erosion tester was constructed in National Key Laboratory for Remanufacturing, which consists of five parts: blasting system, heating system, abrasive blender, erosion room and temperature controlling system. The erosion temperature was monitored by two thermocouples. The sand velocity was modulated by air pressure as the velocity at different air pressure had been determined by velocimeter. The impingement angle was adjusted by rotating the sample. The diagrammatic sketch of the elevated temperature erosion apparatus is shown in Fig. 1. The elevated temperature erosion testing conditions were listed in Table 1. The experimental surfaces of the samples were ground with mesh 800 emery papers before erosion test. The samples were preheated for about 10 min and the sand and gas were heated instantaneously. Each test was repeated three times. Before and after the tests, the specimens were cleaned in acetone solution and the mass loss was measured using a precision electronic balance with an accuracy of 0.1 mg. The erosive rate was defined as the mass loss of samples per abrasive quantity. After erosion tests, the eroded surfaces were characterised by SEM. The phase structure of the eroded coatings was analysed by XRD.

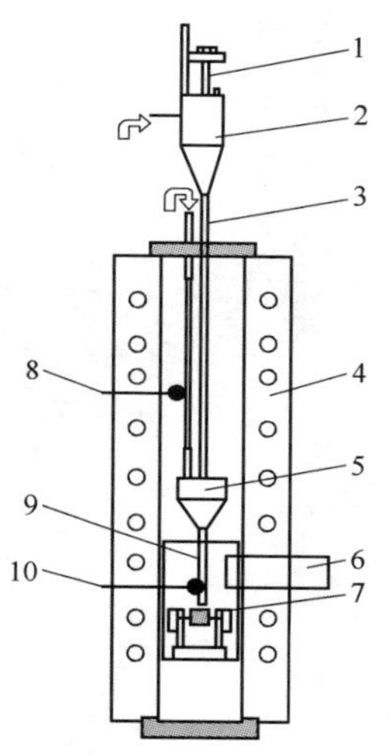

Fig. 1 Sketch map of elevated temperature erosion apparatus

1—Sands regulating stem; 2—Sand hopper; 3—Inlet sand pipe; 4—Electrical resistance furnace; 5—Sand and air mixing chamber; 6—Exhaust pipe; 7—Sample; 8, 10—Thermocouples; 9—Nozzle

Table 1 Parameters of erosion test

Test variable	Test	Test variable	Test
Air pressure/kPa	100	Test time t/h	0.5
Particle velocity v/m · s^{-1}	62	Loading L/g	375
Temperature T/℃	25, 300, 450, 650	Erodent material	150 – 180 μm quartz sand (irregular shape)
Impact angle α/(°)	30, 90		

3 Results and Discussion

3.1 Microstructure characterisation

Fig. 2 shows the SEM image of cross-section in the coatings. It can be seen that the coatings are fully dense smooth, adhering well, and with no cracking. Only some pores exist in the coatings. Both coatings have little change of grey contrast regions, indicating that little oxidation occurred during deposition. Compared with the Fe-Cr-B-Si-Mn-Nb-Y coating, the Fe-B-Si-Nb-Cr coating has a better compact structure. The two coatings have low porosity and their values are 1.7% for the Fe-Cr-B-Si-Mn-Nb-Y coating and 1.5% for the Fe-B-Si-Nb-Cr coating by image analysis respectively. The chemical compositions of the coatings by EDAX analysed are 80.3Fe-10.4Cr-4.9B-1.1Si-1.2Mn-1.7Nb-0.4Y (wt%) (Fig. 2a) and 88.3Fe-5.6B-0.6Si-2.2Nb-3.3Cr(wt%) (Fig. 2b) respectively. However, those chemical compositions derived by EDAX have an indicative, semiquantitative significance.

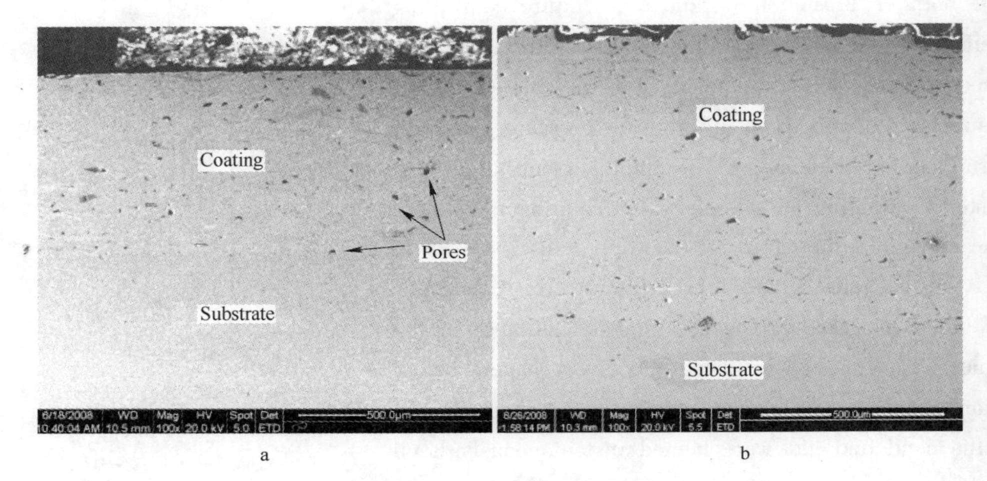

Fig. 2 Images(SEM) of coatings

a—Fe-Cr-B-Si-Mn-Nb-Y; b—Fe-B-Si-Nb-Cr

Fig. 3 illustrates the XRD patterns of the coatings. It can be seen that broad halo peak appears at $2\theta = 43.5°$, confirming that amorphous coatings were formed by wire arc spraying process. The characteristic diffraction peaks of α-(Fe, Cr) are present for the coated sample. The nanocrystalline grains in the Fe-Cr-B-Si-Mn-Nb-Y coating is 17 nm and the Fe-B-Si-Nb-Cr coating is 11 nm calculated by Sherrer formula. Any other oxidised materials peaks are not detected in the patterns, confirming that little oxides exist in the coatings. The formation mechanism of amorphous and nanocrystalline can be interpreted as follows.

During wire arc spraying, individual splat was estimated to cool at a rate of about 10^5 K/s[16], which is suitable for forming amorphous phase. On the other hand, both alloy systems have very strong glass forming ability, which satisfies the three empirical rules proposed by Inoue et al.[18], i. e.:

(1) Multicomponent alloy systems consisting of more than three constituent elements;

(2) Significantly different atomic size ratios above approximately 13% ;

(3) Suitable negative heats of mixing among the constituent elements.

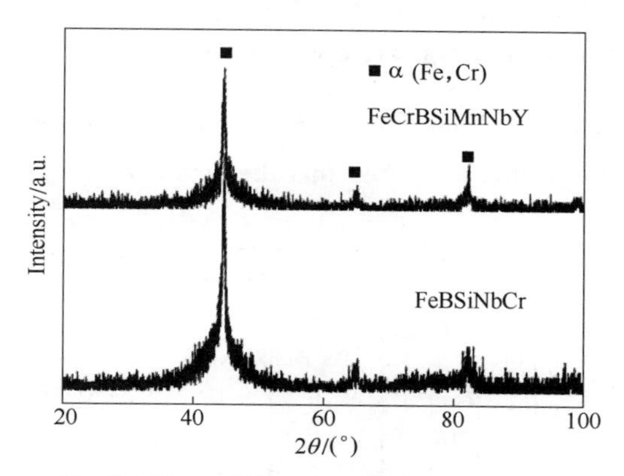

Fig. 3 X-ray diffraction patterns of coatings

The atomic radius and mix enthalpy play important roles in glass formation[18]. The different atomic size in the order of Y > Nb > Cr > Mn > Fe > Si > B in the two alloy systems facilitate the packed local structure. The mixing enthalpies values for Fe-B, Fe-Si, Fe-Cr and Fe-Nb atomic pairs are −11, −18, −1 and −16 kJ/mol respectively. And the mixing enthalpy values of the Cr-B, Cr-Nb, Nb-Si, B-Nb, Y-Mn, Y-Si and Y-Fe atomic pair are −16, −7, −31, −39, −8, −57 and −11 kJ/mol respectively[19]. Large negative mix heat among the constituent elements strengthens the interaction among the components and promotes the chemical short range ordering in the liquids[20]. The formation of the amorphous phase was attributed to the high cooling rates of moulted droplets and the proper materials composition.

However, the amorphous phase is a metastable phase, which would be translated to a stable phase in suitable conditions. As mention above, the cooling rate of individual splat is so high that allows the formation of large number of small nuclei and fine nanoprecipitates. The higher nucleation rate and fine nanostructures are highly dependent on the chemical composition, especially in the presence of the glass forming elements[21]. During the devitrification process, the glass precursor, when heated to its crystallisation temp-erature, readily transforms into nanostructured structure[17]. This refinement is due to the uniform nucleation and extremely high nucleation frequency during crystallisation, resulting in little time for grain growth before impingement between neighbouring grains. By this route, it is possible to develop very stable nanostructures that resist coarsening at elevated temperatures during spaying[22].

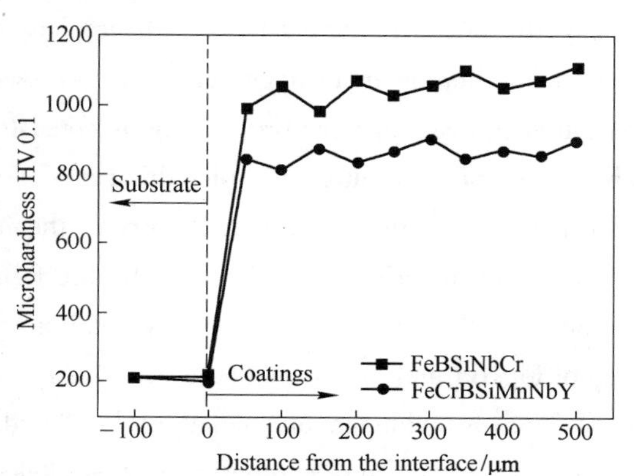

Fig. 4 Microhardness profiles of coatings

Fig. 4 shows the variation of microhardness across the interface from substrate to the coating. The Vickers hardness of the Fe-B-Si-Nb-

Cr coatings is higher than that of the Fe-Cr-B-Si-Mn-Nb-Y coating. The Fe-B-Si-Nb-Cr coating has a higher density structure and lower porosity, so its microhardness is relatively high.

3.2 Effect of impingement angle on erosion wastage

Fig. 5 depicts the effect of impact angle (30° and 90°) on erosion rate at room temperature for the substrate and the coatings. It can be seen that the coatings present a lower erosion rate than

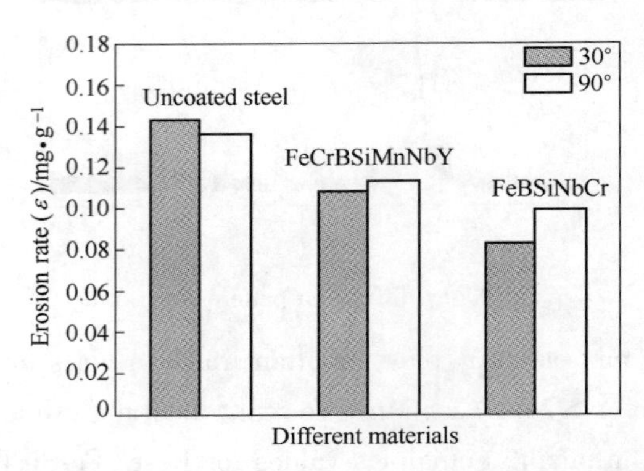

Fig. 5 Erosion rates of amorphous coatings and 45 steel as function of impingement angles at room temperature

that of the substrate. The erosion rate of the substrate tested at a 30° impact angle is higher than that tested at a 90° impact angle. In contrast, the composite coatings exhibit better erosion resistance at lower impact angle. And the Fe-B-Si-Nb-Cr coating has better erosion resistance than that of the Fe-Cr-B-Si-Mn-Nb-Y coating. The following formula is mostly used to express the erosion rate at different angle[23]:

$$\varepsilon = A\cos^2\beta\sin(n\beta) + B\sin^2\beta \quad (1)$$

where ε is the erosion rate; β is the impingement angle; n is a constant; A, B are also constants to describe the brittle and plastic behaviour respectively. For typical brittle material $A = 0$, for typical plastic material $B = 0$, and for the other material, the plastic item displays the main effects at low impingement angle while that is the brittle item at high impingement angle. That is to say, the brittle materials have high erosion rate at high impingement angle, but low erosion rate at low impingement angle, and it is reversed for plastic materials. The eroded results indicate that the coatings are brittle. This is consistent with the high hardness of the coating and its ability to resist the cutting or ploughing mechanism of impinging particles impacting at low angle. At higher angle, the kinetic energy of the impinging particles is transferred directly to the coating, and material removal occurs by the formation of cracks.

The eroded surfaces of the tested coating specimens and the substrate were presented in Fig. 6. Fig. 6a and b shows the eroded morphologies of the substrate at different impact angles. There is evidence of "ploughing" or "cutting" on the erosion surface (see arrows as in Fig. 6a and b). It can be seen that the eroded sample at 90° impact angle has a smoother surface morphology than that of 30°. This corresponds to its lower erosion loss.

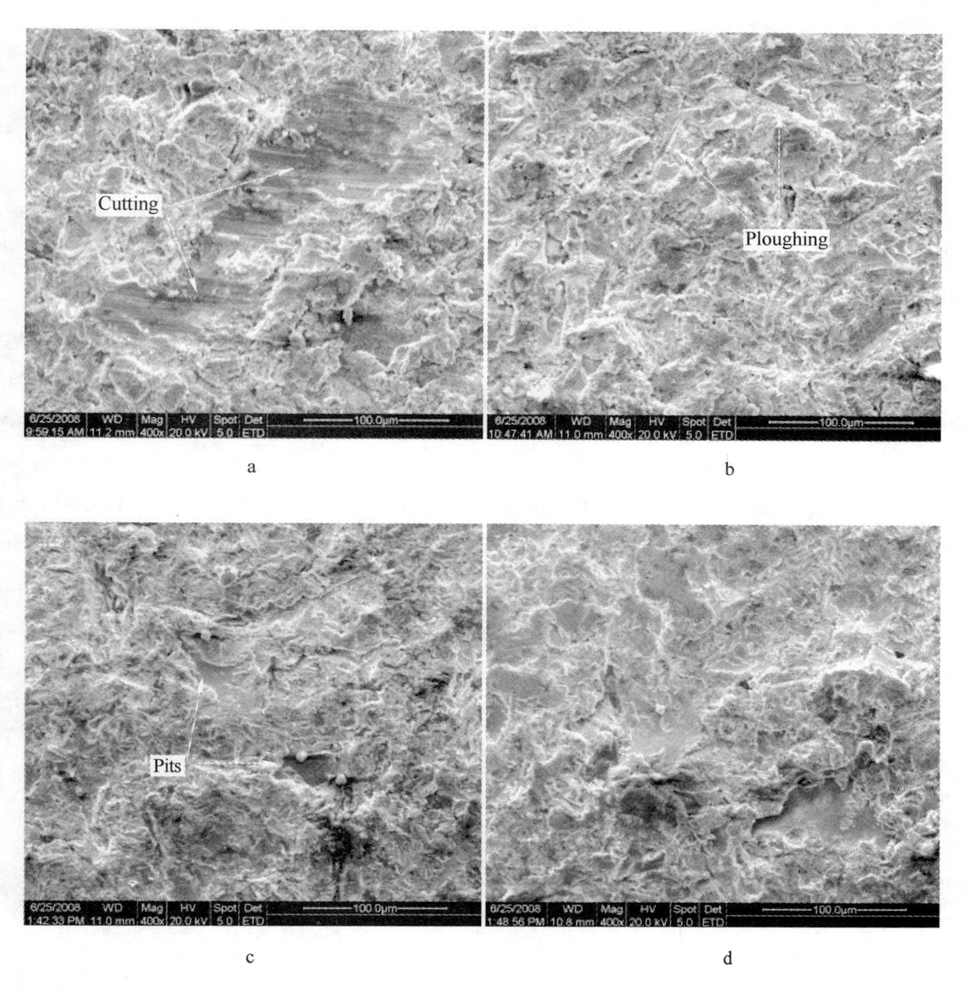

a b

c d

Fig. 6 Surface images(SEM) of substrate and coatings(Fe-B-Si-Nb-Cr) at different impingement angles at room temperature

a—Substrate, 30°; b—Substrate, 90°; c—Fe-B-Si-Nb-Cr coating, 30°; d—Fe-B-Si-Nb-Cr coating, 90°

Fig. 6c and d shows the eroded surface of the Fe-B-Si-Nb-Cr amorphous/nanocrystalline coating at 30°and 90° impact angles. There are typical craters or pits morphology in the eroded surface of the coatings(see arrows as in Fig. 6c). Compared with the coating eroded at 30° impact angle, a large amount of pieces chipped off and a large mass of removed material caused a coarse and rough surface morphology at 90° impact angle. And it accounts for their higher erosion wastage than that eroded at lower impact angle. During erosion, the deformation of coating takes place inside of coating splats initially, which forms small craters and microcracks. And it caused slight erosion wastage. As the test proceeded, propagation of the initial cracks occurs during subsequent attack by erodent particles[24]. And then fractured and loosened pieces were chipped off. Finally many small voids and pits formed[25]. This is also termed as cracking and chipping brittle mechanism.

3.3 Effect of temperature on erosion wastage

The results of the elevated temperature erosion tests conducted at different temperatures with 30° impact angle are given in Fig. 7. When the erosion temperature increases from 25 to 650℃, the erosion rates of the coatings decrease. It indicates that erosion resistance is improved as function of temperatures. The erosion rates of the coatings are superior to the substrate. And the Fe-B-Si-Nb-Cr coating has the highest erosion resistance. When the temperature is 650℃, the erosion rate of the Fe-B-Si-Nb-Cr coating is a negative value. This is attributed to the following factors. First, with increasing the temperatures, more and more nanocrystalline grains are precipitated in the coatings. Fig. 8 shows the XRD patterns of the Fe-B-Si-Nb-Cr coatings after erosion testing. The intensity of the XRD peaks was increased and the XRD curves are narrowed with increasing erosion temperature. This means that the grain size in coatings increases at different erosion temperatures. The sizes of nanoparticles calculated by Scherrer's formula are 11, 17, 32 nm as a function of temperature respectively. Those nanocrystalline grains act as dispersion strengthening to prevent the materials removal. Kim et al. [26] have pointed out that the nanocrystalline grains are too small to contain defects (such as dislocations, stacking faults) and ultra high strength would explain the greater resistance to deformation than the amorphous phase itself. Multiplication of interfaces would play an important role as it is the case with conventional precipitation hardened nanoscale particles. Those nanocrystals acted as dispersion strengthening in the amorphous matrix and could increase the mechanical strength such as fracture stress and could impede crack propagation[27,28]. Second, the other explanation for the behaviour is related to the formation of a thick oxide at elevated temperatures, which reduced the erosion rate of the material. When a thick oxide scale exists, erosion takes place from the scale only[29]. From Fig. 8, it can be seen that the rate of formation of the oxide scale is greater at 600℃ than at 450℃, yielding more oxide scales at the higher temperature.

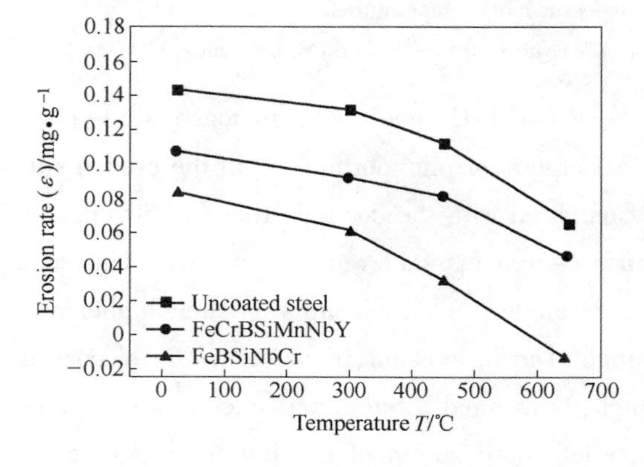

Fig. 7 Erosion rates of amorphous coatings and 45 steel as function of temperatures at 30° impingement angles

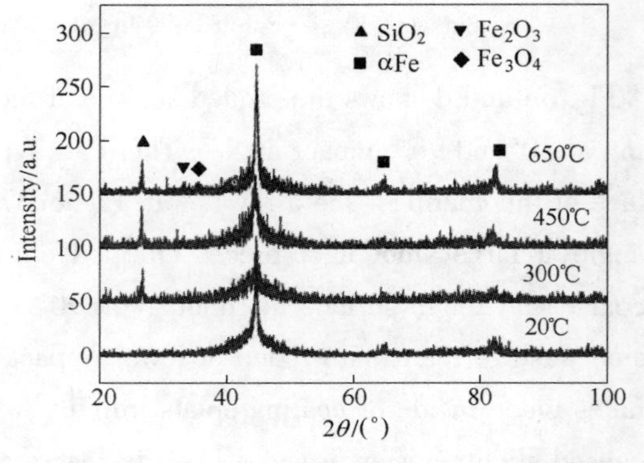

Fig. 8 X-ray diffraction patterns of Fe-B-Si-Nb-Cr coatings

When the hard particles impact the surface of the sample, the surface of eroded sample shows a typical flaky appearance because of the cracking and spalling of the oxide scale. The erosion process of the sample is a repeat process of formation-crack-spalling-formation of oxide scale[30,31]. Third, the reason for the Fe-B-Si-Nb-Cr coatings with excellent erosion resistance is mainly related to the change in the strength and ductility of material with the temperature increase. Generally, the strength decreases and the ductility increases for metal materials as the temperature is increased[29]. Thus, a significant amount of erodent particles are embedded in the surfaces of the coatings. From Fig. 8, the peaks of SiO_2 exist in XRD patterns as a function of erosion temperature. Such embedment would modify the surface of the coatings, reducing their erosion rates after an initial "running in". Therefore, the coatings provide significant erosion resistance and protection over a wide temperature range.

Fig. 9 shows the eroded surface of the Fe-B-Si-Nb-Cr coatings with different erosion temperatures. The craters morphologies exist in the eroded surface of the coatings. With increasing the erosion temperature, the craters are becoming smaller and smaller and the eroded morphologies

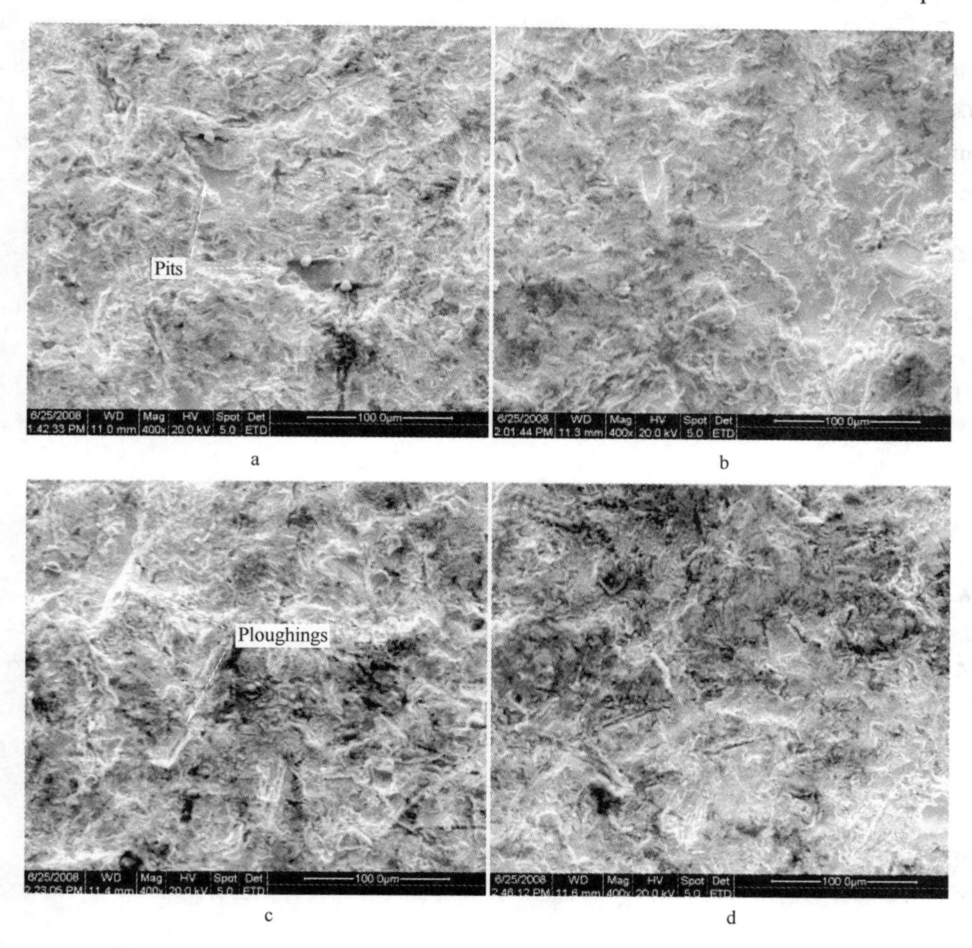

Fig. 9 Images (SEM) of Fe-B-Si-Nb-Cr coatings as function of temperatures at 30° impingement angles
a—25℃; b—300℃; c—450℃; d—650℃

are smoother and smoother. It indicates that the erosion resistance is improved at elevated temperature. When the erosion temperature reaches 450℃, ploughing morphologies appear in the eroded surface (Fig. 9c). It has caused the transition of erosion mechanism from a purely "brittle" to a relatively "ductile" behaviour[32]. This transition is mainly related to the change in the strength and ductility of material with increasing the temperature[32]. And this phenomenon exists in the eroded surface at 650℃ (Fig. 9d). So the erosion mechanism of the coating at elevated temperature is the ductile mechanism and brittle fracture mechanism coexistence. But the predominant erosion mechanism should be brittle fracture mechanism.

4 Conclusions

Fe-Cr-B-Si-Mn-Nb-Y and Fe-B-Si-Nb-Cr amorphous/nanocrystalline coatings were fabricated by wire arc spray process. The microstructure of both coatings consist of amorphous phase matrix containing α-(Fe, Cr) nanocrystalline grains. The coatings are fully dense with low porosity. The porosity of the Fe-Cr-B-Si-Mn-Nb-Y amorphous coating is 1.7% and that of the Fe-B-Si-Nb-Cr amorphous coatings is 1.5%. Little oxides are detected in the both coatings. Both coatings exhibit a lower erosion rate at 30° impact angle. The erosion rates decrease as function of erosion temperatures. The main failure mechanism of the coatings is brittle breaking and fracture mechanism. The Fe-B-Si-Nb-Cr amorphous/nanocrystalline coating has the best elevated erosion resistance.

Acknowledgements

The authors are grateful for the funds support provided by Key Natural Science Foundation of China (grant No. 50735006), National Key Laboratory for Remanufacturing (grant No. 914OC850205080C85) and Key laboratory for Advanced Materials Processing Technology, Ministry of Education, China (200802).

References

[1] C. A. Schuh, T. C. Hufnagel and U. Ramamurty: *Acta Mater.*, 2007, 55, 4067-4109.

[2] A. Inoue: Bulk amorphous alloys, 1; 1998, Uetikon/Zuerich, Trans Tech Publications.

[3] J. Jayaraj, D. J. Sordelet, D. H. Kim, Y. C. Kim and E. Fleury: *Corros. Sci.*, 2006, 48, 950-964.

[4] Z. L. Long, Y. Shao, X. H. Deng, Z. C. Zhang, Y. Jiang, P. Zhang, B. L. Shen and A. Inoue: *Intermetallics*, 2007, 15, 1453-1458.

[5] T. Gloriant: *J. Non-Crsyst. Solids*, 2003, 316, 96-103.

[6] B. H. Kear and G. Skandan: *Nanostruct. Mater.*, 1997, 8, (6), 765-769.

[7] S. J. Pang, T. Zhang, K. Asami and A. Inoue: *Corros. Sci.*, 2002, 44, 1847-1856.

[8] M. E. Mchenty, M. A. Willard and D. E. Laughlin: *Prog. Mater Sci.*, 1999, 44, 291-433.

[9] E. Fleury, S. M. Lee, H. S. Ahn, W. T. Kim and D. H. Kim: *Mater. Sci. Eng. A*, 2004, A375-377, 276-279.

[10] Z. P. Lu, C. T. Liu, J. R. Thompson and W. D. Porter: *Phys. Rev. Lett.*, 2004, 92, 245503.

[11] S. J. Pang, T. Zhang, K. Asami and A. Inoue: *Acta Mater.*, 2002, 50, 489-497.

[12] A. Kobayashi, S. Yano, H. Kimura and A. Inoue: *Surf. Coat. Technol.*, 2008, 202, 2513-2518.

[13] A. Kobayashi, S. Yano, H. Kimura and A. Inoue: *Mater. Sci. Eng. B*, 2008, B148, 110 - 113.
[14] H. S. Ni, X. H. Liu, X. C. Chang, W. L. Hou, W. Liu and J. Q. Wang: *J. Alloys Compd*, 2009, 467, 163 - 167.
[15] P. Georgieva, R. Thorpe, A. Yanski and S. Seal: *Adv. Mater. Process.*, 2006, 8, 68 - 69.
[16] A. P. Newbery, P. S. Grant and R. A. Neiser: *Surf. Coat. Technol.*, 2005, 195, 91 - 101.
[17] D. J. Branagan, M. Breitsameter, B. E. Meacham and V. Belashchenko: *J. Therm. Spray Technol.*, 2005, 14, (2), 196 - 204.
[18] A. Inoue, T. Zhang and T. Masumoto: *Mater. Trans. JIM*, 1989, 30, 965 - 972.
[19] F. R. de Boer, R. Boom, W. C. M. Mattens, A. R. Miedema and A. K. Niessen: Cohesion in metals: transition metal alloys, 217 - 399; 1989, Amsterdam, Elsevier Science Publishing Company, Inc.
[20] V. Ponnambalam, S. J. Poon and G. J. Shiflet: *J. Mater. Res.*, 2004, 19, 1320 - 1323.
[21] P. Georgieva, R. Thorpe, A. Yanski and S. Seal: *Adv. Mater. Process.*, 2006, 8, 68 - 69.
[22] D. J. Branagan, M. J. Kramer and R. W. McCallum: *J. Alloys Compd*, 1996, 244, 27 - 39.
[23] G. Jin, B. S. Xu, H. D. Wang, L. Yin, Q. F. Li, S. C. Wei and X. F. Cui: *Appl. Surf. Sci.*, 2008, 254, 5470 - 5474.
[24] P. J. Hoop and C. Allen: *Wear*, 1999, 233 - 235, 334 - 341.
[25] B. Q. Wang and K. Luer: *Wear*, 1994, 174, 177 - 185.
[26] Y. H. Kim, K. Higara, A. Inoue and T. Masumoto: *Mater. Trans. JIM*, 1994, 35, 293 - 302.
[27] A. Leonhard, L. Q. Xing, M. Heilmaier, A. Gebert, J. Eckert and L. Schultz: *Nanostruct. Mater.*, 1998, 10, 805 - 817.
[28] L. Q. Xing, J. Eckert and L. Schultz: *Nanostruct. Mater.*, 1999, 12, 503 - 506.
[29] E. H. Saarivirta, F. H. Stott, V. Rohr and M. Schütze: *Wear*, 2006, 261, 746 - 759.
[30] X. Q. Yu, M. Fan and Y. S. Sun: *Wear*, 2002, 253, 604 - 609.
[31] M. Roy, K. K. Ray and G. Sundararajan: *Wear*, 1998, 217, 312 - 320.
[32] Y. Wang, Y. Yang and M. F. Yan: *Wear*, 2007, 263, 371 - 378.

Finite Element Modeling of Coating Formation and Transient Heat Transfer in the Electric Arc Spray Process*

Abstract The electric arc sprayed coating can be described as a superposition of Gaussian profile particles whose overlapping depends on the movement of spray gun. The heat transfer behavior during the deposition has a significant influence on the performance of the process. In this paper, simulation of the coating formation and analysis of the transient heat transfer were performed based on a newly developed finite element model, in which the dynamic stochastic multiple particles deposition characteristic of the process was taken into account. In order to investigate the effects of the kinematics and dimensional aspects on the coating/substrate temperature distribution, a traditional layer - by - layer finite element model without consideration of gun movement and particles Gaussian profile was also performed as a comparison. The stochastic deposition model provided a more objective result of the transient heat transfer of the coating/substrate than that of the layer - by - layer model, especially the severely inhomogeneous temperature distribution characteristics in different locations and spraying conditions. Finally, the molding results were experimentally compared with the temperature measurements on the coating surface and substrate back face using an infrared thermal imaging video camera, which shows that most of the modeling findings are consistent with that of the experiment.

Key words thermal spraying, finite element analysis, arc sprayed coating, transient heat transfer

1 Introduction

The twin - wire electric arc thermal spray process has been widely used as a cost - effective method in industry, especially in the fields such as components corrosion protection, wear resistance, and tools spray forming[1]. In the spraying process, two metal wires are connected the anode and cathode of a direct current power supply, and guided to the core of a high velocity gas jet where an electric arc is formed. Subsequently, the feedstock wires are melted, atomized and propelled out of the jet to impinge upon the substrate, resulting in solidification. Heat of the newly deposited particles is transferred to the ambient and lower splat layers or substrate by means of convection, radiation and conduction. In addition, phase change and variation of latent heat may take place during the solidification. The thermal transfer characteristic of the spraying process is therefore very complicated and highly transient with large amplitude excursions. Numerous studies have shown that, the temperature of coating/substrate during the deposition has a significant influence on the properties of the coating, such as residual stress, microstructure, oxidation, and adhesion/cohesion strength[2-6].

---

* Copartner: Chen Yongxiong, Liang Xiubing, Liu Yan, Bai Jinyuan. Reprinted from *International Journal of Heat and Mass Transfer*, 2010, 53(9 - 10): 2012 - 2021.

Thermal analysis and measurements of the electric arc sprayed coatings and other thermal spraying processes by using the instruments such as thermocouple, pyrometer and infrared thermal imager were much concerned in pervious literatures[7-11]. However, precisely measuring the transient temperature variation of the coating/substrate is a real challenge, because the hot particles are added on the substrate or previously formed coating surface dynamically and randomly with a simultaneous movement of the spray gun[12-14]. Numerical simulation provides another way to understand the details of temperature for the thermal sprayed coatings[15]. For example, Finite element and finite difference simulations of temperature field during the thermal spraying processes[16,17], where coating buildup was simplified as a layerby – layer deposition model and the layer thickness was derived from the cross – section profile of an actual coating. Literatures[18,19] also reported a layer – by – layer finite element model (FEM) used to simulate the transient temperature distribution of plasma thermal sprayed coatings, and calculate the associated residual stresses of the coatings. However, most of them mainly concerned the high velocity oxygen flame (HVOF) or plasma spray process, instead of electric arc spray process.

Generally, the thermal sprayed coating can be described as a superposition of Gaussian profile splats whose overlapping depends on the process kinematics[20,21]. These kinematics and dimensional aspects ought to be considered when modeling the process[22,23], because the thermal histories of the individual particles may relate with the coating stress status, deformation and phase transformation. Zhu et al. [24,25] developed a finite difference model (FDM) to analyze the temperature/time history of the electric arc sprayed particles deposition process, where the individual particles adding process was modeled by assuming a Gaussian distribution of particles around the axis of the spraying jet and the movement of the spray gun was taken into account. This study may represent the latest development of the applications of numerical modeling to the electric arc spray process.

Nomenclature

C	specific heat (J/(kg · K))
f	fraction solid or liquid of the deposit
h	heat transfer coefficient (W/(m^2 · K))
ΔH_f	latent heat of fusion (J/kg)
i, j	mesh integers
K	thermal conductivity (W/(m · K))
P	temperature load array (K)
$\dot{q}$	heat flux (W/m^2)
r	radial distance (m)
R	outer radius of the specimen (m)
T	temperature (K)
t	time (s)
x, y	displacement in the x, y direction (m)

Greek symbols

Δ	finite change
δ	Stefan – Boltzmann constant (J/(m^2 · s · K))
ε	emissivity
θ	difference time discrete factor
ρ	density (kg/m^3)

Subscripts

a	ambient
dep	deposit
e	element
eq	equal
H	convection
inter	interface between deposit and substrate
l	liquid
n	step number
q	heat flow
Q	internal heat flow
rad	radiation
s	solid
sub	substrate

This paper is mainly aimed to develop a stochastic multiple particle deposition model (in the following text, as called "stochastic deposition model") to simulate the transient heat transfer and coating formation of electric arc spray process using the finite element method. The finite element method has many advantages in the applications such as thermal and theomechanical coupled analyses[26,27]. In order to investigate the effects of the kinematics and dimensional aspects on the coating/substrate temperature distribution, a traditional layer – by – layer finite element model without consideration of gun movement and particles Gaussian profile was also developed. Comparisons between the stochastic deposition model and the layer – by – layer model were subsequently carried out under different electric arc spraying conditions, such as different gun traverse speeds and interruption spraying modes. Furthermore, an infrared thermal imaging video camera was used to monitor the sample surface temperature during the spraying, and the temperature histories at some specified positions were recorded and compared with the modeling results.

2 Finite Element Model

2.1 Governing equation and boundary conditions

During the electric arc spraying, as schematically shown in Fig. 1, metal droplets with high initial temperature are flattened and deposited on the substrate surface to form coating, the heat of the deposit is therefore transferred to the substrate and surrounding gas. The governing heat transfer equation for the coating structure in two dimensional Cartesian coordinates (x, y) can be written as

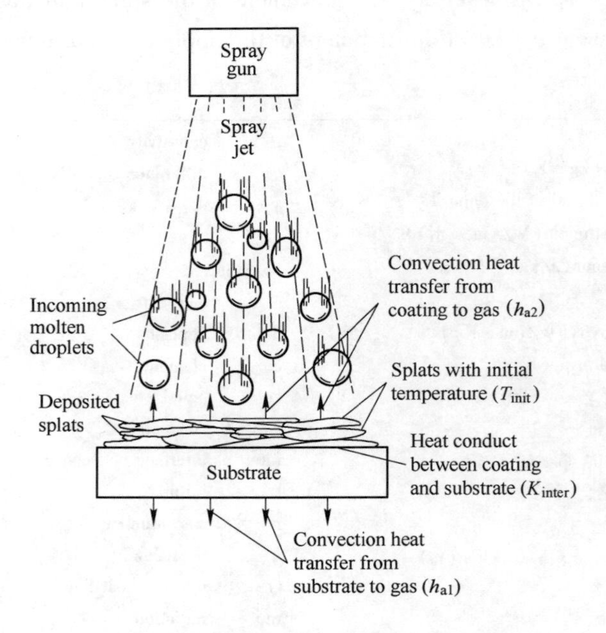

Fig. 1 Schematic physical and thermomechanical description of the electric arc spray process

$$\rho C \frac{\partial T}{\partial t} = K\left(\frac{\partial^2 T}{\partial x^2} + \frac{\partial^2 T}{\partial y^2}\right) + \rho \Delta H_f \frac{\partial f_s}{\partial t} \tag{1}$$

where T is the material temperature; t is time; x and y are the displacements in x and y directions, respectively; ρ is the material density; C is the specific heat; K is the heat conductivity; ΔH_f is the latent heat, and f_s is the solid fraction.

Eq. (1) can be transformed to

$$\rho\left(C - \Delta H_f \frac{\partial f_s}{\partial t}\right)\frac{\partial T}{\partial t} = K\left(\frac{\partial^2 T}{\partial x^2} + \frac{\partial^2 T}{\partial y^2}\right) \tag{2}$$

Defining a term "equivalent specific heat" C_{eq} as

$$C_{eq} = C - \Delta H_f \frac{\partial f_s}{\partial t} \tag{3}$$

A standard heat transfer equation without the last term of latent heat releasing in Eq. (1) is therefore obtained

$$\rho C_{eq} \frac{\partial T}{\partial t} = K\left(\frac{\partial^2 T}{\partial x^2} + \frac{\partial^2 T}{\partial y^2}\right) \tag{4}$$

For the substrate, since the temperature variation is very low, the latent heat change can be ignored. Therefore, the equivalent specific heat C_{eq} in Eq. (4) is equal to the specific heat of the substrate material C_{sub}.

For the coating material, the variation of latent heat due to phase change should be taken into account. Assume that the correlation between the solid fraction and temperature is linear

$$T = T_l - (T_l - T_s) f_s \tag{5}$$

$$\frac{\partial f_s}{\partial T} = \frac{-1}{T_l - T_s} \tag{6}$$

where T_l and T_s are the liquidus and solidus of the deposited material, respectively. The equivalent specific heat C_{eq} in Eq. (4) is equal to

$$C_{eq} = C_{dep} - \Delta H_f \frac{-1}{T_l - T_s} \tag{7}$$

Therefore, the variation of latent heat corresponding to phase change of the coating material was considered in the form of specific heat.

The boundary conditions to specify the above governing equations can be expressed as follows:

(1) At the free surface of the deposit and substrate, the heat flux due to gas convective can be written as

$$\dot{q}_a = h_a (T - T_a) \tag{8}$$

where T_a is the ambient temperature; h_a is the convective heat transfer coefficient. The heat flux due to radiation is considered in the form of equivalent convection heat transfer

$$\dot{q}_{rad} = \varepsilon\delta(T^4 - T_a^4) = h_{rad}(T - T_a) \tag{9}$$

$$h_{rad} = \varepsilon\delta(T^2 + T_a^2)(T + T_a) \tag{10}$$

where ε is the emissivity; δ is the Stefan – Boltzmann constant (6.78×10^{-8} J/(m^2 · s · K));

h_{rad} is the equivalent convection coefficient. It should be noticed that, as marked with arrows in Fig. 1, the value of the convection heat transfer coefficient for the deposit and substrate is usually different.

(2) At the deposit – substrate interface, the heat flux between the deposit and substrate can be described as

$$\dot{q}_{inter} = h_{inter}(T - T_{sub}) \tag{11}$$

where h_{inter} is the heat transfer coefficient between the deposit and the substrate; T_{sub} is the surface temperature of the substrate. If the deposit and the substrate are assumed to bond completely, conduction becomes the dominant heat transfer mechanism. The rate of heat transfer at the deposit – substrate interface can be expressed as follows

$$\dot{q}_{inter} = K_{inter} \left.\frac{\partial T}{\partial r}\right|_{r=R} \tag{12}$$

where K_{inter} is the equivalent heat conductivity of the interface, which can be given by the formula

$$K_{inter} = 2K_{dep} \cdot K_{sub}/(K_{dep} + K_{sub}) \tag{13}$$

K_{dep} and K_{sub} are the heat conductivities of the depositing material and the substrate, respectively; R is the radius of the substrate.

2.2 Finite element algorithm

Finite – element form of the heat conduction equation (Eq. (4)) can be expressed as

$$C\frac{dT}{dt} + KT = P \tag{14}$$

where C is the heat capacity matrix: $C_{ij} = \sum_{re} C^e_{ij}$; K is the thermal conduction matrix: $K_{ij} = \sum_e K^e_{ij} + \sum_e H^e_{ij}$ (H_{ij} is the correction factor matrix); P is the temperature load array: $P_i = \sum_e P^e_{Q_i} + \sum_e P^e_{q_i} + \sum_e P^e_{H_i}$; T is the node temperature array, and dT/dt is the derivative array of temperature to time. Assume that the heat capacity matrix C and thermal conduction matrix K are independent of time, and that the same interpolation function is used for P and T, Eq. (14) can be transformed to

$$(C/\Delta t + K\theta)T_{n+1} - [C/\Delta t + K(1-\theta)T_n] = P_{n+1}\theta + P_n(1-\theta) \tag{15}$$

where θ is the difference time discrete factor. Eq. (13) is then the finite element iteration function calculating the temperature of the coating during deposition. When an appropriate initial time step Δt_1 and subsequent time step Δt_n are used, the temperature in this step will be gained by iterative solution, and results are used as the initial temperature for the next calculation step.

2.3 Description of the layer – by – layer numerical model

Watanable et al.[28] noted that, for the deformation and solidification of a single spraying droplet, solidification occurred immediately after the deformation of the droplet; the solidification time is about two orders of magnitude higher than the deformation time. Therefore, it is reasonable to some degree, to simplify the deposition process of electric arc spraying as that one layer of hot material with certain thickness is instantaneously added to the solidified surface of the substrate or previously deposited layers. The heat flux of the layer is transferred to the lower layers, substrate and ambient air, and there is no mass added until the coming of the next layer.

To allow the problem to be reduced to a two - dimensional axisymmetric case, a columnar specimen substrate was utilized in the present study, as shown in Fig. 2. During the computation, the continuous spraying process must be broken into separate time steps for the study, so discretization is very necessary. Fig. 3 schematically shows the deposition process (step by step) based on the layer - by - layer model. The terminology "period" is used to define the time between a new coming layer and previous layer. Period 1 (Fig. 3b), therefore, corresponds to the time from the impact of the first layer till the arrival of the second layer. The setting of the period is based on the spraying pass, where the spray gun moves from the left to right side of the specimen, with simultaneous rotation of the specimen. Fig. 4 schematically shows the finite element mesh of the above layer - by - layer coating formation model, where a refined mesh is performed in the coating domain.

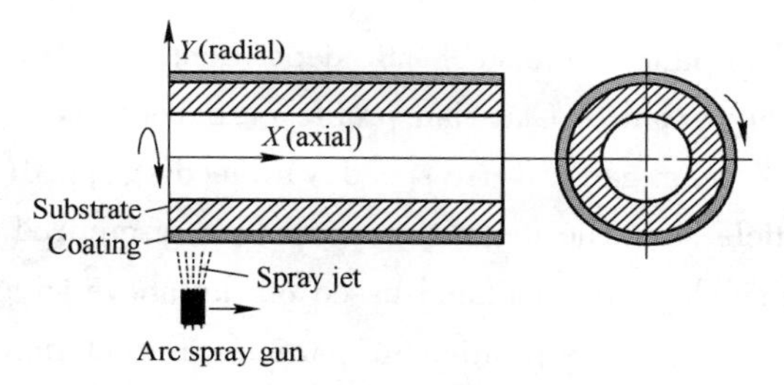

Fig. 2 Schematic of the electric arc sprayed columnar specimen

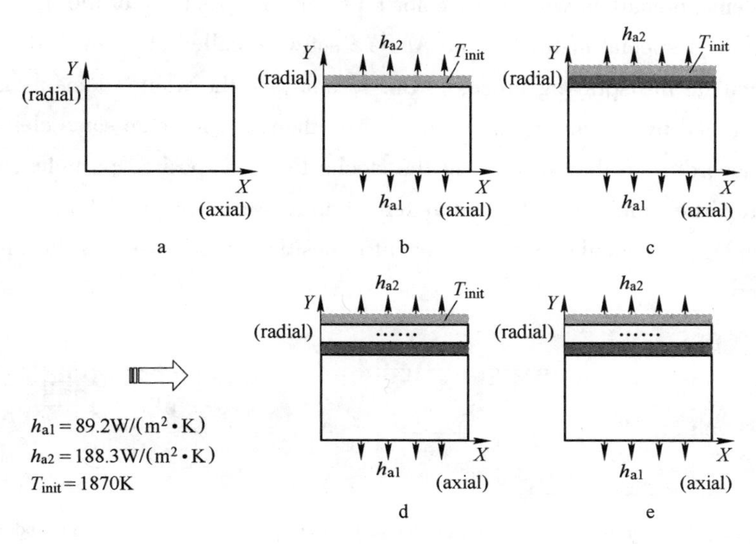

Fig. 3 Schematic description of the layer - by - layer deposition model

a—Zero layer (substrate); b—One layer (period 1); c—Two layers (period 2); d—N layers (period N); e—Cooling down

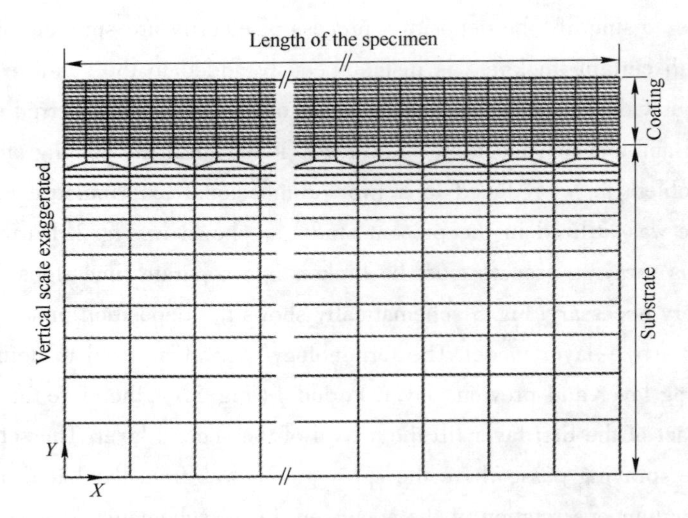

Fig. 4 Mesh of the layer – by – layer model for finite element analysis

2.4 Description of the stochastic deposition numerical model

This section gives the description of the stochastic deposition model, in which assumed that the sprayed coating is a superposition of Gaussian profile particles whose overlapping depends on the spraying parameters[20], e. g. gun traverse speed, wire feeding speed (corresponds to material deposition rate) and particle size. The deposition process is discretized step by step with total number of particles in a single step calculated based on the above spraying parameters. At the beginning of each deposition step, the positions of particles are determined by a Gauss random number generator in the program. Fig. 5 shows the schematic finite element mesh for this model, where the elements marked with black color represent the positions of the deposited particles in this single step. A special method of the ANSYS software called element "birth" and "death" was used to simulate the spraying process. Prior to analysis, the whole elements in the domain that may be occupied by the coating were killed. And then, in each step some elements were activated (birth) according to the positions of the stochastically deposited particles, and the particle material properties including the initial temperature were specified for the activated elements. The temperature calculated in the current deposition step is used as the initial condition for the next step.

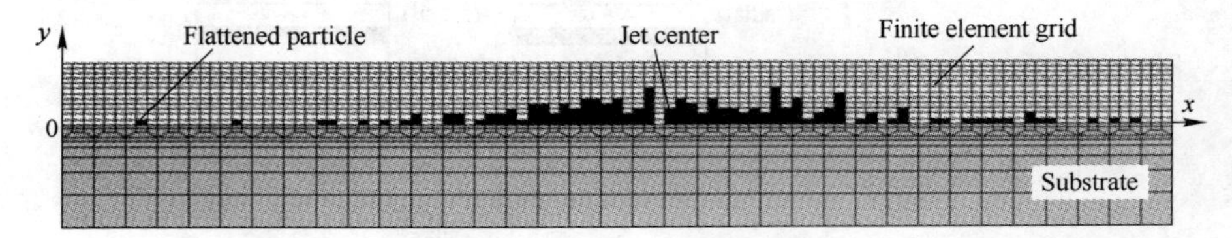

Fig. 5 Schematic of the mesh and stochastic multiple particles deposition model

This model can be used to investigate the effect of various spraying conditions on the temperature field. For example, gun traverse speed, pre – and in situ cooling or heating, specimen dimension, material deposition rate, material properties, jet pass mode, jet divergence, particle size, coating thickness, and droplet temperature, etc.

2.5 Case studied

To perform the comparison between the layer – by – layer model and the stochastic deposition model, the material properties, specimen dimension, element size, boundary conditions, and spraying parameters are prescribed the same. Table 1 shows the material properties of the coating and substrate for the two models. The outer diameter, thickness, length of the substrate was specified, 40 mm, 5 mm and 120 mm, respectively. The initial temperature of the substrate for both models is 298 K. The thickness per layer activated in every step for the layer – by – layer model is calculated based on the stochastic model. In other words, under the same spraying conditions, the total volume of the deposited particles of the stochastic model in every circulation that the gun moves from the left beginning to the right end of the specimen is equal to that of the layer – by – layer model.

Table 1 Material properties

Materials	Temperature/℃	Enthalpy/J · m^{-3}	Thermal conductivity /W · (m · K) $^{-1}$	Density/kg · m^{-3}	Specific heat /J · (kg · K) $^{-1}$
Coating/steel	0	0	29.91	7833	—
	1468	82.476×10^8	31.98	7833	—
	1528	105.461×10^8	25.34	7833	—
	1597	112.156×10^8	25.34	7833	—
Substrate/copper	—	—	106	8960	377

Experimentally, the parameters such as temperature and size of individual particles may vary within particular ranges. But the layer – by – layer model cannot simulate the temperature variation for every layer composed of individual particles, so, a unified initial temperature of 1870 K was specified to all deposited particles for both models. And in the stochastic deposition model, the flattened particle is considered as a thin disk of the same volume with that of the original particle and all the particles are assumed being equal in size (160 μm of diameter and 6 μm of thickness[25]). Another assumption is that, the coating formed by the electric arc spray process is considered of isotropy, and porosity and oxide in the coating are not taken into account.

Eight cases were calculated in the finite element analysis, where "No. " denotes "numerical analysis" (shown in Table 2). No. 1 – No. 3, and No. 5 – No. 7 give the analysis with varying gun traverse speeds for the layer – by – layer model and the stochastic deposition model, respectively. Generally, a unidirectional spraying method is believed to decrease the heat variation to the coating/substrate, and then improve the spray quality. A short pause is therefore needed in the simulations after the spray gun moved form the left beginning to the right end of the specimen, namely, there is no material added on the specimen during the gun returning to the left beginning point. The above cases (No. 1 – No. 3, and No. 5 – No. 7) considered this "standard" spraying mode and specified the pause time with 0. 4 s. Another spraying mode of 3 s spraying and cooling periods was also modeled with No. 4 and No. 8.

Table 2 Model descriptions and spraying parameters

Model No.	Constitutive model	Gun transverse move speed/mm · s^{-1}
1①	Layer – by – layer deposition	100
2①		200
3①		400
4②		200
5①	Stochastic multiple particles deposition	100
6①		200
7①		400
8②		200

①Spray with the standard mode, every period includes the time for the gun moving from the left to right ends of the specimen, and the time for pause (i. e. time for the gun returning to left beginning point, no material added on the specimen, specified 0. 4 s for all models).

②Spray with periodical interruptions, except for the pause described above, another interruption is that after every three spray periods of mode a(3 s), a pause of 3 s implemented for cooling.

3 Finite Element Analysis Results and Discussion

3.1 Comparison between the layer – by – layer model and the stochastic deposition model

For all cases described in Section 2. 5, the coating/substrate temperature contours in every calculating step were picked out and compiled with form of animation by using the software post – processing module. Remarkable differences between the two models are shown by the contours. Fig. 6a and b give the temperature contours of No. 2 for the layer – by – layer model at 34. 5 s, and No. 6 for the stochastic deposition model at 34. 55 s, respectively. It is obvious that, for the layer – by – layer model presented in Fig. 6a, temperature in the x (axial) direction has no variation, while in the y (radius) direction it presents a "layered" typical distribution. However, for the stochastic deposition model shown in Fig. 6b, the temperature gradient at this transient time is very great both in x and y directions, especially in the middle zone underneath the coating surface. In other words, Fig. 6b indicates that the spray gun has reached the middle zone of the specimen, new particles are deposited on the coating surface with high initial temperature, and then the heat instantaneously transfers to the surrounding regions. Furthermore, a rough coating surface profile was gained by the stochastic deposition model, which is consistent with the experimentally obtained coating. The average coating thickness of No. 6 at 34. 55 s is about 0. 47 mm. Fig. 6b additionally shows a small detail that the thickness at the left and right end of the simulated coating is a little less than that in the middle. The reason is that, in every spraying circulation, once the jet center reaches the end of the specimen, the gun will reverse direction, resulting in that the jet passed the specimen end for only half time of that passed the middle.

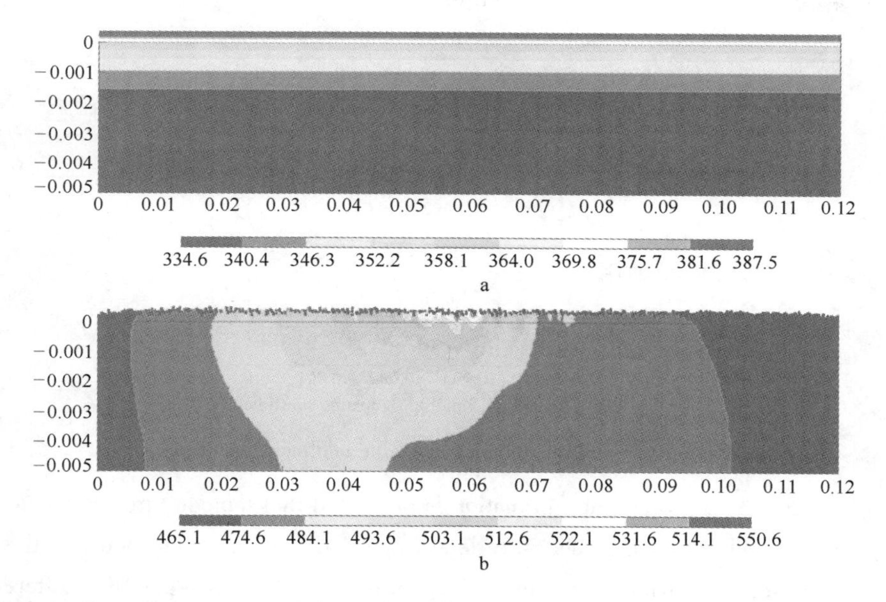

Fig. 6 Temperature contour of the specimen cross – section simulated with different models
a—No. 2 (layer – by – layer model) at 34. 5 s; b—No. 6 (stochastic deposition model) at 34. 55 s
(the thickness dimension was magnified for better view of the coating surface profile)

Fig. 7 shows the temperature distributions along the coating/substrate interface (in x direction) at different time for No. 2 and No. 6. For the layer – by – layer model (No. 2), the temperature values do not vary along the interface, only increase with the spray time, for example, 242. 3℃ at 14. 03 s and 373. 8℃ at 29. 03 s. But, for the stochastic deposition model, the temperature distribution along the interface shows a noticeable variation, the difference between the maximum and minimum temperature is 64℃ at 14. 75 s and 55℃ at 30. 2 s. The two time points (14. 75 s and 30. 2 s) represent the transients that the jet center reached just the midline of the specimen. The inhomogeneous distribution of the temperature simulated by the stochastic model can well explain the frequently observed phenomena such as specimen debonding and distortion due to mismatch of thermal expansion between the coating and substrate. The temperature distributions along the midline of the specimen at the time points described above are shown in Fig. 8. It can be found that, for all the models and corresponding time points, the temperature along the coating/substrate thickness direction presents near – linear distribution, and the coating/substrate interface is the knee point of the temperature gradient. When spraying time is prolonged, the temperature increases slightly with the increase of

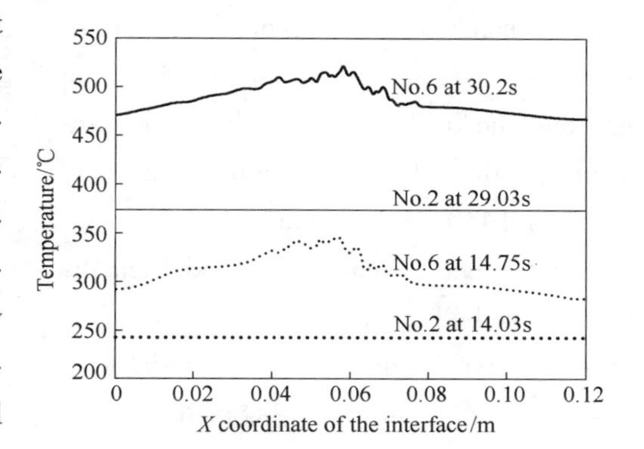

Fig. 7 Temperature distributions along the coating/substrate interface

coating thickness.

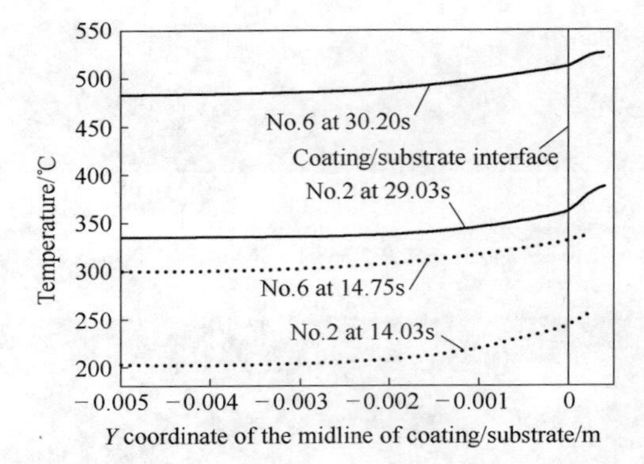

Fig. 8 Temperature distributions along the midline of the specimen

From Fig. 6 – Fig. 8, an important information is presented that temperatures calculated by the layer – by – layer model and the stochastic deposition model at the same point or the near – same deposition time are obviously different. To quantificationally investigate this difference, the temperature/time histories at different points (interface left end, interface midpoint, and substrate back surface midpoint) for No. 2 and No. 6 are shown in Fig. 9. Because the temperature at the interface midpoint and the left end of No. 6 is equal in every deposition time, there is only one curve shown in Fig. 9 for the two points. When deposition time increased, the temperature difference at the interface midpoint between the layer – by – layer model (No. 2) and the stochastic deposition model (No. 6) increased correspondingly. The same regularity is presented at the interface left end and the back surface midpoint. For instance, at time of 30. 9 s, the temperature of No. 2 is 145℃ lower than that of No. 6. This difference is mainly resulted from the loading method for the two models. The layer – by – layer model loaded the initial temperature of the particles prior to their deposition, namely, for every circulation the gun passed the specimen, one layer of elements was activated, which indicates the particles corresponding to these elements were deposited on the surface of the substrate or previous deposited layers. In reality, these elements ought to be activated when the spray gun reaches the corresponding positions. Therefore, the layer – by – layer model assumed the spraying material cools down in advance, but this dynamical process has been respected in the stochastic deposition model. The result evaluated by using the stochastic deposition model is therefore more objective than that of the layer – by – layer model.

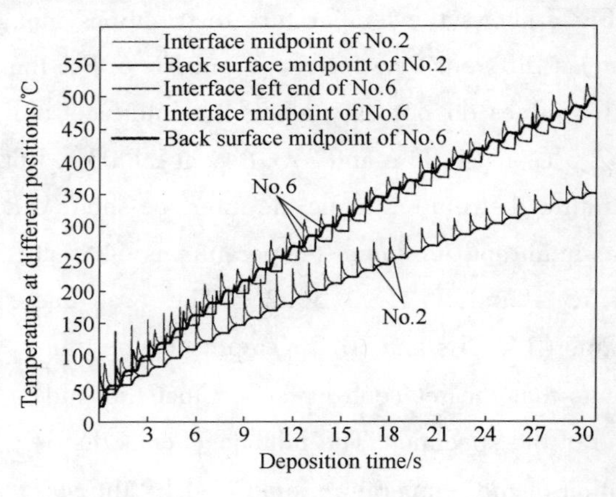

Fig. 9 Thermal histories at different positions of the specimen

3.2 Effect of gun traverse speed on the heat transfer

As described in section 3. 1, the temperature calculated by the layer - by - layer model is lower than that of the stochastic deposition model at the same time and location. But the difference value may change under different spraying conditions, for example, the gun traverse speed, interruption time during spraying, particles initial temperature, wire feeding speed (i. e. deposition rate), etc. This study only simulated the conditions of different gun traverse speeds and interruption spray modes nevertheless typical regularities have been obtained. Fig. 10a - c shows the temperature via time changes for both models under the gun traverse speed of 100 mm/s, 200 mm/s and 400 mm/s, respectively. It is obvious that the difference between the layer - by - layer model and stochastic deposition model decreased when gun traverse speed increased. This can be explained by that, when the traverse speed of the gun increased, the layer thickness per path (i. e. the circulation step that the gun moves from the left to right end of the specimen) decreased correspondingly, and then the "advance time" to load the initial temperature (as described in section 3. 1) became shorter. For instance, when the gun traverse speed for the layer - by - layer model is 100 mm/s (No. 1), the advance time for the particles added on the rightmost of the specimen is 1. 2 s, i. e. the value of gun speed divided by specimen length. When

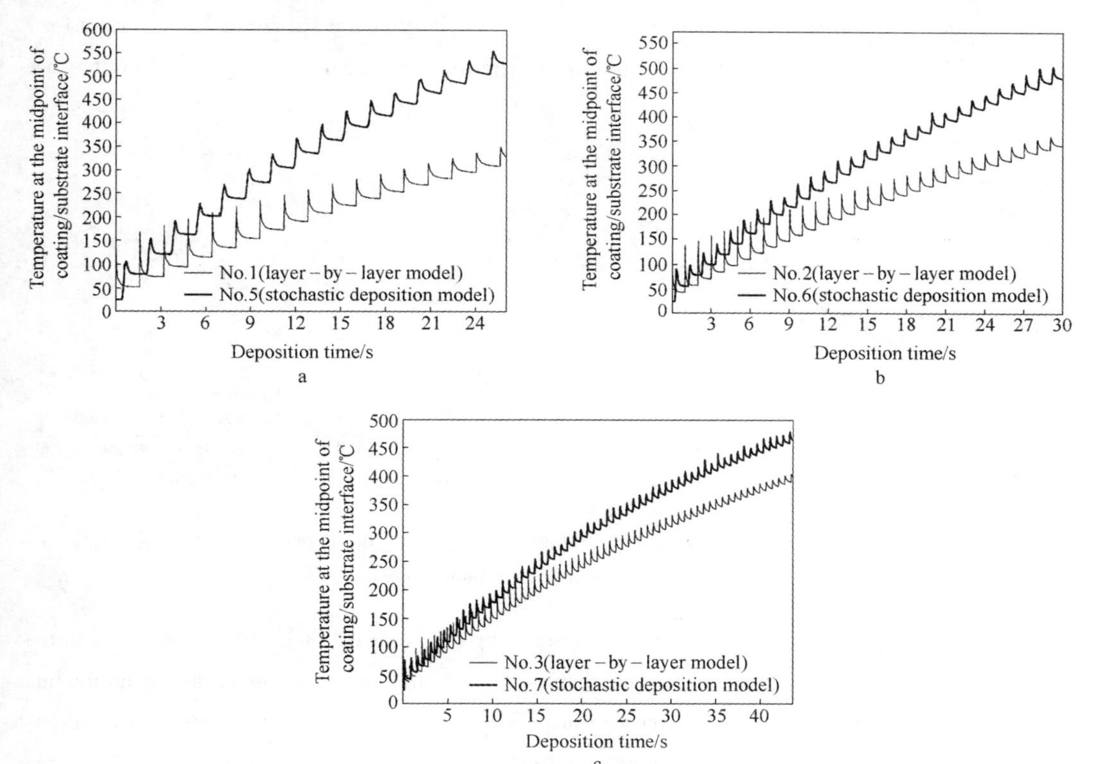

Fig. 10 Effect of the gun transverse move speed on the thermal history at the midpoint of coating/substrate interface

a—100 mm/s; b—200 mm/s; c—400 mm/s

the gun speed increases to 400 mm/s, the advance time will decrease to 0. 3 s, then the difference between the layer – by – layer model and stochastic deposition model weakens noticeably.

Another important discovery from the simulations is that higher traverse speed of the spray gun would result in lower thermal cycling gradient during the deposition. Both of the layer – by – layer model and the stochastic deposition model present a "zigzag" temperature curve due to the continuous circulation of heating and cooling, and the total temperature increases with spraying time prolonged. While the traverse speed of the spray gun increased, the difference value between the peak and valley of the "zigzag" decreased correspondingly. Therefore, the degree of circulation from heating to cooling and then heating was small down, which is useful to improve the coating quality, such as formation of relatively homogenous microstructure and decrease in coating stress during the deposition. That is the reason why a relative higher traverse speed of the gun was first respected in a real thermal spray experiment[29].

3. 3 Effect of interruption spraying mode on the heat transfer

Fig. 11a and b shows the temperature/time history (No. 4 and No. 8, both are interruption spraying modes) at the midpoint of coating/substrate interface and back surface of the substrate, respectively. The temperature difference still exists at the same time points between the layer – by – layer model and the stochastic deposition model, especially at the substrate back surface (refer to Fig. 11b). With the spraying time prolonged, the tendency of increasing the difference becomes clear.

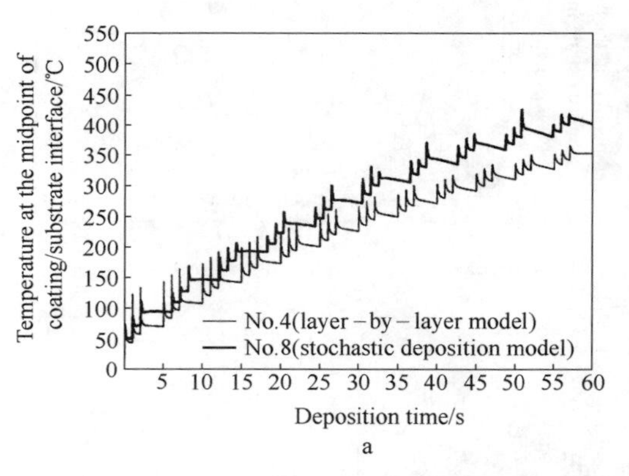

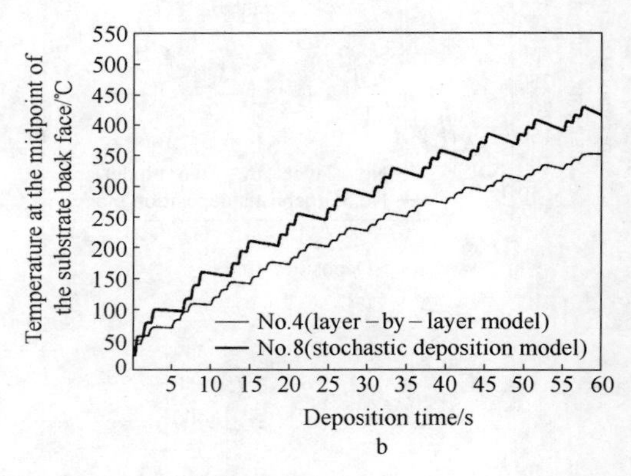

Fig. 11 Evaluation of temperature/time history of the periodical interruption spray mode at different positions

When comparing the periodical interruption spraying mode (Fig. 11) with the standard spraying mode (refer to Fig. 9 or Fig. 10b, correspondingly), the interruption spraying mode has the trend to reduce the overall temperature due to longer convection cooling time, however, interruption spraying mode results in greater temperature fluctuation. In fact, the standard spraying mode described above is another type of interruption spray, while the alternating interval is very short. In other words, a longer alternating interval results in a greater temperature fluctuation.

4 Experimental Verification

To verify the modeling results of the temperature distribution presented in the study, the surface temperature of the sample was monitored during the spraying using a ThermoVision™ A20 M infra – red (IR) thermal imaging video camera. The camera operating at 50 Hz at a wavelength of 8 – 14 μm was fixed on a tripod with a distance of about 1. 0 m to the sample, as shown in Fig. 12. 2. 0 mm diameter carbon steel wire, with composition of 0. 82 wt% C, 0. 6 wt% Mn, 0. 1 wt% Cr and balance Fe, was sprayed onto a rotating tubular copper substrate (120 mm long, 40 mm outer diameter and 5 mm wall thickness). The spraying parameters are given in Table 3. The gun traverse speed and spraying mode are consistent with the modeling settings listed in Table 2. Consider the usual temperature range for the IR camera of 0 – 250℃ is used, the actual deposition length, i. e. the product that multiplying the gun traverse speed with the actual deposition time when the spray jet is sprayed on the specimen, is relatively shorter than that of the modeling, but it is the same for all the measurements with different gun speeds in order to gain the similar coating thickness. Thermal images, of which a typical example is shown in Fig. 13, were recorded. ThermaCAM Researcher Pro 2. 7 software was used to analyze the thermal images and obtain temperatures at two positions: point 1 – near the left end of the tubular inner surface (i. e. the back face of the substrate) and point 2 – near the midpoint of the coating sample top surface, as shown in Fig. 13.

Fig. 12 Electric arc spraying process setup for tubular sample temperature measurement

Table 3 Parameter settings for the electric arc spraying of the carbon steel onto rotating tubular copper substrate

Parameter	Setting			
Spraying mode	Standard spraying		Periodical interruption	
Gun transverse speed	100 mm/s	200 mm/s	400 mm/s	200 mm/s
Arc voltage	34 V			
Arc current	160 A			

Table 3 (continued)

Parameter	Setting
Air pressure	0. 7 MPa
Spray distance①	200 mm
Substrate rotation	30r/min

①Distance along the spray axis from the wire tips to the top surface of the substrate.

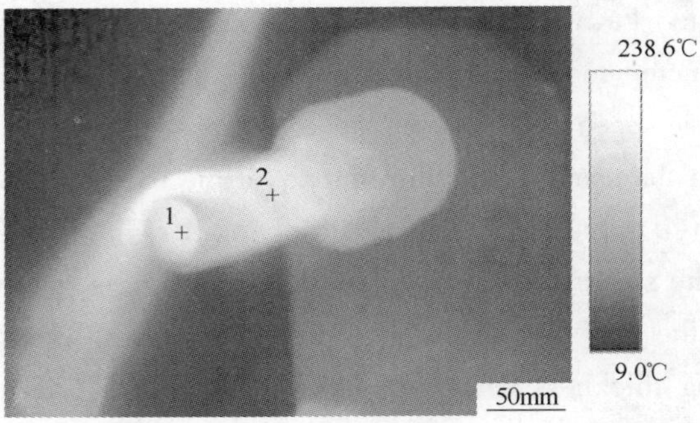

Fig. 13 Example of an IR thermal image, recorded using a standard spraying mode with 200 mm/s gun traverse speed at 8. 76 s, showing the two positions used for temperature measurement: point 1 – near the left end of the tubular inner surface and point 2 – near the midpoint of the coating sample top surface

Fig. 14a – c shows the temperature/time histories at point 1 and point 2 obtained with the gun traverse speed of 100 mm/s, 200 mm/s and 400 mm/s, respectively. Similar to the results obtained from numerical modeling, "zigzag" temperature – time curves are presented in all the measurements performed with different gun traverse speeds, especially for the substrate back face measuring point (point 1). For the coating surface measuring point (point 2), there are some great peaks in the curve which is much higher than other small peaks around them, and with the increase of gun traverse speed, the amount of the great peaks is increased correspondingly. This may result from the measurement conditions that the tubular sample was rotated with a special speed. When the spray jet reached close to the measuring point, the temperature increased remarkably, but before next imaging record was finished, the point heated by the spray jet had rotated to other side.

It should be noted that the overall temperature measured by the IR camera is relatively lower than that of the modeling results under the same spraying conditions (e. g. the same total deposition time). This manifests that, on one side, accurate temperature measurement is difficult to obtain since the derived temperature depended on the conditions such as thermal emissivity, ε, of the material surface. The temperature reported here assumed a constant $\varepsilon = 0.7$ for oxidized copper and steel[30]. In practice, the value of ε will change with respect to temperature, roughness, liquid fraction and amount of oxide. Measurements made in the present study were likely to be lower than the true temperature, especially for point 2, because an oblique observation of

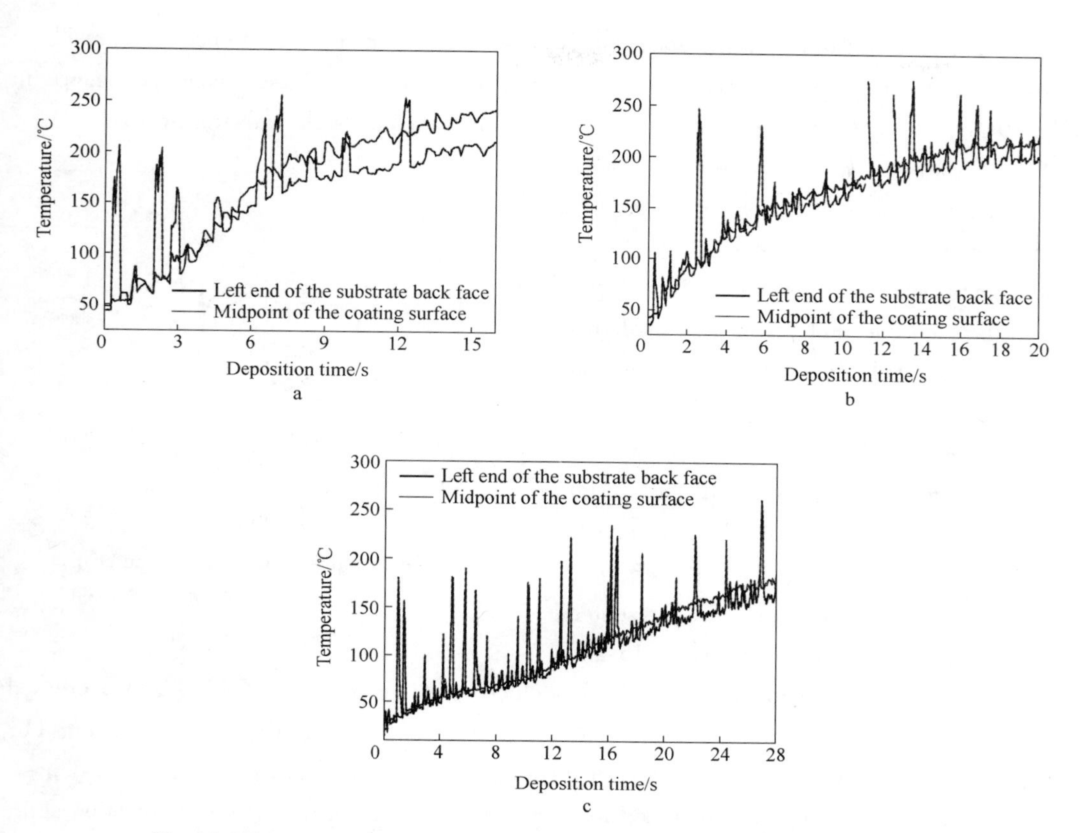

Fig. 14 Temperature histories measured at the two points shown in Fig. 13 with the gun traverse speed of 100 mm/s(a),200 mm/s(b) and 400 mm/s(c)

the IR camera was performed, namely the measuring points are moving along with the rotation of the sample. In addition, existence of the spray jet in IR thermal images also lowered the temperature measurement results for the deposit structure since the temperature of the spray jet is much higher than the deposit. On the other side, the parameters specified in the numerical model, such as materials enthalpy, thermal conductivity, and convection heat transfer coefficient, were specified based on reference literature, most of which were assumed to be temperature independent. In fact, most parameters used in the numerical model are changed with respect to temperature. If more accurate modeling result was expected to obtain, some parameter measurements, including the material properties and convection heat transfer coefficient, should be done first.

Although the overall temperature difference is existed between the modeling and IR measurement results, the present models, especially the stochastic model, are valid to some degree, because most results obtained from the numerical modeling are similar to that of the IR measurement. For example, as shown in Fig. 14, when the gun traverse speed is increased from 100 mm/s to 400 mm/s, the overall temperatures at the measuring points are decreased correspondingly, which is also shown in Fig. 10 for the stochastic deposition model. Another important consistency between the model and experiment is that higher gun traverse speed leads to lower thermal cycling gradient during the deposition, namely increasing the gun traverse speed will result in decrease of difference value between the peak and valley of the "zigzag" curve, as shown in Fig. 10 and Fig. 14 (especially for the measurement at the point of the substrate back face).

Fig. 15 shows the temperature/time history measured at points 1 and 2 under the periodical interruption spraying mode operated with 200 mm/s gun traverse speed. It can be found that, the interruption spraying mode reduced the overall temperature as compared with Fig. 14b, the standard spraying model, if the same coating thickness (i. e. the same actual deposition time) is expected. This result is consistent with the finite element modeling. When comparing Fig. 15 with Fig. 11, some similar phenomena are also presented between them. For instance, both Fig. 15 and Fig. 11 show that the temperatures at the modeling points or measuring points are decreased during every interruption period, in other words, from the curves shown in Fig. 15 and Fig. 11, we can clearly distinguish the regions corresponding to the interruption period and the deposition period. The above findings obtained from the IR measurement under the interruption spraying condition additionally verified the numerical model presented in the study.

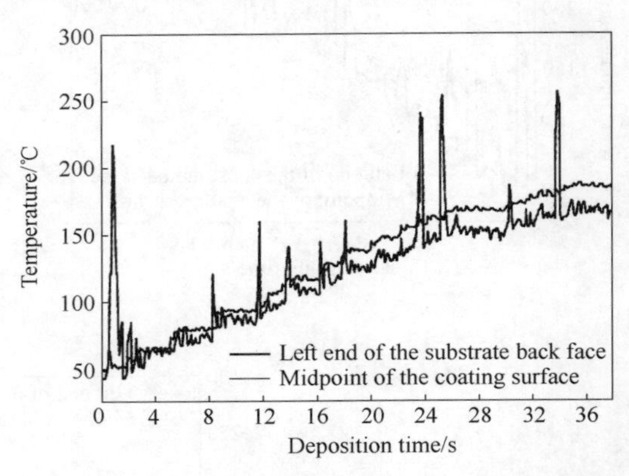

Fig. 15 Temperature histories measured at the two points shown in Fig. 13 for the periodical interruption spraying mode

5 Conclusions

A dynamic stochastic multiple particle deposition model based on finite element method (FEM) was developed to simulate the transient heat transfer of the thermal spray process, and a traditional layer – by – layer deposition model was also introduced as comparison. By using the two FEM models of this paper, transient thermal analysis of electric arc spray process under different spraying conditions was performed. Comparing the stochastic deposition model with the layer – by – layer model, it shows a noticeable difference of temperature, and the difference varies with deposition time, particles locations, gun traverse speeds and interruption spray modes, and so on. Contrary to the layer – by – layer model, the stochastic deposition model concerns the dynamic process of the gun movement and the random feature of particles deposition. The stochastic deposition model reproduced the rough surface of the coating when the coating formation was simulated, more important, it provided a more objective result of the transient temperature distribution of the coating/substrate than that of the layer – by – layer model, especially the severely inhomogeneous temperature distribution in x direction.

By investigating the temperature/time histories with different gun traverse speed and interruption spraying modes, the regularity was confirmed that relatively higher gun traverse speed and shorter interruption interval are helpful for the coating formation, because slower gun speed or longer alternating interval results in greater temperature fluctuation but rapid movement or shorter interval produces smaller temperature variation. Temperature/time histories near the coating surface midpoint and the left end of the substrate back face were recorded using an IR camera, which shows that, as compared with the numerical models (especially for the stochastic deposition model), most of experimental findings under different gun transverse speed and spraying mode conditions are consistent with that of the modeling.

The severely inhomogeneous and dynamic temperature distribution of the coating/substrate demonstrates the necessity of modeling the stochastic multiple particle deposition to build up coating in thermal spray processes, because the temperature/time history during coating build-up has a significant influence on the stress/time history and then on the finial residual stress state.

Acknowledgements

The authors are grateful to the priority support by China Natural Science Foundation (50735006, 50905185) and National "863" Project of China (2009AA032342), and would also like to thank Prof. Y. X. Wu and Dr. J. B. Cheng of the School of Material Science and Engineering of Shanghai Jiaotong University for supply of the ANSYS software simulation.

References

[1] D. I. Wimpenny, G. J. Gibbons, Metal spray tooling for composite forming, J. Mater. Proc. Technol. 138 (2003) 443 - 448.

[2] A. K. Srivastava, R. C. Anandani, A. Dhar, A. K. Gupta, Effect of thermal conditions on microstructural features during spray forming, Mater. Sci. Eng. A 304/306 (2001) 587 - 591.

[3] G. Jandin, H. Liao, Z. Q. Feng, C. Coddet, Correlations between operating conditions microstructure and mechanical properties of twin wire arc sprayed steel coatings, Mater. Sci. Eng. A 349 (2003) 298 - 305.

[4] A. Vaidya, T. Streibl, L. Li, S. Sampath, O. Kovarik, R. Greenlaw, An integrated study of thermal spray process - structure - property correlations: a case study for plasma sprayed molybdenum coatings, Mater. Sci. Eng. A 403 (2005) 191 - 204.

[5] A. P. Newbery, P. S. Grant, Oxidation during electric arc spray forming of steel, J. Mater. Proc. Technol. 178 (2006) 259 - 269.

[6] H. Zhang, X. Y. Wang, L. L. Zheng, X. Y. Jiang, Studies of splat morphology and rapid solidification during thermal spraying, Int. J. Heat Mass Transfer 44 (2001) 4579 - 4592.

[7] L. Moulla, Z. Salhi, M. P. Planche, M. Cherigui, C. Coddet, On the measurement of substrate temperature during thermal spraying, in: E. Lugscheider (Ed.), Thermal Spray Connects: Explore its Surfacing Potential, DVS, Düsseldor, DE, 2005, pp. 679 - 683.

[8] P. D. A. Jones, S. R. Duncan, T. Rayment, P. S. Grant, Control of temperature profile for a spray deposition process, IEEE Trans. Control Syst. Technol. 11 (2003) 656 - 667.

[9] J. Matejicek, S. Sampath, In situ measurement of residual stresses and elastic moduli in thermal sprayed coatings: Part 1: apparatus and analysis, Acta Mater. 51 (2003) 863 - 872.

[10] A. McDonald, C. Moreau, S. Chandra, Thermal contact resistance between plasma - sprayed particles and flat surfaces, Int. J. Heat Mass Transfer 50 (2007) 1737 - 1749.

[11] K. Nagashio, K. Kodaira, K. Kuribayashi, T. Motegi, Spreading and solidification of a highly undercooled $Y_3Al_5O_{12}$ droplet impinging on a substrate, Int. J. Heat Mass Transfer 51(2008)2455 – 2461.

[12] R. Ghafouri – Azar, J. Mostaghimi, S. Chandra, M. Charmchi, A stochastic model to simulate the formation of a thermal spray coating, J. Therm. Spray Technol. 12(2003)53 – 69.

[13] Y. Chen, G. Wang, H. Zhang, Numerical simulation of coating growth and pore formation in rapid plasma spray tooling, Thin Solid Films 390(2001)13 – 19.

[14] J. Mostaghimi, S. Chandra, R. Ghafouri – Azar, A. Dolatabadi, Modeling thermal spray coating processes: a powerful tool in design and optimization, Surf. Coat. Technol. 163/164(2003)1 – 11.

[15] X. Sheng, C. Mackie, T. C. A. Hall III, Heat transfer characterization of the solidification process resulting from a spray forming process, Int. Commun. Heat Mass Transfer 32(2005)872 – 883.

[16] R. Bhardwaj, J. P. Longtin, D. Attinger, A numerical investigation on the influence of liquid properties and interfacial heat transfer during microdroplet deposition onto a glass substrate, Int. J. Heat Mass Transfer 50(2007)2912 – 2923.

[17] F. Hugot, J. Patru, P. Fauchais, L. Bianchi, Modeling of a substrate thermomechanical behavior during plasma spraying, J. Mater. Proc. Technol. 190(2007)317 – 323.

[18] Z. Gan, H. W. Ng, A. Devasenapathi, Deposition – induced residual stresses in plasma – sprayed coatings, Surf. Coat. Technol. 187(2004)307 – 319.

[19] H. W. Ng, Z. Gan, A finite element analysis technique for predicting as – sprayed residual stresses generated by the plasma spray coating process, Finite Elem. Anal. Des. 41(2005)1235 – 1254.

[20] S. C. Gill, Residual stresses in plasma sprayed deposits, Ph. D. Thesis, University of Cambridge, The U. K. ,1991.

[21] A. P. Newbery, P. S. Grant, R. A. Neiser, The velocity and temperature of steel droplets during electric arc spraying, Surf. Coat. Technol. 195(2005)91 – 101.

[22] C. Aumund – Kopp, D. H. Müller, Influencing the contour of spray – formed sheets and billets by changing the scanning kinematics of the atomiser, Mater. Sci. Eng. A 326(2002)176 – 183.

[23] S. Markus, C. Cui, U. Fritsching, Analysis of deposit growth in spray forming with multiple atomizers, Mater. Sci. Eng. A 383(2004)166 – 174.

[24] Y. Zhu, H. Liao, C. Coddet, Transient thermal analysis and coating formation simulation of arc spray process, in: E. Lugscheider (Ed.), Thermal Spray Connects: Explore its Surfacing Potential, DVS, Düsseldor, DE, 2005, pp. 1376 – 1381.

[25] Y. Zhu, H. Liao, C. Coddet, Transient thermal analysis and coating formation simulation of thermal spray process by finite difference method, Surf. Coat. Technol. 200(2006)4665 – 4673.

[26] Z. Gan, H. W. Ng, Experiments and inelastic finite element analyses of plasma sprayed graded coatings under cyclic thermal shock, Mater. Sci. Eng. A 385(2004)314 – 324.

[27] M. Y. He, J. W. Hutchinson, A. G. Evans, Simulation of stresses and delamination in a plasma – sprayed thermal barrier system upon thermal cycling, Mater. Sci. Eng. A 345(2003)172 – 178.

[28] T. Watanabe, I. Kuribayashi, T. Honda, A. Kanzawa, Deformation and solidification of a droplet on a cold substrate, Chem. Eng. Sci. A 47(1992)3059 – 3065.

[29] S. Hoile, T. Rayment, P. S. Grant, A. D. Roche, Oxide formation in the sprayform tool process, Mater. Sci. Eng. A 383(2004)50 – 57.

[30] ThermoVision™ A20 M operator's manual. FLIR Systems, Danderyd, Sweden, 2004.

Tribological Behaviors of Surface-coated Serpentine Ultrafine Powders as Lubricant Additive*

Abstract The effect of surface – coated ultrafine powders (UFPs) of serpentine suspended in lubricants on the tribological behaviors of a mated 1045 steel contact was investigated. Through the addition of serpentine UFPs to oil, the wear resistance ability was improved and the friction coefficient was decreased. The addition of 1.5wt% serpentine to oil is found most efficient in reducing friction and wear. The nano – hardness and the ratio of hardness to modulus of friction surface are observably increased. Such effects can be attributed to the formation of a tribofilm of multi – apertured oxide layer, on which the micrometric alumina particles embedded and serpentine nano – particles adsorbed.

Key words serpentine, ultrafine powder, additive, tribological property

1 Introduction

Over the last few years, interest in the synthesis and tribological properties of ultrafine powders (UFPs) as lubricant additives has steadily grown due to their efficacies in reducing friction and wear[1-4]. There have been many investigations on the behaviors of inorganic or organic UFPs as extreme pressure (EP) and anti – wear (AW) additives for liquid lubricants. It is found that the lubricating properties of oils were significantly improved when nano – sized particles were used, while micro – sized particles had much smaller effects[5-7].

Compared with nano – particles, the micrometric UFPs in liquid media are more thermodynamically unstable and tend to spontaneously subside. The large – scale particles, on the other hand, can act as abrasive particles on contacts, as accordingly result in severe wear. Therefore, the previous and current studies on UFPs additives mainly focused on the tribology testing of nanoscale particles of metals[8-13], carbon materials[14-17], oxides[18-20], sulfides[21-23], borates[24], RE compounds[25-27], polymers[28], etc. Besides BN particles[29,30], few studies were carried out on micro – sized UFPs due to their small contribution towards friction – reducing and wear resistance.

Serpentine group describes a group of common rock – forming hydrous magnesium iron phyllosilicate ($(Mg, Fe)_6Si_4O_{10}(OH)_8$) minerals. They may contain minor amounts of other elements including chromium, manganese, cobalt and nickel. Recent researches indicate the micro – sized serpentine ($Mg_6[Si_4O_{10}](OH)_8$) UFPs present excellent tribological properties when added to liquid lubricants[31,32]. Jin et al. [33] investigated the tribological behaviors of crankcase

* Copartner: H. L. Yu, Y. Xu, P. J. Shi, H. M. Wang, Y. Zhao, Z. M. Bai. Reprinted from *Tribology International*, 2010, 43 (3): 667 – 675.

oil suspended serpentine particles(≤2. 0μm) in railway diesel engines under field trial conditions. They found that a super – hard and super – lubricious oxide layer formed on the worn ferrous surface, as accordingly lowers the friction and wear. Yu et al. [34] studied the lubricating effect of natural mineral admixtures(size:0. 3 –3μm) that mainly composed of serpentine(90% –95%) and schungite(4. 8% –9. 8%). They found that a DLC film with Si or Si – O structures doped formed on the worn steel contacts, as contributes to the excellent mechanical and tribological properties of the friction surface.

The present study aimed to further clarify the mechanisms responsible for the effect of serpentine minerals used as additives. The effect of surface – coated serpentine UFPs, with an average particle size of 1. 0 μm, suspended in lubricating oils on the tribological behaviors of steel contact was reported. The morphologies and element distributions of the tribofilm formed by the serpentine were studied. In particular, the nano – mechanical properties of the tribofilm were measured by a nano – indentation tester.

2 Experimental

2. 1 Materials

The UFPs used in the present study were prepared by mechanical crushing and ball – milling the serpentine mineral (Liaoning Province, China). Table 1 lists the chemical compositions of the starting materials, its crystal formula can be expressed as $Mg_{5.70}Al_{0.13}Fe_{0.02}Ca_{0.06}K_{0.04}Mn_{0.03}$-$[Si_{4.05}O_{10}](OH)_8$, which is close to the ideal formula of serpentine, i. e. $Mg_6[Si_4O_{10}]$-$(OH)_8$. Fig. 1 shows the X – ray diffraction pattern of the ball – milled serpentine UFPs. The diffraction peaks of d = 0. 7282, 0. 3617 and 0. 2519 nm can be indexed to those of antigorite, corresponding to the (0 0 1), (1 0 2), and (16 0 1) planes, respectively. Its lattice parameters are calculated as follows: a = 0. 536 nm, b = 0. 928 nm, c = 0. 732 nm, $\alpha = \gamma = 90°$ and β = 91. 38°, further proving the antigorite structure. To provide good stabilization in viscous liquid, a mixture of boric acid ester and Span 60 (mol. ratio = 1∶1) was mixed with the UFPs in a globe mill for 6 h operation (rotating speed = 300 r/min) to produce an organic coating layer. Fig. 2 shows the SEM image and size distribution of the surface – coated serpentine UFPs. The particle size is mostly in the range of 0. 1 –5 μm and the average size approximates 1. 0μm. The final surface – coated UFPs can be dispersed well in some organic solvents, such as chloroform, benzene, methylbenzene and lubricating oil.

Table 1 Chemical composition of the serpentine minerals

Oxides	Content/wt%	Oxides	Content/wt%
SiO_2	43. 49	CaO	0. 64
Al_2O_3	1. 18	K_2O	0. 33
FeO	0. 25	P_2O_5	0. 085
MnO	0. 32	H_2O^+	12. 37
MgO	41. 0	H_2O^-	0. 29

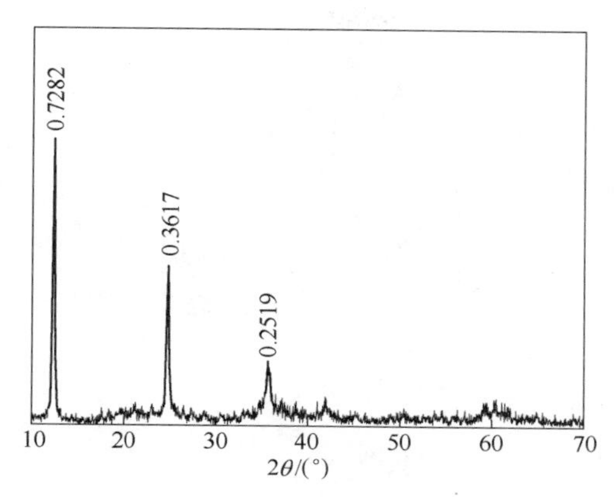

Fig. 1 XRD pattern of untreated serpentine UFPs

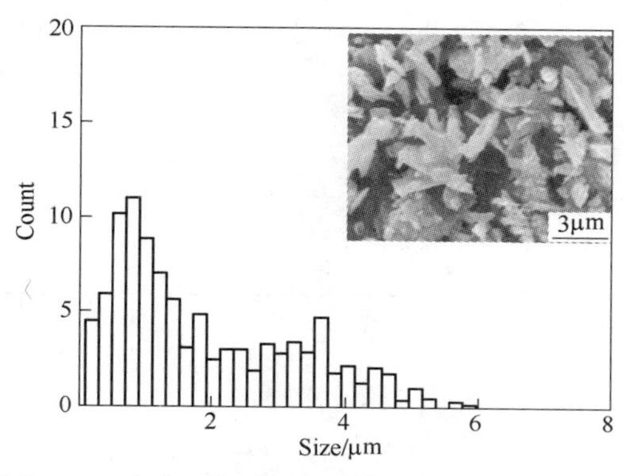

Fig. 2 SEM image and size distribution of surface - coated serpentine UFPs

2. 2 Friction and wear test

An MM - 10 W sliding friction tribotester was employed to study the friction - reduction and anti - wear abilities. As shown schematically in Fig. 3, the MM - 10 W comprises an upper rotating ring specimen, which came into contact with a lower disk specimen fixed in an oil bath. The friction coefficient and temperature of the disk specimen were then measured. Diesel engine oil (grade: 50 CC) was used as a lubricant for the friction specimens of the rings and disks (1045 steel, hardness: 210 HB, surface roughness: 0. 20 - 0. 25μm). Table 2 gives viscosity of the oil with and without serpentine UFPs. High - energy mechanical ball - milling agitation (rotating speed = 120 r/min, duration = 60 min) and ultrasonic dispersion (power = 200 W, temperature = 40℃, duration = 30 min) were used to provide good dispersion stability of the surfacecoated UFPs in oil. The experimental conditions were: atmospheric environment, room temperature, normal load = 100 N, 200 N, 300 N and 400 N, sliding speed = 1. 51 m/s and test duration = 120 min. The corresponding initial mean Hertzian pressure at the contacts was 0. 554 MPa (100N), 1. 11 MPa (200 N), 1. 66 MPa (300 N) and 2. 21 MPa (400 N). After the test, the steel disks were cleaned in petroleum ether and absolute ethyl alcohol. The wear rates of the disks were then calculated by measuring the weight loss of the disks using a mass balance to an accuracy of 0. 1 mg. All the experimental results were the average of three sets of the experi-

mental data.

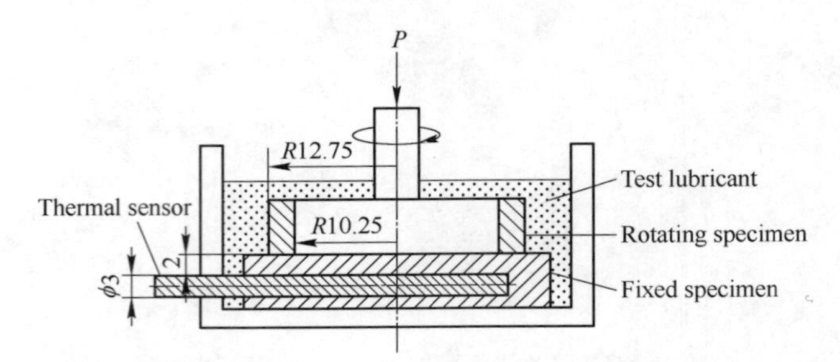

Fig. 3 Schematic illustration of friction and wear test

Table 2 Viscosity of oil with and without serpentine UFPs

Item	Kinematic viscosity(100℃)/$mm^2 \cdot s^{-1}$	Viscosity(150℃,10^6 s)/mPa · s
Pure oil	18.6	3.7
0.5 wt% UFPs	18.6	3.7
1.0 wt% UFPs	18.6	3.7
1.5 wt% UFPs	18.7	3.7
2.0 wt% UFPs	18.7	3.8

2.3 Worn surface analysis

Scanning electronic microscope (SEM) equipped with energy dispersive X – ray spectroscopy (EDS) was utilized to analyze the morphologies and element distributions of the worn steel disks. Nano – indentation tester (Nano Test 600, Micro Materials Ltd.) was utilized to investigate the mechanical properties of the rubbing surfaces. A three sided pyramidal diamond indenter, with a diameter of 50 nm, was used throughout the test. In a constant maximal load test, the initial and maximum load was 0.03 mN and 15 mN, respectively. The loading and unloading rates were both 0.3 mN/s, and the dwell time at maximum load was 60 s. In constant maximal indentation depth tests, the maximal depth of indentation was controlled as follows: 50 nm, 100 nm, 200 nm, 500 nm, 1000 nm, 1500 nm and 2000 nm. The initial load was 0.03 mN. The loading and unloading rates were both 0.3 mN/s. In order to avoid the interaction between two test points, distance between any two points was not shorter than 10 μm.

3 Results

3.1 Tribological properties

Fig. 4 shows the variation of friction coefficient with applied load for oil with and without serpentine UFPs. The coefficient of pure oil increases linearly along with the increasing normal load. With the addition of serpentine to oil, the friction decreases at all applied loads. The best friction – reduction property can be obtained when the UFPs concentration is 1.5 wt%. Such concentration decreases the friction coefficient by 50.6% at 100 N, 58.1% at 200 N, 56.3%

at 300 N and 56.1% at 400 N, as compared to pure oils. When the concentration is higher than that, the friction – reduction property is weakened, but still superior to pure oil.

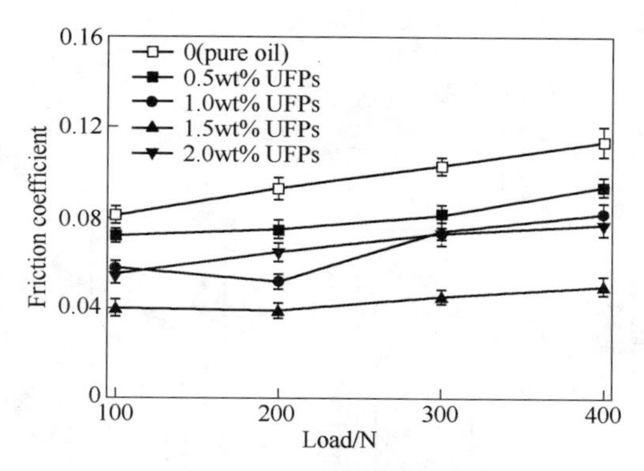

Fig. 4 Variation of friction coefficient with normal load

During the friction test, the specimen temperature can be measured by the sensor that inserted in the disk. Fig. 5 demonstrates time dependency of specimen temperature at applied load of 200 N. It shows that the temperature of disks lubricated with oil containing UFPs is lower than that of pure oil. Similar with the results of friction coefficient, the lowest specimen temperature is obtained when serpentine concentration is 1.5 wt%. The corresponding temperature decreases by 13.5% compared to pure oil. It is known that when surfaces slide relatively, almost all the energy dissipated in friction appears in the form of heat at the interface. This frictional heat raises the temperature of specimen and oil. It is noted that the 2.0 wt% suspension causes a higher temperature with regard to the rest of samples during the first half of friction tests. That may be caused by local breakage of oil film initiated by increasing solid particles of serpentine in the beginning compared with lower concentration, as resulted in higher friction and specimen temperature.

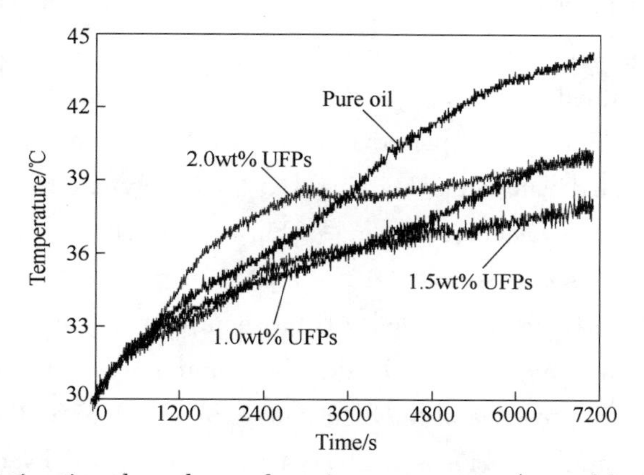

Fig. 5 Friction time dependency of specimen temperature (normal load = 200 N)

Fig. 6 shows the anti – wear property of oil with and without serpentine UFPs versus applied load. Wear of disk specimens under the lubrication of oil with and without serpentine increases with the increasing load. With the addition of the UFPs to oil, the wear of steel disk is reduced. Similar with the variation of friction coefficient, the best anti-wear property can be obtained when serpentine concentration is 1.5 wt%. As compared to pure oils, such concentration decreases wear rate of disk specimen by 82.3% at 100 N, 89% at 200 N, 82.5% at 300 N and 74.1% at 400N.

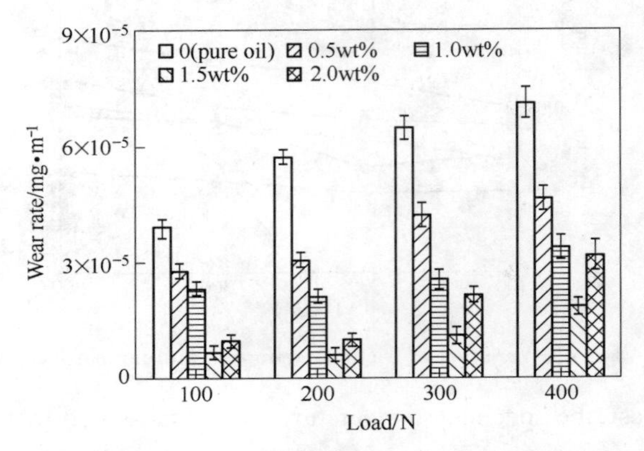

Fig. 6 Wear rate of disk specimens under the lubrication of oil with and without serpentine UFPs

3.2 Morphology and elementary analysis of the worn surface

Typical SEM images of the worn surfaces under the lubrication of oil with and without serpentine (normal load = 200 N) are shown in Fig. 7. The obvious furrows and grooves in sliding direction, formed by the wear debris, are wide and deep for the disks lubricated with pure oil, showing severe abrasive wear. The furrows become shallow and narrow when adding 1.0 wt% serpentine to oil. A large number of micro-apertures with a uniform distribution appear on the worn surface. The rubbing surface lubricated with 1.5 wt% suspension is smoother and multiapertured, few grooves can be found. This result is in accordance with the best tribological behaviors of oil containing 1.5 wt% serpentine.

Fig. 8 shows the magnified morphology and corresponding elemental distribution map of the image shown in Fig. 7c. Four typical patterns marked as A, B, C and D can be clearly seen on the worn surface lubricated with oil containing 1.5 wt% serpentine UFPs. It is obvious that most areas of the surface are smooth, e.g. zone A. Furthermore, there exist a lot of micro-apertures in which the micro-sized particles (0.5 - 1 μm), e.g. particle B, and their aggregations (2 - 3 μm), e.g. particle C, are embedded. At the same time, a large number of nano-sized adsorbates (100 - 200 nm), e.g. particle D, distribute uniformly on the surface. The elemental distribution maps show the presence of the elements of Fe, O, Al, Si and Mg on the rubbing surface. The relative concentration of Fe is highest, as may be caused by the electronic beam of EDS which penetrated the worn surface and hit directly the substrate steel. Mg and Si distribute uniformly across the surface. While Al exists only in the regions corresponding to the apertures embedded micro-sized particles, where the relative concentration of O is also higher. That indicates the micro-sized particles mainly consist of Al and O.

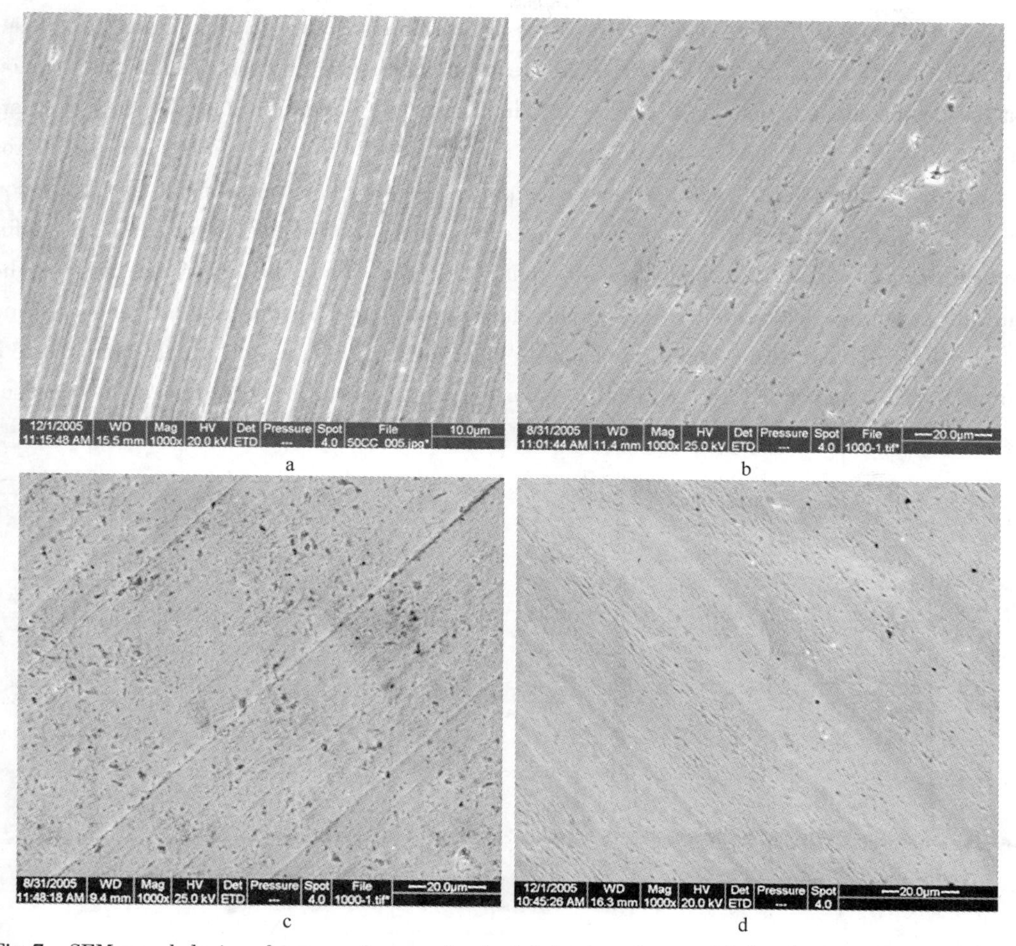

Fig. 7 SEM morphologies of worn surfaces under the lubrication of pure oil(a), oil + 1.0 wt% UFPs(b), oil + 1.5 wt% UFPs(c) and oil + 2.0 wt% UFPs(d)(normal load = 200 N)

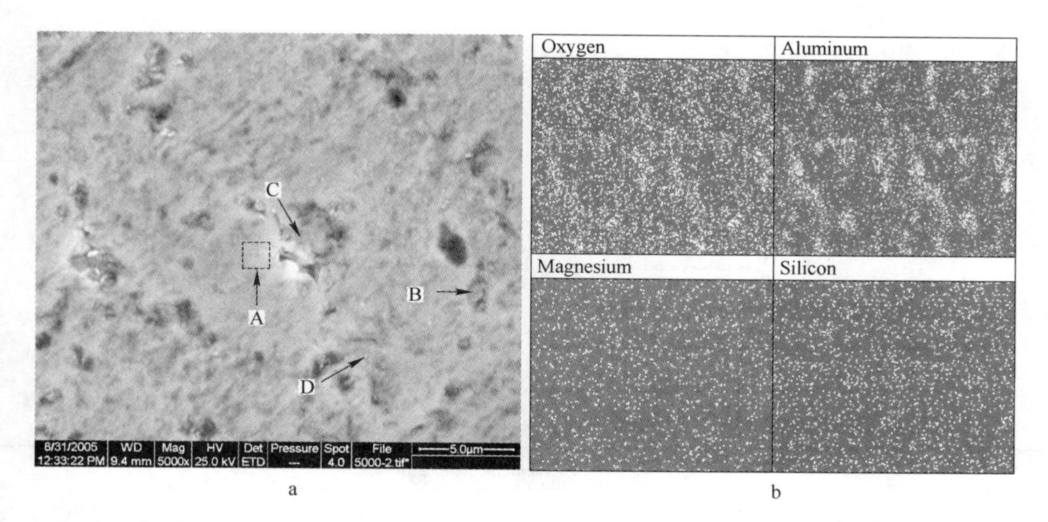

Fig. 8 Enlarged SEM image(a) and elemental distribution(b) of the morphology shown in Fig. 7c (Surface lubricate with 1.5 wt% UFPs suspension at a normal load of 200 N)

Fig. 9 shows the EDS patterns of(1) the different worn areas shown in Fig. 8a and(2) the surface lubricated with pure oil. Table 3 gives the semiquantitative analysis results. It is obvious that the compositions of regions B and C are similar, mainly containing the elements of O and Al. The ratio of oxygen atom to aluminum atom is about 3 to 2. So, it is inferred that the microsized particles embedded in the apertures are alumina(Al_2O_3) particles. Particle D mainly contains O, Si and Mg, showing it is possible the serpentine particles that porphyrized by sliding contacts during friction. It is seen in Fig. 9c that the oxygen content of zone A was much higher than that of surface lubricated with pure oil, showing an oxide layer forms. The above – mentioned results indicate that a tribofilm of multi – apertured oxide layer, on which micrometric alumina particles embedded and serpentine nano – particles adsorbed, has formed on worn surface with oil containing serpentine.

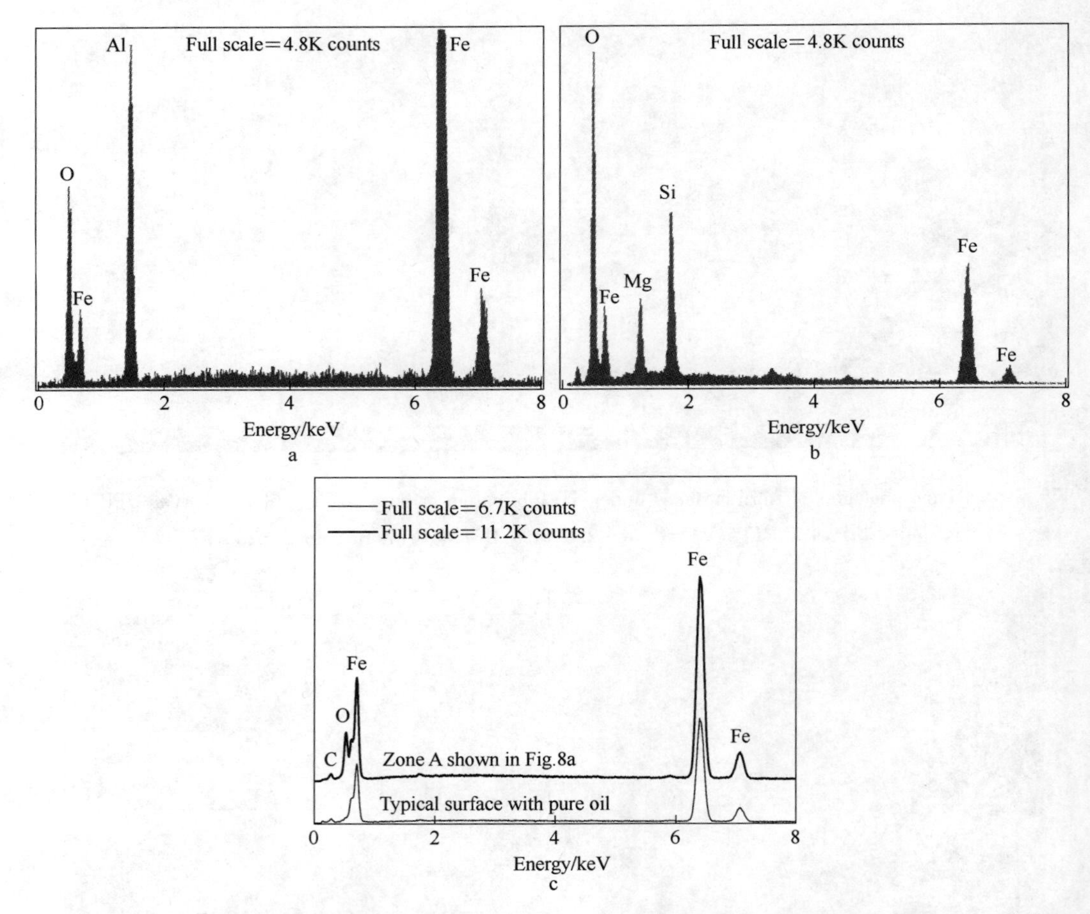

Fig. 9 EDS analysis of particle B(a), particle D(b), zone A and surface lubricated with pure oil(c)

Table 3 Elemental compositions of the different worn zones shown in Fig. 8a

Zone	Element composition/at%					
	Fe	O	Si	Mg	Al	Others
A	84.51	9.68	0.26	0.38	0.25	0.42

Table 3(continued)

Zone	Element composition/at%					
	Fe	O	Si	Mg	Al	Others
B	37. 34	33. 52	0. 60	0. 82	27. 52	0. 20
C	41. 92	33. 96	0. 15	0. 20	23. 40	0. 37
D	77. 96	12. 40	3. 60	5. 42	0. 17	0. 45

3. 3 Nano – mechanical properties of the worn surface

It is generally difficult to avoid the effects of substrate when measuring the mechanical properties of thin films. Therefore, mechanical characterization of tribofilms is generally not easy. Many works have been carried out to do that using nanoindentation method[33 – 38]. Here, two nano – indentation testing methods, namely controlling maximal applied load and maximal indentation depth, were performed to study mechanical properties of normal worn surface (pure oil, normal load = 200 N) and the tribofilm (1. 5% UFPs suspension, normal load = 200 N).

Controlling maximum load indentation tests were performed to compare the mechanical properties of different surfaces. Thirty points were chosen at random on the surface of every friction sample to measure their nano – hardness (H) and elastic modulus (E). Fig. 10 shows the typical and total 30 load – depth curves of normal worn surface and the tribofilm tested with a maximal applied load of 15 mN. The maximum indentation depth and plastic deformation for the tribofilm are much smaller than those of normal worn surface, indicating the tribofilm was a hard thin layer. Table 4 lists the nano – hardness and modulus of different worn surfaces. The nano – hardness (H), elastic modulus (E), and their ratio (H/E) of the tribofilm are superior to normal worn surface. The addition of 1. 5 wt% serpentine increased the average H of the worn surface by 94%, and increased the mean value of H/E by 75%.

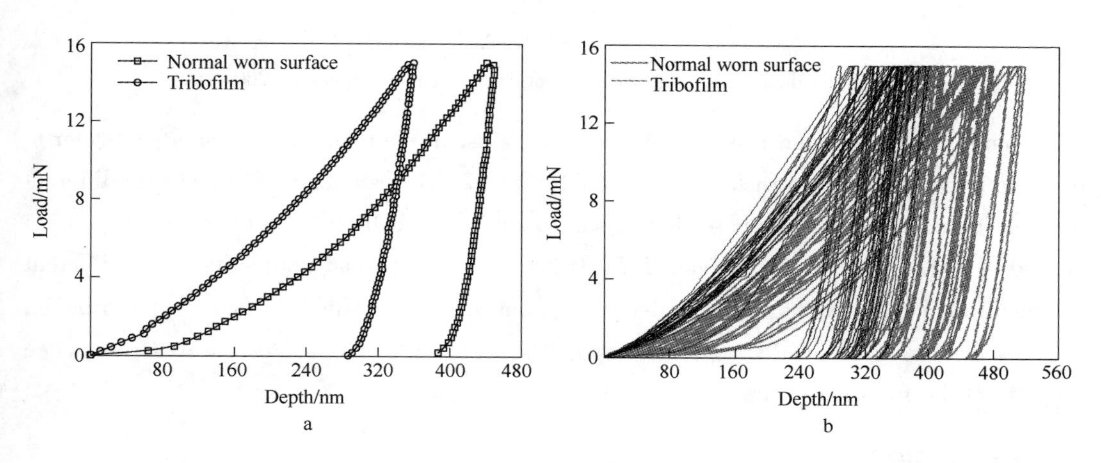

Fig. 10 Typical (a) and total (b) load – depth curves of different surfaces tested by nano – indentation tester with a maximal applied load of 15 mN

Table 4 Nano – mechanical properties of normal worn surface and the tribofilm measured by controlling the maximal load at 15 mN

Item	Nano – mechanical properties				
	Max depth/nm	Plastic depth/nm	H/GPa	E/GPa	H/E
Normal surface	438.80 ±59.70	393.50 ±54.47	3.45 ±0.85	215.53 ±32.10	1.60×10^{-2}
Tribofilm	342.55 ±42.41	289.65 ±37.57	6.68 ±0.65	238.52 ±29.65	2.80×10^{-2}

In order to clarify the influence of substrate on the mechanical properties of the tribofilm, indentation tests were performed with different maximal indentation depths. For a given maximal indentation depth, five points were chosen at random on the tribofilm (worn surface lubricated with 1.5 wt% UFPs suspension at load of 200 N). Fig. 11 shows the dependency of H and E on maximal indentation depth. When tested with a maximal depth of 50 nm, 100 nm and 200 nm, H of the tribofilm approximates 8.2 GPa. The value decreases rapidly with the increasing maximal indentation depth when it is deeper than 200 nm, showing the effects of substrate increase. When the depth reaches 2000 nm, H approximates 3.6 GPa which is in accord with the 1045 steel. The results indicate the real nano – hardness of the tribofilm is approximately 8.2 GPa and its thickness is possible between 200 nm and 2000 nm.

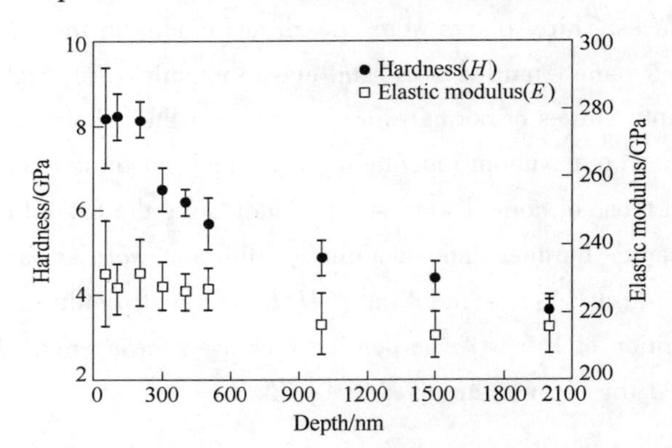

Fig. 11 Dependency of hardness (H) and elastic modulus (E) on maximal indentation depth measured on the tribofilm (1.5 UFPs suspension, 200 N)

The ratio of hardness and elastic modulus, H/E, is usually introduced as a main parameter to estimate the relative wear resistance of materials. Fig. 12 demonstrates H/E tested with different maximal indentation depth on normal worn surface and the tribofilm (1.5 wt% UFPs suspension at load of 200 N). The value is about $(1.5 \pm 0.1) \times 10^{-2}$ for normal worn surface with different maximal indentation depths, while it decreases from 3.41×10^{-2} to 1.79×10^{-2} for tribofilm with the increasing maximal indentation depth. The result further indicates the wear resistance property of the tribofilm is superior to the normal worn surface.

4 Discussion

According to the present experimental results, it can be concluded that an oxide tribofilm has

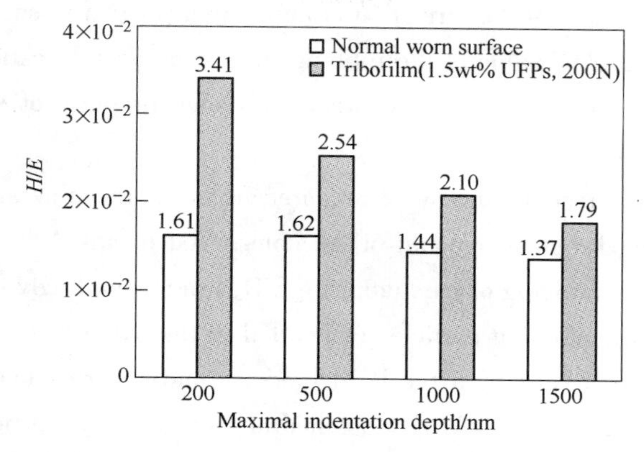

Fig. 12 Dependency of ratio of hardness to elastic modulus on maximal indentation depth

formed on the worn surface under the lubrication of oil with surface - coated serpentine UF-Ps. The result is similar with what has been reported by Jin et al. [33]. But, what is different from the previous studies is that the tribofilm is a multi - apertured oxide layer on which micrometric alumina particles embedded, and serpentine nano - particles adsorbed. It is obvious that the film consists of three typical structures: (1) oxide layer with excellent nano - mechanical properties, (2) micro - apertures with alumina particles embedded, and (3) third bodies formed by nano - scale serpentine particles.

Serpentine as a typical phyllosilicate has a basic structure of [SiO_4] which is connected by three bridging oxygen atoms between each other, as accordingly forms Si—O tetrahedron layers in two - dimensional space. The non - bridging oxygen atoms in [SiO_4] point to the same direction and connect with Mg^{2+}, and consequently form Mg—O octahedral layers. The two layers are connected by weak molecular bond and/or hydrogen bond to form the typical layer structure of serpentine, which displays good capability of oxygen release due to the weak bond between the Si—O tetrahedral layer and Mg—O octahedral layer. Mechanical action caused by the direct contact of friction surfaces initiate the crystal fracture of the layered structure of serpentine particles that suspended in oil. Then oxygen atoms and oxygen - containing species were released for the bond breakage of Si—O tetrahedron and Mg—O octahedral. Finally, the mechanochemical/thermochemical oxidation reaction of iron atoms and oxygen atoms occurred[33], and a hard oxide layer was accordingly formed on the worn ferrous surface.

In general, a hard oxide layer can provide a sufficient resistance not only to the embedding of abrasive particles but the plastic deformation of materials in sliding friction as well, the abrasive wear can be accordingly reduced. But, there is no inevitable dependence between the increasing of hardness and the reduction of abrasive wear. Practically, the reduction of elastic modulus usually results in the reducing of wear. That means, for a metal surface with lower elastic modulus, elastic deformation would more easily occur when contact stress applied, as increases the contact area between the sliding surfaces, and reduces the contact stress. In this case, it is more easily for the abrasive particles to pass the contacts, and consequently reduce the abrasive wear. Physically, the higher is the H/E value, the better is the wear resistance. Nano - indentation test results demonstrate that the formation of an oxide layer increased significantly the average nano - hardness and *H/E* value of the rubbing surface under lubrication of oil with serpentine. That may be the main reason why the wear resistance property of worn surface was improved.

The UFPs used in the present study were prepared from natural minerals of serpentine which was impure and contained a small amount of Al atoms. That means Al^{3+} substituted for Mg^{2+}, connected with the non - bridging oxygen atomsin[SiO_4] and accordingly formed Al—O octahedral. It is obvious that the alumina particles embedded in the oxide layer come form the Al—O octahedral impurity. It is possible that the local pressure and shearing between the sliding surfaces initiate the bond breakage of Al—O octahedral, and alumina particles were accordingly released.

As a ceramic phase, the hardness of the alumina particles is much higher than that of carbon steels and their oxides. It is easy for them to be embedded in the contact surfaces, and accordingly form a ceramic particle reinforced phase. Moreover, alarge number of micron apertures were formed in the case of alumina particles embedded in the contacts. The benefit arising from these apertures on worn surface is a combination of several effects improving oil supply and reducing abrasion in the sliding contact[39,40]. (1) Apertures can act as oil reservoirs, which transport or retain oil to be released in emergency situations. (2) Apertures in friction surface can disarm wear particles by entrapping them, thereby suppressing abrasion and plowing friction. (3) When surface irregularities appear at sufficient density, they can improve the wetting of the surface by oil, and thereby support the lubricating oil film formation.

Physically, occurring dehydration reaction at certain temperature and pressure is one of the most important properties of silicate minerals. So, it is possible for serpentine particles in oil to partly dehydrate due to the local overheating, high flash temperature and high contact stress during the sliding of contact surfaces[41]. Once dehydration reaction occurred, the size of serpentine particles was refined and its hardness was reduced [42]. The particle size would be further decreased by the grinding that initiated by the contact sliding surfaces. That may be the reason why a large number of nano - scale serpentine particles, which acted as the third bodies in friction, were observed on the worn surface lubricated with oil containing surface - coated serpentine UFPs. Traditionally, inorganic silicates such as talcum($Mg_6Si_4O_{10}(OH)_2$) are among solid lubricants[43]. This is attributable to the four - member - ring stacked layers in certain silicates that are formed from the structural skeleton of silicon and oxygen, $[Si_4O_{10}]_n$. Therefore, the friction and wear could be further reduced due to the uniformly distributed third bodies that formed by refined serpentine particles with a nano - scale diameter.

5 Conclusion

(1) Surface - coated serpentine UFPs suspended in oil present excellent anti - wear and friction - reducing properties. The addition of 1. 5 wt% serpentine into oil is found most efficient in reducing friction and wear.

(2) A tribofilm of multi – apertured oxide layer, on which the micrometric alumina particles embedded and serpentine nano – particles adsorbed, is found on surface under the lubrication of oil with serpentine UFPs. It possesses excellent mechanical properties and contributes to the excellent tribological behaviors of serpentine UFPs additive.

Acknowledgements

This research was supported by the Key Basic Research and Development Program of China (2007CB607601) and National Natural Science Foundation of China (50735006, 50805146).

References

[1] Xue QJ, Liu WM, Zhang ZJ. Friction and wear properties of a surface – modified TiO_2 nanoparticle as an additive in liquid paraffin. Wear 1997;213:29 – 32.

[2] Hu ZS, Dong JX, Chen GX. Study on antiwear and reducing friction additive of nanometer ferric oxide. Tribol Int 1998;31:355 – 360.

[3] Zhou JF, Yang JJ, Zhang ZJ, Liu WM, Xue QJ. Study on the structure and tribological properties of surface – modified Cu nanoparticles. Mater Res Bull 1999;34:1361 – 1367.

[4] Bakunin VN, Suslov AY, Kuzmina GN, Parenago OP. Recent achievements in the synthesis and application of inorganic nanoparticles as lubricant components. Lubr Sci 2005;17(2):127 – 145.

[5] Ye PP, Jiang XX, Li S, Li SZ. Preparation of $NiMoO_2S_2$ nanoparticle and investigation of its tribological behavior as additive in lubricating oils. Wear 2002;253:572 – 575.

[6] Radice S, Mischler S. Effect of electrochemical and mechanical parameters on the lubrication behaviour of Al_2O_3 nanoparticles in aqueous suspensions. Wear 2006;261:1032 – 1041.

[7] Hu KH, Liu M, Wang QJ, Xu YF, Schraube S, Hu XG. Tribological properties of molybdenum disulfide nanosheets by monolayer restacking process as additive in liquid paraffin. Tribol Int 2009;42:33 – 39.

[8] Tarasov S, Kolubaev A, Belyaev S, Lerner M, Tepper F. Study of friction reduction by nanocopper additives to motor oil. Wear 2002;252:63 – 69.

[9] Sun L, Zhang ZJ, Wu ZS, Dang HX. Synthesis and characterization of DDP coated Ag nanoparticles. Mater Sci Eng A 2004;379:378 – 383.

[10] Qiu SQ, Zhou ZR, Dong JX, Chen GX. Preparation of Ni nanoparticles and evaluation of their tribological performance as potential additives in oils. J Tribol 2001;123:441 – 443.

[11] Zhao YB, Zhang ZJ, Dang HX. A simple way to prepare bismuth nanoparticles. Mater Lett 2004;58:790 – 793.

[12] Kolodziejczyk L, Martínez – Martínez D, Rojas TC, Fernández A, Sánchez – López JC. Surface – modified Pd nanoparticles as a superior additive for lubrication. J Nanoparticle Res 2007;9:639 – 645.

[13] Ma JQ, Mo YF, Bai MW. Effect of Ag nanoparticles additive on the tribological behavior of multialkylated cyclopentanes (MACs). Wear 2009;266:627 – 631.

[14] Peng YT, Hu YZ, Wang H. Tribological behaviors of surfactant – functionalized carbon nanotubes as lubricant additive in water. Tribol Lett 2007;25:247 – 253.

[15] Pei XW, Hu LT, Liu WM, Hao JC. Synthesis of water – soluble carbon nanotubes via surface initiated redox polymerization and their tribological properties as water – based lubricant additive. Eur Polym J 2008; 44:2458 – 2464.

[16] Joly – Pottuz L, Matsumoto N, Kinoshita H, Vacher B, Belin M, Montagnac G, et al. Diamond – derived

carbon onions as lubricant additives. Tribol Int 2008;41:69 – 78.

[17] Lee JK, Cho SW, Hwang YJ, Cho HJ, Lee CG, Choi YM, et al. Application of fullerene – added nano – oil for lubrication enhancement in friction surfaces. Tribol Int 2009;42:440 – 447.

[18] Hernandez Battez A, Fernandez Rico JE, Navas Arias A, Viesca Rodriguez JL, Chou Rodriguez R, Diaz Fernandez JM. The tribological behaviour of ZnO nanoparticles as an additive to PAO6. Wear 2006;261: 256 – 263.

[19] Wu YY, Tsui WC, Liu TC. Experimental analysis of tribological properties of lubricating oils with nanoparticle additives. Wear 2007;262:819 – 825.

[20] Hernández Battez A, González R, Viesca JL, Fernández JE, Díaz Fernández JM, Machado A, et al. CuO, ZrO_2 and ZnO nanoparticles as antiwear additive in oil lubricants. Wear 2008;265:422 – 428.

[21] Chen S, Liu WM, Yu LG. Preparation of DDP – coated PbS nanoparticles and investigation of the antiwear ability of the prepared nanoparticles as additive in liquid paraffin. Wear 1998;218:153 – 158.

[22] Rapoport L, Leshchinsky V, Lvovsky M, Nepomnyashchy O, Volovik YU, Tenne R. Friction and wear of powdered composites impregnated with WS_2 inorganic fullerene – like nanoparticles. Wear 2002; 252: 518 – 527.

[23] Moshkovith A, Perfiliev V, Verdyan A, Lapsker I, Popovitz – Biro R, Tenne R, et al. Sedimentation of IF – WS_2 aggregates and a reproducibility of the tribological data. Tribol Int 2007;40:117 – 124.

[24] Dong JX, Hu ZS. A study of the anti – wear and friction – reducing properties of the lubricant additive, nanometer zinc borate. Tribol Int 1998;31:219 – 223.

[25] Zhou JF, Wu ZS, Zhang ZJ, Liu WM, Dang HX. Study on an antiwear and extreme pressure additive of surface coated LaF_3 nanoparticles in liquid paraffin. Wear 2001;249:333 – 337.

[26] Wang LB, Zhang M, Wang XB, Liu WM. The preparation of CeF_3 nanocluster capped with oleic acid by extraction method and application to lithium grease. Mater Res Bull 2008;43:2220 – 2227.

[27] Yao YL, Wang XM, Guo JJ, Yang XW, Xu BS. Tribological property of onion – like fullerenes as lubricant additive. Mater Lett 2008;62:2524 – 2527.

[28] Fernández Rico E, Minondo I, García Cuervo D. The effectiveness of PTFE nanoparticle powder as an EP additive to mineral base oils. Wear 2007;262:1399 – 1406.

[29] Kimura Y, Wakabayashi T, Okada K, Wada T, Nishikawa H. Boron nitride as a lubricant additive. Wear 1999;232:199 – 206.

[30] Pawlak Z, Kaldonski T, Pai R, Bayraktar E, Oloyede A. A comparative study on the tribological behaviour of hexagonal boron nitride (h – BN) as lubricating micro – particles – an additive in porous sliding bearings for a car clutch. Wear 2009;267:1198 – 1202.

[31] Alexandrov SN. Method of treatment of friction surfaces of friction units. World Patents, WO01/38466, 2001 – 05 – 31.

[32] Alexandrov SN. Method of treatment of friction surfaces of friction units. Chinese Patents, CN1317041, 2001 – 10 – 10.

[33] Jin YS, Li SH, Zhang ZY, Yang H, Wang F. In situ mechanochemical reconditioning of worn ferrous surfaces. Tribol Int 2004;37:561 – 567.

[34] Yu Y, Gu JL, Kang FY, Kong XQ, Mo W. Surface restoration induced by lubricant additive of natural minerals. Appl Surf Sci 2007;253:7549 – 7553.

[35] Bec S, Tonck A. Nanometer scale mechanical properties of tribochemical films. Tribol Ser 1996;31:173 – 184.

[36] Demmou K, Bec S, Loubet JL, Martin JM. Temperature effects on mechanical properties of zinc dithio-

phosphate tribofilms. Tribol Int 2006;39:1558 - 1563.

[37] Pereira G, Munoz - Paniagua D, Lachenwitzer A, Kasrai M, Norton PR, Capehart TW, et al. A variable temperature mechanical analysis of ZDDP - derived antiwear films formed on 52100 steel. Wear 2007;262:461 - 470.

[38] Yu HL, Xu Y, Shi PJ, Xu BS, Wang XL, Liu Q, et al. Characterization and nanomechanical properties of tribofilms using Cu nanoparticles as additives. Surf Coat Technol 2008;203:28 - 34.

[39] Parry AO, Swain PS, Fox J. Fluidad sorption at a non - planar wall: roughnessinduced first - order wetting. J Phys Condens Matter 1996;8:L659 - 666.

[40] Chow TS. Wetting of rough surfaces. J Phys Condens Matter 1998;10:L445 - 451.

[41] Tatsumi Y. Migration of fluid phases and genesis of basalt magmas in subduction zones. J Geophys Res 1989;94:4697 - 4707.

[42] Xie HS, Zhou WG, Li YW, Guo J, Xu ZM. The elastic characteristics of serpentinite dehydration at high temperature - high pressure and its significance. Chin J Geophys 2000;43(6):806 - 811(in Chinese).

[43] Savan A, Pflüger E, Voumard P, Schröer A, Simmonds M. Modern solid lubrication: recent developments and applications of MoS_2. Lubr Sci 2000;12:185 - 203.

再制造工程的现状与前沿*

摘　要　再制造是维修工程和表面工程发展的高级阶段，是先进制造的重要组成，是废旧产品高技术修复、改造的产业化；再制造具有“两型社会、五六七”的特征。中国经济社会发展对再制造具有十分迫切的需求，再制造的初步应用也已取得了非常明显的节能减排效果。国际上，美国的再制造体系较完善，近10年来，中国的再制造发展迅猛，在再制造的政策法规、产业实践及基础研究方面成绩斐然，已成为世界再制造中心之一。再制造的发展前沿可概括为“探索再制造的科学基础，创新再制造的关键技术，制定再制造的行业标准”。

关键词　再制造工程　研究现状　发展前沿

1　再制造的内涵与研究范围

再制造以废旧产品的零部件为毛坯，主要以先进的表面工程技术为修复手段（即在损伤的零件表面制备一层薄的耐磨、耐蚀、抗疲劳的表面涂层），因此无论是毛坯来源还是再制造过程，对能源和资源的需求、对废物废气的排放都是极少的，具有很高的绿色度。再制造具有如下重要特征：再制造产品的质量和性能不低于原型新品，有的还超过原型新品，成本只是原型新品的50%，节能60%，节材70%，对环境的不良影响显著降低，有力促进了资源节约型、环境友好型社会的建设[1]。上述特征可概括为：“两型社会”、“五六七”。

再制造的出现，完善了全寿命周期的内涵，使得产品在全寿命周期的末端，即报废阶段，不再成为固体垃圾。再制造不仅可使废旧产品起死回生，还可很好地解决资源节约和环境污染问题。因此，再制造是对产品全寿命周期的延伸和拓展，赋予了废旧产品新的寿命，形成了产品的多寿命周期循环。这是面向循环经济的再制造的重要理论成果。

再制造的研究内容非常广泛，贯穿产品的全寿命周期，体现着深刻的基础性和科学性。在产品设计阶段，要考虑产品的再制造性设计；在产品的服役至报废阶段，要考虑产品的全寿命周期信息跟踪；在产品的报废阶段，要考虑产品的非破坏性拆解、低排放式物理清洗，要进行零部件的失效分析及剩余寿命演变规律的探索，要完成零部件失效部位的具有高结合强度和良好摩擦学性能的表面涂层的设计、制备与加工，以及对表面涂层和零部件尺寸超差部位的机械平整加工及质量控制等。

2　再制造的迫切性和重要意义

我国已进入机械装备和家用电器报废的高峰期，再制造势在必行。目前全国役龄

* 原发表于《材料热处理学报》，2010，31(1)：10~14。

10 年以上的传统旧机床超过 200 万台，80% 的在役工程机械超过保质期；年报废汽车约 500 万辆，报废电脑、电视机、电冰箱 1600 万台，报废手机 2000 万部，每年产生约 8 亿吨固体废物。

我国的装备运行损失十分惊人，再制造也势在必行。仅以腐蚀和磨损为例，2003 年中国工程院发布腐蚀调查报告[2]：2002 年我国由腐蚀造成的损失近 6000 亿元，占当年 GDP 的 5%。2007 年中国工程院发布摩擦学调查报告[3]：2006 年全国由摩擦磨损造成的损失高达 9500 亿元，占当年 GDP 的 4.5%。两项损失合计 1.55 万亿元，粗略估算占 GDP 的 9.5%，而发达国家仅为 4% ~5%。若能采取有效措施挽回 10% 的损失，则每年可节约 1550 亿元。

再制造的社会效益十分巨大。与相关制造业相比，再制造业的就业人数是其 2 ~3 倍。2005 年，美国再制造业的年产值为 750 亿美元，雇佣员工 100 万人，同年，美国计算机制造业的产值与此相当，雇佣员工只有 35 万人，说明再制造业具有显著的创造就业与再就业的能力。

再制造的资源与环境效益同样十分巨大。据美国 Argonne 国家重点实验室统计，新制造 1 辆汽车的能耗是再制造的 6 倍，新制造 1 台汽车发电机的能耗是再制造的 7 倍，新制造 1 台汽车发动机的能耗是再制造的 11 倍。据对我国第一家再制造领域的循环经济示范试点企业济南复强再制造公司的数据统计，若每年再制造 5 万台斯泰尔发动机，则可节省近 4 万吨金属，实现利税 16.15 亿元，节电 7250 万度，实现利税 1.45 亿元，减少 CO_2 排放 3000t[4]。

中国特色的再制造来源于维修工程和表面工程，是维修工程和表面工程发展的高级阶段；同时，再制造是先进制造的组成部分，属于绿色制造。但是，再制造又明显区别于维修和制造，具有独立的学科方向。与制造相比，再制造有更多的科学和技术基础问题需要独立解决：（1）加工对象更苛刻，制造的对象是经铸锻焊、车铣磨、热处理后的新毛坯，性能均质、单一，而再制造的对象是旧毛坯，即报废的成型零件，存在着尺寸超差、残余应力、内部裂纹和表面变形等一系列缺陷；（2）前期处理更繁琐，制造的毛坯是基本清洁的，很少需要前处理，而再制造的毛坯必须去除油污、水垢、锈蚀层及硬化层；（3）质量控制更困难，制造过程的质量控制已趋成熟，再制造毛坯的寿命预测和质量控制，因毛坯损伤的复杂性和特殊性而变得非常困难；（4）工艺标准更严格，制造过程非常规范，再制造过程中废旧零件的尺寸变形和表面损伤程度各不相同，必须采用更高技术标准的加工工艺。

上述特殊的基础理论和工程需求催生了再制造工程新学科。近年来我国大力推进再制造学科的发展，建立了专门从事再制造研究的国家级重点实验室——装备再制造技术国防科技重点实验室。学科的发展也有力地推动了国家政策的发展。2009 年 1 月，《中华人民共和国循环经济促进法》生效，该法在第 2、第 40 及第 56 条中六次阐述再制造，标志着再制造已进入国家法律。

3 国内外再制造的发展现状

3.1 国外再制造

再制造在欧美发达国家已形成了巨大的产业。2005 年全球再制造业产值已超过

1000 亿美元，美国的再制造产业规模最大，达到 750 亿美元，其中汽车和工程机械再制造占 2/3 以上，约 500 亿美元。

美军高度重视再制造。隶属于美国国家科学研究委员会的“2010 年后国防制造工业委员会”制定了 2010 年国防工业制造技术的框架，将武器系统的再制造列为国防工业的重要研究领域。美军也是再制造的最大受益者。美空军 B-52 战略轰炸机，1962 年生产，1980 年、1996 年两次再制造，到 1997 年时平均自然寿命还有 13000 飞行小时，可服役到 2030 年；2005 年，美空军完成了 269 架阿帕奇直升机的再制造，2015 年前还将完成 750 架的再制造。再制造后的阿帕奇直升机成为美军现役武装直升机中战斗力最强的一种机型[5]。

近年来，日本加强了对工程机械的再制造，至 2008 年，再制造的工程机械中，58% 由日本国内用户使用，34% 出口到国外，其余的 8% 拆解后作为配件出售。至 2004 年，德国大众汽车公司已再制造汽车发动机 748 万台，变速器 240 万台，公司销售的再制造发动机及其配件和新机的比例达到 9∶1。

欧美国家的再制造，在再制造设计方面，主要是结合具体产品，针对再制造过程中的重要设计要素如拆卸性能、零件的材料种类、设计结构与紧固方式等进行研究；在再制造加工方面，对于机械产品，主要通过换件修理法（将失效件更换为新件以完成再制造的方法）和尺寸修理法（将失配的零部件表面尺寸加工修复到可以配合的范围，如缸套－活塞环磨损失效后，通过镗缸的方法恢复缸套的尺寸精度，再配以大尺寸的活塞环以完成再制造的方法）来恢复零部件的尺寸，如英国 Lister Petter 再制造公司，他们每年为英、美军方再制造 3000 多台废旧发动机，再制造时，对磨损超差的缸套、凸轮轴等关键零件都予以更换新件，并不修复。对于电子产品，再制造的内涵就是对仍具有使用价值的零部件予以直接的再利用。如德国柏林工业大学[6]对平板显示器的再制造就是先将液晶显示器 LCD、印刷线路板 PCB、冷阴极荧光灯 CCFL 等关键零部件进行拆解，经检测合格后进行再利用；德国 ReMobile 公司[7]对移动电话的再制造也是先拆解再检测最后再利用；此外还有对数码相机（日本柯达公司）、打印机墨盒（美国施乐公司）、品牌电脑（美国 HP 公司）等的再制造也都是以再利用为主。

3.2 国内再制造

我国再制造产业发展虽晚，但势头非常好，目前已成为世界上重要的再制造中心之一，而且在基础理论研究与技术应用开发方面走在了世界前列。

3.2.1 *在再制造政策方面*

1999 年 6 月，笔者在中国西安召开的“先进制造技术国际会议”上作了《表面工程与再制造技术》的特邀报告，在中国率先提出“再制造”的概念。

2000 年，“再制造工程技术及理论研究”被国家自然科学基金委机械学科列为“十五”优先发展领域，标志着再制造的基础研究已经得到了国家的重视和认可；2005 年，“资源循环型制造与再制造”又被国家自然科学基金委机械学科列为“十一五”优先发展领域，再制造的学术地位得到进一步巩固。

2005 年 7 月，国务院颁布的 21、22 号文件明确表示国家将支持废旧机电产品再制造，并把“绿色再制造技术”列为“国务院有关部门和地方政府加大经费支持力度的关键、共性项目之一”；11 月，国家发改委等 6 部委联合公布了国家首批循环经济示范试点领域及企业名单，再制造成为 4 个重点领域之一。

2006 年 4 月，时任国务院副总理曾培炎在国家发改委上报的《关于汽车零部件再制造产业发展及有关对策措施建议的报告》上批示：“同意以汽车零部件为再制造产业试点，探索经验，研发技术。同时要考虑定时修订有关法律法规”。

2008 年 3 月，国家发改委批准全国 14 家企业作为新一轮“汽车零部件再制造产业试点企业”，其中包括一汽、东风、上汽、重汽等整车制造企业和潍柴、玉柴等发动机制造企业。

2009 年 1 月，《中华人民共和国循环经济促进法》正式生效。该法指出：“国家支持企业开展机动车零部件、工程机械、机床等产品的再制造”。

2009 年 4 月，国务院召开全国循环经济座谈会，中共中央政治局常委、国务院副总理李克强，全国政协副主席、科技部部长万钢，中国工程院院长徐匡迪等领导出席，特邀 9 名循环经济领域的院士和企业家出席会议。笔者在会上做了《中国特色的再制造产业发展现状与对策建议》的发言，重点介绍了中国特色的再制造区别于国外再制造的主要特征，自主研发的再制造高新技术在循环经济试点企业的应用情况，再制造对节能减排的贡献，再制造在抵御全球金融危机方面发挥的重要作用等内容。笔者的发言受到李克强副总理的重视，不断插话、提问题并讨论，表现出对再制造的高度关注。

2009 年 5 月起，为落实李克强副总理的指示，国家发改委和中国工程院先后组织相关院士、专家赴北京、天津、山东、贵州、重庆考察循环经济试点企业。

2009 年 9 月，国家发改委组织了“循环经济专家行”的再制造专项活动，对 2008 年 3 月立项的 14 家汽车零部件再制造试点企业的工作进度、生产状况及技术应用情况进行考察。

3.2.2 在再制造实践方面

仅以装备再制造技术国防科技重点实验室为例，采用等离子喷涂技术，完成了某型主战坦克转向机构重要薄壁零件“行星框架”易热变形的再制造难题，对 6 辆坦克的实车考核结果表明，再制造行星框架的使用寿命达到新品的 3 倍，成本仅为新品的 10%，材料消耗为 1%；英国路虎汽车（Land - Rover）的铝合金发动机缸盖，服役后出现环形压槽，造成气密性下降，英方无法修复，委托重点实验室研究解决。采用材料成型与制备一体化技术成功完成了路虎汽车铝合金发动机缸盖的再制造，突破了铝合金材料零件再制造的国际难题，再制造的发动机已投入实车考核，性能稳定。

3.2.3 在再制造产业化方面

我国已基本构建了再制造产业，越来越多的专业化再制造企业不断出现。仅 2008 年一年，在机械产品领域，就有近 30 家再制造企业挂牌，如二汽康明斯发动机再制造公司、广西玉柴发动机再制造公司等。目前，发动机再制造企业济南复强动力有限公司是我国最大的再制造企业，专门从事斯太尔、康明斯、三菱等种类型号，尤其是重型汽车发动机的再制造。在 2005 年成为国家循环经济示范试点企业后，该公司加强了与装备再制造技术国防科技重点实验室的合作，将最新的纳米表面工程技术和自动化表面工程技术应用于生产线，显著提升了废旧发动机的再制造水平和再制造率，现已达到年产再制造发动机 25000 台的能力。我军对再制造非常重视，近年来在某军工厂建立了全军第一条军用汽车发动机再制造生产线，单班年再制造能力近1万台。

3.2.4　*在再制造基础研究方面*

国内许多单位，如装甲兵工程学院装备再制造技术国防科技重点实验室、上海交通大学、合肥工业大学、山东大学、中科院兰州化物所等，深入开展了再制造的基础研究。在理论基础方面，完善了涂层残余应力的计算方法，探索并初步建立了寿命预测评估模型。如研究并初步提出了再制造零部件涂层中残余应力的计算方法；以废旧柴油机曲轴为对象，研究了非线性动力学分析模型，探讨了废旧零部件疲劳试验数据与模型分析数据的映射关系，初步建立了剩余寿命预测模型[8]；基于金属磁记忆原理和疲劳寿命评定准则，初步提出了废旧零部件的剩余寿命评估和再制造零部件服役寿命预测的方法[9]。在技术基础方面，发展、创新了多项再制造关键技术，并深入研究了相关的基础理论。如研究了高温条件下 Fe－Al 金属间化合物的形成机理，首次将高速电弧喷涂技术与粉芯丝材相结合的方法应用于再制造零部件的表面修复，实现了 Fe－Al（基）金属间化合物的制备与涂层成形一体化技术[10]；发明了一种“双通道、双温区”的超声速等离子喷涂新工艺，解决了涂层熔滴的过熔、夹生及烧损问题[11]；利用具有自主知识产权的高能机械化学法，解决了纳米颗粒在多离子溶液体系中的均匀分散与悬浮稳定的难题，实现了纳米电刷镀过程中非导电的纳米颗粒与导电的基质金属镍的高效共沉积[12]。

4　再制造的发展前沿

放眼未来，中国的再制造应从三个方面予以重点突破，即“探索再制造的科学基础、创新再制造的关键技术、制定再制造的行业标准”：

（1）探索再制造的科学基础，即深入探索研究以产品全寿命周期理论、废旧零件和再制造零件的寿命评估预测理论等为代表的再制造基础理论，以揭示产品寿命演变规律的科学本质。再制造是来自实践的工程科学，经验性更强。废旧零件的剩余寿命是否足够，再制造零件的使用寿命是否可保持一个完整的服役周期？这样一些重大问题，由于缺少理论依据，有时仅凭简单的检测设备，甚至只靠工人师傅的目测或经验判断来完成。为解决这个重大难题，必须探索研究更多更有效的无损检测及寿命预测理论与技术。如应加快研究声发射、交流阻抗、三维辐射 CT、金属磁记忆、超声相控阵等先进的无损检测技术，研究基于断裂力学、弹塑性力学的力学损伤评价理论，以及模拟仿真和虚拟现实技术，准确把握裂纹的萌生征兆，深刻理解裂纹的扩展规律，科学建立零件的寿命模型。

（2）创新再制造的关键技术，即不断创新研发用于再制造的先进表面工程技术群，使再制造零件表面涂层的强度更高、寿命更长，确保再制造产品的质量不低于或超过新品。先后开发成功纳米表面工程技术和自动化表面工程技术，前者包括纳米颗粒复合电刷镀技术、纳米热喷涂技术、纳米减摩自修复添加剂技术等，后者包括自动化电弧喷涂技术、自动化纳米颗粒复合电刷镀技术等。纳米表面工程技术的核心是利用纳米颗粒材料的小尺寸效应，通过在涂层或添加剂中的均匀、弥散分布，实现纳米颗粒与基质金属间原子尺度的化学键结合，从而显著提高涂层的强度学和摩擦学性能；自动化表面工程技术的核心是利用机器人或操作机来取代手工操作，通过自动控制规划路径，实时反馈调节涂层成型工艺参数，实现表面涂层制备的自动化、智能化。上述技术已应用于发动机再制造生产线，如纳米颗粒复合电刷镀技术成功修复了进口飞机发动机压气机叶片，300h 台架试验满足要求，突破了对国外进口产品的国产化维修技术瓶颈；自动化电弧喷涂技术用于重载汽车发动机缸体、曲轴箱体等零件的再制造，单件发动机箱体的再制造时间由 90min 缩短为 20min，且材料消耗仅为零件本体质量的 0.5%，费用投入不超过新品价格的 10%。下一步除了继续完善纳米表面工程技术和自动化表面工程技术外，还需研发生物表面工程技术等新的方向。

（3）制定再制造的行业标准，即尽早建立系统、完善的再制造工艺技术标准、质量检测标准等体现再制造走向规范化的标准体系。国内再制造因起步较晚，再制造企业的技术积累少，再制造的标准缺乏，因而一定程度上阻碍了再制造的广泛应用。2008 年，国家标准化管理委员会批准成立了“全国绿色制造标准化技术委员会再制造分技术委员会”，该委员会正陆续制定并有望近期出台“再制造概念、术语”和“再制造率的概念及评估方法”等共性基础标准。同时，国内相关高等院校和再制造企业正在联合制定“再制造技术工艺标准、再制造质量检测标准、再制造产品认证标准”等多类标准草案，包括再制造发动机工艺流程标准、发动机再制造产品性能评价与质量检测标准、废旧发动机零件剩余寿命评估标准、再制造的关键零件（曲轴、缸体、凸轮轴、连杆轴等）质量检测标准、再制造发动机试车考核标准等。下一步应深化标准内涵，制定出具有良好通用性和可操作性的标准方案。

5 结论

（1）再制造是废旧产品高技术修复、改造的产业化。再制造的重要特征是再制造产品质量和性能不低于新品，有些还能超过新品，成本只是新品的 50%，节能 60%，节材 70%，对环境的不良影响显著降低。可概括为：“两型社会”、“五六七”。

（2）中国特色的再制造来源于维修工程和表面工程，是维修工程、表面工程发展的高级阶段，也是先进制造的重要组成，但是与维修和制造相比，再制造蕴含着更深的科学理论和更高的技术基础。再制造已创造出了巨大的经济和社会效益。

（3）再制造在中国得到快速发展，再制造不仅进入了国家法律，而且在产业化实践和基础研究等方面均取得了良好的阶段性成果，中国已成为国际再制造中心之一，在国际再制造领域发挥着重要作用。

（4）为推动再制造的进一步发展，今后可围绕“探索再制造的科学基础、创新再制造的关键技术、制定再制造的行业标准”等发展前沿展开研究工作。

参 考 文 献

[1] 徐滨士，刘世参，王海斗．大力发展再制造产业［J］．求是，2005，(12)：46～47.
[2] 柯伟．中国腐蚀调查报告［M］．北京：化学工业出版社，2003：10.
[3] 谢友柏，张嗣伟．摩擦学科学及工程应用现状与发展战略研究——摩擦学在工业节能、降耗、减排中地位与作用的调查［M］．北京：高等教育出版社，2009：3.
[4] 徐滨士．再制造工程基础及其应用［M］．哈尔滨：哈尔滨工业大学出版社，2005：10.
[5] 徐滨士．装备再制造工程的理论与技术［M］．北京：国防工业出版社，2007：7.
[6] Seliger G，Frank C，Ciupek M. Process and facility planning for mobile phone remanufacturing［J］.

Annals of the CIRP, 2004, 53 (1): 9 ~ 12.

[7] Frank C, Yakut E. Process design to the mobiles in the remanufacturing network [C]. Proceedings Global Conference on Sustainable Product Development and Life Cycle Engineering, 2004: 191 ~ 198.

[8] 张国庆，荆学东，浦耿强，等. 产品可再制造性评价方法与模型 [J]. 上海交通大学学报，2005，39(9)：1431 ~ 1436.

[9] Dong Lihong, Xu Binshi, Dong Shiyun, et al. Monitoring fatigue crack propagation of materials with spontaneous abnormalmagnetic signals [J]. International Journal of Fatigue, 2008, 30(8): 1599 ~ 1605.

[10] XU Binshi, ZHU Zixin, Ma Shining, et al. Sliding wear behavior of Fe - Al and Fe - Al/WC coatings prepared by high velocity arc spraying [J]. Wear, 2004, 257(11): 1089 ~ 1095.

[11] HAN Zhihai, XU Binshi, WANG Haijun, et al. A comparison between the thermal shock behavior of currently plasm a spray and recently supersonic plasm a spray $CeO_2 - Y_2O_3 - ZrO_2$ graded TBCs [J]. Surface and Coatings Technology, 2007, 201(9 ~ 11): 5232 ~ 5235.

[12] XU Binshi, WANG Haidou, DONG Shiyun, et al. Electrodepositing nickel silica nano - composites coatings [J]. Electrochemistry Communications, 2005, 7(6): 572 ~ 575.

State of the Art and Future Development in Remanufacturing Engineering

Xu Binshi

(Academy of Armored Forces Engineering, National Key Laboratory for Remanufacturing, Beijing, 100072, China)

Abstract Remanufacturing engineering is the senior stage of maintenance engineering and surface engineering, and the important part of advanced manufacturing, and the industrialization for the waste products with repair and reformation. The character, named as "two - oriented society and 567", has been held by remanufacturing engineering. The demand from the rapid economy and society development to remanufacturing in China is very urgent, and the preliminary application of remanufacturing has obtained the obvious effect of energy - saving and discharge - reducing. Around the world, the relatively perfect remanufacturing system has been set up in USA. In the recent 10 years, remanufacturing in China develops quickly, especially in policy, industry and theory of remanufaturing, and now China becomes one of the international remanufacturing centers. The front of remanufacturing could be generalized as "exploring the foundational theory, innovating the key technology, establishing the industrial standard".

Key words remanufacturing engineering, state of the art, future development

Elastoplastic Analysis of Process Induced Residual Stresses in Thermally Sprayed Coatings*

Abstract The residual stresses induced from thermal spraying process have been extensively investigated in previous studies. However, most of such works were focused on the elastic deformation range. In this paper, an elastoplastic model for predicting the residual stresses in thermally sprayed coatings was developed, in which two main contributions were considered, namely the deposition induced stress and that due to differential thermal contraction between the substrate and coating during cooling. The deposition induced stress was analyzed based on the assumption that the coating is formed layer-by-layer, and then a misfit strain is accommodated within the multilayer structure after the addition of each layer (plastic deformation is induced consequently). From a knowledge of specimen dimensions, processing temperatures and material properties, residual stress distributions within the structure can be determined by implementing the model with a simple computer program. A case study for the plasma sprayed NiCoCrAlY on Inconel 718 system was performed finally. Besides some similar phenomena observed from the present study as compared with previous elastic model reported in literature, the elastoplastic model also provides some interesting features for prediction of the residual stresses.

1 Introduction

In the process of coating production by thermal spraying, the material to be deposited is introduced as the form of powder or wire into the core of a high velocity gas jet, where it is melted by the heater such as plasma, electric arc and high velocity oxygen flame (HVOF). The molten particles are atomized and accelerated towards the substrate, resulting in flattening, solidification (within a very short time, about few milliseconds), and then coating formation. Due to the large temperature difference during the process, residual stress is unavoidably generated in the coating/substrate structure, which strongly affects the coating properties and therefore the service life. Generally, the residual stress can be divided into two main sources[1,2]: "quenching" (primary) and "thermal" (secondary) stresses. During the deposition, the temperature of molten particles striking on the substrate drops instantaneously, thermal contraction is constrained by the underlying solid, which results in tensile stress being developed in the splat layer, known as "quenching", "intrinsic" or "deposition" stress. After spraying, the deposit and substrate cool down to room temperature, accompanied with thermal cooling stress (tensile or compressive) due to mismatch of the coefficient thermal expansion (CTE) between the materials. As a result, the

* Copartner: Chen Yongxiong, Liang Xiubing, Liu Yan. Reprinted from *Journal of Applied physics*, 2010, 108, 013517: 1-9.

residual stress is locked into the coating at the room temperature and any subsequent usage of the coating components. In most cases, residual stresses have adverse effects on coating properties, such as bonding strength, resistance to thermal shock, fatigue life under bending and erosion resistance. Understanding of the development of stresses during spraying is therefore of utmost importance, in order to devise counter measures and to have desirable coating properties.

As an important factor controlling the coating properties, the residual stress has been widely investigated in previous works. Among them the Stoney equation[3] can be considered as one of the earliest achievements, which can be conveniently used to determine the residual stress in very thin films from the knowledge of curvature. Based on the Stoney equation[4,5], some ex situ and in situ curvature experimental methods were developed to measure stresses. However, this equation may result in considerable error if applied in thick films, and only an average stress can be obtained in the whole film structure. Besides the curvature methods, other experimental methods to determine stresses in coatings were also developed and modified, e. g. X - ray and neutron diffractions, material removal methods[6,7]. Each method has specific requirements on the specimens, instruments, measurement and evaluation procedures, and they also differ in the "richness" and nature of provided information[4]. For the X - ray and neutron diffraction stress measurement methods, as an example, one main limitation is that only on the coating surface the stress can be detected. This is a significant problem in many applications, because the stress status within the coating/substrate structure may be more important than that on the coating surface if the maximum stress locates within the structure.

Besides the experimental methods for prediction of coating residual stresses, numerical and analytical modeling can also be considered as useful candidates[8-10], because both the stress on the coating surface and that within the structure can be determined by the models. For example, the finite element (FE) modeling technique has been widely used to simulate the residual stress generation during the thermal spraying process. But some models considered the stresses only due to differential thermal contraction and neglect the quenching contribution, and others assumed that deformation is limited in elastic range[11-14]. These assumptions are often totally unjustified. Recently some FE models were developed based on the elastoplastic assumption for thermally sprayed coatings, especially for the graded or multilayered composite coatings[15-18]. The numerical results obtained from the FE models presented detailed stress distributions in the whole coating and substrate. Some of them can be used to predict the coating residual stress distributions induced from the thermal spraying process. But it is always unaffordable for the computer ability when a thick coating is to be considered. Analytical modeling provides another way in understanding the stress distributions in thermally sprayed coatings, which is more cost - and - time efficient for calculation as compared with the FE method. Tsui and Clyne et al. developed analytical models to estimate the generation of stresses in coatings deposited onto planar[6,19,20] and cylindrical[21] specimens, assuming that the coating is formed with a large amount of thin layers continuously deposited one by one, and the quenching stresses are taken into account. In their studies, the models were limited in the elastic range and the intrinsic (quenching) stress should be pre - determined before calculation of the final residual stress was performed. In reality, the thermal contraction of a thermal - sprayed splat during quenching is usually large, which may initiate plastic deformation in the structure, and it's difficult to obtain the real value of intrinsic stress for the thin layers. Therefore, the present study is aimed at developing an elastoplastic analytical model to investigate the residual stresses in thermally sprayed coatings in the whole elastic - plastic range, where the intrinsic stress need not be pre - specified. Section Ⅱ presents the quenching stress formation, in which the coating system with planar geometry is considered as a multilayer composite beam which is formed with progressively deposited thin layers. Section Ⅲ describes the formation of thermal stress in the coating system during cooling down to the room temperature. Section Ⅳ gives the result of the final residual stresses. Finally a case study is presented in Section Ⅴ.

2 Formation of Deposition Induced Stress

The geometry of the multilayered material considered here is such that the problem can be reduced to one dimension, and that analytically tractable solutions can be used. For analytical simplicity, attention is confined to those cases where only thermal, elastic and plastic deformations are considered, while other factors affecting the stresses such as creeping, microcracks and micropores in the material can be ignored. The interfaces between the layers (including the coating/substrate interface) are assumed to be perfectly bonded at all times. For most thermal spraying processes, the feedstock can be heated upper its melting point, and deposited on the substrate in the form of small molten droplets. So the initial temperature of each depositing layer is assumed to be the melting point of the material, and the layered material is consequently considered to be initially stress - free. It should be noted that in some cases the spraying particles may have an initial temperature lower than the melting temperature (e. g. for the process of HVOF), where significant peening stresses are generated during the impact of the semi - molten particles on the substrate[22]. Prediction of the residual stresses for these cases is therefore become more complex, which is not taken into account in the present study.

2.1 Deposition of the first layer (bilayer structures)

Fig. 1 schematically illustrates the geometry of the layered structure. Consider that the sample is long enough as compared with its total thickness. The analysis developed for the plane stress problem can be easily extended to the biaxial case by simply replacing Young's modulus E by $E/(1-v)$ where v is Poisson's ratio; the equal biaxial stress state represents the most realistic geometrical condition in the layered material.

Assume that the first layer impinging on the substrate is quenched from its initial temperature T_{in} to the deposition temperature T_d, and that the temperature change of the substrate induced by the deposition is so little that can be ignored, namely the temperature of the substrate is equal to T_d which is assumed to be fixed during the process. The misfit strain $\Delta\varepsilon$ between the

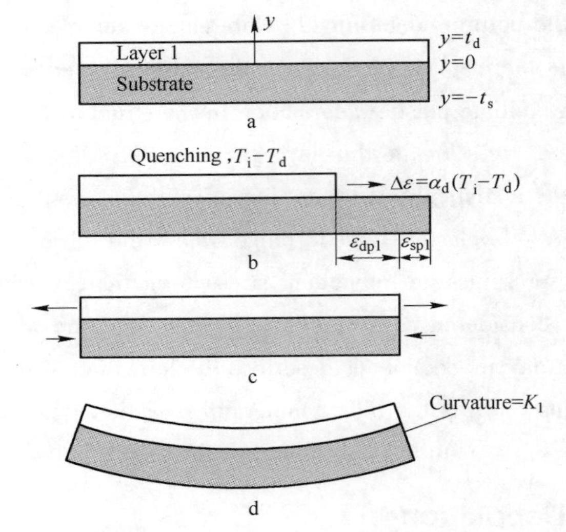

Fig. 1 Schematics depiction of the generation of strains due to the deposition of the first layer on the substrate surface

a—Stress - free condition; b—Quenching induced misfit strain (unconstrained) $\Delta\varepsilon$;

c—Planar strain component of the layer ε_{dpl} and the substrate ε_{spl};

d—Asymmetric stresses induced bending

first layer and substrate is given by

$$\Delta\varepsilon = \alpha_d(T_{in} - T_d) = \alpha_d \Delta T \tag{1}$$

where α_d is the deposit CTE. Imposition of this misfit strain sets up a tensile force acting on the deposit and a compressive force acting on the substrate. Because of this pair of equal and opposite forces, a bending moment is generated in the structure. The bending strain ε_K can be defined as

$$\varepsilon_K = (K_1 - K_0)(y - y_{el})(-t_s \leqslant y \leqslant t_d) \tag{2}$$

where K_1, K_0 and y_{el} is the curvature due to the deposition of the first layer, the initial curvature and the position of the neutral axial, respectively. Normally the initial curvature K_0 would be equal to zero. The curvature is defined to be negative in this case, as shown in Fig. 1. For the deposit and substrate, the total strain is therefore composed of two parts, one is correlated with the planar force, and the other is correlated with the bending, which can be expressed as

$$\varepsilon_{dl} = \varepsilon_{dpl} + (K_1 - K_0)(y - y_{el})(0 \leqslant y \leqslant t_d) \tag{3}$$

$$\varepsilon_{sl} = \varepsilon_{spl} + (K_1 - K_0)(y - y_{el})(-t_s \leqslant y \leqslant 0) \tag{4}$$

Note that $K_1 - K_0$ and y_{el} are independent of y, and that $\varepsilon_{dpl} - \varepsilon_{spl} = \Delta\varepsilon = \alpha_d\Delta T$. Eq. (3) and Eq. (4) can be redefined as

$$\varepsilon_{dl} = (K_1 - K_0)y + \delta_1 + \alpha_d\Delta T(0 \leqslant y \leqslant t_d) \tag{5}$$

$$\varepsilon_{sl} = (K_1 - K_0)y + \delta_1(-t_s \leqslant y \leqslant 0) \tag{6}$$

where $\delta_1 = \varepsilon_{spl} - (K_1 - K_0)y_{el}$. For convenience in the discussion followed, it is assumed that the material constants are temperature independent and that the substrate is always in the elastic deformation range, the stress versus strain relationship of the substrate is therefore given by

$$\sigma_{sl} = E_s\varepsilon_{sl} \tag{7}$$

In most cases, the temperature difference $T_{in} - T_d$ is significant and the contraction of the layer is usually large (at least several millistrain), which results in that the deposited layer may be beyond its elastic deformation range. The stress versus strain relationship of most coating materials in a range slightly over the elastic limit may be estimated using a bilinear model as illustrated in Fig. 2, where E_d, H_d, and σ_Y are the young's modulus, strain hardening (in the tensile direction), and initial yield stress, respectively. Here, providing the temperature change of the deposit is higher than the critical value (this critical value, denoted by ΔT_1, will be solved later), which leads to the whole layer reaches the plastic state, the stress versus strain relationship of the deposit can be expressed as[23,24]

$$\sigma_{d1} = H_d\left(\varepsilon_{d1} - \frac{\sigma_Y}{E_d}\right) + \sigma_Y \tag{8}$$

For the deposit/substrate structure, the following equilibrium conditions should be met

$$\int_{-t_s}^{0} \sigma_{s1} dy + \int_{0}^{t_d} \sigma_{d1} dy = 0 \tag{9a}$$

$$\int_{-t_s}^{0} \sigma_{s1}(y - y_{e1}) dy + \int_{0}^{t_d} \sigma_{d1}(y - y_{e1}) dy = 0 \tag{9b}$$

Rearranging Eq. (9b) results in

$$\int_{-t_s}^{0} \sigma_{s1} y dy + \int_{0}^{t_d} \sigma_{d1} y dy - y_{e1}\left(\int_{-t_s}^{0} \sigma_{s1} dy + \int_{0}^{t_d} \sigma_{d1} dy\right) = 0 \tag{9c}$$

Referring to Eq. (9a), Eq. (9c) can be reduced to

$$\int_{-t_s}^{0} \sigma_{s1} y dy + \int_{0}^{t_d} \sigma_{d1} y dy = 0 \tag{9d}$$

Combining Eq. (5) – Eq. (9a), and Eq. (9d), the explicit expressions for K_1, δ_1, σ_{d1}, and σ_{s1} can be conveniently obtained as

$$K_1 = -\frac{6E_s t_s t_d (t_s + t_d)[H_d(\alpha_d \Delta T - \sigma_Y/E_d) + \sigma_Y]}{H_d^2 t_d^4 + E_s^2 t_s^4 + 2H_d E_s t_d t_s (2t_d^2 + 3t_d t_s + 2t_s^2)} \tag{10}$$

$$\delta_1 = K_1 \frac{H_d t_d^3 + 4E_s t_s^3 + 3E_s t_s^2 t_d}{6E_s t_s (t_s + t_d)} \tag{11}$$

$$\sigma_{d1} = K_1 \frac{H_d t_d [6(t_s + t_d) y - 3t_d t_s - 4t_d^2] - E_s t_s^3}{6t_d (t_s + t_d)} \tag{12}$$

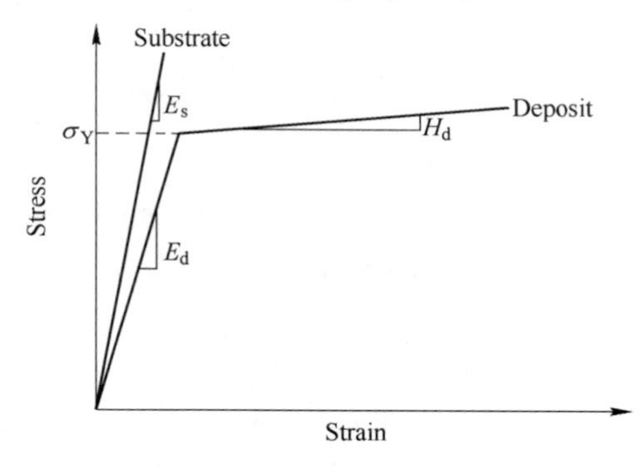

Fig. 2　Schematic representation of the stress – strain curve for the coating and substrate

$$\sigma_{s1} = K_1 \frac{H_d t_d^3 - 2E_s t_s^3 - 3E_s t_s^2 t_d + 6E_s t_s (t_s + t_d)(y + t_s)}{6t_s(t_s + t_d)} \tag{13}$$

It should be noted that Eq. (10) – Eq. (13) are identical to that reported in literature[23,25], but there are some differences between them. The solution in literature[23,25] is based on the linear strain assumption while the presents study is based on the generation of force and moment. Moreover, the final objective of the present study is to obtain an elastoplastic solution for thermal sprayed multilayer structures, while the discussions in previous literature[23,25] are only limited in bilayer or tri – layer structures.

If the above solutions are compared with that performed in literature[19], it can be found that, since there are some similarities of the solution procedure used between the two models, the present study deduced the generation of stresses in the structure based on the assumption of temperature variation induced misfit strain, where using the fixed quenching stress is avoided (while it is used in literature[19]), moreover, plastic deformation of the layer during deposition is taken into account in the present study.

2.2 The critical temperature variation required for full plastic deformation

From the above discuss, it can be found that with the increase of temperature difference (i. e. ΔT), the stress in the coating layer will increase accordingly. If the temperature difference is increased to a critical value, plastic deformation will generate in the coating. According to Eq. (12), it can be found that the maximum and minimum stresses in the coating layer appear at the coating/substrate interface and the top surface of the coating layer, respectively. The minimum stress can be expressed as

$$\begin{aligned}\sigma_{d1}^{min} &= K_1 \frac{H_d t_d (3t_d t_s + 2t_d^2) - E_s t_s^3}{6t_d(t_s + t_d)} \\ &= \frac{E_s t_s (H_d \alpha_d \Delta T - \sigma_Y H_d/E_d + \sigma_Y)[E_s t_s^3 - H_d t_d(3t_d t_s + 2t_d^2)]}{H_d^2 t_d^4 + E_s^2 t_s^4 + 2H_d E_s t_d t_s (2t_d^2 + 3t_d t_s + 2t_s^2)}\end{aligned} \tag{14}$$

When plastic deformation reaches the whole coating layer, the minimum stress in the layer must be equal to the initial yield stress. As defined previously, ΔT_1 is the critical temperature at which plastic deformation just reaches the whole layer. Therefore, substituting $\sigma_{d1}^{min} = \sigma_Y$ into Eq. (14), ΔT_1 can be obtained as

$$\Delta T_1 = \frac{\sigma_Y}{E_d \alpha_d} + \frac{\sigma_Y}{E_s \alpha_d} \cdot \frac{H_d (t_d/t_s)^4 + 6E_s (t_d/t_s)^3 + 9E_s (t_d/t_s)^2 + 4E_s (t_d/t_s)}{E_s - 2H_d (t_d/t_s)^3 - 3H_d (t_d/t_s)^2} \tag{15}$$

In most thermal spraying cases, the thickness of the deposited layer (t_d) is about 2 ~ 3 levels lower than that of the substrate (t_s). Therefore, Eq. (15) can be estimated by

$$\Delta T_1 \approx \frac{\sigma_Y}{E_d \alpha_d} \tag{16}$$

Since σ_Y/E_d is equal to the elastic strain limit of the coating material, as shown in Fig. 2, Eq. (16) manifests that the critical temperature ΔT_1 corresponds to the condition that misfit strain initiated from temperature variation reaches the material elastic limit. For the spraying process, as mentioned before, the temperature difference $T_{in} - T_d$ is significant, and the misfit strain $\Delta\varepsilon = \alpha_d \Delta T$ is much higher than the elastic strain limit, therefore the condition of Eq. (15) can be easily satisfied for most thermal sprayed materials. The following study will only focus on the case in which this condition is satisfied. Since the minimum stress is higher than the initial yield stress and plastic deformation is generated in the whole layer, discussion about the maximum stress is somewhat unnecessary here.

2.3 Deposition of the second layer (tri – layer structures)

Consider the second layer impinging on the coated substrate (as shown in Fig. 3). The initial temperature of the second layer and the deposition temperature of the substrate (together with the first layer) are the same as before, so the misfit strain between the second layer and the coated substrate can also be expressed by Eq. (1). The strains in this stage for the two coating layers and the substrate are given as

$$\varepsilon_{d2} = \varepsilon_{dp2} + (K_2 - K_1)(y - y_{e2})\ (t_d \leqslant y \leqslant 2t_d) \quad (17)$$

$$\varepsilon_{d1} = K_1 y + \delta_1 + \alpha_d \Delta T + \varepsilon_{sp2} + (K_2 - K_1)(y - y_{e2})\ (0 \leqslant y \leqslant t_d) \quad (18)$$

$$\varepsilon_{s2} = K_1 y + \delta_1 + \varepsilon_{sp2} + (K_2 - K_1)(y - y_{e2})\ (-t_s \leqslant y \leqslant 0) \quad (19)$$

where ε_{dp2} and ε_{sp2} is the strain component correlated with the planar force of the second layer

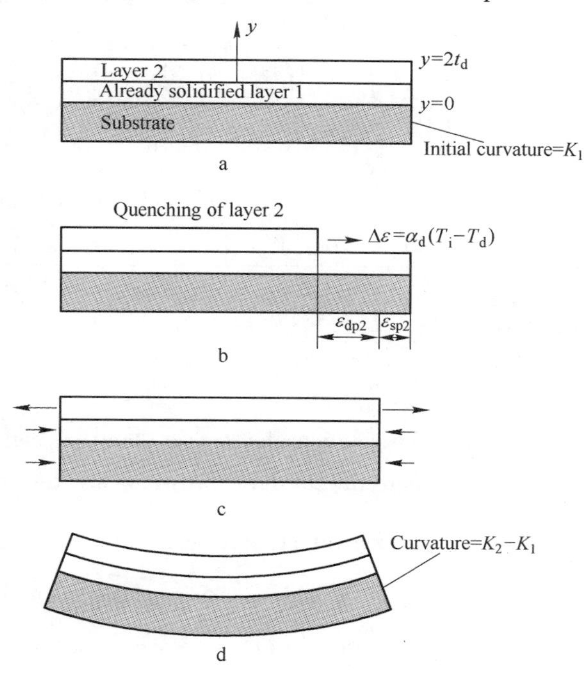

Fig. 3 Schematics depiction of the generation of strains due to the deposition of the second layer on the coated substrate surface

a—Stress – free condition; b—Quenching induced misfit strain (unconstrained) between the second layer and the coated substrate $\Delta\varepsilon$;

c—Planar strain component of the layer ε_{dp2} and the substrate ε_{sp2};

d—Asymmetric stresses induced bending

and the coated substrate (together with the first layer), respectively. Their relationship is

$$\varepsilon_{dp2} - \varepsilon_{sp2} = \Delta\varepsilon = \alpha_d \Delta T \tag{20}$$

If it is assumed that $\delta_2 = \varepsilon_{sp2} - (K_2 - K_1) y_{e2}$, Eq. (17) – Eq. (19) can be rearranged as

$$\varepsilon_{d2} = (K_2 - K_1) y + \delta_2 + \alpha_d \Delta T (t_d \leqslant y \leqslant 2t_d) \tag{21}$$

$$\varepsilon_{d1} = K_1 y + \delta_1 + \alpha_d \Delta T + (K_2 - K_1) y + \delta_2 (0 \leqslant y \leqslant t_d) \tag{22}$$

$$\varepsilon_{s2} = K_1 y + \delta_1 + (K_2 - K_1) y + \delta_2 (-t_s \leqslant y \leqslant 0) \tag{23}$$

Consider the elastic deformation for the substrate and plastic deformation for the whole second layer. Their stress versus strain relationships are given by

$$\sigma_{s2} = E_s \varepsilon_{s2} \tag{24}$$

$$\sigma_{d2} = H_d \left(\varepsilon_{d2} - \frac{\sigma_Y}{E_d} \right) + \sigma_Y \tag{25}$$

For the first layer, consider that the stress component resulted from deposition of the second layer is always compressive (the tensile possibility will be discussed later), and that the modulus in the compressive direction of the plastically deformed coating material can be set equal to its elastic modulus, the stress versus strain relationship for the first layer is then given as

$$\sigma_{d1} = H_d \left(K_1 y + \delta_1 + \alpha_d \Delta T - \frac{\sigma_Y}{E_d} \right) + \sigma_Y + E_d [(K_2 - K_1) y + \delta_2] \tag{26}$$

The stress and momentum equilibrium conditions are

$$\int_{-t_s}^{0} \sigma_{s2} \mathrm{d}y + \int_{0}^{t_d} \sigma_{d1} \mathrm{d}y + \int_{t_d}^{2t_d} \sigma_{d2} \mathrm{d}y = 0 \tag{27}$$

$$\int_{-t_s}^{0} \sigma_{s2} y \mathrm{d}y + \int_{0}^{t_d} \sigma_{d1} y \mathrm{d}y + \int_{t_d}^{2t_d} \sigma_{d2} y \mathrm{d}y = 0 \tag{28}$$

Combining Eq. (21) – Eq. (28), the explicit expressions of δ_2 and K_2 can be obtained as

$$\delta_2 = -t_d \frac{(4t_s^3 E_s + 9t_s^2 E_s t_d - 4H_d t_d^3)(-H_d \sigma_Y + \sigma_Y E_d + \alpha_d \Delta T H_d E_d)}{E_d (32 t_s t_d^3 H_d E_s + 24 t_s^2 t_d^2 H_d E_s + 8 t_s^3 t_d H_d E_s + E_s^2 t_s^4 + 16 H_d^2 t_d^4)} \tag{29}$$

$$K_2 = 6\delta_2 \frac{[2t_d^6 H_d^3 + 2t_s^6 E_s^3 + 4t_s^5 E_s^3 t_d + t_s t_d H_d E_s (12 t_s^4 E_s + 52 t_s^3 t_d E_s + 8 t_s^2 t_d^2 H_d + 78 t_s^2 t_d^2 E_s + 44 t_s t_d^3 E_s + 29 t_s t_d^3 H_d + 27 t_d^4 H_d)]}{(E_s^2 t_s^4 + 6 t_s^2 t_d^2 H_d E_s + H_d^2 t_d^4 + 4 t_s t_d^3 H_d E_s + 4 t_s^3 t_d H_d E_s)(4 t_s^3 E_s + 9 t_s^2 E_s t_d - 4 H_d t_d^3)} \tag{30}$$

With the expressions of δ_2 and K_2, the solutions for σ_{d2}, σ_{d1} and σ_{s2} can be conveniently obtained, while they are not presented here due to the expressions are very long.

2.4 Deposition of the nth layer (multilayer structures)

The above procedure can be extended to analysis deposition of the nth layer. The strain for n layers of the coating as well as the substrate can be expressed as follows

$$\varepsilon_{dn} = (K_n - K_{n-1}) y + \delta_n + \alpha_d \Delta T ((n-1) t_d \leqslant y \leqslant n t_d) \tag{31}$$

$$\varepsilon_{dj} = (K_j - K_{j-1}) y + \delta_j + \alpha_d \Delta T + \sum_{i=j+1}^{n} [(K_i - K_{i-1}) y + \delta_i] ((j-1) t_d \leqslant y \leqslant j t_d) \tag{32}$$

$$\varepsilon_s^{ns} = K_n y + \sum_{i=1}^{n} \delta_i (-t_s \leqslant y \leqslant 0) \tag{33}$$

where $1 \leqslant j \leqslant n-1$. Assume that, for all previously deposited $n-1$ layers, the stress component resulted from deposition of the nth layer is still compressive. The strain versus stress relationship for the coating/substrate structure can be obtained as

$$\sigma_{dn} = H_d[(K_n - K_{n-1})y + \delta_n + \alpha_d\Delta T - \sigma_Y/E_d] + \sigma_Y((n-1)t_d \leqslant y \leqslant nt_d) \tag{34}$$

$$\sigma_{dj} = H_d[(K_j - K_{j-1})y + \delta_j + \alpha_d\Delta T - \sigma_Y/E_d] + \sigma_Y + E_d\left\{\sum_{i=j+1}^{n}[(K_i - K_{i-1})y + \delta_i]\right\}((j-1)t_d \leqslant y \leqslant jt_d) \tag{35}$$

$$\sigma_{sn} = E_s\left(K_n y + \sum_{i=1}^{n}\delta_i\right)(-t_s \leqslant y \leqslant 0) \tag{36}$$

The equilibrium conditions become

$$\int_{-t_s}^{0}\sigma_{sn}\mathrm{d}y + \sum_{j=1}^{n-1}\int_{(j-1)t_d}^{jt_d}\sigma_{dj}\mathrm{d}y + \int_{(n-1)t_d}^{nt_d}\sigma_{dn}\mathrm{d}y = 0 \tag{37}$$

$$\int_{-t_s}^{0}\sigma_{sn}y\mathrm{d}y + \sum_{j=1}^{n-1}\int_{(j-1)t_d}^{jt_d}\sigma_{dj}y\mathrm{d}y + \int_{(n-1)t_d}^{nt_d}\sigma_{dn}y\mathrm{d}y = 0 \tag{38}$$

With the above equations, expressions of $K_1, K_2, \cdots, K_n$ and $\delta_1, \delta_2, \cdots, \delta_n$ as well as the stresses set up due to deposition are determined. This can be done by writing a simple computer program, which is straightforward and time – effective.

It should be noted that it is possible that, for some previously deposited layers, the stress (strain) component resulted from the deposition of nth layer becomes tensile in some cases. But it is clear that the resulted planar strain (due to the thermal contraction of nth layer) acting on the previously deposited layers and substrate is still compressive. The sign of the total strain component (the sum of planar strain and the bending strain) is therefore determined by the bending component, namely only when the bending strain becomes positive (tensile) and its absolute value is higher than that of the planar strain, the total strain may become tensile. As defined in Eq. (2), the bending strain for the material located lower than the neutral axis is positive and that higher than the neutral axis is negative. As a result, if the bending strain component for some deposited layers changes from compressive to tensile, it must starts from the layer/substrate interface (i. e. at point $y = 0$). For a coating/substrate structure in pure elastic range, the neutral axis position y'_{en} can be given as[19]

$$y'_{en} = \frac{(nt_d)^2E_d - t_s^2E_s}{2(nt_dE_d + t_sE_s)} \tag{39}$$

If $y'_{en} > 0$, the neutral axis will located in the deposit, and the layers located lower than y'_{en} will become tensile, and it is possible that some of them have a tensile total strain component mentioned above. While for a coating/substrate structure in an elastic – plastic range, if plastic deformation is initiated in the layers, the neutral axis position will lower than that of pure elastic coating/substrate structure under the same conditions (e. g. the same dimensions, load and material). Therefore, an approximatively maximum thickness of nt_d can obtained if Eq. (39) is equal to zero, which can be used to predict the critical status in which the total strain component changes from compressive to tensile. This value is very conservative because when the neutral axis position reaches to the substrate/coating interface, only the bending strain becomes positive (tensile) in the layers below the neutral axis, but the total stain (sum of the bending and planar strain components) in these layers is still negative. In another side, this prediction method is enough because few industrial coatings have the thickness beyond this critical value.

3 Formation of CTE Mismatch Induced Stress During Cooling

When the whole coating/substrate structure cools down to room temperature, stresses due to CTE mismatch will generate in the structure. Therefore, a misfit strain between the two bonded plates, $\Delta\varepsilon = (\alpha_d - \alpha_s)(T_d - T_r)$, is created. Similar to that described in section 2, a planar strain/stress and a bending strain/stress are acted on the whole coating and substrate due to the misfit strain between the coating and substrate. The final curvature after cooling is K_c, while K_n is the curvature adopted after the last layer has been deposited. For convenience in expressions followed, the coating thickness is defined as h (i. e. $h = nt_d$). The total strain of the system is given as

$$\varepsilon_{dc} = \varepsilon_{dp} + (K_c - K_n)(y - y_{ec})(0 \leqslant y \leqslant h) \tag{40}$$

$$\varepsilon_{sc} = \varepsilon_{sp} + (K_c - K_n)(y - y_{ec})(-t_s \leqslant y \leqslant 0) \tag{41}$$

Consider $\delta_c = \varepsilon_{sp} - (K_c - K_n)y_{ec}$. Eq. (40) – Eq. (41) can be rearranged as

$$\varepsilon_{dc} = (K_c - K_n)y + \delta_c + (\alpha_d - \alpha_s)(T_d - T_r)(0 \leqslant y \leqslant h) \tag{42}$$

$$\varepsilon_{sc} = (K_c - K_n)y + \delta_c(-t_s \leqslant y \leqslant 0) \tag{43}$$

As discussed previously, in the progressive deposition process, the tensile component of stress in each layer is progressively reduced by deposition of successive layers on top of it (see Eq. (35)). Therefore, a rebound process to some degree has been generated in the deposited layers, which may result in that the stress acting on some layers is below the material yielding strength again. Consider the tensile deformation ($\varepsilon_{dc} > 0$ or $\varepsilon_{sc} > 0$) as a loading process and the compressive deformation ($\varepsilon_{dc} < 0$ or $\varepsilon_{sc} < 0$) as an unloading process for the plastically deformed deposit. The stress status will become very complex in the deposit during cooling down because the sign of the deformation is determined by the parameters such as the deposition temperature, CTE difference between the coating and substrate, and sample dimensions as well as the pre-existed stress magnitude during the deposition. In other words, besides the deposition induced stresses generated in the layers, there are several possibilities that an additional loading or unloading process acting on the coating, therefore, it is difficult to list out every possibility in the analytical model, and to list out whether elastic modulus or strain hardening modulus should be used to describe the stress versus strain relationship for the coating. For convenience in discussion followed, the CTE mismatch induced strain in the coating is only limited in the case that $\varepsilon_{dc} < 0$, the relationship of stress versus strain generated in the cooling process for the substrate and coating is given as

$$\sigma_{sc} = E_s\varepsilon_{sc} \tag{44}$$

$$\sigma_{dc} = E_d\varepsilon_{dc} \tag{45}$$

Consider the equilibrium conditions below

$$\int_{-t_s}^{0}\sigma_{sc}\mathrm{d}y + \int_{0}^{h}\sigma_{dc}\mathrm{d}y = 0 \tag{46}$$

$$\int_{-t_s}^{0} \sigma_{sc} y \mathrm{d}y + \int_{0}^{h} \sigma_{dc} y \mathrm{d}y = 0 \tag{47}$$

Combining Eq. (42) – Eq. (47), the solutions for $K_c - K_n$, δ_c, σ_{dc}, and σ_{sc} can be obtained as

$$K_c - K_n = -\frac{6ht_s E_s E_d (t_s + h)(\alpha_d - \alpha_s)(T_d - T_r)}{E_d^2 h^4 + E_s^2 t_s^4 + 2ht_s E_d E_s (2h^2 + 3ht_s + 2t_s^2)} \tag{48}$$

$$\delta_c = (K_c - K_n) \frac{E_d h^3 - 2E_s t_s^3 - 3E_s t_s^2 h}{6E_s t_s (t_s + h)} \tag{49}$$

$$\sigma_{dc} = (K_c - K_n) \frac{E_d h[6(t_s + h)y - 3ht_s - 4h^2] - E_s t_s^3}{6h(t_s + h)} \tag{50}$$

$$\sigma_{sc} = (K_c - K_n) \frac{E_d t_d^3 - 2E_s t_s^3 - 3E_s t_s^2 t_d + 6E_s t_s (t_s + t_d)(y + t_s)}{6t_s (t_s + t_d)} \tag{51}$$

4 Final Residual Stresses

From the above knowledge, the final residual stresses distribution in the substrate and coating layer can be conveniently obtained by adding the contributions represented by Eq. (51) to Eq. (36), and adding Eq. (50) to Eq. (34) – Eq. (35), respectively.

$$\sigma_{sn}^{res} = (K_c - K_n) \frac{E_d t_d^3 - 2E_s t_s^3 - 3E_s t_s^2 t_d + 6E_s t_s (t_s + t_d)(y + t_s)}{6t_s (t_s + t_d)} + $$

$$E_s \left(K_n y + \sum_{i=1}^{n} \delta_i\right) (-t_s \leqslant y \leqslant 0) \tag{52}$$

$$\sigma_{dn}^{res} = (K_c - K_n) \frac{E_d h[6(t_s + h)y - 3ht_s - 4h^2] - E_s t_s^3}{6h(t_s + h)} + $$

$$H_d[(K_n - K_{n-1})y + \delta_n + \alpha_d \Delta T - \sigma_Y / E_d] + \sigma_Y ((n-1)t_d \leqslant y \leqslant nt_d) \tag{53}$$

$$\sigma_{dj}^{res} = (K_c - K_n) \frac{E_d h[6(t_s + h)y - 3ht_s - 4h^2] - E_s t_s^3}{6h(t_s + h)} + $$

$$H_d[(K_j - K_{j-1})y + \delta_j + \alpha_d \Delta T - \sigma_Y / E_d] + \sigma_Y + $$

$$E_d \left\{ \sum_{i=j+1}^{n} [(K_i - K_{i-1})y + \delta_i] \right\} ((j-1)t_d \leqslant y \leqslant jt_d) \tag{54}$$

Therefore, the solution of the residual stresses is just a simple addition of the deposition induced quenching stresses and the cooling induced thermal stresses solved in the above sections.

5 A Case of Elastoplastic Analysis

Once the dimensions, material properties of the substrate and coating layer are determined, implementation of the model is fairly straightforward. In order to illustrate the use of the present model, typically atmospheric plasma sprayed (APS) NiCoCrAlY coating on the Ni alloy substrate is considered, because this coating system is often used in industry and widely investigated in previous literature. The properties required for the model are listed in Table 1. The flow chart giving the sequence of the residual stress calculation is shown in Fig. 4. It should be noted that the current model does not consider the variation of material properties with temperature, which would affect the accuracy of the predictions if the deposition temperature is very high and the properties varies considerably with temperature.

Table 1 Material properties used in the analytical model[1,26,27]

Property	Material	
	Inconel 718	NiCoCrAlY
CTE/M · K^{-1}	14.4	11.6
Yong's modulus/GPa	200	204
Poisson's ratio	0.3	0.3
Yield start stress/MPa	1185	270
Strain hardening/GPa	—	5
Melting point/K	—	1673

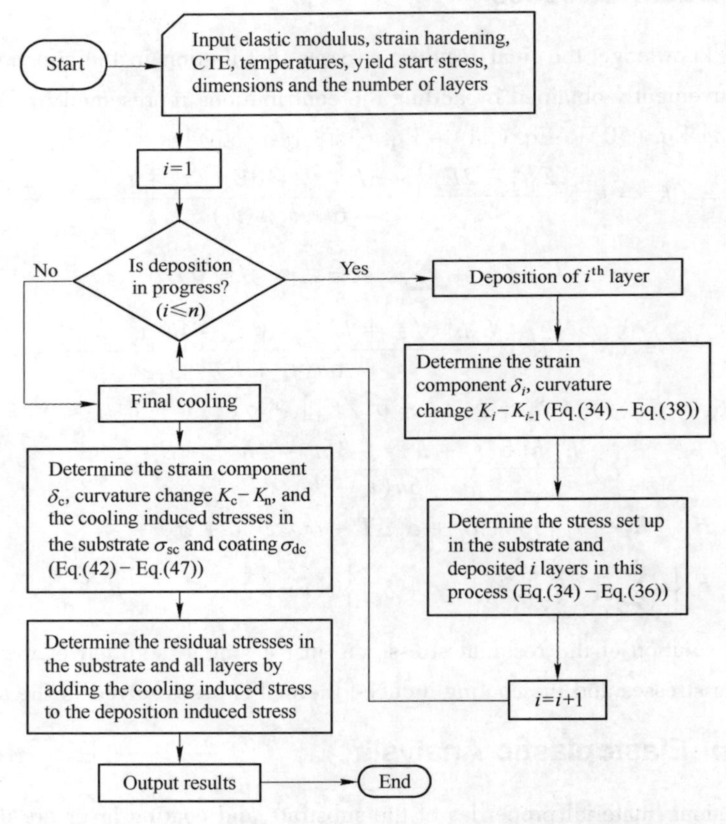

Fig. 4 Flow chart showing the main calculation processes for the present model

When using the model, selection of n (hence h) is arbitrary. Fig. 5 shows the effect of vary n on the predicted residual stress distribution due to quenching only for the APS NiCoCrAlY on Inconel 718 system. For the case when n is equal to 1, i. e. block deposition, the stress level and the stress gradient in the structure (especially in the coating) are significantly different to those when n is equal to 5 or 100. This is due to the fact that, in the progressive deposition process, the tensile stress in each layer, as mentioned previously, is progressively reduced by deposition of successive layers on top of it, because the deposition induced stress component in the successive layers is always compressive. Therefore, the stress at the coating top surface should be more tensile as compared to that at the interface (i. e. a positive stress gradient). But for the case that the process is considered as a block deposition, a negative stress gradient is obtained. This is because the stress component originated from the curvature in each layer is compressive. In fact, such a negative stress gradient is generated within every layer, which is very apparent for the $n = 5$ plot. However, as n becomes large, this effect becomes insignificant because the thickness of each layer is too small to be identified from the plot, and the real trend is that of the stress rising from the bottom to the top surface of the coating. Therefore, using a suitable large value for n will be more identical to the reality and will give a fairly accurate prediction of the stress distribution. It is important to be noted that the above results obtained from the present model are similar to that reported by Y. C. Tsui and T. W. Clyne[19], which is solved based on the elastic deformation assumption. However, details of the plots indicate some remarkable differences. For example, for the case when n is equal to 1, the negative gradient of the stress in the present model is less obvious than that in the literature, while for the case when n is equal to 100, the positive gradient in the former is much higher than the later. This is mainly due to the different assumptions of the deformation range and that of the intrinsic stress. In the previous literature[19], as mentioned before, the material properties are assumed to be of pure elastic deformation, and a fixed quenching (intrinsic) stress has to be pre – specified before calculation. In practice, it is difficult to obtain the real value of intrinsic stress for the thin layers, even if by some experimental apparatuses or numerical methods. In the present model, all the material properties (see Table 1), including the parameters correlated with the plastic deformation, can be directly obtained from experiments or referring previous works.

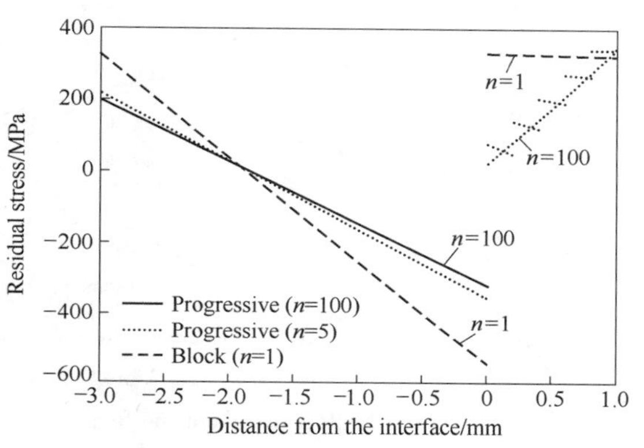

Fig. 5 Plots of deposition induced stress for APS NiCoCrAlY on Inconel 718, by treating the deposition as a block process ($n = 1$) or as a progressive process with n ($n = 5, 100$)

Fig. 6 shows the stress distributions for the APS NiCoCrAlY on Inconel 718 system with and without the final cool – down process. It is found that the final residual stress is close to that due to quenching only, which manifests that the stresses arising from quenching of splats account for most of the final stress levels. For the case shown in Fig. 6, a constant deposition temperature of 700K was assumed and room temperature is assumed to be 300K. Whereas in fact the deposition temperature can be varied somewhat. So it is worthwhile to investigate the effect of assuming different deposition temperatures on the final residual stresses. Fig. 7 shows the residual stress plots with three different deposition temperatures (covered the actual range during spraying), i. e. 473K, 623K and 773K, respectively. It can be found that, the residual stress in the coating decreases with the increase of deposition temperature. For the substrate, as the deposition temperature increases, the curvature slope (absolute value) of the residual stress decreases correspondingly. Therefore, it can be concluded that using a relatively higher deposition temperature is helpful to improve the residual stress distribution in the coating system.

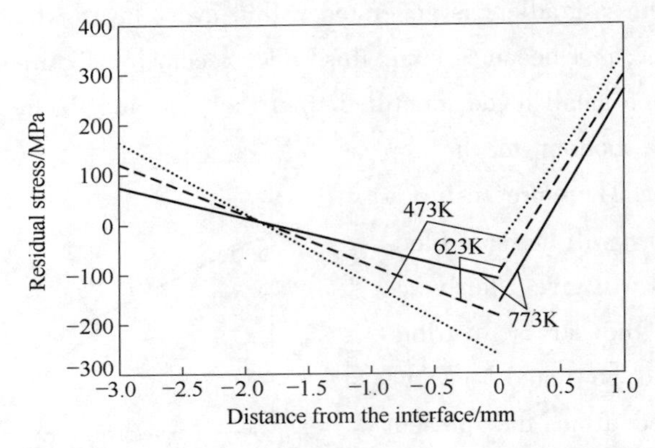

Fig. 6 Predicted stress distributions from the present model ($n = 100$) for APS NiCoCrAlY on Inconel 718 with and without the final cool – down process (from 700K)

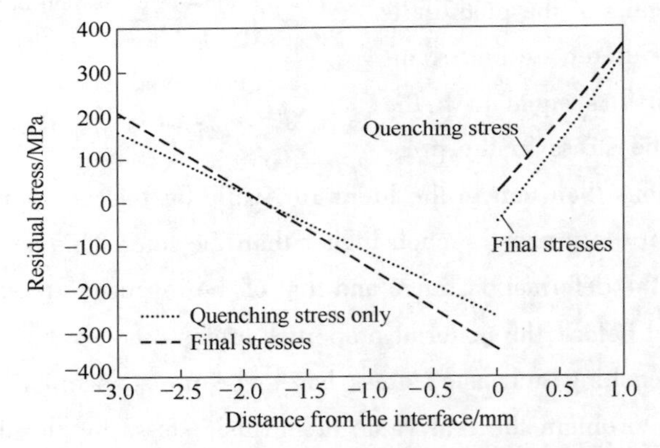

Fig. 7 Predicted residual stress distributions from the present model ($n = 100$) for APS NiCoCrAlY on Inconel 718 with different deposition temperatures (473K, 623K and 773K)

The effect of coating thickness on the residual stress distribution was also investigated. As shown in Fig. 8, with the increase of coating thickness, the residual stress on the coating surface (tensile) increases correspondingly, and the magnitude of residual stress (compressive stress in most cases) at the interface also increases. For the substrate, a higher curvature slope (absolute value) of the residual stress is obtained when the coating thickness increased. In addition, it can be found that, in the coating systems with three different thicknesses, the coating residual stresses at the same distance from the interface are different from each other, and that, at the same position, the stress value in the structure with higher coating thickness is always lower than that with lower thickness.

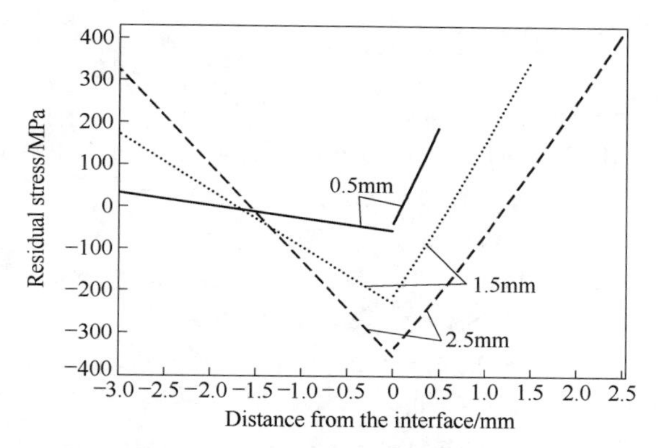

Fig. 8 Predicted residual stress distributions from the present model($n = 100$) for APS NiCoCrAlY on Inconel 718 with different coating thicknesses

6 Conclusions

This paper presents an elastoplastic analysis of the residual stress in thermally spraying coating systems. Two main residual stress generation mechanisms are taken into account in the developed elastoplastic model, i. e. the quenching stress during deposition and thermal mismatch between coating and substrate during cooling. The deposition induced stress was analyzed based on the layer – by – layer deposition assumption which clearly illustrated how stresses build up during the spraying process. Although only one possibility for the thermal mismatch stress generation was discussed in the present study, the approach presented here can be extended to other possibilities. As compared with other numerical or analytical models, the main advantage of the present model lies in that, without predetermining the intrinsic stress, the stress distribution in the coating can be easily determined by knowing the specimen dimensions and material properties. The case study for the APS NiCoCrAlY on Inconel 718 system shows that the stresses resulting from quenching of splats account for most of the final stress levels, and that controlling the residual stress levels can be easily performed based on the model by optimizing the process parameters, e. g. deposition temperature and coating thickness.

Although the material properties considered in the present model are temperature independent, in most cases, this does not lead to large errors, at least in the applications of qualitative prediction of the residual stress level in the coating system. In addition, the thermally sprayed coating concerned in the elatoplastic model is considered as a multilayer structure, so the model can be easily applied into other multilayer thin films, such as CVD and PVD films.

Acknowledgements

The authors are grateful to the priority support by China Natural Science Foundation(50735006,

50905185), National Science and Technology Support Program(2006BAF02A19), and National "863" Project(2009AA03Z342).

References

[1] S. Kuroda and T. W. Clyne, Thin Solid Films 200, 49(1991).

[2] J. Matejicek, S. Sampath, P. C. Brand, and H. J. Prask, Acta Mater. 47, 607(1999).

[3] G. G. Stoney, Proc. Roy. Soc. A A82, 172(1909).

[4] J. Matejicek and S. Sampath, Acta Mater. 51, 863(2003).

[5] S. Kuroda, T. Kukushima, and S. Kitahara, Thin Solid Films 164, 157(1988).

[6] T. W. Clyne and S. C. Gill, J. Therm. Spray Technol. 5, 401(1996).

[7] Y. Y. Santana, J. G. La Barbera - Sosa, M. H. Staia, J. Lesage, E. S. Puchi - Cabrera, D. Chicot, and E. Bemporad, Surf. Coat. Technol. 201, 2092(2006).

[8] M. Buchmann, R. Gadow, and J. Tabellion, Mater. Sic. Eng. A288, 154(2000).

[9] C. H. Hsueh, Thin Solid Films 418, 182(2002).

[10] M. Wenzelburger and M. Escribano, Surf. Coat. Technol. 180 - 181, 429(2004).

[11] Z. C. Feng and H. D. Liu, J. Appl. Phy. 54, 83(1983).

[12] C. H. Hsueh, J. Appl. Phy. 91, 9652(2002).

[13] P. H. Townsend, D. M. Barnett, and T. A. Brunner, J. Appl. Phy. 62, 4438(1987).

[14] G. P. Nikishkov, J. Appl. Phy. 94, 5333(2003).

[15] A. E. Giannakopoulos, S. Suresh, M. Finot, and M. Olsson, Acta Metall. Mater. 43, 1335(1995).

[16] Z. H. Gan and H. Wah Ng, Mater. Sci. Eng. A385, 314(2004).

[17] J. Stokes and L. Looney, J. Mater. Eng. Perf. 18, 21(2009).

[18] J. Stokes and L. Looney, J. Therm. Spray Technol. 17, 908(2008).

[19] Y. C. Tsui and T. W. Clyne, Thin Solid Films 306, 23(1997).

[20] T. W. Clyne, Key Eng. Mater. 116 - 117, 307(1996).

[21] Y. C. Tsui and T. W. Clyne, Thin Solid Films 306, 34(1997).

[22] P. Bansal, P. H. Shipway, and S. B. Leen, Acta Mater. 55, 5089(2007).

[23] Y. Y. Hu and W. M. Huang, J. Appl. Phy. 96, 4154(2004).

[24] H. Shames, *Introduction to solid mechanics*(Prentice - Hall, Englewood - Cliffs. NJ., 1989).

[25] X. C. Zhang, J. Appl. Phy. 103, 073505(2008).

[26] S. Widjaja, A. M. Limarga, and T. H. Yip, Thin Solid Films 434, 216(2003).

[27] Z. H. Gan, H. W. Ng, and A. Devasenapathi, Surf. Coat. Technol. 187, 307(2004).

自动化电弧喷涂路径对涂层残余应力的影响*

摘　要　喷涂路径是影响电弧喷涂层残余应力的重要因素之一。利用自行开发的自动化电弧喷涂系统设计了四种典型的喷涂路径，采用X射线残余应力测试仪测量了不同路径下制备的82B高碳钢涂层的表面残余应力分布，并利用金相显微镜和X射线衍射仪分析了涂层的组织形貌与相组成。试验显示，“环形”喷涂路径制备的涂层的残余应力分布最均匀，最大应力值最低；“平行对称”路径的情况最差。组织分析表明涂层主要由氧化物和块状马氏体组织组成。最后探讨了高碳钢涂层骤冷应力、热应力和相变应力三种因素的影响机理。

关键词　电弧喷涂　高碳钢涂层　残余应力　路径规划

1　引言

电弧喷涂制备涂层技术因具有设备简单、成本低、效率高、质量可靠等特点而被广泛用于表面磨损、腐蚀等失效零件的维修与再制造。近年来，结合自动化技术的优势和快速成型的思想，开发了电弧喷涂快速成型技术[1~3]，并应用Zn、Zn－Al合金、巴氏合金等低熔点喷涂材料进行模具制造[4]。低熔点材料沉积温度低、塑性好、残余应力低，进而使得模具热变形小、精度高。但这些材料涂层的硬度只有40~50HV，模具寿命较低。针对这些不足，尝试使用高熔点的碳钢、不锈钢等作为喷涂材料[5~7]。高熔点涂层具有高的硬度和强度，可在较高的温度和压力条件下工作，但是其工艺复杂性和成型难度相应增大，因为高熔点的熔滴沉积到基体上会产生很大的应力，使基体和涂层发生变形，并可能导致成型层的失效[8]。

导致涂层产生残余应力的因素较多，其中，因喷涂路径而产生热失配应力是较重要的因素之一。张海鸥等人[9,10]研究了不同的等离子喷涂路径下涂层残余应力的分布规律，并认为喷涂方向的不断改变和叠加会减小涂层的应力峰值。关于电弧喷涂路径与残余应力之间关系的研究，主要是通过数值模拟和测温试验的方法观察喷涂路径对涂层温度场分布的影响[3,11]，进而认为相对均匀的温度场分布会降低涂层的残余应力。然而，喷涂路径对涂层残余应力的影响程度的定量评估报道比较少。为此，规划设计了四种电弧喷涂路径，采用X射线应力分析技术研究高碳钢电弧喷涂层的残余应力行为，并分析其分布规律及作用机理。

2　试验方法

自动化电弧喷涂的路径规划关系到沉积涂层的厚度、涂层表面温度场分布是否均匀，进而影响到残余应力的分布。使用自行开发的操作机自动化电弧喷涂系统制备涂

* 本文合作者：陈永雄、梁秀兵、刘燕。原发表于《焊接学报》，2010，31(8)：97~100。国家科技支撑计划基金资助项目（2006BAF02A19）；中国机械工程学会焊接学会创新思路预研奖学金资助项目（08－12－001）。

层。首先设计了如图 1a ~ d 所示四种不同的喷涂路径，通过 Workbench V5 离线编程软件编写自动控制程序，最后在操作机上进行喷涂。

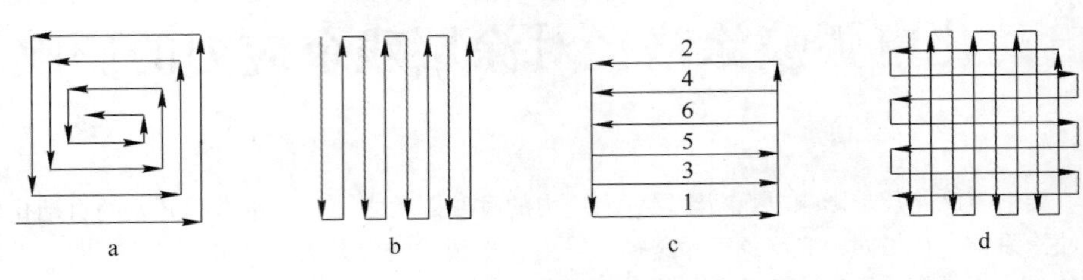

图 1　四种喷涂路径示意图

Fig. 1　Schematic of four spraying paths

a—环形路径；b—“Z”字形路径；c—平行对称路径；d—“田”字形路径

通过试验发现，测试 HAS - 2 型喷枪在 210mm 的喷涂距离时沉积斑点的直径约为 40mm，据此四种路径的喷涂间距都设定为 20mm。选用长 200mm，宽 200mm，厚 6mm 的 45 钢板作为喷涂成型基体，喷涂前进行喷砂处理。喷涂材料选用 2mm 直径的 82B 丝材（Fe - 0. 82% C - 0. 6% Mn）。喷涂的工艺参数在整个对比试验中都固定，分别为喷涂电流 70A，喷涂电压 36V，喷涂距离 210mm，压缩空气压力 0. 7MPa，喷枪移动速度 1. 4m/min。采用间歇喷涂方式制备涂层，即上述路径每喷涂 1 个循环后，停止送丝，吹气冷却 1 个循环，保证喷涂层的温度不超过 300℃。

当涂层厚度达到 0. 2mm 时，停止喷涂，使用 X - 350A 型 X 射线应力测试仪进行涂层残余应力的测量，表面不做任何处理，分别测试 5 个点，各点的坐标位置如图 2 所示。第一次测量完之后，分别按照前述工艺进行第二次喷涂，待涂层厚度达到 2mm 时，停止喷涂并进行第二次表面残余应力测试，测量位置与第一次相同。X 射线残余应力测试的原理是，通过 X 射线以一定的入射角照射试样表面，当材料存在残余应变时，晶面间距将发生变化，相应的 X 射线谱的衍射峰发生移动，这样就可以根据残余应变与 X 射线的入射角及衍射角之间的关系计算出残余应力值。由于文中试样的长度和宽度远大于涂层/基体的厚度，可考虑为平面应力情形，此时 X 射线应力 σ 的计算公式为：

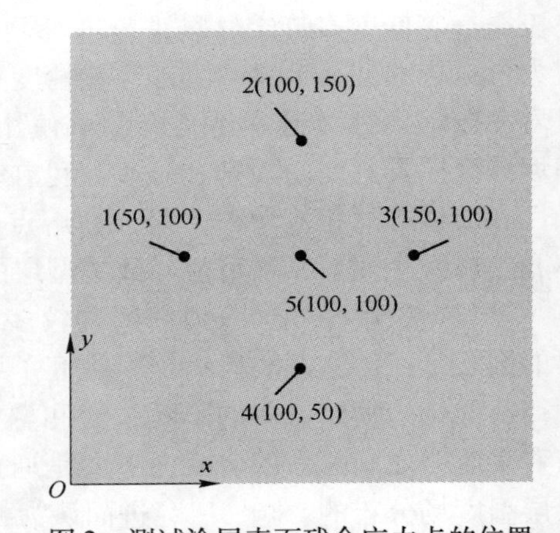

图 2　测试涂层表面残余应力点的位置

Fig. 2　Test positions for coating residual stress

$$\sigma = -[E/1(1+\nu)]\cot\theta_0(\pi/180)[\partial(2\theta)/\partial(\sin^2\psi)] \tag{1}$$

$$k = -[K/2(1+\nu)]\cot\theta_0(\pi/180) \tag{2}$$

$$M = \partial(2\theta)/\partial(\sin^2\psi) \tag{3}$$

式中，E、ν、θ_0 分别为材料的弹性模量、泊松比和无应力状态时的 θ 角；k 为应力常数（文中定为 - 318MPa/(°)）；M 为不同入射角 ψ 测得的衍射线 2θ 角与 $\sin^2\psi$ 直线关系的斜率。试验时采用侧倾固定法进行 X 射线应力测量，衍射线 2θ 角的扫描范围为151°~162°，扫描步距为0.1°，计数时间1.0s，使用 CrK_α 靶辐射，分析晶面为（211），入射角 ψ 定为0°，30°和45°，X 光管电压为22.0kV，电流6.0mA，准直管直径为2mm。理想的平面应力状态下，x 和 y 方向（图2）的应力值是相等的，但实际中由于喷涂路径的影响，会产生较大的差别，为此，在每点处分别测量了图示 x、y 两个方向的残余应力值，即 ψ 角的变动方向分别与 x、y 轴平行。

为了探讨涂层的残余应力形成机理，用2%的硝酸酒精溶液腐蚀试样，使用 Olympus PMG3 型光学显微镜观察涂层截面的组织形貌，并结合 Bruker D8 Advance X 射线衍射仪分析涂层的相组成。

3　试验结果与分析

3.1　喷涂路径之间的比较

表1、表2 分别为四种路径下涂层厚度为 0.2mm，2mm 时的残余应力测试结果。比较可以发现，环形路径在两种厚度下的最大残余应力测试值都最小，分别为 69MPa（0.2mm）和 86MPa（2mm），波动范围也最小，而平行对称路径在 2mm 时应力最大值达到了 172MPa，平行对称路径在两种厚度下的应力波动范围也最大。另外，当涂层厚度从 0.2mm 增大到2mm 时，大部分测试点的应力值都有不同程度的增加。田字形路径在厚度为 0.2mm 时应力波动相对较小，但当厚度增大后波动变得剧烈。Z 字形路径的应力值和波动范围都处于中间水平。从表1 和表2 还可看出，大多数测试点处 x 方向和 y 方向的残余应力值差别较大，这说明喷涂路径、喷枪扫描方向等因素对涂层的残余应力产生了影响。

表1　涂层厚0.2mm 时残余应力测试结果

Table 1　Residual stress measurement data of coating with 0.2mm thickness

测量位置	应力方向	残余应力 σ/MPa			
		环形	Z 字形	平行对称	田字形
1	x	46	53	90	90
	y	46	71	116	88
2	x	11	73	66	98
	y	28	46	98	30
3	x	14	131	50	81
	y	34	59	87	58
4	x	19	89	71	83
	y	69	44	128	131
5	x	11	27	19	90
	y	64	49	39	66

表 2 涂层厚 2mm 时残余应力测试结果

Table 2 Residual stress measurement data of coating with 2mm thickness

测量点	方向	残余应力 σ/MPa			
		环形	Z 字形	平行对称	田字形
1	x	61	119	123	59
	y	15	70	90	68
2	x	86	98	6	128
	y	71	116	172	106
3	x	11	65	121	11
	y	25	42	74	55
4	x	50	72	105	49
	y	39	54	120	65
5	x	1	76	137	51
	y	17	30	75	123

通过涂层表面残余应力值可以大致预测涂层内部的应力范围，因为在喷涂过程中，最上面一层涂层沉积后，由于快速凝固收缩，最上一层会产生骤冷拉应力，而下面的涂层和基体的束缚会阻碍最上一层进一步收缩，这样系统内的受力会达到一个平衡状态，使得最上一层涂层产生一定量的拉应力，而下面的涂层和基体会产生额外的压应力作用[12]，因此，每当沉积一层新的涂层，就会对已沉积的涂层产生附加的压应力，涂层沉积完之后，最表层的残余应力势必要大于基体界面处的应力值。

涂层的变形往往是由涂层的残余应力分布不均匀造成的，当残余应力的峰值超过涂层的极限值时会萌生裂纹而导致开裂。综上可以判定，环形路径最有利于喷涂成型，其残余应力值最低，分布最均匀；而平行对称路径的情况最差，其最大表面应力值最高，波动的幅度也最大。

除了喷涂路径外，文中的其他喷涂工艺都相同，因此，涂层残余应力的差别主要是不同的喷涂路径造成的。对比四种喷涂路径，环形路径的温度场分布最均匀，Z 字形路径次之，这一点已在张海鸥等人[9]研究等离子喷涂的结果中得到证实；平行对称路径虽以对称的方式进行喷涂，但从第一个平行移动到第二个平行移动所经历的时间间隔要高于其他几种路径（例如图 1c 中从路线“1”到路线“2”），这样导致涂层受热和冷却的幅度增大，温度变化剧烈；田字形路径虽然涂层的厚度分布最均匀，但从粒子沉积的循环过程到吹气冷却的时间间隔延长（相当于 Z 字形路径的 2 倍），温度波动照样剧烈。温度波动越大，由此引起的热应力也越大，所以，利用 X 射线测试不同路径的应力分布规律与涂层温度场分析的结果是一致的。

3.2 涂层残余应力的影响因素分析

对于常规热喷涂层，残余应力主要有两种，即骤冷应力和热应力。无论是骤冷应力还是热应力，都是由温度引起了材料的失配[13]。文中由于采用了间歇喷涂的方式，涂层的沉积温度较低，且涂层与基体的线膨胀系数相差不大，因此喷涂完成后冷却过程中的热应力可忽略。而且，如果不考虑涂层内部应力释放，涂层在弹性范围内的残余应力可用下式估算，即：

$$\sigma_r = \left(\frac{E}{1-\nu}\right)\alpha(T_m - T_d) \tag{4}$$

式中，E、ν 为涂层的弹性模量和泊松比；T_m 和 T_d 分别为涂层自由应变点温度和沉积温度；α 为涂层的线膨胀系数。对于碳钢喷涂材料[11]，E 约 100GPa，$\nu = 0.33$，$\alpha = 12 \times 10^{-6}K^{-1}$，$T_m - T_d$ 约 623K，则 σ_r 约 1116MPa，显然这与测试结果相差甚远。实际上涂层材料在喷涂过程中会发生屈服变形，而且涂层中孔隙和微裂纹也会释放部分应力。因此，涂层的残余应力值应该在材料的屈服点附近，块体碳钢材料在 250℃ 时屈服强度约为 170MPa[11]。

另外，对于高碳钢材料，除了骤冷应力和热应力外，还会产生相变应力，这是它区别于其他热喷涂层的显著特征。熔融的碳钢粒子在沉积过程中，发生快速凝固，粒子会由奥氏体迅速转变为贝氏体或马氏体[7,11]，诱发体积膨胀，进而产生较大的压应力。图 3 和图 4 为采用环形路径制备的涂层的 X 射线衍射图谱和金相形貌。可以看出，由于采用间歇式喷涂，涂层的沉积温度比较低，喷涂材料会发生奥氏体向马氏体的转变，使得涂层中含有大量的马氏体（图 3），图 4 中灰色条状区域主要为氧化物相，白色区域主要是由块状的 Fe－C 合金组织组成。通过 X 射线残余应力的测试结果也可以看出，涂层中的大部分测试点应力值都远小于 170MPa，这进一步说明，涂层必然产生了相变压应力，抵消部分拉应力，使得总体残余应力较低。如果要对涂层的显微组织特征进行更深入的分析，还需借助 TEM 等手段。

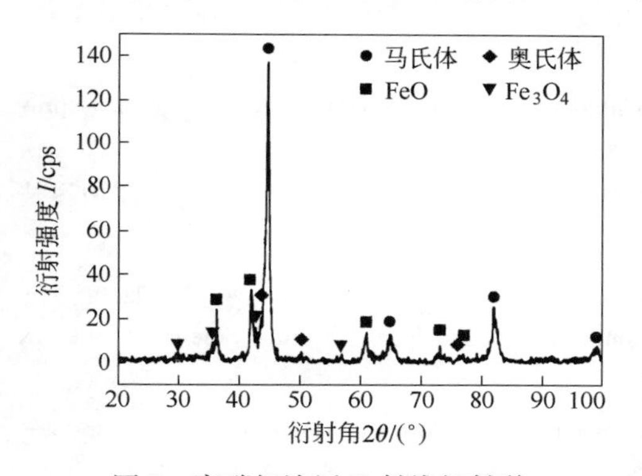

图 3　高碳钢涂层 X 射线衍射谱

Fig. 3　XRD pattern of high carbon steel coating

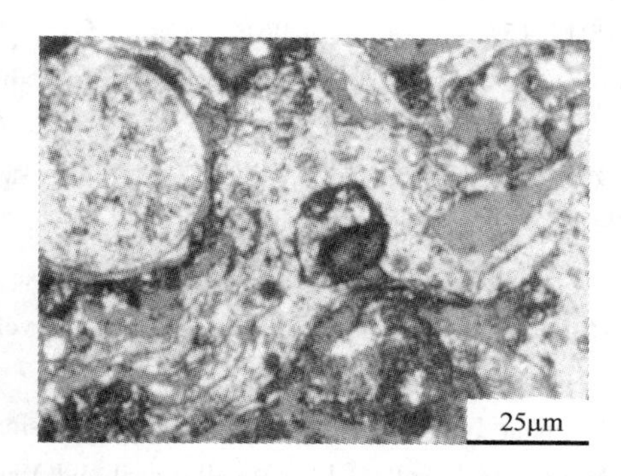

图 4　高碳钢涂层截面组织形貌（2% 硝酸酒精腐蚀）

Fig. 4　Cross－section metallograph of high－carbon steel coating etched by 2% nital

4　结论

（1）环形路径最有利于喷涂成型，涂层温度场分布最均匀，对应的残余应力最低，波动范围最小；平行对称路径的应力分布情况最差。

（2）随着涂层厚度的增加，涂层的表面残余应力有增大的趋势，且表面的残余应力值要高于涂层/基体界面处的应力，因此 X 射线方法可以大致预测涂层内部的残余应力分布范围。

（3）除了骤冷应力（拉应力）和热应力外，相变也是高碳钢喷涂层产生残余应力的重要因素之一，马氏体相变产生的压应力可以部分抵消拉应力，同时涂层通过塑性变形、孔隙和微裂纹等方法释放部分应力，使得涂层中的总体残余应力较小。

参考文献

[1] Chua C K, Hong K H, Ho S L. Rapidtooling technology part 2: a case study using arc spray metal tooling [J]. International Journal of Advanced Manufacturing Technology, 1999, 15: 609~614.

[2] Fang J C, Xu W J, Zhao Z Y. Arc spray forming [J]. Journal of Materials Processing Technology, 2005, 164~165: 1032~1037.

[3] Wimpenny D I, Gibbons G J. Metal spray tooling for composite forming [J]. Journal of Materials Processing Technology, 2003, 138: 443~448.

[4] 王伊卿. 电弧喷涂制造模具关键技术及材料开发研究 [D]. 西安: 西安交通大学, 2001.

[5] Zhao Z Y, Fang J C, Wang H, et al. Arc spray forming of stainless steel mould [J]. Key Engineering Materials, 2005, 291~292: 603~608.

[6] 陈正江. 电弧熔射成形法快速制造模具技术研究 [D]. 大连: 大连理工大学, 2005.

[7] New bery A P, Grant P S. Oxidation during electric arc spray forming of steel [J]. Journal of Materials Processing Technology, 2006, 178: 259~269.

[8] 程江波, 梁秀兵, 陈永雄, 等. 再制造电弧喷涂成形层的残余应力分析 [J]. 焊接学报, 2008, 29(6): 17~20.

Cheng Jiangbo, Liang Xiubing, Cheng Yongxiong, et al. Residual stress in electric arc sprayed coatings for remanufacturing [J]. Transactions of the China Welding Institution, 2008, 29(6): 17~20.

[9] 张海鸥, 刘辉祥, 王桂兰. 等离子熔射制模过程中皮膜温度场实验研究 [J]. 中国机械工程, 2004, 15(21): 1958~1961.

Zhang Haiou, Liu Huixiang, Wang Guilan. Experimental research on temperature field of plasma spray coating [J]. China Mechanical Engineering, 2004, 15(21): 1958~1961.

[10] 刘辉祥, 张海鸥, 韩光超, 等. 机器人等离子熔射路径对皮膜残余应力的影响 [J]. 华中科技大学学报（自然科学版）, 2004, 32(11): 21~23.

Liu Huixiang, Zhang Haiou, Han Guangchao, et al. Effect of plasma spray path on coating residual stress [J]. Journal of Huazhong University of Science and Technology (Natural Science Edition), 2004, 32(11): 21~23.

[11] Rayment T, Hoile S, Grant P S. Phase transformations and control of residual stresses in thick spray-formed steel shells [J]. Metallurgical and Materials Transactions, 2004, 35B: 1113~1122.

[12] Tsui Y C, Clyne T W. An analytical model for predicting residual stresses in progressively deposited coatings, Part 1: Planar geometry [J]. Thin Solid Films, 1997, 306: 23~33.

[13] 张显程. 面向再制造寿命预测的等离子喷涂涂层结构完整性研究 [D]. 上海: 上海交通大学, 2007.

Cu Nanoparticles Effect on the Tribological Properties of Hydrosilicate Powders as Lubricant Additive for Steel-steel Contacts*

Abstract The effect of Cu nanoparticles(NPs) on the tribological behaviors of serpentine powders (SPs) suspended in diesel oil was investigated. Results show that the optimum mass ratio of Cu NPs to SPs is 7.5:92.5. With the addition of the above mixture to oil, the tribological properties can be significantly improved compared with those of the oil containing SPs alone. A more smooth and compact tribofilm has formed on the worn surface, which is responsible for the further reduced friction and wear, mainly with iron oxides, silicon oxides, species enriched in Si—O structures, graphite, organic compounds, and Cu^0, Cu^+ and Cu^{2+} species.

Key words serpentine, friction and wear, lubricant additive, nanoparticles

1 Introduction

Friction and wear, as one of the main causes of materials failure, has attracted more and more attentions all over the world[1]. In recent years, numerous studies have reported that the addition of nanoparticles, such as metal[2], metal oxide[3], metal sulfides[4,5], carbonate[6], borate[7], carbon materials[8,9], organic material[10] and rare-earth compound[11], to lubricants is effective in reducing friction and wear. The friction-reduction and anti-wear behaviors are dependent on the characteristics of nanoparticles, such as size, shape and physicochemical nature[12]. The mechanisms for the reduction in friction and wear of nanoparticles are mainly associated with the following aspects: (1) the colloidal effect, i. e. the nanoparticles penetrate the elastohydrodynamic contacts by mechanical entrapment to form a boundary lubricating film[13]; (2) the spherical nanoparticles act as nano-bearings between the rubbed surfaces to reduce the friction and wear by surface polishing and the increase in surface hardness effects[8]; (3) the nanoparticles serve as spacers to eliminate the metal-to-metal contact between the asperities of the two mating surfaces[5]; and (4) various boundary films with excellent mechanical and lubricating properties are generated on the rubbed surfaces[9-11]. The above cited mechanisms take place mostly in the region of mixed lubrication or boundary lubrication. It is found that Cu NPs often display good friction and wear reducing characteristics by the effect of nano-bearings and the formation of a metallic film and/or a tribo-sintered film with low shearing strength on rubbed surfaces[2].

Compared with nanoparticles, the micro-scale particles are more intrinsically unstable and

---

* Copartner: Zhang Baosen, Xu Yi, Gao Fei, Shi Peijing, Wu Yixiong. Reprinted from *Tribology International*, 2011, 44: 878-886.

easy to subside in liquid lubricants. Moreover, the large - scale particles possibly act as abrasive particles to aggravate the scuffing of rubbed surfaces[14]. Therefore, few of previous studies have been reported focusing on the use of micro- scale particles as additives of liquid lubricants. However, some recent researches show that the micro - scale mineral powders as lubricant additive exhibit excellent tribological properties[14-17]. Yu et al. [14] found that the serpentine powders (≤10μm) can help generate an oxide layer with high mechanical properties on the worn surface. Yu et al. [15] showed that the ophite powders (0. 3 - 3. 0 μm) as oil additive can present restoration effects on steel contacts with an amorphous diamond like carbon (DLC) film. Jin et al. [16] evaluated the reconditioning effect of the ART packages on ferrous surfaces under field trial conditions and found a cermet layer on the cast iron cylinder bore. Wang[17] clarified that in the presence of serpentine powders (1. 0 - 10μm), the protective layer formed on the cast iron cylinder bore was mainly composed of Fe_3O_4 nanoparticles and carbyne. The above studies demonstrate that the mineral powders may be a promising candidate for lubricating machines due to the low cost and high efficacy in reducing friction and wear.

Due to the different physicochemical characteristics of the Cu NPs and SPs, and their great potential in practical applications, it is necessary to investigate how the two particles behave together as lubricant additives. The aim of the present study is to discuss the effect of Cu nanoparticles on the tribological properties of serpentine powders as oil additive for steel contacts and disclose the mechanism for the reduction in friction and wear by analyzing the worn surface.

2 Experimental

2.1 Materials

A commercial diesel engine oil (grade: CD 15w/40) was used as the base oil without further treatment. The typical physicochemical parameters of the CD 15w/40 are listed in Table 1.

Table 1 Typical physicochemical parameters of the diesel engine oil

Grade	Item value (SAE)				
	Density /g · cm^{-3}	Kinematic viscosity /mm^2 · s^{-1}	Viscosity index	Pour point /℃	Flashing point/℃
CD 15w/40	0. 794	15. 02, 100℃ /110. 6, 40℃	141	-27	228

The SPs were prepared by mechanical crushing and ball - grinding the serpentine mineral (Liaoning Province, China). The chemical composition of serpentine was determined with an X-ray fluorescence spectrometer and shown in Table 2. Its chemical formula can be expressed as $Mg_{5.82}Al_{0.02}Fe_{0.05}Ca_{0.01}K_{0.07}Mn_{0.01}(Si_{4.21}O_{10.36})(OH)_8$, which is close to the ideal chemical formula of serpentine group, i. e. $Mg_6Si_4O_{10}(OH)_8$. Fig. 1 and Fig. 2 show the SEM morphology and X-ray diffraction pattern of SPs, respectively. The SPs reveals laminar with the particles size mostly less than 0. 5μm. The characteristic diffraction peaks at $2\theta = 12.1°, 24.5°$ and $35.6°$, corresponding to the crystal planes of (0 0 1), (1 0 2) and (16 0 1), respectively, are indexed to those of antigorite (JCPDS No. 22-1163).

Table 2 Chemical composition of serpentine

Oxides	Content/wt%	Oxides	Content/wt%
SiO_2	44.57	CaO	0.13
MgO	41.10	K_2O	0.60
Al_2O_3	0.20	MnO	0.07
FeO	0.62	H_2O	12.71

Fig. 1 SEM morphology of SPs

Fig. 2 X-ray diffraction pattern of SPs

The Cu NPs synthesis was based on the procedure described in Ref. [18]. Fig. 3 reveals that most of the Cu NPs show agglomerated and keep a nearly-elliptical shape with size around 50 nm. The diffraction peaks shown in Fig. 4 corresponding to the (111), (2 0 0) and (2 2 0) planes of Cu NPs indicate that the product is highly pure without oxidization.

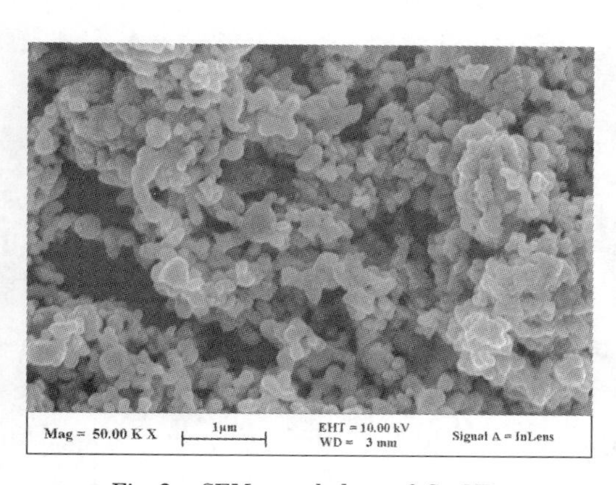

Fig. 3 SEM morphology of Cu NPs

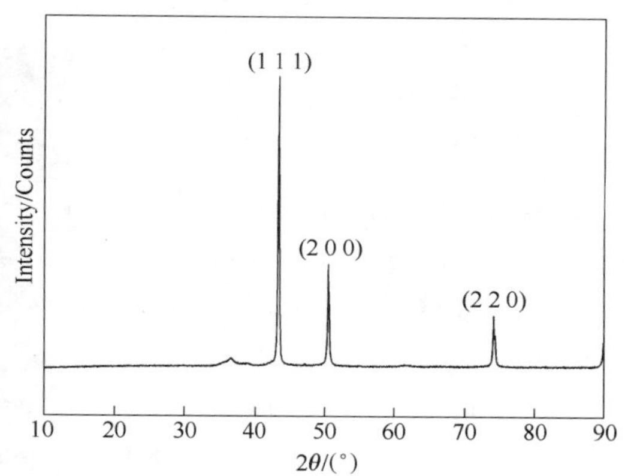

Fig. 4 X-ray diffraction pattern of Cu NPs

To gain good dispersion stabilization in oil, the 5.0wt% mass fraction of oleic acid was mixed with the SPs or Cu NPs in a ball mill (rotating speed = 350rpm, duration = 16h) to produce an organic coating layer. TEM micrographs of the SPs and Cu NPs (see Fig. 5) show that they are monodisperse with clear grain boundaries, and can be dispersed well in nonpolar organic solvents and a number of mineral and synthetic oils. For the tribological tests, the addition of additives to oil was fixed at 0. 5wt% and the lubricants shown in Table 3 were comparatively evaluated. Before tests, high-speed mechanical agitation (rotating speed = 5000r/min, duration = 30min) and ultrasonic dispersion (power = 400W, temperature = 35℃ and duration = 30min) were used to obtain the uniform suspensions.

Table 3 The lubricants used in the present study

ID	Constituent
SR0	CD 15w/40
SR1	Oil + 0. 5wt% SPs
SR2	Oil + 0. 5wt% (97. 5% SPs-2. 5wt% Cu NPs)
SR3	Oil + 0. 5wt% (95. 0% SPs-5. 0wt% Cu NPs)
SR4	Oil + 0. 5wt% (92. 5% SPs-7. 5wt% Cu NPs)
SR5	Oil + 0. 5wt% (90. 0% SPs-10. 0wt% Cu NPs)

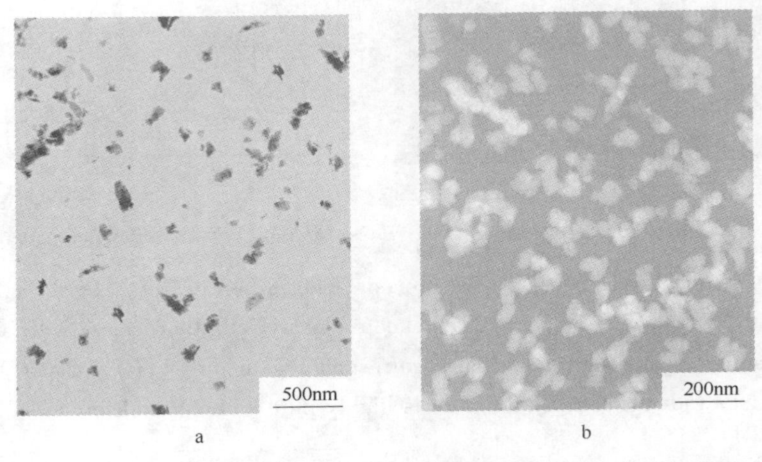

Fig. 5 TEM micrographs of the surface – modified SPs(a) and Cu NPs(b)

2. 2 Friction and wear test

An Optimal SRV-Ⅳ friction and wear tester with a schematic illustration shown in Ref. [6] was employed to examine the tribological properties of the oil samples. An AISI 52100 steel ball (ϕ10mm, HRC61- 63) was used as a rider, which was pressed against the stationary AISI 1045 steel disk (ϕ24mm × 8mm, HRC42- 45) to reciprocate by a horizontal oscillating rod. The chemical constituents of the ball and disk were shown in Table 4. Prior to tests, the disks were mechanically ground and polished to a mirror finish surface, and the specimens were cleaned in petroleum ether.

Table 4 Chemical constituents of the specimens

Specimens	Chemical composition/wt%						
	C	Si	Mn	Cr	S	P	Fe
Disk AISI 1045	0. 40 – 0. 50	0. 15 – 0. 40	0. 50 – 0. 80	<0. 025	<0. 035	<0. 035	Balance
Ball AISI 52100	0. 95 – 1. 05	0. 15 – 0. 35	0. 20 – 0. 40	1. 30 – 1. 65	<0. 027	<0. 027	Balance

The tests were carried out in atmospheric environment(room temperature, about 30% humidity) under load of 50N (corresponds to the mean Hertzian pressure of 1.15GPa), sliding frequency ramp from 10 to 30Hz, oscillating amplitude of 1mm, and a duration of 30min. The friction coefficient was automatically recorded with a computer attached to the test rig. The wear scar diameter of the ball was observed by an optical microscope. The wear volume loss of the disk was measured with a MicroXAM profilometer. The final results were quoted from the averages of the three sets of experimental data. The relative errors measured are of the order of ±5%.

2.3 Worn surface analysis

The morphology and element distribution of the worn surface on the disk were examined using scanning electron microscopy (SEM) equipped with energy dispersive X-ray spectroscopy (EDS). The chemical analysis of the worn surface on the disk was performed with a PHI-5702 multifunctional X-ray photoelectron spectrometer(XPS).

3 Results and discussion

3.1 Friction and wear behavior

Fig. 6 shows the variation of friction coefficient and wear loss with the lubricating conditions. When the base oil(SR0) was used as lubricant, the friction coefficient shows high value with much fluctuation and the corresponding wear loss also shows great improvement. While with the addition of additive, both of them significantly decrease compared with those of the pure oil. The addition of Cu NPs presents remarkable effects on the tribological properties of SPs, which shows the most efficient ones enhancing the friction and wear reducing properties at the content of 7.5wt%. The friction coefficient and wear loss decrease as the content of Cu NPs increases below 7.5wt%, but when the content is higher than that, the tribological properties are weakened on the contrary, which is even worse than those of the SPs alone but still superior to those of the base oil. This may be attributed to the competitive interplays between the SPs and the Cu NPs. The Cu NPs often reduce the friction and wear by rolling like miniature bearings, forming a deposited or tribo-sintered boundary lubricating film due to the low microhardness and melting point [2]. While for the SPs, as one of the phyllosilicates, they possess laminar shape and high surface hardness. During the sliding process, the SPs tend to provide filling, grinding and polishing effects on the contact surface, and then the SPs are refined and a protective layer with lower roughness, higher hardness and lubricating ability may be generated on the worn surface for the decrease in friction and wear[14-16]. As the Cu NPs content increases below 7.5wt%, more particles are possibly adsorbed onto the contact areas to present compensating and synergistic effects with the SPs, and thus the friction and wear decrease. However, as the Cu NPs content further increases above that, the refining effect on the SPs and the interactions between the SPs and the friction interface may be weakened, implying that some coagulation of the micro – scale SPs was formed owing to the friction effect, which made the sliding process unstable and resulted in an increase in friction and wear[14].

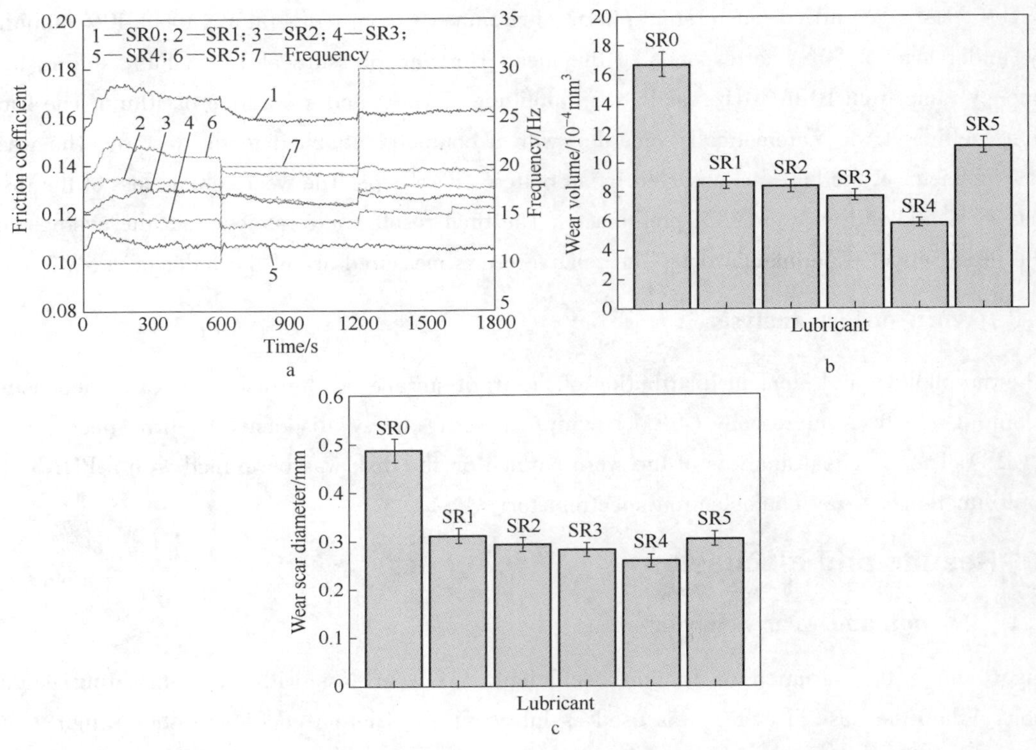

Fig. 6 Friction coefficient(a), wear volume of the disk(b) and wear scar diameter of the ball(c) as a function of lubricating condition

Fig. 6a also shows the influence of sliding frequency as a representation of sliding speed on the friction coefficient. It is found that in the case of SR4, the friction process shows steady tendency, while those of other conditions show a similar change tendency, i. e. they gradually decrease as the frequency increases below 20Hz, while as it further increases up to 30 Hz, the case is the contrary. This result means that excessively high frequency, i. e. sliding speed results in a decrease in tribological properties. The friction behavior can be discussed by considering the two aspects: (1) Stribeck curves[19,20], which present the relationship between the friction coefficient and the test parameters such as sliding speed(V), normal force(P) and kinematic viscosity of oil(η), i. e. $\eta V/p$. Under the boundary lubricating conditions, the friction coefficient decreases as the value of $\eta V/p$ increases. When the normal load and lubricant are definite, the value of $\eta V/p$ increases as the sliding speed appropriately increases, thus the friction coefficient decreases. (2) the competition between the formation and the removal of the protective film. The appropriate increase in speed may provide more energy and promote the formation of protective film, while as the speed further increases, the total sliding distance remarkably increases, which enhances the removal of the protective film, and therefore the friction and wear increase[21].

3.2 Worn Surface Analysis

3.2.1 Morphology and elementary analysis

Fig. 7 shows the SEM micrographs of the worn surfaces on the disks under various lubricating conditions. In the presence of SR1 or SR4, the wear scar diameter exhibits much smaller value than that of SR0. A number of small holes and deep grooves coupled with plastic deformation were found on the worn surface for SR0, indicating that severe scuffing happened (see Fig. 7a and b). While for SR1, the worn surface was relatively smooth and porous with tiny pits and embedded particles, and the scuffing was relatively mild (see Fig. 7c and d). In the case of SR4, the worn surface was very smooth and compact with only slight signs of wear and many scattered tiny dark regions (see Fig. 7e and f). The observed results accord well with the corresponding tribological behaviors.

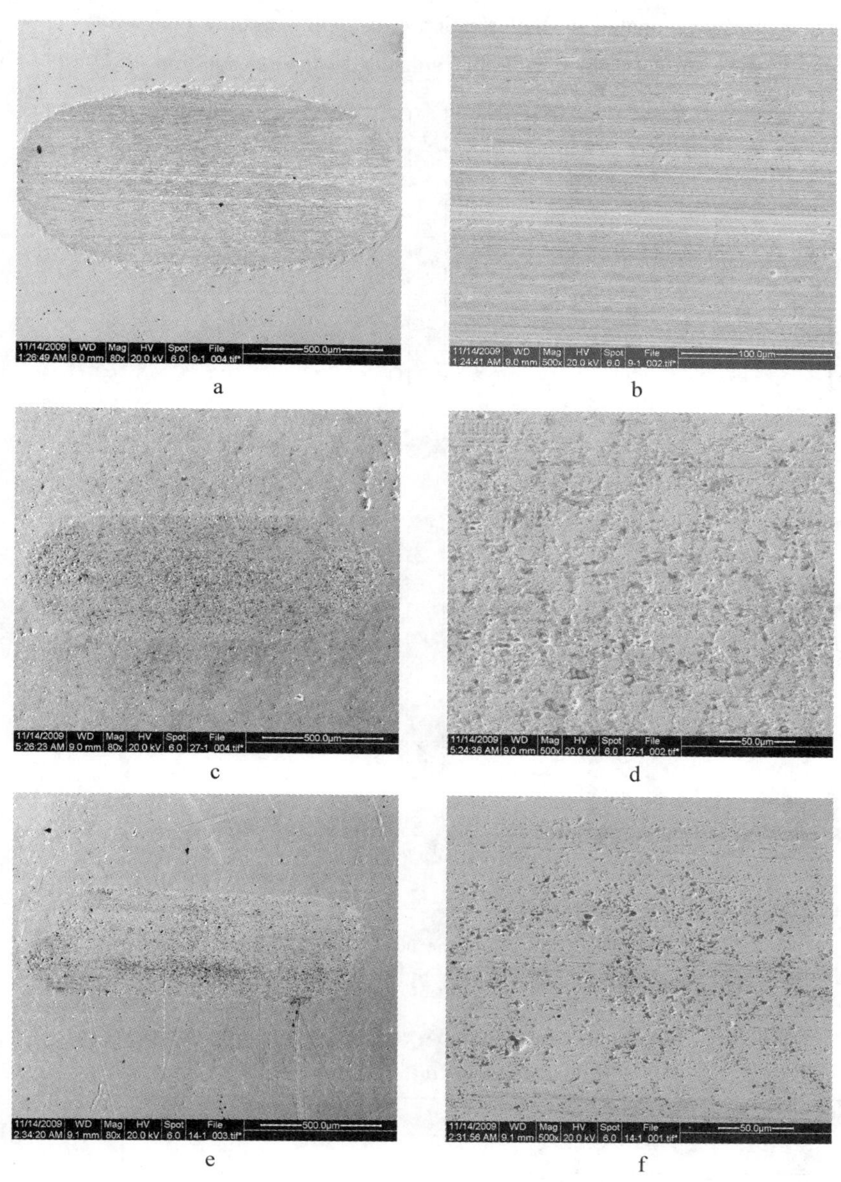

Fig. 7 SEM micrographs of the worn surfaces lubricated with SR0(a,b), SR1(c,d) and SR4(e,f)

Fig. 8 and Table 5 show the elementary analysis of the non – contact region, and the contact regions lubricated with different lubricants. The content of C and O of the contact region for SR0 relatively increases compared with that of the virgin substrate. While for SR1, the contact region shows a significantly increased content of C, O and Si besides a very small quantity of Mg. In the case of SR4, besides a small content of Mg and Cu, a further increased content of C, O and Si is found on the contact region compared with that of SR1. The Mg, Si and Cu come from the additives, the C possibly derives from the substrate and the oil decomposition, and the O may be attributed to the air, SPs and organic compounds. The results indicate that the SPs and Cu NPs have experienced and participated in the formation of tribofilm, and the interactions between the additives and the friction surfaces are remarkably enhanced with the addition of 7.5wt% Cu NPs.

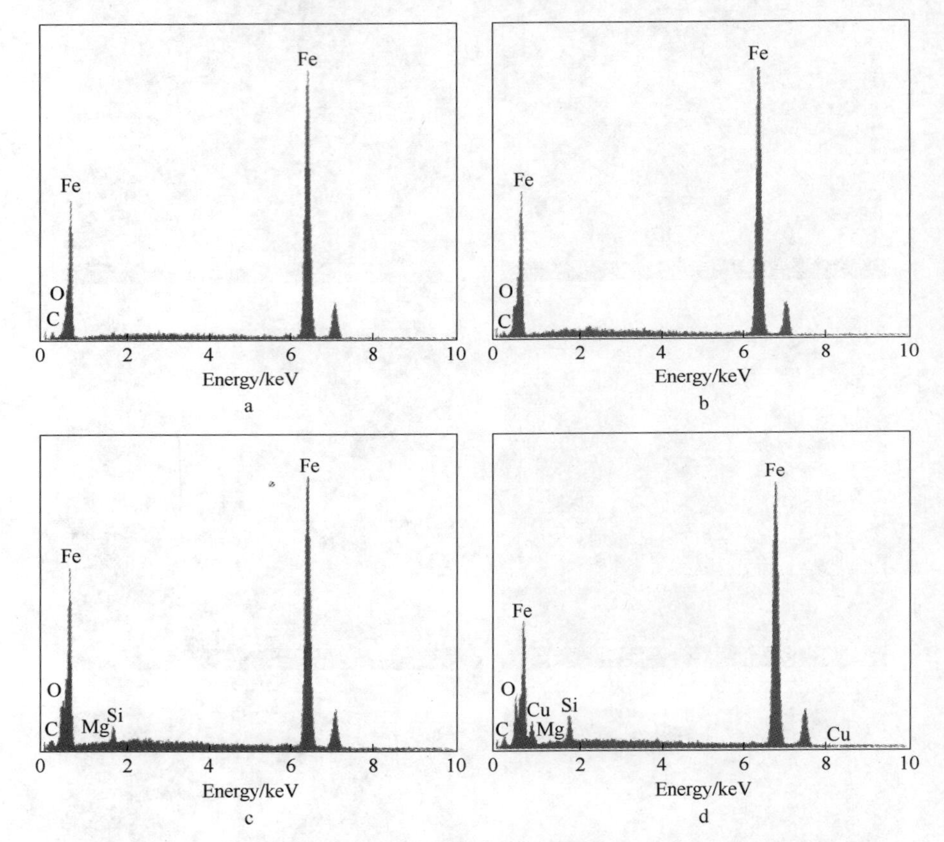

Fig. 8　EDS patterns of the non – contact region(a) and the contact regions shown in Fig. 7b(b), Fig. 7d(c) and Fig. 7f(d)

Table 5　EDS analysis of the non – contact region and contact regions lubricated with different lubricants

Item	Elemental content/at%					
	Mg	Si	Fe	O	C	Cu
Non – contact region of the disk			95.13	3.42	1.45	
The contact region shown in Fig. 7b			86.37	5.95	7.68	

Table 5 (continued)

Item	Elemental content/at%					
	Mg	Si	Fe	O	C	Cu
The contact region shown in Fig. 7d	0.11	0.94	76.38	10.26	12.31	
The contact region shown in Fig. 7f	0.17	3.46	63.75	13.49	18.22	0.91

Fig. 9 shows the magnified morphology and the corresponding elemental distribution maps of the micrograph shown in Fig. 7f. Five characteristic regions marked as A, B, C, D and E can be clearly seen. Most areas of the surface are smooth and there exist some spalling pits, dark regions, aggregates and micrometric sheets thereon (see Fig. 9a). As shown in Fig. 9b, Mg, Cu and C distribute uniformly while Fe, Si and O distribute heterogeneously across the surface. The

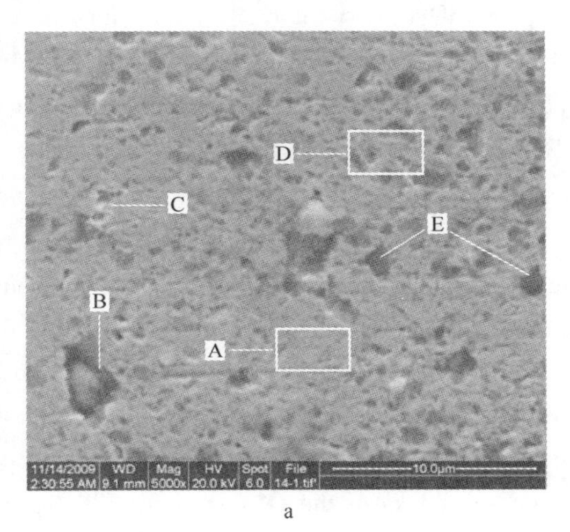

a

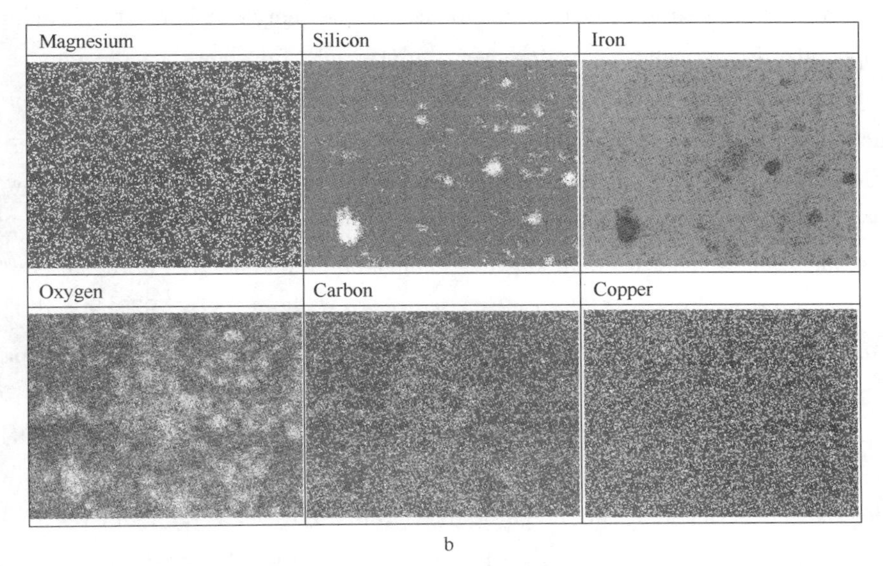

b

Fig. 9 Magnified morphology (a) and elemental distribution maps (b) of the micrograph shown in Fig. 7f

smooth area as region A and spalling pits as region C show a high content of Fe and O, suggesting the formation of oxide layer. The regions B, D and E enriched in Si and O mean that there are many Si-O structures, micro-scale particles and their aggregations in the protective layer. The results demonstrate that a tribofilm of oxide layer, on which Cu – containing species deposited, Si-O structures, refined SPs particles and their aggregations embedded, has formed on the worn surface in the presence of SR4.

3.2.2 XPS analysis of worn surface

Fig. 10 presents the XPS analysis results of the worn surfaces lubricated with SR0, SR1 and SR4. The Mg2p spectrum with only noise signal means that almost no Mg exists on the worn surfaces (see Fig. 10a). In the presence of SR1, the Si2p peak at 101.3 eV with much noise is identified as silicon oxides, i.e. $SiO_x (0 < x < 2)$[22,23], while for that of SR0, only noise signal is found. In the case of SR4, the Si2p with two sub-peaks at 101.3 and 102.1 eV indicates that SiO_x and species enriched in Si-O structures coexist on the worn surface (see Fig. 10b)[22-24]. Under the condition of SR0, the $Fe2p_{3/2}$ with five sub-peaks around 707.3, 709.1, 710.3, 711.5 and 713.6 eV suggests that Fe, FeO, Fe_3O_4, Fe_2O_3 and Fe-containing organic compound exist on the worn surface[22,25]. In the presence of additives, the above Fe-containing species are also found on the worn surfaces and the high oxides show more intensity than those of SR0 especially under the condition of SR4 (see Fig. 10c), which means that the SPs can enhance the oxidation reactions at the friction interface especially with the addition of Cu NPs. The O1s for the worn surface lubricated with SR0 presents the sub-peaks at 529.8, 530.2, 531.2, 532.3 and 533.2 eV suggesting the coexistence of iron oxides and organic compounds, while in the presence of SR1 or SR4, the sub-peaks of the O1s at about 529.9, 530.3, 531.3, 532.3 and 533.3 are assigned to the iron oxides, silicon oxides, species enriched in Si-O structures and organic compounds, respectively[22-25]. Moreover, the O1s for SR4 shows a more intense component at high binding energy than that of SR1, which means that the SPs with addition of Cu NPs can promote the oxidizing actions at the friction interface (see Fig. 10d). The C1s for SR0 with peaks at 284.8 and 285.3 eV indicates the existence of organic compounds on the worn surface[22]. For both of SR1 and SR4, the C1s with sub-peaks at 284.4, 284.8, and 285.4eV shows that the graphite and residual organic compounds exist on the worn surface[22] (see Fig. 10e). In the presence of SR4, the $Cu2p_{3/2}$ with sub-peaks at 932.6, 932.9 and 933.9 eV suggests that the Cu^0, Cu^+ and Cu^{2+} species exist on the worn surface[26-28], while for both of SR0 and SR1, only noise signals are found (see Fig. 10f). The results mean that the Cu NPs have deposited on the worn surface and most of them have experienced and participated in the chemical reactions at the friction interface.

From the above experimental data, it can be inferred that the SPs as oil additive exhibit excellent friction and wear reducing properties especially with addition of 7.5wt% Cu NPs. This may be associated with the friction conditions, crystal structure and physicochemical characteristics of the additives. The SPs of the present study possess a corrugated structure with the Si-O tetra-

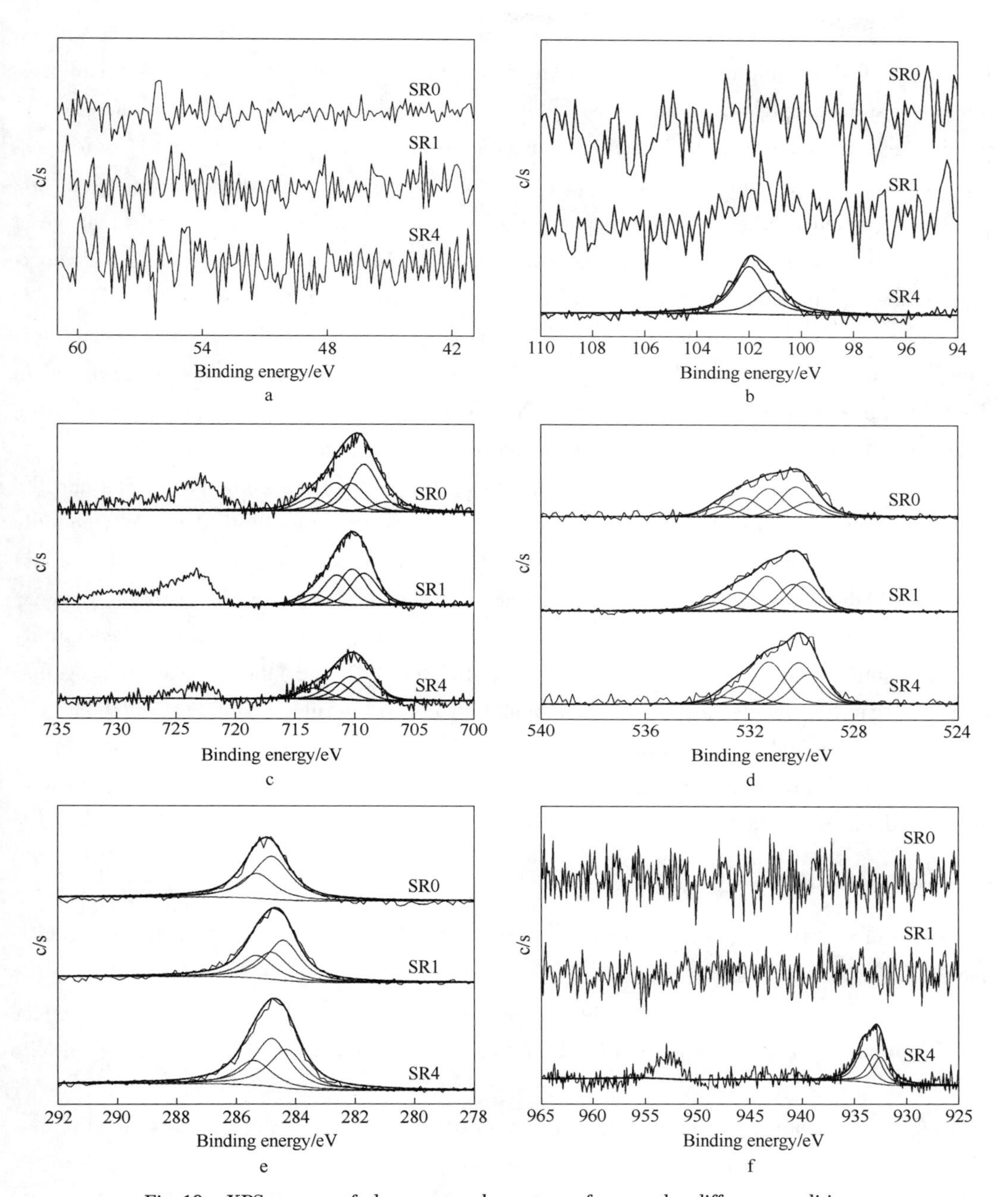

Fig. 10 XPS spectra of elements on the worn surfaces under different conditions
a—Mg2p; b—Si2p; c—Fe2p; d—O1s; e—C1s; f—Cu2p

hedral layer and brucite - like octahedral layer interconnected by vertices, hydrogen bonds and Van der Waals forces. The binding forces between the layers are very weak and there exist some dimensional disproportions and residual stresses in the crystals, which makes them good basal cleavage and capability of oxygen release[14,16]. Moreover, numerous unsaturated or dangling bonds, such as Si-O-Si, O-Si-O and Mg-OH/O, provide the SPs with high polarity, chemical activity and affinity to the metal surface[14,16]. Under the friction forces, the following processes may occur at the friction interface: (1) the SPs are delivered onto the contact areas and present mechanical filling, grinding and polishing effects on the rubbed surface, which consequently reduces the surface roughness and increases the surface hardness[14,15]; (2) the SPs may be refined by experiencing the dehydration reactions due to the local high pressure and flash temperature caused by the collision and rupture of the asperities between the mating metal surfaces[14,17], which results in the decomposition of oil, the crystal fracture and bond breakage of the SPs, contributing to the release of secondary particles, Si-O structures, free water, active oxygen atoms and organic compounds[15,16]; (3) the new products with more unsaturated bonds and higher chemical activity tend to deposit on and react with the freshly exposed contact surface[15,16]. Thus a boundary lubricating film of oxide layer doped with the species enriched in Si-O structures is generated on the worn surface, and it can provide a conjunct effect with the graphite and organic compounds to result in the reduction of friction and wear[14,15].

Furthermore, the enhancement effect of Cu NPs on the interactions between the SPs and the friction surfaces may be attributed to the following aspects: (1) the elliptical Cu NPs possibly prefer rolling like nano-bearings between the contact surfaces to reduce the sliding friction and wear[2]; (2) the Cu NPs, which possess high chemical activity, low hardness and melting point, can deposit on the contact areas and react with other constituents to form a tribofilm containing Cu^0, Cu^+ and Cu^{2+} species with lower shearing strength[2,3,26]; (3) the soft Cu NPs together with the organic molecules may have the colloidal effects on the rubbed surfaces, which can enhance the mechanical entrapment of the refined SPs[2,13]. Thus a more smooth and compact tribofilm with higher content of C, O, Si and Cu elements has formed during the sliding process to further reduce the friction and wear.

4 Conclusions

The effect of Cu NPs on the tribological properties of the SPs suspended in diesel oil was investigated. On the basis of above results and discussion, it can be concluded that:

(1) The Cu NPs can act as a promoter of the SPs as oil additive. The most synergistic effect on friction and wear reducing properties of the SPs can be obtained when the content of Cu NPs is 7.5wt%.

(2) A more smooth and compact tribofilm enriched in Fe, C, O, Si and Cu has formed on the worn surface in the presence of SPs with addition of 7.5wt% Cu NPs.

(3) The outmost surface of the protective film for SR4 mainly contains iron oxides, silicon oxides, species enriched in Si-O structures, graphite, organic compounds, and Cu^0, Cu^{1+} and Cu^{2+} species, which possesses high mechanical properties and lubricating ability, and contributes to the further reduced friction and wear.

Acknowledgements

The authors are grateful for the financial support sponsored by China National 973 Program (2007CB607601) and NSFC (50735006, 50805146, and 50904072).

References

[1] Sun J S. Wear of Metals. Beijing: Metallurgy Industrial Press; 1992(in Chinese).

[2] Liu G, Li X, Qin B, Xing D, Guo Y, Fan R. Investigation of the mending effect and mechanism of copper nano – particles on a tribologically stressed surface. Tribol Lett 2004; 17: 961 – 966.

[3] Hernández Battez A, González R, Viesca J L, Fernández J E, Díaz Fernández J M, Machado A et al. CuO, ZrO_2 and ZnO nanoparticles as antiwear additive in oil lubricants. Wear 2008; 265: 422 – 428.

[4] Chen S, Liu W M, Yu L G. Preparation of DDP – coated PbS nanoparticles and investigation of the antiwear ability of the prepared nanoparticles as additive in liquid paraffin. Wear 1998; 218: 153 – 158.

[5] Rapoport L, Feldman Y, Homyonfer M, Cohen H, Sloan J, Hutchison J L et al. Inorganic fullerene – like material as additives to lubricants: structure – function relationship. Wear 1999; 225 – 229: 975 – 982.

[6] Zhang M, Wang X B, Fu S H, Xia Y Q. Performance and anti – wear mechanism of $CaCO_3$ nanoparticles as a green additive in poly – alpha – olefin. Tribol Int 2009; 42: 1029 – 1039.

[7] Hu Z S, Dong J X. Study on antiwear and reducing friction additive of nanometer titanium borate. Wear 1998; 216: 87 – 91.

[8] Xu T, Zhao J Z, Xu K. The ball – bearing effect of diamond nanoparticles as an oil additive. J Phys D 1996; 29: 2932 – 2937.

[9] Huang H D, Tu J P, Gan L P, Li C Z. An investigation on tribological properties of graphite nanosheets as oil additive. Wear 2006; 261: 140 – 144.

[10] Fernández Rico E, Minondo I, García Cuervo D. The effectiveness of PTFE nanoparticle powder as an EP additive to mineral base oils. Wear 2007; 262: 1399 – 1406.

[11] Zhang Z F, Yu L G, Liu W M, Xue Q J. The effect of LaF_3 nanocluster modified with succinimide on the lubricating performance of liquid paraffin for steel – on – steel system. Tribol Int 2001; 34: 83 – 88.

[12] Wu Y Y, Tsui W C, Liu T C. Experimental analysis of tribological properties of lubricating oils with nanoparticle additives. Wear 2007; 262: 819 – 825.

[13] Chinas – Castillo F, Spikes H A. Mechanism of action of colloidal solid dispersions. Trans ASME 2003; 125: 552 – 527.

[14] Yu H L, Xu Y, Shi P J, Wang H M, Zhao Y, Xu B S et al. Tribological behaviors of surface – coated serpentine ultrafine powders as lubricant additive. Tribol Int 2010; 43: 677 – 685.

[15] Yu Y, Gu J L, Kang F Y, Kong X Q, Mo W. Surface restoration induced by lubricant additive of natural minerals. Appl Surf Sci 2007; 253: 7549 – 7553.

[16] Jin Y S, Li S H, Zhang Z Y, Yang H, Wang F. In situ mechanochemical reconditioning of worn ferrous surfaces. Tribol Int 2004; 37: 561 – 567.

[17] Wang F. Research on microstructure of the auto – restoration layer of worn surface of metals. Mater Sci Eng A 2005; 399: 271 – 275.

[18] Wang X L, Xu B S, Xu Y, Yu H L, Shi P J, Liu Q. Preparation of nano – copper as lubricant oil additive. J Cent South Univ Technol 2005; 12: 203 – 206.

[19] Talke Frank E. A review of ‘contact recording’ technologies. Wear 1997; 207: 118 – 121.

[20] Lee J, Cho S, Hwang Y, Lee C, Kim SH. Enhancement of lubrication properties of nano – oil by controlling the amount of fullerene nanoparticle additives. Tribol Lett 2007; 28: 203 – 208.

[21] Huang W J, Dong J X, Wu G F, Zhang C Y. A study of S – [2 – (acetamido) benzothiazol – 1 – yl] N, N – dibutyl dithiocarbamate as an oil additive in liquid paraffin. Tribol Int 2004; 37: 71 – 76.

[22] Wagner C D, Riggs W M, Davis L E, Moulder J F, Muilenburg G E. Handbook of X – ray photoelectron

spectroscopy. Eden Prairie: Perkin – Elmer Corporation; 1979.

[23] Lamontagne B, Semond F, Roy D. X – ray photoelectron spectroscopic study of Si(111) oxidation promoted by potassium multilayers under low O_2 pressures. J Electron Spectrosc Relat Phenom 1995; 73: 81 – 88.

[24] Koshizaki N, Umehara H, Oyama T. XPS characterization and optical properties of Si/SiO_2, Si/Al_2O_3 and Si/MgO co – sputtered films. Thin Solid Films 1998; 325: 130 – 136.

[25] McIntyre N S, Zetaruk D G. X – ray photoelectron spectroscopic studies of iron oxides. Anal Chem 1977; 49: 1521 – 1529.

[26] Hernández Battez A, Viesca J L, González R, Blanco D, Asedegbega E, Osorio A. Friction reduction properties of a CuO nanolubricant used as lubricant for a NiCrBSi coating. Wear 2010; 268: 325 – 328.

[27] Paschoalino M, Guedes N C, Jardim W, Mielczarski E, Mielczarski J A, Bowen P et al. Inactivation of *E. coli* mediated by high surface area CuO accelerated by light irradiation > 360 nm. J Photochem Photobiol A 2008; 199: 105 – 111.

[28] Zhang T Y, Wang S P, Yu Y, Su Y, Guo X Z, Wang S R et al. Synthesis, characterization of $CuO/Ce_{0.8}Sn_{0.2}O_2$ catalysts for low – temperature CO oxidation. Catal Commun 2008; 9: 1259 – 1264.

Effect of Surface Nanocrystallization on the Tribological Properties of 1Cr18Ni9Ti Stainless Steel*

Abstract Surface nanocrystallization of 1Cr18Ni9Ti austenite stainless steel was conducted by the supersonic fine particles bombarding (SFPB) technique. The friction coefficients and wear losses in air and vacuum were tested to analyse the effect of surface nanocrystallization on the tribological properties of 1Cr18Ni9Ti steel. The results show that the microstructure of the surface layer was refined into nano - grains successfully by SFPB treatment; furthermore, strain - induced martensitic transformation occurred during the treatment. The tribological properties of SFPB treated samples enhanced greatly, The dominant wear mechanism of the original 1Cr18Ni9Ti stainless steel is abrasive wear and adhesive wear, while it transfers to the combined action of fatigue wear, abrasive wear and adhesive wear after surface nanocrystallization by SFPB.

Key words 1Cr18Ni9Ti, nanomaterials, surface nanocrystallization, microstructure, tribological properties

1 Introduction

Nanomaterials with unique structure and excellent properties have attracted great scientific attention since last century[1]. Recently, surface nanocrystallization is considered the most likely breakthrough in the engineering applications of nanotechnology[2]. The surface nanocrystallization of metal materials can improve their physical and chemical properties notably, especially the tribological properties[3]. Applying impact energy of certain frequency and intensity to a component surface to induce severe plastic deformation is the main method of surface nanocrystallization[4]. Researchers have developed mechanical attrition treatment (SMAT), ultrasonic shot peening (USSP), supersonic fine particles bombardment (SFPB) and other surface nanocrystallization process, and fabricated nanocrystalline surface layer on pure iron, carbon steel, stainless steel and other materials successfully[5]. Among them, the SFPB technology is very suitable for the surface nanocrystallization of components with complex shape or large dimension, making it have large engineering application potentiality.

In this study, the 1Cr18Ni9Ti steel was treated by SFPB. The microstructure and mechanical properties of the SFPBed (SFPB treated) 1Cr18Ni9Ti steel were studied to reveal the effect of surface nanocrystallization on the tribological properties.

---

* Copartner: Ma Guozheng, Wang Haidou, Si Hongjuan, Yang Daxiang. Reprinted from *Materials Letters*, 2011, 65 (9): 1268 - 1271.

2 Experimental Details

The material used in this study was a commercial 1Cr18Ni9Ti austenite stainless steel. Before the surface nanocrystallization treatment, the 1Cr18Ni9Ti samples were polished to an average roughness of 0.08μm. And the SFPB process was described in detail in our previous reports[6].

The surface microstructure of the SFPBed samples was characterized by the H – 800 transmission electron microscopy (TEM, operated at 120kV). The thin foil specimens for TEM observation were prepared by polishing and single – sided ion – milling from the backside.

The mechanical properties of the SFPBed and original samples were measured on a Nano Test 600 type nano – hardness tester (NHT). The tribology tests were carried out both in air and vacuum (1×10^{-5} Pa) at room temperature using a УТИ – 1000 type ball – on – disk vacuum tribometer. The morphologies and compositions of the worn scars were examined by a Quanta 200 type scanning electron microscopy (SEM) and EDAX energy dispersive spectrometer (EDS).

3 Results

3.1 Microstructure and mechanical properties

Fig. 1 shows the bright – field TEM images of the top surface layer and of the layer about 20 μm deep from the treated surface. It can be seen from Fig. 1a that the microstructure of the top surface layer consists of roughly equiaxed nanograins, and the inserted selected area electron diffraction (SAED) pattern indicates that the nanograins with random crystallographic orientations can be exactly indexed as bcc α phase. It can be concluded that the SFPB treatment induced transformation from γ austenite to α phase. At about 20μm deep from the treated surface, the grain size is still in nanometer range. It's clear that a parallel finger – like martensite structure appeared in this layer; the diffraction rings can be identified as α martensite and γ austenite, as shown in Fig. 1b. This provides an unarguable evidence for the strain induced martensite phase transformation during SFPB treatment.

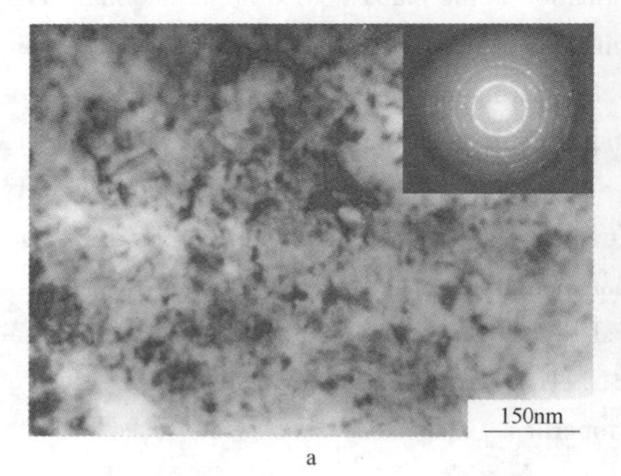

a

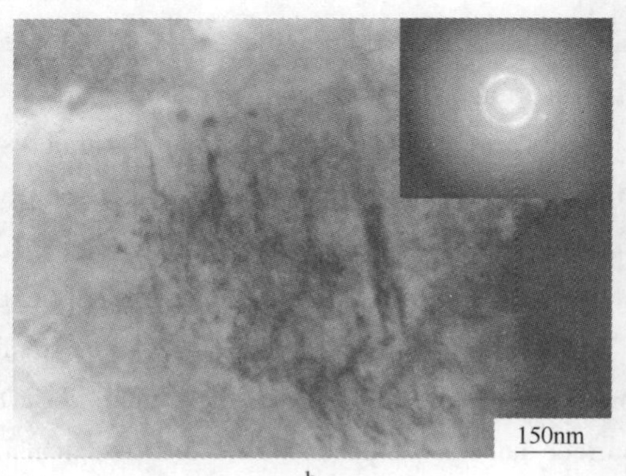

b

Fig. 1 Bright – field TEM images and corresponding SAED patterns of the SFPBed samples

a—At the top surface layer; b—At about 20μm deep from the treated surface

Fig. 2 illustrates the mechanical properties of the samples before and after SFPB treatment by NHT. Each cure represents an average value of five indentations. The hardness of the original sample is 4.68GPa, while that of the SFPBed sample is 7.57GPa. Within a grain size range of 15 – 100nm, the relationship between the hardness and the grain size is still consistent with the Hall – Petch relationship[7], so the notable increase in hardness is mainly attributed to grain refinement. Besides, the martensite phase transformation can also increase the surface hardness.

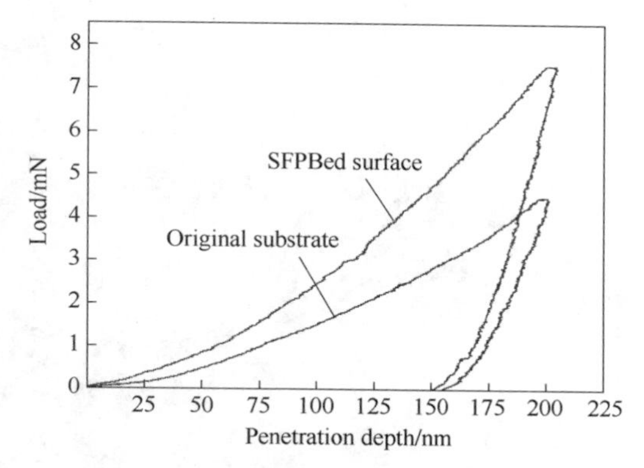

Fig. 2 Load – depth curves of nanoindentation for samples before and after SFPB treatment

3.2 Tribological behaviour

Fig. 3 shows the tribological properties of the original and SFPBed samples in air and vacuum, respectively. The friction coefficients of the SFPBed samples are always lower than those of the original samples. The friction coefficient and wear loss of the SFPBed sample don't change with atmospheric pressure obviously, but those for the original sample are very sharp. The friction coefficient of the original sample shows a violent fluctuation in vacuum, while that of the SFPBed sample is low and stable. The wear mass losses of the SFPBed sample are much lower, and even down to 1/3 of that of the corresponding original sample in vacuum.

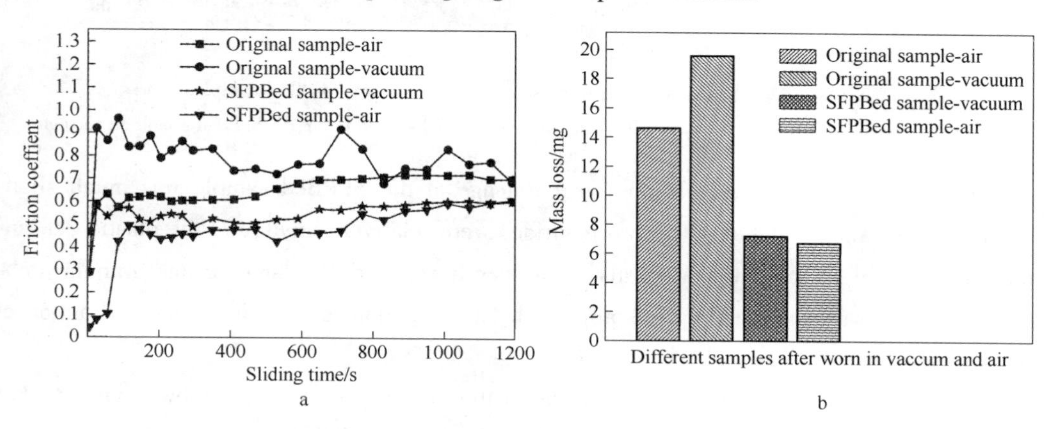

Fig. 3 Tribological properties of original and SFPBed samples

a—Friction coefficient versus friction time; b—Comparison of wear mass loss

Fig. 4 shows the SEM morphologies and the corresponding EDS analysis results of the two kinds of samples after wearing in air and vacuum. The original sample suffered severe wear both in air and vacuum, bulk of materials were removed, leaving deep and wide parallel grooves, and an amount of material accumulated on the sides of the grooves. The main wear damage mechanisms were micro – cutting, adhesion – tearing and material transfer caused by severe plastic deformation.

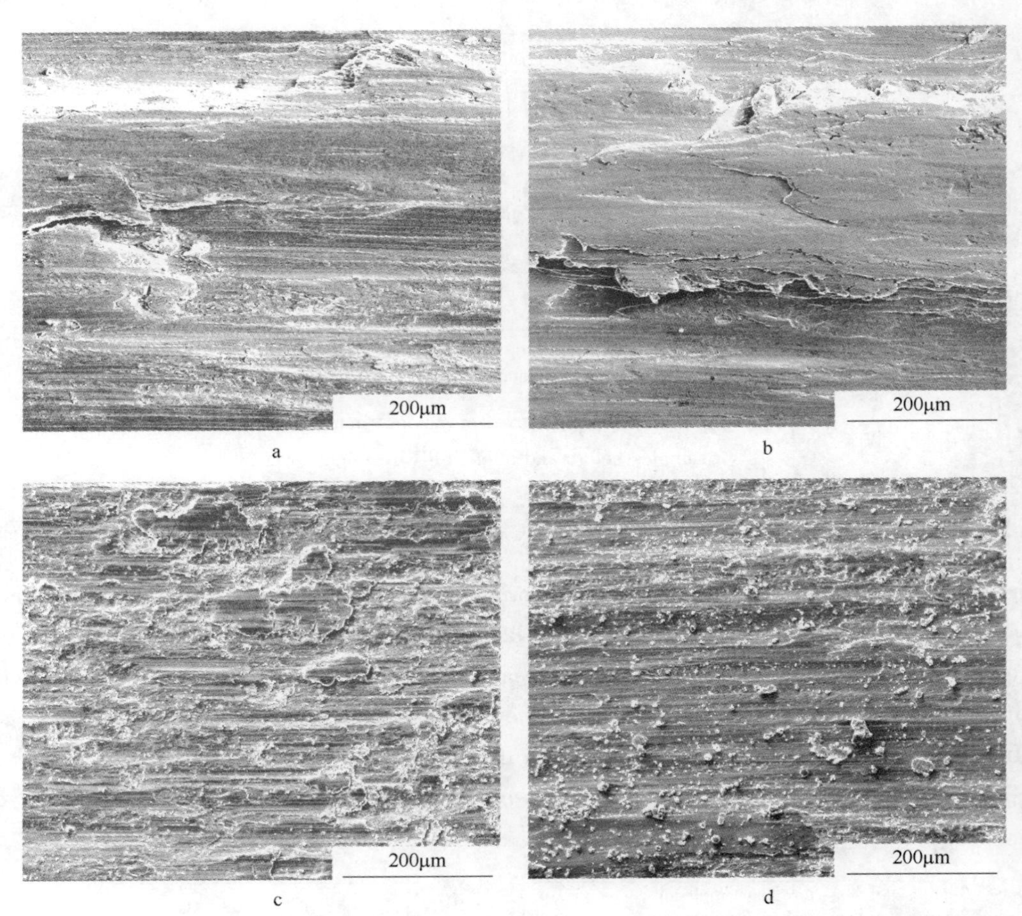

Fig. 4 Morphologies of the worn scars of original and SFPBed samples

a—Original sample in air; b—Original sample in vacuum; c—SFPBed sample in air; d—SFPBed sample in vacuum

Compared to the original sample, the wear damage of the SFPBed sample was much slighter. The grooves and scratches on the worn surface were relatively small, and the plastic deformation and material accumulation were also significantly reduced, but lamellar and granular debris increased, and a lot of spalling pits appeared. The wear damage was the combined action of fatigue wear, abrasive wear and adhesive wear.

The corresponding EDS composition analysis of the above worn surfaces shows that: (1) For both kinds of samples, the worn surfaces in air contain more O element and less Fe element than those in vacuum, which indicates that the tribochemical oxidation reaction occurred during wearing in air; (2) In the same air conditions, the O content on the worn surface of the SFPBed sample was much higher (30.29, at%) than that of the original sample (19.07, at%). This may be because the nanocrystalline layer with high chemical activity can absorb much more external O atoms and form iron oxide more easily during the wearing. The oxide film has an excellent friction – reducing property and can prevent the direct contact among fresh metals effectively[8]. So the friction coefficients and wear losses of both kinds of samples in air were slightly less than those in vacuum, and the tribological properties in air of the SFPBed sample were much better than the original sample.

4 Discussion

Experimental results indicated that, in both vacuum and air conditions, the friction coefficients and wear losses of the surface nanocrystallized 1Cr18Ni9Ti decreased obviously. And the tribological properties didn't change with the atmospheric pressure notably, which is significantly different from the original 1Cr18Ni9Ti. The nanocrystalline layer was formed and the martensitic phase transformation was also induced on the SFPBed surface, resulting in a dramatic increase in surface hardness and strength. The nanocrystallized layer with high resistance to plastic deformation and plastic flow can reduce the adhesive wear effectively. Besides, the actual contact area of the friction interface and the depth of the abrasive particles penetrating into the high hardness nanocrystalline layer were both reduced, and the frictional force and removal of material were reduced correspondingly. So, the friction coefficient and wear loss of the SFPBed sample reduced significantly. In addition, the nanocrystalline surface with high activity is much easier to form continuous oxide film during the wearing, which is also an important reason for the enhancement of its tribological properties in air.

The dominant wear mechanisms of the original 1Cr18Ni9Ti steel were adhesive wear and abrasive wear based on plastic deformation, while the wear damage of SFPBed samples was the co-actions of fatigue wear, adhesive wear and abrasive wear. SFPB treatment formed a specific structure of "external hard and internal soft", and the friction interface was in the condition of elastic-plastic contact. After SFPB treatment, the increase of surface hardness reduced the plowing and adhesion effects, but high density of dislocations and other crystal defects gathered in the subsurface during the repetitive bombardments process[9]; under the periodic friction stress, crack source initiated in the location of defects and ultimately resulted in the spalling of the surface layer.

5 Conclusions

The 1Cr18Ni9Ti stainless steel was surface nanocrystallized successfully by SFPB technology, and the surface hardness increased notably resulting from grain refinement and martensite phase transformation after the SFPB treatment. The nanocrystalline surface layer with high hardness can reduce the adhesive wear and abrasive wear effectively. The higher surface activity of the nanocrystallized layer is helpful for forming oxide film during wearing, which can enhance its

tribological properties in air. The wear damage of the original 1Cr18Ni9Ti steel is plastic deformation – induced adhesive wear and abrasive wear, while the dominant wear mechanisms of the SFPBed samples change to the coactions of fatigue wear, adhesive wear and abrasive wear.

Acknowledgements

This research was financially supported by the 973 Project (2007CB607601), the NSFC(50975285), and the Advanced Maintenance Research Project(9140A270304090C8501).

References

[1] Gleiter H. Progress of nanocrystalline materials. Mater Sci 1989;33:223.

[2] Huang L, Lü J, Troyon M. Nanomechanical properties of nanostructured titanium prepared by SMAT. Surf Coat Technol 2006;201:208 – 213.

[3] Tong W P, Han Z, Wang L M, Lü J, Lu K. Low – temperature nitriding of 38CrMoAl steel with a nanostructured surface layer induced by surface mechanical attrition treatment. Surf Coat Technol 2008;202:4957 – 4963.

[4] Dai K, Shaw L. Comparison between shot peening and surface nanocrystallization and hardening processes. Mater Sci Eng A 2007;463:46 – 53.

[5] Liu G, Wang S C, Lou X F, Lu K. Low carbon steel with nanostructured surface layer induced by high – energy shot peening. Script Mater 2001;8:1791 – 1795.

[6] Ma G Z, Xu B S, Wang H D, Si H J. Effects of surface nanocrystallization pre – treatment on low – temperature ion sulfurization behavior of 1Cr18Ni9Ti stainless steel. Appl Surf Sci 2010;257:1204 – 1210.

[7] Farhat Z N, Ding Y, Northwood D O, Alpas A T. Effect of grain size on friction and wear of nanocrystalline aluminum. Mater Sci Eng A 1996;206:302 – 313.

[8] Li X Y, Tandon K N. Microstructural characterization of mechanically mixed layer and wear debris in sliding wear of an Al alloy and an Al based composite. Wear 2000;245:148 – 161.

[9] Guan X S, Dong Z F, Li D Y. Surface nanocrystallization by sandblasting and annealing for improved mechanical and tribological properties. Nanotechnology 2005;16:2963 – 2971.

Zn - Al 涂层腐蚀电化学行为研究*

摘　要　通过电化学交流阻抗谱（EIS）研究了 Zn - Al 涂层在铜加速醋酸盐雾试验中的腐蚀行为，建立了等效电路模型并分析了相关电化学参数随时间的变化规律。采用扫描电镜及能谱仪分析了涂层腐蚀后的形貌及成分组成，阐释了电化学参数的演变规律。结果表明，盐雾试验中 Zn - Al 涂层的腐蚀由电化学反应控制逐渐转变为扩散控制，最后进入基体腐蚀阶段。

关键词　Zn - Al 涂层　腐蚀　电化学交流阻抗

1　引言

Zn - Al 涂层的腐蚀电位介于锌涂层和铝涂层之间，兼有锌涂层和铝涂层的优势，能够对钢结构提供有效保护。Zn - Al 涂层在很多环境下具有良好的耐蚀性，美国 TAFA 公司的测试表明，涂层经盐雾试验 3000h 仍未出现锈斑，试件表面的腐蚀产物较少[1]，因此，Zn - Al 涂层成为海洋防腐领域常用的涂层之一。热浸镀和热喷涂是制备 Zn - Al 涂层的常用方法[2~6]。热浸镀获得的镀层厚度有限，而热喷涂则可以沉积厚涂层，且成本低廉，因而获得了广泛应用。

通常情况下热喷涂涂层含有少量的孔隙。但 Zn - Al 涂层能够发挥一定的自封闭性，即涂层在腐蚀过程中形成致密的腐蚀产物，堵塞涂层中的孔隙。近几年有学者开展了涂层自封闭性的研究[7,8]，但是有关 Zn - Al 涂层的腐蚀行为及腐蚀产物自封闭性的研究还有待进一步探讨。本文在前人研究工作的基础上，通过高速电弧喷涂技术制备了 Zn - Al 涂层，采用交流阻抗谱研究了涂层在铜加速醋酸盐雾试验后的腐蚀行为，并辅以微观组织分析加以论证。

2　实验

基体选用 25mm × 20mm × 3mm 的 Q235 钢。实验前使用丙酮将试样超声清洗 10min，然后用 25 目棕刚玉砂对基体进行喷砂处理。使用 HAS - 02 型高速喷枪及装备再制造技术国防科技重点实验室研制的自动化高速电弧喷涂系统，在基体上制备 Zn - Al 涂层。喷涂工艺参数为：喷涂电压 34V。喷涂电流 100A，喷涂距离 200mm。雾化气压力 0.7MPa。

盐雾试验采用铜加速醋酸盐雾试验（CASS）。试验条件参照 GB/T 10125—1997《人造气氛腐蚀试验——盐雾试验》，试验周期为 240h。

交流阻抗谱（EIS）使用德国 ZAHNER IM6e 电化学工作站测量，溶液为 3.5% NaCl 水溶液。试样先在 3.5% NaCl 溶液中浸泡 0.5h。待开路电位稳定后再测试交流阻

---

* 本文合作者：刘毅、魏世丞，王玉江。原发表于《功能材料》，2011（42）：226 ~ 229。

抗谱。测试温度为20℃，采用三电极体系，试样为工作电极，参比电极选用饱和甘汞电极。辅助电极为铂电极。交流信号的扰动幅值为5mV。频率范围为$10^{-2}\sim10^{5}$Hz，试样的工作面积为$1cm^2$，阻抗数据采用Zsimp Win软件进行拟合。

采用FEI Quanta 200型扫描电子显微镜（SEM）观察涂层腐蚀后的形貌。采用EDAX公司的Genesis型能谱仪分析涂层成分。

3 实验结果及分析

3.1 Zn－Al涂层的交流阻抗谱

图1a给出了Zn－Al涂层在盐雾试验不同时期的Nyquist图。

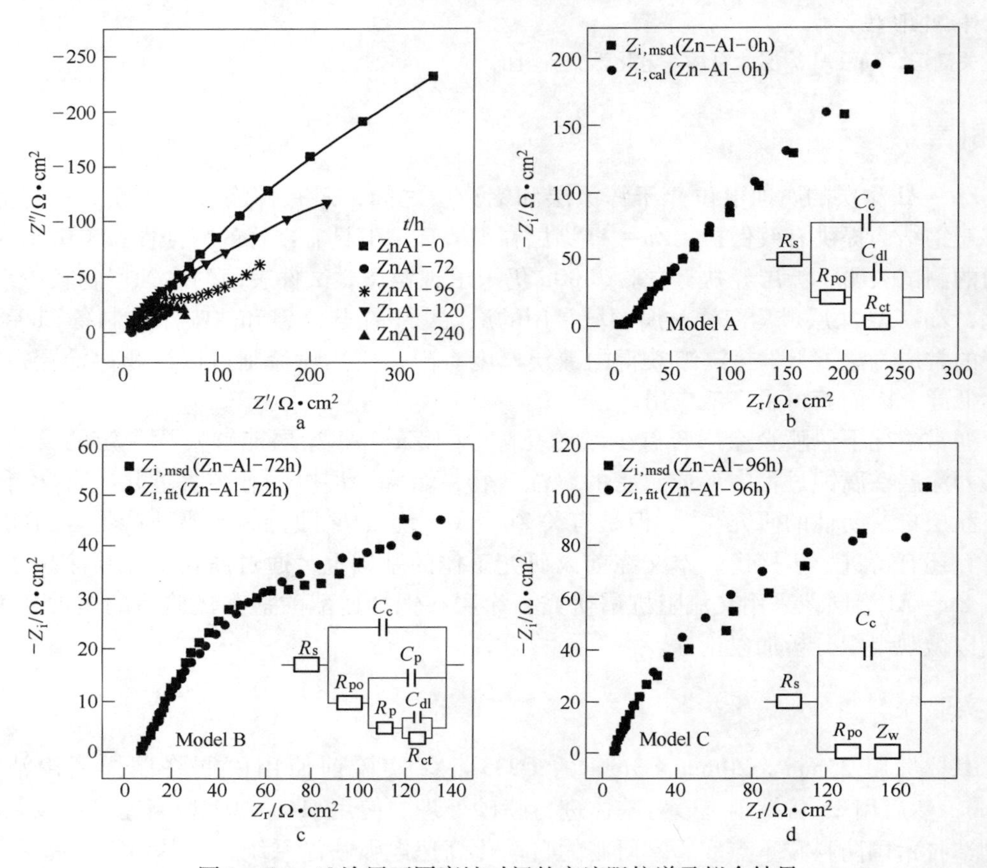

图1　Zn－Al涂层不同腐蚀时间的交流阻抗谱及拟合结果

由图1可知，盐雾试验前涂层阻抗谱的容抗弧较大，盐雾试验开始后，容抗弧经历了先减小后增大再减小的过程。为了更清晰地了解盐雾试验中涂层阻抗谱随时间的变化规律，分别对不同腐蚀时间下涂层的阻抗谱进行拟合分析。

盐雾试验前，图1b的阻抗复平面图出现两个容抗弧，说明Zn－Al涂层阻抗谱含有两个时间常数，高频区容抗弧对应涂层电容及表面微孔电阻，低频区容抗弧对应双电层电容和电化学溶解阻抗[9~11]。低频区容抗弧的半径较大，表明涂层能够有效抑制基体金属腐蚀的发生和发展[12]。采用$R_s(C_c(R_{po}(C_{dl}R_{ct})))$的等效电路进行拟合

（记为模型 A）。各参数物理意义如下：R_s 表示溶液电阻，R_{po}为涂层表面的微孔电阻，C_c 为涂层电容，R_{ct}表示电荷转移电阻，C_{dl}为双电层电容。由于图中阻抗谱的圆心偏向第四象限，考虑到弥散效应，用常相位角元件 CPE 表示电容。拟合结果如图 1b 所示，拟合结果与测试结果较为吻合。虽然阻抗谱中两个时间常数的界限并不清晰，但模型 A 的等效电路对实验数据拟合得较好。这可能是由于电极表面的电化学反应面积较小，导致高低频段的时间常数 $\tau_1(R_{po}C_c)$ 与 $\tau_2(R_{ct}C_{dl})$ 在同一数量级而不易区分[13]。

盐雾试验进行 72h，Zn－Al 涂层的容抗弧与试验前相比大幅减小，说明涂层此时处于活性溶解阶段。相应的阻抗谱图出现了 3 个时间常数，分析认为第 3 个时间常数来源于腐蚀产物电阻 R_p 和腐蚀产物电容 C_p（用常相位角元件表示）的贡献，采用 $R_s(C_c(R_{po}(C_p(R_p(C_{dl}R_{ct})))))$ 等效电路模型对其进行拟合（记为模型 B），拟合结果如图 1c 所示。该腐蚀过程阴极反应是氧的去极化反应：

$$\frac{1}{2}O_2 + H_2O + 2e \longrightarrow 2OH^-, \ \varphi^0 = 0.401V \tag{1}$$

阳极反应是 Zn 的溶解反应[9]：

$$Zn + 2OH^- \longrightarrow Zn(OH)_2 + 2e, \ \varphi^0 = -1.248V \tag{2}$$

Zn－Al 涂层盐雾试验进行 96h 的阻抗谱如图 1d 所示。其低频区出现斜线，呈现 Warburg 阻抗特征，表明材料的腐蚀过程由电化学反应控制转变为扩散控制。这可能是由于 Zn－Al 涂层经历较长时间的电化学腐蚀，在涂层/基体界面上的腐蚀产物发生聚集，在有涂层覆盖的情况下很难疏散，从而使传质成为法拉第过程的控制步骤。图 1d 的阻抗谱图只有一个容抗弧，说明只含一个时间常数，故用 $R_s(C_c(R_{po}Z_w))$ 的等效电路进行拟合（记为模型 C），拟合结果如图 1d 所示。从 EIS 谱图解析结果看，此时腐蚀已经到达基体界面，该物理模型反映了基体表面的腐蚀状况[10]。为验证腐蚀是否到达基体，采用能谱仪分析了铜加速醋酸盐雾试验 72h 和 120h 涂层表面的成分，如图 2 所示，腐蚀 120h 涂层表面出现了 Fe 元素，证明腐蚀已到达基体。Zn－Al 涂层盐雾试验进行 120h，阻抗谱仍可使用模型 C 的等效电路拟合。但是到盐雾试验进行 240h 时，阻抗谱再次出现两个时间常数，这可能是由于涂层/基体界面处的部分腐蚀产物进一步参与化学或电化学反应而被消耗。[13]。

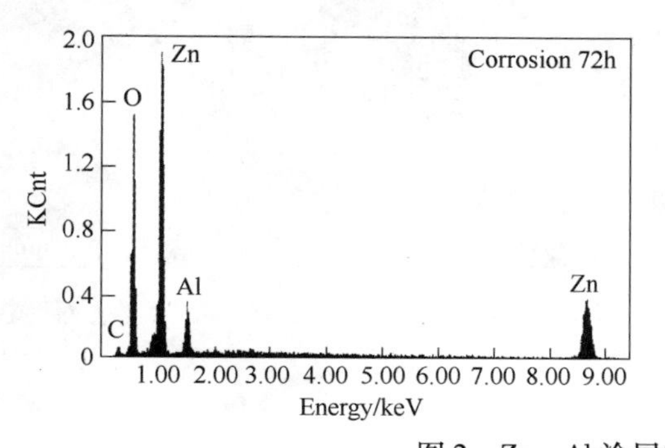

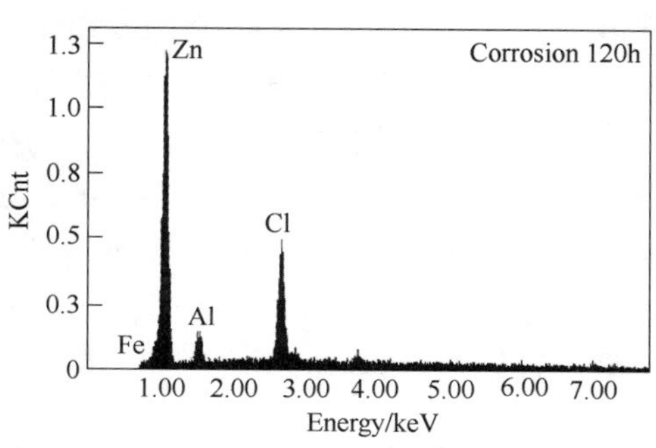

图 2　Zn－Al 涂层表面的 EDS 谱图

3.2　电化学参数的演化规律

图 3 给出了 Zn－Al 涂层的 R_{po}和 C_{dl}随时间的变化曲线。

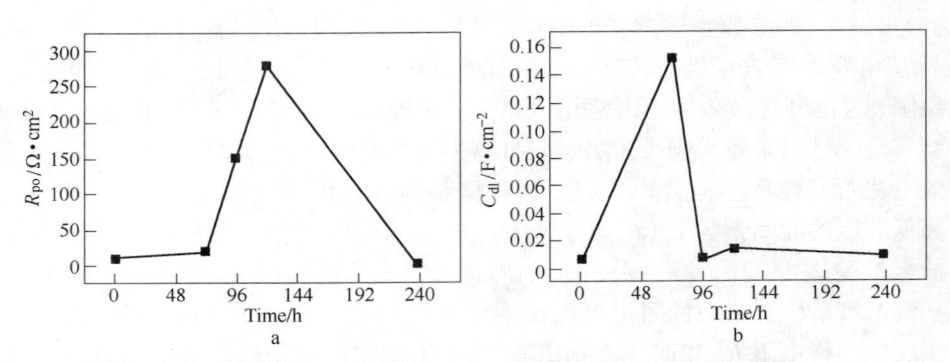

图 3　涂层孔隙电阻和双电层电容随时间的变化曲线

a—涂层孔隙电阻；b—双电层电容

如图 3a 所示，涂层孔隙电阻随时间延长先增大后减小。R_{po}增大是由于涂层的腐蚀产物逐渐增多，微溶性的腐蚀产物积累在涂层表面并逐渐变得致密[14]，腐蚀产物不仅减小了涂层的活性表面，并在一定程度上堵塞了涂层中存在的孔隙，减缓了腐蚀介质向涂层内部的渗透，使得孔隙电阻增大。盐雾试验后期，涂层的孔隙电阻大幅降低，可能是由于 Zn－Al 涂层微溶性的腐蚀产物发生了溶解和脱落以及氯离子的不断渗透形成了贯穿型孔隙，导致孔隙电阻大幅降低。

图 3b 是涂层与基体界面的双电层电容随时间的变化曲线，C_{dl}先增大后减小最后趋于稳定。C_{dl}值与腐蚀反应面积密切相关[11,12]，C_{dl}增大表明腐蚀反应面积扩大，引起这种变化趋势的原因是电解质溶液的渗入增大了活性表面，与组成涂层的物质及涂层中的孔隙相比，电解质溶液具有较小的电阻值。C_{dl}减小是由于涂层表面覆盖了腐蚀产物以及腐蚀产物堵塞涂层中的孔隙导致活性表面减少。但是锌的腐蚀产物微溶于水，部分腐蚀产物的溶解使涂层重新露出活性表面。当电解质溶液的渗透及腐蚀产物的形成与溶解达到平衡时，C_{dl}趋于稳定。图 4 给出了 Zn－Al 涂层腐蚀 120h 的 SEM 照片，涂层表面由大量絮状腐蚀产物构成，这些腐蚀产物减小了涂层的活性表面，并在一定程度上封闭了涂层中的孔隙，表现出自封闭性。但是图 4 也显示出腐蚀产物之间存在着裂纹，这些裂纹是由于腐蚀产物的生长引起的局部应力得到释放而产生的，随腐蚀时间的延长，裂纹不断变宽变深[11]。这些裂纹提供了腐蚀介质向涂层内部渗透的通道，这与 R_{po}和 C_{dl}的拟合结果相吻合。

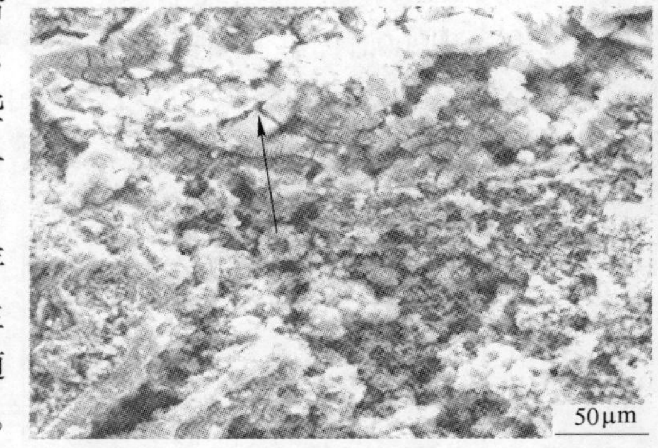

图 4　Zn－Al 涂层盐雾试验 120h 涂层表面形貌

3.3　低频阻抗模值

D. P. Schmidt[15]、谢德明等人[16]认为低频阻抗模值（maximum impedance at lowest frequency，简称 LF）是 EIS 的一个重要参数，能够灵敏地反映涂层耐蚀性的差异，常被用来评价涂层的抗腐蚀性能。Zn－Al 涂层的 LF 值如图 5 所示，LF 值随时间处于波动状态，但总体呈现下降趋势。LF减小是由于腐蚀介质对涂层的溶胀作用使得孔径加大，并且由于腐蚀介质的渗入，通道内的电导率增加及形成新微观通道。而腐蚀产物对涂层的屏蔽以及对涂层微孔的堵塞则可能导致LF增加，LF的这种波动现象即为“自封闭性”。Zn - Al涂层LF值的变化规律说明其具有一定的自封闭性，但自封闭性不足。

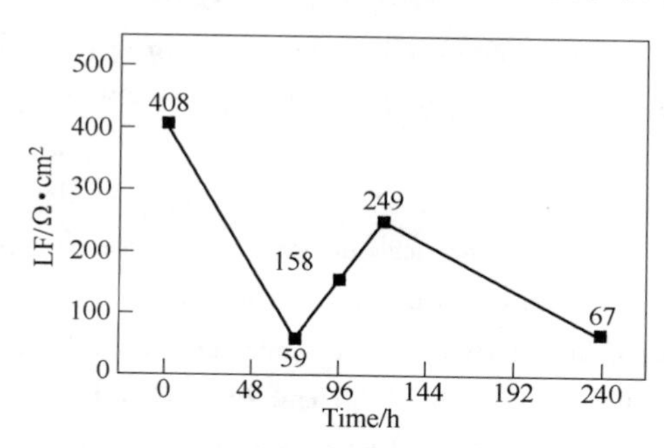

图5　Zn - Al涂层的低频阻抗模值LF

4　结论

（1）Zn - Al涂层在铜加速醋酸盐雾试验中的交流阻抗谱，由电化学反应控制逐渐转变为扩散控制，最后进入基体腐蚀阶段。

（2）涂层孔隙电阻R_{po}、双电层电容C_{dl}及低频阻抗模值LF的变化反映了氯离子的渗透以及微溶性的腐蚀产物不断形成与溶解并达到平衡的过程。

（3）腐蚀后涂层表面覆盖了大量絮状腐蚀产物，表现出一定的自封闭性，但腐蚀产物之间存在着裂纹，这些裂纹成为氯离子向涂层内部渗透的通道。盐雾加速试验120h的涂层表面出现了Fe成分，说明腐蚀到达了基体。

参考文献

[1] 张忠礼．钢结构热喷涂防腐蚀技术［M］．北京：化学工业出版社，2004.

[2] Wang S M, He M Y, Zhao X J. Chinese Journal of Mechanical Engineering, 2009, 22(4): 608 ~ 613.

[3] Wallinder I, Leygraf C, Karlén C. Corrosion Science, 2001, 43(5): 809 ~ 816.

[4] 潘应君，张恒，黄宁．腐蚀科学与防护技术，2003，15(4)：231 ~ 233.

[5] de Rincón O, Rincón A, Sánchez M. Construction and Building Materials, 2009, 23(3): 1465 ~ 1471.

[6] Souto R M, Fernández - Mérida L, González S.［J］. Corrosion Science, 2006, 48(5): 1182 ~ 1192.

[7] 刘燕，朱子新，马洁．中国表面工程，2005，(2)：27 ~ 30.

[8] Zhao X H, Zuo Y, Zhao J M. Surface & Coatings Technology, 2006, 200(24): 6846 ~ 6853.

[9] Chung S C, Cheng J R, Chiou S D. Corrosion Science, 2000, 42(7): 1249 ~ 1268.

[10] 曹楚南，张鉴清．电化学阻抗谱导论［M］．北京：科学出版社，2002.

[11] Hu H L, Li N, Cheng J N. Journal of Alloys and Compounds, 2009, 472(1 ~ 2): 219 ~ 224.

[12] 张金涛，胡吉明，张鉴清．金属学报，2006，42(5)：528 ~ 532.

[13] 胡吉明，张鉴清，谢德明．物理化学学报，2003，19(2)：144 ~ 149.

[14] 章小鸽．腐蚀与防护，2006，27(2)：98 ~ 108.

[15] Schmidt D P, Shaw B A, Sikora E. Progress in Organic Coatings, 2006, 57(4): 352 ~ 364.

[16] 谢德明，童少平，胡吉明．金属学报，2004，40(7)：749 ~ 753.

Electrochemical Behavior of Zn – Al Coating in CASS

Liu Yi, Wei Shicheng, Wang Yujiang, Xu Binshi

(National Key Laboratory for Remanufacturing, Academy of Armored Forces Engineering, Beijing, 100072, China)

Abstract The corrosion processes of Zn – Al coating in copper accelerated acetic acid salt spray test (CASS) was studied with electrochemical impedance spectroscopy (EIS), and equivalent circuits were proposed, which was based on the physicochemical process of the corrosion. The evolution features of the corresponding electrochemical parameters were analyzed. Scanning electron microscope and electron spectrometer were used to analyze the microstructure and compositions of Zn – Al coating after corrosion, which was used to explain the evolution features of the corresponding electrochemical parameters. The results revealed that corrosion process of Zn – Al coating changed from charge transfer control to diffusion control and finally got into matrix corrosion period in CASS.

Key words Zn – Al coating, corrosion, electrochemical impedance spectrum

绿色再制造工程的发展现状和未来展望*

摘　要　再制造工程是废旧机电产品高技术修复改造的产业化，是循环经济和节能环保产业的重要技术支撑。中国的再制造工程经历了产业萌生、科学论证和政府推进三个阶段。中国特色的再制造主要基于尺寸恢复和性能提升，并以先进的寿命评估技术、纳米表面工程和自动化表面工程技术为支撑，其重要特征是再制造产品的质量和性能不低于原型新品，成本为新品的50%、节能60%、节材70%，取得显著环保效果。具有中国特色的再制造模式已逐渐形成，并取得了重要成果。

关键词　再制造工程　中国特色　产业化　关键技术

1　引言

进入21世纪，保护地球环境、构建循环经济、保持社会经济可持续发展已成为世界各国共同关心的话题。目前大力提倡的循环经济模式是追求更大经济效益、更少资源消耗、更低环境污染和更多劳动就业的一种先进经济模式[1,2]。

再制造工程，作为我国21世纪重点发展起来的新方向，以节约资源、节省能源、保护环境为特色，以综合利用信息技术、纳米技术、生物技术等高技术为核心，充分体现了具有中国特色自主创新的特点。再制造可使废旧资源中蕴含的价值得到最大限度的开发和利用，缓解资源短缺与资源浪费的矛盾，减少大量的失效、报废产品对环境的危害，是废旧机电产品资源化的最佳形式和首选途径，是节约资源的重要手段。再制造工程高度契合了国家构建循环经济的战略需求，并为其提供了关键技术支撑，大力开展再制造工程是实现循环经济、节能减排和可持续发展的主要途径之一[3,4]。

2　中国特色的再制造工程的内涵与特征

欧美等国的再制造是在原型产品制造工业基础上发展起来的，目前主要以尺寸修理法和换件修理法为主。随着科技的迅速发展，这种再制造模式存在以下四方面的问题：一是旧件再制造率低，节能节材的效果差；二是再制造次数受限；三是难以提升再制造产品的性能；四是加工量大，环保效果不佳[5]。

中国特色的再制造工程可以简单概括为：再制造是废旧产品高技术修复、改造的产业化。中国特色的再制造工程是在维修工程、表面工程基础上发展起来的，主要基于寿命评估技术、复合表面工程技术、纳米表面技术和自动化表面技术，这些先进的表面技术是国外再制造时所不曾采用的。其重要特征是再制造产品的质量和性能不低于新品，成本只有新品的50%、节能60%、节材70%，对环境的不良影响与制造新品

* 原发表于《中国工程科学》，2011，13(1)：4~10。国家自然科学基金重点项目（50735006）、国家“973”计划项目（2007CB607601）、国家科技支撑计划项目（2008BAK42B03）资助。

相比显著降低。先进表面工程技术在再制造中的应用，可将旧件再制造率大幅度提高（以斯太尔发动机为例，再制造率可提高到92%），使零件的尺寸精度和质量性能标准不低于原型新品水平，而且在耐磨、防腐、抗疲劳等性能方面达到原型新品水平，并最终确保再制造装备零部件的性能质量不低于原型新品。

3 中国再制造的发展历程及面临的任务

笔者1999年起大力宣传并深入研究再制造工程。10年期间，我国仅有的几个再制造企业由最初面临重重困难、各项工作难以开展，发展到目前国家政府机关、行业领域和社会各界广泛认可与大力支持。由于再制造是落实国家节约资源、节省能源和发展循环经济的重要举措，前景广阔；特别是再制造在落实全球可持续发展战略方面发挥了重要作用，备受发达国家关注。

我国的再制造发展经历了3个主要阶段。

第一阶段是再制造产业萌生阶段。

自20世纪90年代初开始，我国相继出现了一些再制造企业，如中国重汽集团济南复强动力有限公司（中英合资）、上海大众汽车有限公司的动力再制造分厂（中德合资）等，分别在重型卡车发动机、轿车发动机等领域开展再制造。产品均按国际标准加工，质量符合再制造的要求。但是，为取缔汽车非法拼装市场，2001年国务院307号令规定旧汽车五大总成一律回炉，切断了这些企业的再制造毛坯来源，产量严重下滑。

第二阶段是学术研究、科研论证阶段。

1999年6月，笔者在西安召开的“先进制造技术”国际会议上发表了《表面工程与再制造技术》的学术论文，在国内首次提出了“再制造”的概念；同年12月，在广州召开的国家自然科学基金委员会机械学科前沿及优先领域研讨会上，“再制造工程技术及理论研究”被列为国家自然科学基金机械学科发展前沿与优先发展领域[6]。

2000年3月，在瑞典哥德堡召开的第15届欧洲维修国际会议上，笔者发表了题为《面向21世纪的再制造工程》的会议论文，这是我国学者在国际维修学术会议上首次发表“再制造”论文；同年12月，中国工程院咨询项目“绿色再制造工程在我国应用的前景”研究报告引起了国务院领导的高度重视，并被批转国家计委（现发改委）、经贸委、科技部、教育部、国防科工委、铁道部、信息产业部、环保总局、民航总局等国务院领导机关参阅[7]。

2001年5月，总装备部批准立项建设我国首个再制造领域的国家级重点实验室——装备再制造技术国防科技重点实验室，于2003年6月正式投入使用。

2002年9月及2007年9月，国家自然科学基金委员会先后批准了两项关于再制造基础理论与关键技术研究的重点项目；2003年8月起，国务院总理温家宝组织了2000多位科学家从国家需求、发展趋势、主要科技问题及目标等方面对“国家中长期科学和技术发展规划”进行了论证研究，其中第三专题《制造业发展科学问题研究》将“机械装备的自修复与再制造”列为19项关键技术之一。

2003年12月，中国工程院咨询报告《废旧机电产品资源化》完成，研究结果表明，废旧机电产品资源化的基本途径是再利用、再制造和再循环，其目标是使再利用、再制造的部分最大化，使再循环的部分最小化，使安全处理的部分趋零化。2004 年 9 月，美国再制造产业网站报道了一条题为《再制造全球竞争——中国正在迎头赶上》的新闻，介绍了再制造在中国的发展状况，并且预言中国将成为美国在再制造领域最强劲的全球竞争对手[8]。

2006 年 12 月，中国工程院咨询报告《建设节约型社会战略研究》中把机电产品回收利用与再制造列为建设节约型社会的 17 项重点工程之一。

上述多角度的深入论证，为政府决策提供了科学依据。

第三阶段是人大颁布法律、政府全力推进阶段。

2005 年 6、7 月间，国务院颁发的 21、22 号文件均明确指出国家“支持废旧机电产品再制造”，并“组织相关绿色再制造技术及其创新能力的研发”。同年 11 月，国家发改委等 6 部委联合颁布了《关于组织开展循环经济试点（第一批）工作的通知》，其中再制造被列为四个重点领域之一，我国发动机再制造企业“济南复强动力有限公司”被列为再制造重点领域中的试点单位。

2006 年，前国务院曾培炎副总理就发展我国汽车零件再制造产业做出批示：“同意以汽车零部件为再制造产业试点，探索经验，研发技术。同时要考虑定时修订有关法律法规”。2008 年，国家发改委组织了“全国汽车零部件再制造产业试点实施方案评审会”，对从全国各省市 40 余家申报单位中筛选出来的 14 家汽车零部件再制造试点企业进行了评审，包括一汽、东风、上汽、重汽、奇瑞等整车制造企业和潍柴、玉柴等发动机制造企业纷纷开始实施再制造项目。

2009 年 1 月，《中华人民共和国循环经济促进法》正式生效，第 2、第 40、第 56 条中 6 次阐述再制造，为推进再制造产业发展提供了法律依据。2009 年 4 月，国家发改委组织“全国循环经济座谈会暨循环经济专家行启动仪式”，笔者向李克强副总理汇报了我国再制造产业发展现状与对策建议，受到李克强副总理的高度重视，他指出“今后要大力推进再制造新兴产业，建议把汽车零部件再制造进一步扩大到机床、工程机械等领域，同时注重再制造与改造相结合；并建议实施汽车下乡工程与再制造生产相结合，促进形成新的产业链”。

2009 年 11 月，工业与信息化部启动了包括工程机械、矿采机械、机床、船舶、再制造产业集聚区等在内的 8 大领域 35 家企业参加的再制造试点工作。

2009 年 12 月，中共中央政治局常委、国务院总理温家宝对再制造作出重要批示，高度肯定再制造产业的重要性，认为再制造不仅关系到循环经济的发展，而且关系到扩大内需（如家电、汽车以旧换新）和环境保护；同时温总理还认为再制造产业链条长，涉及政策、法规、标准、技术和组织，是一项比较复杂的系统工程。

2010 年 2 月 20 日，国家发改委和国家工商管理总局确定启用汽车零部件再制造产品标志，目的在于更好地加强对再制造产品的监管力度，进一步推进汽车零部件再制造产业的健康发展。

2010 年 3 月 13 日，第十一届全国人大三次会议新闻中心专门安排了主题为“再制造与汽车产业的可持续发展”的集体采访活动。

2010 年 5 月，国家发改委、科技部、工信部、公安部、财政部、商务部等 11 个部委联合下发《关于推进再制造产业发展的意见》，指导全国加快再制造的产业发展，并将再制造产业作为国家新的经济增长点予以培育。

2010 年 10 月，国务院 32 号文件《国务院关于加快培育和发展战略性新兴产业的决定》指出：要加快资源循环利用关键共性技术研发和产业化示范，提高资源综合利用水平和再制造产业化水平。

上述法律条款以及党和国家领导人的指示精神，为再制造的发展注入了强大动力。可以说，我国已进入到以国家目标推动再制造产业发展为中心内容的新阶段，国内再制造的发展呈现出前所未有的良好发展态势。

机遇与挑战并存。在国家宏观政策为再制造的发展提供重大机遇的同时，今后的任务仍很繁重。在大众认识层面，再制造作为一个新的理念还没有被人们广泛认识，消费者没有真正认识到使用再制造产品的好处，不少制造企业对发展再制造产业的积极性不高，没有看到再制造对企业可持续发展的深远影响；在物流回收层面，目前我国仍然强制规定报废汽车五大总成必须回炉冶炼，这些政策切断了再制造企业原旧件的来源，是当前制约再制造产业发展的"瓶颈"；在关键技术层面，当前以发动机再制造为主要应用对象的关键技术无法完全满足短期内迅速扩展的各个再制造行业（如工程机械、机床等再制造）对再制造关键技术的多层次需求；在再制造模式层面，国内还有相当多的再制造企业，因引进国外再制造生产线，故仍简单地套用着国外的尺寸修理模式和换件修理模式，对具有中国特色的"尺寸恢复、性能提升"模式认识不足；在中试平台层面，由于缺乏中间转换环节，如国家工程中心、再制造中试基地等，实验室里研发的再制造关键技术未经中试，直接应用于再制造工厂的生产线，影响了生产效率和可靠性；在监管方面，还没有相应的市场准入制度来对再制造企业进行管理，尚未全面实行再制造企业认证、产品标识、产品信息备案等制度。未来的几年中，必须迎难而上，应对挑战，突破瓶颈，切实落实好温总理的指示精神，牢牢抓住再制造发展难得的大好机遇，实现再制造的重大突破。

4　自主创新的中国特色的再制造模式及成果

再制造的基础理论和关键技术研究主要从 20 世纪末开始研究与实践，目前已经形成了"以高新技术为支撑，以恢复尺寸、提升性能的表面工程技术为手段，产学研相结合，既循环又经济"的中国特色的再制造模式。

该模式注重基础研究与工程实践结合，创新发展了中国特色的再制造关键技术，构建了废旧产品的再制造质量控制体系，保证了再制造产品的性能质量和可靠性；注重企业需求与学科建设融合，提升企业与实验室核心竞争力；注重社会效益与经济效益兼顾，促进国家循环经济建设。

4.1　再制造模式的主要特色

（1）技术手段的集约性。再制造是由维修工程和表面工程发展而来的，又结合了力学、摩擦学、材料学等多学科理论，因此再制造的技术手段体现了集约性，既有传统的作为主体的维修技术、表面工程技术，又有新兴的无损检测、寿命评估预测、质量控制等先进技术。

（2）节能环保的实效性。中国特色再制造模式不同于国外换件修理和尺寸修理模式的主要创新在于引入了先进的表面工程技术作为再制造的主要技术手段，通过表面工程技术对零件的局部损伤进行“加法”修复，以恢复并提升零件的性能，最大限度地挖掘了废旧零件中蕴含的附加值，避免了回炉和再成型等一系列加工中的资源能源消耗和环境污染。

（3）基础研究的前瞻性。采用超声波、涡流检测、金属磁记忆等无损检测技术与模拟评估手段，创新性地进行了国际前沿的再制造寿命评估基础研究，为再制造产品性能达到或超过原型新品奠定了坚实的理论基础。

（4）关键技术的先进性。将自主研发的先进表面工程、纳米技术和自动化技术用于再制造生产，大大提升了再制造的品质，不仅使再制造产品的性能达到甚至超过新品，而且对资源、能源的节约和对环境的保护效果更为优异。现已成功开发了再制造寿命评估仪器及软件、自动化纳米电刷镀设备、自动化高速电弧喷涂设备、自动化等离子熔覆设备和智能纳米减摩自修复添加剂技术等，应用效果表明，先进表面工程技术在发动机再制造中的推广应用，大大提高了旧件利用率，降低了再制造成本，不仅使工厂获得了经济效益，还为国家节能、节材及保护环境做出了重要贡献。

（5）工程应用的先导性。通过产学研的联合攻关为我国再制造企业发展提供了重要技术支撑。目前已形成了具有中国特色的再制造工程，引领着我国再制造技术的发展方向，并在国际上占有重要的一席之地。

4.2 再制造模式的基础理论研究成果

4.2.1 拓展了产品全寿命周期理论，提出了再制造循环寿命周期理论

产品的全寿命周期是指产品从设计、制造、使用、维修到报废所经历的全部时间，其特征是“研制—使用—报废”，其物流是一个开环系统；再制造的出现，完善了全寿命周期的内涵，使得产品在全寿命周期的末端，即报废阶段，不再作为废品报废，而是依靠高新技术恢复性能、重新焕发生命力，此时全寿命周期的特征已转变为“研制—使用—报废—再生”，其物流已成为一个闭环系统[9]。因此，再制造是对产品全寿命周期的延伸和拓展，赋予了废旧产品新的寿命，形成了再制造产品的循环寿命周期。再制造不仅可使废旧产品起死回生，还可很好地解决资源能源节约和环境污染问题。

4.2.2 创新了再制造寿命评估理论，确保了废旧产品的再制造质量基础前提和再制造产品的质量保证体系

再制造寿命评估包含再制造前的再制造毛坯（废旧零部件）寿命评估和再制造后的再制造产品（再制造零部件）寿命预测两部分内容。其中，废旧零部件寿命评估是通过对废旧零件的剩余寿命评估，回答废旧零部件能否再制造、能再制造几次（剩余寿命是否足够）的问题，是保证再制造毛坯质量的重要途径。再制造产品寿命预测是通过对再制造产品表面涂层质量和服役寿命进行评估，保证再制造产品的性能不低于新品。

再制造零部件疲劳损伤规律。研究提出疲劳寿命是机械零部件寿命的核心，深入研究了典型零部件疲劳损伤累积、疲劳应力集中裂纹萌生和扩展规律；并借助涡流检测、超声检测、金属磁记忆检测多种无损检测技术手段，实现零部件内部和表面应力集中与裂纹的无损检测，为再制造毛坯剩余寿命评估提供了检测技术和理论指导[10]。

再制造毛坯剩余寿命无损评估理论。创新性利用金属磁记忆对再制造毛坯的剩余寿命评估进行探索研究，发现了金属磁记忆信号与废旧零部件所受疲劳载荷大小与历史应力、残余应力和应力集中之间的关系以及废旧零部件磁畴与载荷和磁记忆信号之间的关系，初步构建出表征铁磁性废旧零部件疲劳裂纹萌生寿命模型及裂纹扩展寿命模型，并初步实现了发动机气门杆、连杆、曲轴等重要零部件损伤和寿命的检测评估，为再制造质量控制提供了理论基础。

再制造产品寿命预测理论。再制造产品的结构疲劳寿命以原结构件疲劳寿命为基础。重点攻克了再制造涂层接触疲劳、腐蚀、磨损寿命预测理论，并指出再制造涂层接触疲劳寿命与原结构基体材料密切相关。创新性地将实验力学和声发射理论进行综合集成，通过解决典型声发射信号特征参量的甄选及其指代信息分析，获得真实准确地反映再制造零件表面涂层内部微裂纹萌生、扩展及断裂等的实验力学信息，初步实现对再制造零件表面涂层寿命演变规律的把握，建立了再制造零件涂层的抗接触疲劳损伤失效模型。

再制造产品台架试验及实车考核。在实验室研究结果基础上，针对在选定的再制造材料、工艺和技术规范下获得的再制造零部件，通过台架试验和实车考核，对再制造产品进行整体综合评价，获得充足的剩余寿命实车考核数据，确保再制造产品能够重新服役一个完整的寿命周期。

4.2.3 提出了再制造性和再制造率的概念，完善了产品的再制造评价体系

再制造性是废旧产品能否进行再制造的重要属性，它是指在规定的条件及时间内使用的产品退役后，综合考虑技术、环境等因素，在达到规定性能时，通过再制造获取原产品价值的能力。目前已初步构建起了再制造性函数、再制造费用的统计分布模型、系统再制造费用分析计算模型等。

衡量再制造对节能节材的重要指标是废旧零件再制造率的高低，国际上通常采用计重法统计，笔者创新提出重量再制造率、数量再制造率、价值再制造率以及数量比再制造率、价值比再制造率等多维评价指标体系。

4.3 再制造模式的关键技术与工程实践成果

由于再制造使用的是经过长期服役而报废的各种成型零件，其损伤失效形式复杂多样，残余应力、内部裂纹和疲劳层的存在导致寿命评估与服役周期复杂难测，再制造还要在保持废旧零（部）件材质和形状基本不变的前提下，采用高技术恢复原产品的尺寸标准，达到或超过原产品的性能指标，实现原产品的功能升级，同时也采用正规化、规模化的加工手段，因此加工工艺更为复杂。

再制造寿命检测的核心是疲劳寿命，再制造质量控制的关键是裂纹控制，再制造的主要损伤形式是表面磨损。实验室根据再制造产品失效特征和质量性能不低于新品的标准要求，通过多年研究与实践，研发了多项中国自主创新的再制造技术。

（1）再制造无损检测评估技术及其仪器设备。废旧零部件损伤状态无损检测与评估是再制造质量控制体系的重要内容。研究了汽车发动机缸体、曲轴、连杆、气门杆等不同再制造零部件的多种无损检测评估技术（涡流、超声、金属磁记忆、声发射等），并研制出了高频涡流无损检测仪（气门杆、连杆等，通用性强）、高穿透力超声

无损检测仪（曲轴等，通用性强）、缸体涡流/磁记忆综合无损检测评估仪、金属磁记忆寿命评估仪、纳米复合刷镀层无损测厚仪等，初步实现了发动机连杆、曲轴、发动机缸体等重要零部件的损伤和无损检测评估，为再制造产品质量不低于新品的质量控制体系提供了有力保障。相关技术和仪器设备已在济南复强动力有限公司再制造生产线上应用试验，为再制造毛坯质量控制体系提供了技术支撑。

（2）自动化纳米颗粒复合电刷镀技术。这是自主研发的一项先进的再制造技术，纳米刷镀层与不含纳米颗粒的金属刷镀层相比，耐磨性能提高 1.5 倍，抗温性由 200℃ 提高到 400℃，抗接触疲劳性能由 10^5 周次提高到 10^6 周次，显著延长零件的使用寿命，并成功应用于飞机发动机叶片、汽车发动机连杆、凸轮轴和缸体的再制造[11]。但由于手工纳米电刷镀生产效率低、劳动强度大，针对重载汽车发动机连杆和缸体缸筒的再制造难题和产业化生产需求，自主研发了发动机连杆自动化纳米电刷镀专用设备和汽缸筒自动化纳米电刷镀专用设备，实现了镀液连续供应和循环利用、纳米电刷镀再制造工艺过程综合监控。生产应用表明，生产率提高 5 ~ 10 倍，再制造消耗材料仅为该零件本体重量的 1% ~ 2%，费用是新品价格的 1/10，实现了废旧零件再制造的需求。并且镀液循环利用，废水集中处理，可实现全过程的绿色化要求。

（3）纳米减摩智能自修复添加剂技术。研制的摩擦副损伤原位自修复添加剂已在济南复强动力有限公司的设备及发动机台架上进行实车考核试验，同时在安徽定远进行了实车试验。试验结果证明，该技术可实现对早期磨损表面的轻度微损伤的原位动态自修复，对零件表面形貌进行优化，显著提高零件表面硬度和降低粗糙度，进而改善润滑状况，延长零部件的使用寿命，并节约燃油 3% ~ 7%，降低润滑油温度 40%，显著延长换油周期，节能减排效果明显。

（4）自动化高速电弧喷涂技术。自主研发了自动化高速电弧喷涂技术，采用机器人或操作机的操作臂夹持喷枪，通过红外温度场监测和编程控制高速电弧喷枪实现各种规划路径，实时反馈调节喷涂工艺参数，实现自动喷涂作业的智能控制。该技术结合新开发的 FeAl 和 FeAlMn 系粉芯丝材制备出的喷涂层，结合强度高，硬度高，耐磨损性能好，已成功应用于废旧斯太尔发动机缸体的再制造，已完成再制造量 200 多台。采用自动化电弧喷涂技术后，再制造单件发动机缸体的时间由手工的 1.5h 缩短为 20min，喷涂效率提高 4.5 倍。

（5）自动化微束等离子熔覆技术。自主创新设计了 70kHz 高频逆变微束等离子电源，高于目前通常采用的 20kHz 逆变频率，从而减少了设备的体积，提高了系统的响应特性，使得微束等离子弧的工作更加稳定。利用该技术对发动机废旧排气门密封锥面进行再制造后的气门变形量小，表面硬度恢复到新品数值，力学性能满足要求，成本仅为新品的 1/5。

5　再制造产业发展的前沿问题

展望未来，中国的再制造应从 4 个方面予以重点突破，即“研究再制造质量控制的科学基础，创新再制造成型加工的关键技术，制定再制造的行业标准，探索加强国内外再制造技术的交叉融合”。

（1）研究再制造质量控制的科学基础。寿命评估是再制造质量控制的核心研究内容，建立准确的再制造寿命预测模型，需要深入研究探索以产品全寿命周期理论、废旧零件和再制造零件的寿命评估预测理论等为代表的再制造基础理论，以揭示产品寿命演变规律的科学本质。为解决装备寿命评估这一世界性难题，必须研究更多更有效的无损检测及寿命预测评估理论与技术。

（2）创新再制造先进成型加工的关键技术。需要不断创新研发用于再制造的先进表面工程技术群，使再制造零件表面涂层的强度更高、寿命更长，确保再制造产品的质量不低于和超过新品。现已成功开发出纳米表面工程技术和自动化表面工程技术，除需对它们进一步完善外，还需研发生物表面工程技术等新的方向。

（3）制定再制造的行业标准。我国再制造因起步较晚，再制造企业的技术积累少，再制造标准缺乏，在一定程度上阻碍了再制造的广泛应用。应尽早建立系统、完善的再制造工艺技术标准、质量检测标准等体现再制造走向规范化的标准体系。

（4）加强国内外再制造的交叉融合。一是借鉴国外再制造产业发展模式，加快国内再制造产业发展；二是借鉴国外再制造逆向物流与信息化管理手段，完善国内再制造流通管理；三是宣传中国自主创新的表面工程与修复技术，加强再制造工艺手段的交叉融合与应用；四是探索加强中国自主创新的再制造质量控制方法与标准的交叉融合。

6 结语

（1）再制造是循环经济的重要技术支撑，是对废旧产品进行高技术修复、改造的产业化。再制造的重要特征是再制造产品的质量和性能不低于新品，有些能超过新品，成本却只是新品的50%，节能60%，节材70%，对环境的不良影响显著降低。

（2）再制造作为节能环保产业的重要组成，已被列为战略性新兴产业。目前在我国得到快速发展，有关再制造的法律法规、基础理论、关键技术、行业标准等不断完善，尤为重要的是，再制造的产业试点已全面铺开。

（3）产品质量是再制造的生命，先进的表面工程、纳米表面工程、自动化表面工程技术和严格的无损检测及寿命预测技术是再制造产品质量的有力保证。

（4）国外再制造采用的是换件修理法和尺寸修复法；而我国探索形成了“以高新技术为支撑，以恢复尺寸、提升性能的表面工程技术为依托，产学研相结合，既循环又经济”的中国特色的再制造模式。

（5）今后要结合国情，深入研究国外再制造产业模式，加强中国自主创新的再制造模式和国外再制造的交叉融合与发展。

参考文献

[1] 徐匡迪．工程师——从物质财富的创造者到可持续发展的实践者［J］．中国表面工程，2004，17(6)：1～6.

[2] 邢忠，姜爱良，谢建军．汽车发动机再制造效益分析及表面工程技术的应用［J］．中国表面工程，2004，17(4)：1～5.

[3] 徐滨士，朱胜，马世宁，等．装备再制造工程学科的建设与发展［J］．中国表面工程，2003，16(3)：1～6.

[4] 徐滨士，刘世参，王海斗．大力发展再制造产业［J］．求是，2005(12)：46～47.

[5] Xu Binshi, Zhu Sheng. Advanced remanufacturing technologies based on nano-surface engineering [C] //Proc. 3rd Int. Conf. on Advances in Production Eng, 2004: 35 ~ 43.
[6] 徐滨士，马世宁，刘世参，等．绿色再制造工程设计基础及其关键技术 [J]．中国表面工程，2001，14(2)：12 ~ 15.
[7] 徐滨士．绿色再制造工程及其在我国应用的前景 [R]．中国工程院咨询报告，2000.
[8] 国家自然科学基金委员会工程与材料科学部．机械与制造科学 [M]．北京：科学出版社，2006：373 ~ 398.
[9] 徐滨士．装备再制造工程的理论与技术 [M]．北京：国防工业出版社，2007.
[10] 徐滨士，刘世参，史佩京，等．再制造工程的发展及推进产业化中的前沿问题 [J]．中国表面工程，2008，21(1)：1 ~ 6.
[11] Xu Binshi. The remanufacturing engineering and automatic surface engineering technology [J]. Key Engineering Materials, 2008(373 ~ 374): 1 ~ 10.

Development Status and Prospect of Green Remanufacturing Engineering

Xu Binshi

(National Key Laboratory for Remanufacturing, Academy of Armored Force Engineering, Beijing, 100072, China)

Abstract Remanufacturing engineering is the industrialization of high technology maintenance of the waste productions, and it is important technology support of circular economy and energy saving & environmental protection industry. The remanufacturing engineering in China has experienced three periods, namely industrial emergence, scientific demonstration and governmental promotion. Based on performance improvement method and size restitution method, Chinese characteristic remanufacturing engineering takes advanced life assessment technology, nano surface engineering and automated surface engineering technologies as the key remanufacturing technologies, and the main character of which is that the performance and quality of remanufacturing products are at least as good as the new ones of the prototype, with cost at 50%, energy saving 60%, material saving 70%, which will improve environmental protection obviously. The new remanufacturing mode with Chinese character has gradually established and obtained great achievements.

Key words remanufacturing engineering, Chinese character, industrialization, key technology

再制造成型技术发展及展望*

摘　要　再制造作为国家新兴战略性产业，高度契合了国家发展循环经济的战略，得到政府和企业的高度重视。再制造成型技术能够恢复废旧零部件尺寸，恢复甚至提升其服役性能，是再制造工程的核心。综述再制造成型技术的国内外研究和发展现状，比较分析国内外再制造成型技术和再制造成型质量控制标准的差异，并重点从“尺寸恢复法”再制造成型技术、“尺寸恢复与机械加工”融合的再制造成型技术和再制造成型质量控制与评价方法三方面，详细阐述我国再制造成型技术的最新成果。在此基础上，提出我国再制造成型技术的发展趋势和发展策略建议。

关键词　循环经济　再制造成型技术　现状　发展趋势

1　引言

进入21世纪，保护地球环境、构建循环经济、保持社会经济可持续发展已成为世界各国共同关心的话题。

再制造工程以节约资源能源、保护环境为特色，以综合利用信息技术、纳米技术、生物技术等高技术为核心，可使废旧资源中蕴含的价值得到最大限度的开发和利用，缓解资源短缺与资源浪费的矛盾，减少大量的失效、报废产品对环境的危害，是废旧机电产品资源化的有效途径。再制造工程高度契合了国家发展循环经济的战略需求，并为其提供了关键技术支撑，近年来得到了政府和企业的高度重视。再制造产业已被国家列为新兴战略性产业，并且列入了我国社会经济发展“十二五”规划。再制造已成为节能减排、促进循环经济发展的主要途径之一[1~3]。

再制造成型技术是以废旧机械零部件作为对象，恢复废旧零部件原始尺寸，并且恢复甚至提升其服役性能的材料成型技术手段的统称，其是再制造工程的核心。废旧零件再制造成型，主要包含两方面的内容：（1）恢复废旧零件失效部位的原始尺寸；（2）恢复甚至提升废旧零件的性能。

针对废旧件而言，恢复其原始尺寸，可以是恢复表面或表层尺寸，也可以是恢复局部的三维立体尺寸，主要采用能够在零件基体损伤部位沉积成型修复性表面涂层或三维金属体的各种先进表面工程技术、熔焊沉积技术、快速成型技术等先进再制造技术手段；恢复甚至提升其服役性能主要取决于再制造成型所采用的材料和技术工艺。再制造成型技术具有如下重要特征：再制造成型产品的质量和性能不低于新品，成本约为新品的50%，节能60%、节材70%，对环境的不良影响显著降低。随着再制造产业发展，再制造成型技术及其相关理论也获得了快速发展[2,4]。

* 本文合作者：董世运、朱胜、史佩京。原发表于《机械工程学报》，2012，48(15)：96~105。国家重点基础研究发展计划（“973”计划，2011CB013403）和“十二五”预研（51327040401）资助项目。

2 再制造成型技术的研究和发展现状

2.1 国外研究和发展现状

在工业发达国家，废旧产品造成的危害暴露较早，因而相应的对策也被较早提出。再制造在欧美等发达国家已有几十年的发展历史，无论在废品回收责任制、再制造产品质量保证，还是在再制造产品销售和售后服务等方面都已形成了较完整的产业体系。

2005 年全球再制造业产值已超过 1000 亿美元，美国的再制造产业规模最大，达到 750 亿美元，其中汽车和工程机械再制造占 2/3 以上，约 500 亿美元。美军高度重视再制造，也是再制造的最大受益者。美军军费世界第一，但美军仍然认为再制造具有重大作用，尤其是“在财政预算有限、新装备配备不到位、制造新装备费用高昂的情况下，再制造可维持武器装备的战备完好率”。隶属于美国国家科学研究委员会的“2010 年后国防制造工业委员会”制定了 2010 年国防工业制造技术的框架，将武器系统的再制造列为国防工业的重要研究领域。近年来，日本加强了对工程机械的再制造，再制造的工程机械中，58% 由日本国内用户使用，34% 出口到国外，其余的 8% 拆解后作为配件出售。

在机械产品再制造成型技术方面，欧美和日本等工业发达国家在再制造工业生产中，主要采用换件法和基于机械加工方法的再制造成型技术（即尺寸修理法）进行，如英国 ListerPetter 再制造公司，每年为英国、美国军方再制造 3000 多台废旧发动机时，对磨损超差的缸套、凸轮轴等关键零件都予以更换新件，并不修复。美国康明斯发动机公司采用机械加工方法对发动机缸体、缸盖、曲轴、连杆等主要贵重零部件进行尺寸修理，并更换其相关配偶件和附件，对发动机进行再制造生产。

随着社会可持续发展理念的提出，欧美等工业发达国家也开始注意减少更换件的数量，而对基于表面工程的再制造成型技术及其成套设备的研发投入了大量力量，主要研发了基于激光熔覆、等离子堆焊等原理的高能束再制造成型技术，并凭借其良好的工业基础和先进科技成果，把再制造成型技术与工业机器人、自动控制技术、监检测技术等有机集成，研发出了先进的成套设备，在国防装备和工业设备关键零部件再制造成型中获得了成功应用[5~8]。

2.2 中国再制造成型技术的发展现状

1999 年 6 月，在西安召开的“先进制造技术”国际会议上，徐滨士院士发表了《表面工程与再制造技术》的学术论文，在国内首次提出了“再制造”的概念，至今已经过去 10 年有余。当前，我国对发展再制造产业高度重视，鼓励政策和法律法规相继出台，再制造示范试点工作稳步进行，再制造理论与技术的研究已取得重要成果。随着我国再制造产业的发展，高等院校和企业对再制造成型技术和理论的研究不断深入和拓展，促使再制造成型技术近年来快速发展。

装甲兵工程学院装备再制造技术国防科技重点实验室已初步构建起再制造理论体系框架，主要包括再制造产品多寿命周期理论、再制造产品设计基础、再制造毛坯剩余寿命评估、再制造产品寿命预测、再制造关键技术基础和再制造模拟与仿真六个方面，并在自动化高速电弧喷涂再制造成型、等离子熔覆再制造成型、激光熔覆再制造成型、自动化纳米电刷镀再制造成型等技术方面取得了突破。针对不同零件再制造成型，研究了再制造成型机理、材料、工艺成套设备以及再制造控性控形方法。综合机器人堆焊和数控铣削加工，实现了机器人 GMAW——数控铣削复合快速再制造过程，深入研究了堆焊再制造的成型路径规划理论和方法。天津工业大学杨洸臣教授课题组研究了纺织设备、冶金设备等重要零部件的激光熔覆再制造成型方法[9]。西北工业大学黄卫东教授课题组等研究了航空发动机钛合金叶片等零部件的激光再制造成型方法[10]。上海宝冶工程技术公司研究和应用了高能微弧冷焊再制造成型技术。沈阳大陆激光技术有限公司和上海宝钢集团等研究了连轧机牌坊等冶金设备重要零部件的激光熔覆再制造成型，并实现了工程应用。广州有色金属研究院研究了多种热喷涂再制造成型方法。

目前，再制造成型技术的应用领域已涉及冶金、石化、能源、交通、采矿、武器装备等国民经济和国防各工业领域。再制造成型技术解决的装备零部件失效问题包含了磨损、腐蚀、裂纹、疲劳、外物损伤等各种损伤失效形式。针对不同损伤失效形式零部件，根据其服役条件要求，很好地解决了零部件性能恢复与提升、形状原尺寸恢复等再制造成型过程中的形状控制和性能控制问题。

2.3 再制造成型技术途径

按照再制造成型过程中零件尺寸的增减变化情况，再制造成型技术可以分为“尺寸恢复法”再制造成型和“尺寸加工法”再制造成型两种技术途径。其中，“尺寸恢复法”再制造成型不但可以恢复零件的原始设计尺寸，而且可以通过应用新材料提高零件性能，因此，又称为“尺寸恢复与性能提升方法”。纵观国内最近发展，重点在“尺寸恢复法”再制造成型技术方面获得了显著进展，并逐步实现了两种技术途径的融合创新发展。

2.3.1 “尺寸恢复法”再制造成型技术

“尺寸恢复法”再制造成型技术是针对磨损、腐蚀等表面损伤零件和缺损、裂纹等三维体积损伤零件，采用先进表面工程技术、三维沉积成型技术等恢复零件损伤部位的几何尺寸，并通过优化再制造成型所用材料和工艺方法，恢复和提升零部件性能。针对表面损伤零部件的再制造成型，主要采用先进表面工程技术手段，因此，许多表面工程技术可以应用于再制造成型领域。

表面工程技术进步促进了再制造成型技术和再制造产业发展，尤其在高精度高性能表面再制造成型方面取得了显著成绩。例如，采用气相沉积硬膜技术实现了发动机活塞密封环、气门挺杆盘面等的再制造成型，采用激光熔覆技术实现了凸轮轴凸轮的再制造成型，采用等离子喷涂技术再制造成型发动机叶片表面热障涂层等，均实现了高精度高性能表面的再制造成型和零件延寿，解决了困扰装备维修保障人员多年的技术难题。

在“尺寸恢复法”再制造成型技术领域的突破，可归纳为以下几个主要方面。

（1）三维体积损伤机械零部件的再制造成型技术。三维体积损伤机械零部件的损伤一般由应力或者外力作用引起，因此，该部位的再制造成型必须考虑到承受载荷能力。为此，一般对其再制造成型技术的基本要求是沉积成型金属具有优异的力学性能，并且在再制造成型过程中尽可能不降低零件基体材料性能。

在三维体积再制造成型领域，有突破的技术主要是通过各种热源熔化添加材料的能量束再制造成型技术，如激光熔覆再制造成型技术、等离子熔覆再制造成型技术、电弧堆焊再制造成型技术、高速电喷涂再制造成型技术等。

例如，装甲兵工程学院对重载车辆中难修复典型零件的激光再制造成型进行了系统研究，成功再制造成型了齿类件和薄壁件等三维损伤零件[11]。激光再制造成型与等离子喷涂成型等多种再制造成型技术手段复合，实现了不同工业领域燃汽机、烟机以及航空发动机等定子和转子叶片的三维再制造成型。

随着新型材料研制和成型工艺监控技术的提升，高速电弧喷涂再制造成型技术已成为可以实现大厚度再制造成型的一项技术，已由原来主要用于表面涂层制备，发展到了具备厚成型能力的水平，这得益于电弧喷涂成型理论、材料和技术工艺方法方面研究的突破。

（2）自动化、智能化再制造成型技术。再制造成型技术方法已由最初重视废旧零件尺寸和性能恢复的成型技术手段的研究，正在向提高再制造成型效率的自动化再制造成型方法研究方向发展。其中，一个重要突破是把三维反求建模技术和再制造工艺相结合，实现了再制造成型技术的自动化过程。

大连海事大学、华中科技大学等单位针对再制造成型过程中的零件缺损部位的反求建模，在理论和技术研究方面取得了突破性进展。

近两年，针对机器人操作自动化再制造成型过程，在损伤部位再制造路径生成理论和方法以及自动化再制造成型设备系统等方面，均取得了重要进展。装甲兵工程学院系统开展了基于机器人的惰性气体保护焊（metal inertia gas，MIG）熔敷再制造成型技术研究，研发了基于机器人 MIG 堆焊熔敷再制造成型系统，并对缺损零件的非接触式三维扫描反求测量机制，各子系统标定方法和再制造成型建模方法，空间曲面分层方法，成型路径规划，基于 MIG 堆焊再制造成型过程中的备件形变机理和形变规律以及控形机制，基于 MIG 堆焊/铣削复合工艺的近净成型技术，装备备件再制造成型材料的集约化，面向轻质金属的再制造成型技术等进行了广泛深入的研究，成功实现了典型装备备件的制造与再制造成型[12]。结合再制造产业化发展需要，针对纳米复合电刷镀技术和高速电弧喷涂技术的发展，在手工操作的再制造成型技术的基础上，深入研究技术基本原理，优化技术工艺，实现了技术的自动化、智能化工艺过程，研发了适合再制造产业化生产需要的自动化再制造成型设备和工艺[13,14]，解决了镀液连续循环供应、工序切换、刷镀过程控制及工艺过程多参数监控等技术难题，实现了自动化纳米电刷镀再制造成型过程，并已经应用在再制造工业生产中。自动化纳米电刷镀再制造成型技术的应用，解决了原来劳动强度大、再制造生产质量不稳定、生产效率低等困扰再制造生产的实际问题，相关技术成果已获得 2009 年国家技术发明二等奖。

自动化高速电弧喷涂再制造成型技术将智能控制技术、逆变电源技术、红外测温技术、数值仿真技术综合集成创新，通过操作机或机器人夹持高速电弧喷涂枪，采取数控系统控制喷枪在空间进行各种运动，实时反馈控制与调节喷涂工艺参数，保证涂层的精度与质量，最终可实现零件的高性能快速再制造成型。该技术已在工厂汽车发动机再制造生产线上成功应用，技术成果获得了 2011 年国家科技进步二等奖。

（3）再制造成型新材料。目前，再制造成型所用的固态材料主要有粉末和丝材。由于装备机械零部件多种多样，其材质不同、服役工况复杂多样，对不同材质、不同服役工况、不同损伤形式的机械零部件进行再制造时，对再制造所用材料的性能有不同要求，因此，引起再制造所用材料的多样性和复杂性。为了适应再制造成型技术的推广应用和便于现场或野外作业，实现再制造成型材料的集约化具有重要意义，也就是说，用尽可能少的材料适应尽可能广泛的应用需求，或者说用一种粉末材料或丝材实现不同材质零件、不同服役性能的零部件的再制造。

结合再制造成型技术应用需要，装备再制造技术国防科技重点实验室研发了强韧型、耐磨型和耐蚀性等几种不同功能的集约化再制造成型材料。

针对快速堆焊再制造成型技术需要，综合考虑再制造材料成型性、再制造成型件力学性能、材料之间的相容性等问题，研发了铁基强韧型金属芯焊丝和耐磨型金属芯焊丝新型集约化材料，再制造成型了损伤/损毁的轴类零件。

针对自动化高速电弧喷涂再制造成型技术需要，基于自动化智能高速电弧喷涂设备效率高、稳定性好的优点，在材料制备与成型一体化和材料集约化思路指导下，研发了具有防腐、耐磨和抗热腐蚀与冲蚀的三类低成本、高性能的新型喷涂丝材，主要包括：1）具有“自封闭”效应的 ZnAlMgRE 系防海水腐蚀用粉芯丝材；2）具有自熔剂合金特点的非晶纳米晶 FeCrSiBNb 系列耐磨用粉芯丝材；3）陶瓷颗粒增强的 FeAl 基金属间化合物抗热腐蚀与冲蚀用粉芯丝材，在舰船、重载车辆、电厂设备等装备关键零部件再制造成型中获得了成功应用。

集约化材料的研发和应用，简化了再制造成型技术在施工生产中的技术复杂性，同时，提升了再制造成型技术在伴随保障、现场抢修等重要紧急情况下的技术水平和再制造成型零件质量与性能的可靠性。

（4）现场快速再制造成型技术。现场快速再制造主要是实现工业生产中大型设备贵重零部件现场快速抢修，这样可以显著降低设备维修成本，显著减少设备停产造成的损失。现场再制造成型与在实验室和工厂车间生产线进行再制造成型相比，约束条件多，难度更大。随着工业的发展，现场快速高性能再制造成为了再制造成型技术及其应用发展的新领域。

现场快速高性能再制造成型技术的发展主要得益于再制造成型工艺设备的发展。例如，随着结构紧凑、性能可靠性高、便于移动的全固态激光器、大功率半导体激光器和光纤激光器技术水平的提升，激光熔覆再制造成型技术已应用到冶金设备等大型设备关键零部件的现场快速高性能再制造。在国内，可移动式激光再制造成型技术及其设备系统研发一直是近几年的研究热点。中国科学院半导体研究所、北京工业大学、沈阳大陆集团公司、西安钜光公司等国内多家科研院所和企业，针对激光再制造成型技术的应用背景，在全固态激光器、半导体激光器和光纤激光器方面进行了系统研发。其中，沈阳大陆集团柔性制造公司等单位通过产学研结合，成功研发了输出激光功率大于 1kW 的全固态激光器和输出功率为 2.5kW 的半导体激光再制造成型设备系统，并已在钢厂、汽车制造厂、发电厂等不同工业领域的大型设备现场快速高性能再制造中获得成功应用，解决了工业生产中的设备抢修难题，创造了显著经济效益和社会效益。

等离子熔覆再制造成型技术具有设备系统结构紧凑、工艺操作简洁、再制造零件性能优异等优势，逐步发展成为可实现零件现场快速再制造成型的先进技术之一。装备再制造技术国防科技重点实验室已研制出了等离子熔覆再制造成型技术工程车，满足了到野外现场作业的需要。

2.3.2 “尺寸恢复与机械加工”融合的再制造成型技术

在“尺寸加工法”再制造成型技术方面，主要针对如何提高装备零件再制造成型的效率和质量进行研究，其现阶段的主要成果在于新型多功能加工设备以及与“尺寸恢复法”再制造成型技术复合的技术方法。

“尺寸加工法”再制造成型技术主要分为两种途径。一种途径是针对失效的旧零件，采用机械加工的方法，去除零部件表面损伤层或局部部位，然后选择“尺寸加大”的合适配偶件进行配合，或者另外加入一个“衬套”弥补机械加工去除的尺寸，这是现在国内再制造企业主要在采用的方法，通常也称为“尺寸加工与换件方法”。另一重要途径是针对采用“尺寸恢复法”再制造后的零件，由于其尺寸过剩或表面精度无法满足装配要求，而采用机械加工的方法，去除“尺寸恢复法”再制造多余的尺寸。

第一条途径所采用的技术属于传统机械加工和制造技术。第二条途径是再制造成型领域的特色内容。“尺寸恢复法”再制造后进行机械加工，与制造过程中机械零件的机械加工存在明显的区别。制造时的加工，零件的装卡定位相对灵活，因为毛坯有一定的加工余量。而在再制造加工时，再制造零件加工的毛坯是已经用过的零件，其再制造的只是其中的一个面或几个面，而其他没有再制造的面已经没有加工余量了，因此在再制造加工时，不能将其他面破坏，其装卡方式受到限制。对精度要求高的零件，如果其加工基准发生变化，则加工精度难以保证，如对同轴度有较高要求的台阶轴，如果再制造时只加工其中的一段轴，而与它有同轴度要求的面没有再制造的必要，此时，只加工再制造部分，则难以保证同轴度的要求。目前，关于再制造加工的装卡定位问题，还需要深入研究。下面主要针对第二条途径阐述。

（1）再制造后机械加工成型设备。鉴于再制造后机械加工的特殊性，装甲兵工程学院和北京理工大学合作，针对再制造零件机械后加工的复杂性，研制出了多功能复合机床。该机床具有车、铣、磨、车铣、车磨、钻、铰、攻螺纹等多种加工工艺，不仅可以加工回转类零件，还可以加工非回转类零件[15]。该机床可以完成再制造成型加工所需要的多种加工手段，实现了多工艺复合、多工序复合、多种机床类型复合，同时解决了模块化、单机与多机数字化控制以及人机协同交互等多种技术问题，以满足再制造成型加工的适应性、可重构性、敏捷性等要求。

（2）再制造机械加工成型方法。这方面的一个重要突破是把机械加工的材料去除过程和再制造熔积的材料尺寸增加过程进行融合，实现再制造和机械后加工的同工位完成，简化了生产流程，提高了再制造成型生产效率。

把金属熔积再制造工艺与铣削加工工艺相结合的复合快速再制造成型技术成为了研究热点之一，并取得了较好的成果。这种技术思路起源于美国进行快速成型制造的技术思路。现在，国内装甲兵工程学院、沈阳航空工业学院等单位均在此方面取得了突破性进展。装甲兵工程学院成功实现了机器人自动化堆焊再制造和数控铣削加工的复合，有效提高了再制造成型效率。

沈阳航空工业学院把金属熔化沉积工艺与五轴铣削工艺相结合，提出了连续熔积多层后一并铣削加工的复合成型方式，并研究了根据五轴铣削刀具对层的接近性来判断可一并铣削加工的连续熔积层数的算法。先计算出无干涉的刀位点后，再判断该刀位点是否存在与此相对应的无干涉的刀轴方向，确定五轴铣削刀具对再制造沉积金属的接近性和连续熔积层数，从而大幅度减少了工位变换次数，有效提高了再制造成型效率[16]。

但是，由于这种途径的再制造成型过程是两种工艺交替并行的复合过程，需要不断地变换工位，这显然影响成型效率。因而如何把金属熔化沉积的再制造过程与铣削工艺有机地结合起来，实现同时工作，并在提高表面质量的同时提高成型效率，将具有重要意义。

2.3.3 再制造成型质量控制与评价方法

再制造产业得以健康发展的技术保证在于如何确保再制造成型产品的性能不低于新品。近年来，人们在再制造成型质量控制和评价理论与方法方面进行了大量深入研究。从分析再制造生产工艺流程考虑，各环节均会影响再制造成型产品的最终质量。针对再制造生产过程，除了从再制造成型技术方法与材料选择、工艺优化和成型过程监控等方面进行考虑外，再制造成型前的废旧零部件质量检测控制以及再制造成型后的涂层和成型零件的检测评价，对确保再制造成型产品的质量和性能具有“把关”作用，可以让人们对再制造成型产品“心中有数”。

近年来，人们针对再制造前废旧零件的缺陷、残余应力和剩余疲劳寿命等质量与性能指标的无损评价，研究了金属磁记忆、超声、涡流、声发射等多种无损检测评价理论和技术方法，获得了铁磁性金属零件剩余疲劳寿命的金属磁记忆无损评价理论模型和评价方法，并研发出了适用于典型零部件再制造生产线的专用无损检测仪器设备系统，为再制造成型技术的产业化应用提供了有力的技术支撑。

采用不同再制造成型技术所获得的再制造成型零件，其服役性能和服役寿命取决于再制造毛坯（基体）和再制造成型涂层两个方面。由于在再制造之前，毛坯经过严格的无损检测和评价，这样，再制造成型涂层的评价就成为再制造成型零件质量评价的核心。目前，再制造成型涂层质量评价的内容主要包括缺陷、残余应力、接触疲劳寿命以及硬度等方面。

国内大连理工大学等单位针对零件表面再制造成型涂层中的缺陷，系统研究了超声无损检测评价理论和技术方法。针对再制造成型涂层残余应力评价，装甲兵工程学院在残余应力理论计算上取得了突破，分析了现存大量涂层残余应力近似解的误差，建立了涂层内部残余应力及涂层/基体界面残余应力分布的预测模型，并获得了这些应力的闭合解，为再制造成型涂层的材料和工艺的优化提供了理论指导。国内学者对热喷涂、激光熔覆、堆焊等再制造成型工艺过程中涂层与基体残余应力的变化进行了计算和模拟仿真研究，得到了多种材料和工艺下控制涂层残余应力的模型。在残余应力测量技术方面，各种传统的有损和无损方法在我国均得到了较为广泛的研究和对比，多家单位在研究采用金属磁记忆检测、超声波检测等无损检测方法以及微/纳米压痕等微创伤方法对涂层的残余应力进行检测评价。

涂层接触疲劳寿命评价和预测是目前研究的前沿领域。西班牙学者TOBE等对不同材料体系喷涂层的接触疲劳进行研究后发现，喷涂过程所引起的残余应力是影响再制造涂层接触疲劳寿命的关键因素之一，并指出涂层的抗压强度和界面抗剪强度是影响涂层抗接触疲劳性能的关键因素。燕山大学研制了专门用于考核再制造成型涂层接触疲劳寿命的加速试验机，通过模拟轴承的接触形式考核涂层的接触疲劳寿命，为大样本考核再制造涂层接触疲劳寿命规律搭建了良好的试验平台。装甲兵工程学院对热喷涂再制造成型涂层的接触疲劳寿命评估开展了系统研究，引入 Weibull 分布和 $S-N$ 曲线法等数理统计方式，得到了涂层寿命与施加载荷的对应关系，直观地得到再制造成型零件表面在任意接触载荷作用、任意失效概率下的疲劳寿命（即循环次数），实现了再制造成型零件的接触疲劳寿命预测。

随着再制造事业的蓬勃发展，再制造概念的深入人心，再制造零件的寿命评估必将成为专业学者群和再制造产品客户群关注的焦点，而再制造零件的评估研究尚属起步阶段，可靠的理论和技术还需要不断丰富和完善。因此，针对中国特色“尺寸恢复法”再制造产品的寿命评估和预测还有很多工作要做，尚需要再制造研究者和从业者的不懈努力和顽强拼搏。

3 国内外情况比较分析

3.1 再制造成型技术比较分析

国外的再制造是在制造业基础上发展起来的，一般由制造企业开展再制造生产，其再制造生产主要采用基于机械加工技术的尺寸修理方法和换件方法，因此，其再制造成型技术的主体内容还是制造业中的机械加工技术。尺寸修理法再制造成型虽然能恢复零件的出厂性能，但因破坏了零部件的互换性，且使用了非标准件，故达不到原型机新品的使用寿命。

近年来，美国、加拿大、英国和德国等工业发达国家的很多企业，也开始把基于表面工程的修复技术归类为再制造成型技术，并且具有较高的技术含量，实现了装备关键机械零件的高性能再制造。例如，加拿大研发出了微束等离子弧再制造成型技术及其自动化作业生产设备系统，并且其技术工艺和设备系统已在我国部分再制造企业得到应用，实现了飞机发动机叶片等典型零件的高质量再制造成型；俄罗斯研发了高温合金粉末冶金再制造技术、热障涂层再制造技术等，解决了高温部件等关键零部件的高性能再制造难题，这些技术引进到我国，实现了航空发动机重要零部件的再制造成型；美国研发的 LENS 系统及其基于激光熔覆技术的再制造成型技术，实现了多种机械零部件的快速再制造成型，并且已经在舰船、装甲车辆等武器装备的伴随保障和军队战地装备维修保障中获得了成功应用。

中国的再制造是在国家建设资源节约型和环境友好型社会的科技大环境下，随着绿色制造领域的拓展，在维修工程和表面工程基础上发展起来的，其再制造成型技术主要是自主创新的先进表面工程技术和熔积成型技术等，具有中国特色。我国已经探索形成了“以高新技术为支撑，以恢复尺寸、提升性能的表面工程技术为依托，产学研相结合，既循环又经济”的中国特色的再制造模式。

我国自主创新的先进再制造成型技术的发展，可以归纳为以下几个主要技术领域：

（1）纳米复合再制造成型技术。主要指借助纳米科学与技术新成果，把纳米材料、纳米制造技术等与传统维修技术和再制造技术复合，研发出先进的再制造成型技术，例如纳米复合电刷镀再制造成型技术、纳米热喷涂再制造成型技术等。

（2）能束能场再制造成型技术。主要指利用激光束、电子束、等离子束以及电弧等能量束和电场、磁场、超声波、火焰、电化学能等能量实现机械零部件的再制造成型过程，例如激光熔覆技术、等离子熔覆技术、电子束焊接技术、热喷涂技术、电沉积技术等。

（3）自动化再制造成型技术。主要指综合创新测量与控制、工业机器人、先进材料、先进表面工程和快速成型技术等多领域技术成果，研发再制造成型工艺设备和技术工艺，实现废旧零件再制造成型的自动化过程。装备再制造技术国防科技重点实验室自主研制出了用于等离子弧熔覆再制造工艺的微束等离子弧电源系统。基于自动化等离子弧熔覆再制造成型技术，实现了发动机缸体止推面以及发动机排气门密封锥面的再制造，研究表明，由于等离子弧能量密度高，对基体的热输入量低，工件的变形小，可获得高质量的熔覆层，再制造质量优异。

对比分析可以看出：国外尺寸加工法的再制造成型技术重点关注零部件之间的配合，但不能恢复零件原始设计尺寸，从而不得不使用大量非标件，旧件再制造率低，节能节材效果差，难以提升再制造产品性能；我国基于先进表面工程的再制造成型技术重视恢复零件原始尺寸、恢复零件性能，并可通过引入再制造新材料而提升零件性能，因此，再制造成型件仍是标准件，同时，零件可以经历多次再制造，从而，旧件再制造率高，节能节材效果好。

虽然我国的再制造成型技术发展迅速，具有先进性，但在激光、等离子等高能束再制造成型工艺设备可靠性方面，我国与国外还有较大差距。例如，由于我国在半导体靶条制造与封装、光纤制造等基础制造业方面技术落后，适合再制造成型技术工业应用的大功率光纤激光器和半导体激光器的性能稳定性和运行可靠性与国外尚存在很大差距，因此，可满足移动式作业需求或大型设备现场再制造成型需求的大功率光纤激光或大功率半导体激光再制造设备系统的核心部件尚不具有自主知识产权。

3.2 再制造成型质量控制标准比较分析

英美等发达国家非常注重再制造法规方面的建设，相继建立了相对完善的再制造技术标准、法规和指令体系，并对再制造产品品质保证有严格要求。无论再制造的技术和步骤如何，再制造产品必须在产品质量、性能、耐用性和售后服务上达到与新品一样的水平。但是，国外的再制造主要基于制造业，其再制造生产中的技术主要是制造过程中的尺寸加工等技术。在再制造生产中，主要采用尺寸加工和换件方法。因此，其再制造产品质量和技术标准也一般直接采用制造标准。

国内的再制造是基于机械设备维修而发展起来的，其再制造成型技术主要是各种表面工程技术。再制造过程中，废旧件作为“基体”，通过多种高新技术在废旧零部件的失效表面生成涂覆层，恢复失效零件的尺寸并提升其性能，获得再制造产品。因此，再制造产品的质量是由废旧件（即再制造毛坯）的原始质量和再制造恢复涂层的质量两部分共同决定的。其中，废旧件的原始质量则是制造质量和服役工况共同作用的结果，尤其服役工况中含有很多不可控制的随机因素，一些危险缺陷常常在服役条件下生成并扩展，这将导致废旧件的制造质量急剧降低；而再制造恢复涂层质量取决于再制造技术，包含再制造材料、技术工艺和工艺设备等。再制造零件使用过程中，依靠再制造毛坯和再制造涂层共同承担服役工况的载荷要求，控制再制造毛坯的原始质量和再制造涂层的质量就能够控制再制造产品的质量。

由以上可以看出，我国的再制造不能照搬制造业标准或者国外标准。但是，我国再制造因起步较晚，再制造企业的技术积累少，再制造标准缺乏，这在很大程度上阻碍了再制造技术的广泛推广和应用。因此，要发展我国再制造业，需要逐步建立系统、完善的再制造工艺标准和质量检测标准等。国家和行业已充分认识到了再制造标准的紧迫性，已成立了全国绿色制造技术标准化技术委员会再制造分技术委员会、全国产品回收利用基础与管理标准化技术委员会、全国激光修复技术标准化技术委员会等组织，致力于推进再制造标准化工作。具有中国特色的再制造标准已经建立了标准体系。

4 再制造成型技术的发展趋势

鉴于再制造在工业节能减排和发展循环经济中的巨大潜力，国家十分重视再制造产业化发展，提出了提高再制造产业化水平的要求，这将引导今后国内再制造企业不断增加、行业领域不断拓展、产业规模不断扩大、技术水平不断提升，同时，必将推动再制造成型技术快速发展。

4.1 发展趋势

再制造成型技术是再制造生产活动的技术核心。随着再制造产业发展，再制造成型技术的发展趋势，将体现在三个方面，即：（1）正朝着智能化、复合化和专业化等适合再制造批量生产的方向发展；（2）正由宏观尺度再制造成型向微纳观尺度再制造成型发展；（3）正由纯机械零部件的再制造成型技术向机械/电子复合、机械/功能复合等以机械系统为载体的多功能复合再制造成型技术发展。概括起来，再制造成型技术发展趋势也可以归纳为“五化”，即智能化、复合化、专业化、微纳化和功能化。

（1）智能化。再制造成型技术的智能化，主要包含两方面内容。一方面是针对具体零部件，基于专家数据库等信息，实现再制造成型技术方法优化等再制造成型技术方案的智能化设计；另一方面是针对具体的再制造成型技术方法，基于零部件再制造成型过程，在过程参数反馈控制或逻辑程序控制下由工业机器人自动操作完成再制造成型过程。

智能化再制造成型技术将直接提升再制造生产过程的自动化和柔性化水平，适应再制造大批量生产活动的需要。再制造成型技术已基本摆脱了手工操作，其技术设备和技术过程正逐步实现自动化和智能化，再制造生产正在朝着工业机器人操作代替人工操作的方向发展。柔性化主要是指再制造设备系统和再制造工艺可以满足不同零件再制造的需要。由于近阶段再制造批量化生产量相对较小，能够适应不同种类、不同规格型号零件再制造的柔性化的设备和技术工艺将具有广泛适应性和广阔发展前景。

（2）复合化。再制造成型技术的复合化，也主要包含两个方面，即再制造成型技术手段的复合和再制造成型所用材料的复合。例如，电弧与激光两种能量束复合，实现再制造生产中的高效率高质量的再制造成型；电沉积和热喷涂与激光重熔或喷焊重熔等不同涂层制备技术方法的复合，实现高性能零部件再制造成型过程中高质量涂层的成型制备；针对零件再制造涂层，通过具有抗磨、减摩、抗蚀、抗高温等不同性能的材料的复合，制备出同时具有多方面优异性能的涂层，可以赋予再制造成型零件优良的综合性能。

一个具体的废旧零件，其损伤形式和失效机理往往比较复杂，要把它再制造成为性能合格的再制造产品，往往需要采用多种再制造成型技术手段，通过技术集成化的途径实现。复合化再制造成型技术将为复杂失效零件的高性能再制造成型提供技术途径。

（3）专业化。各工业领域的生产实践已经表明，专业化是提高生产效率和产品合格率的有效途径。机械零部件的失效形式虽然多种多样，但正常情况下，同一种零件的主要失效形式一般具有相对普遍的规律特点。针对某一类零件，为了提高再制造生产效率和再制造产品质量稳定性，往往需要研发专用的再制造技术工艺和专用设备。再制造成型技术专业化的内涵之一是提高其技术水平，实现规范化和标准化，做精一个产品或一道工序。随着再制造产业化规模和水平发展，再制造成型技术的专业化将是提升再制造产品质量的一条有效途径。

（4）微纳化。再制造成型技术的微纳化方向，也主要包含两方面内容。一方面，通过微纳加工技术，对宏观机械零部件功能部位进行的再制造处理，提升机械零部件的服役性能。例如，通过采用激光微纳织构化处理，在再制造成型后的发动机活塞表面制备出微纳结构，提高活塞表面的抗磨减摩性能。另一方面，对微纳系统或微纳结构进行再制造成型处理。微纳技术是21世纪的先进技术，近年来发展迅速，微纳系统或结构技术含量高、附加值高，对其进行再制造必然是今后发展的一个前沿方向。

（5）功能化。在21世纪，各工业领域都在朝着信息化方向发展。现代装备已不是单纯的机械装备，而是机电一体化的复杂系统，其中包含着大量具有特殊功能要求的零部件。但是，现阶段，我国的再制造成型技术还主要局限于机械零部件的再制造成型，致力于恢复和提升机械零部件的抗外力、抗腐蚀介质等力学性能，而对以机械为载体的机电一体化系统及其具备电、磁、声、光等特殊功能的器件或零部件的再制造成型技术，研究得很少。可以预料，在21世纪，针对信息化装备及其功能器件的再制造成型技术必将快速发展。

4.2 发展策略

再制造成型技术以节约资源能源、保护环境为特色，以综合利用高科技为核心，充分体现了具有中国特色自主创新的特点。放眼未来发展，结合当前需求，从实际出发，发展中国的再制造成型技术，应采取适当措施，重点突破如下几方面，即探索理论基础、创新关键技术、制定技术标准、培养专业人才。

（1）探索再制造成型的理论基础。再制造成型的基础理论发展是推动再制造成型技术创新和产业化发展的基石。应进一步深入探索研究以产品全寿命周期理论、废旧零件和再制造零件的寿命评估预测理论等为代表的再制造基础理论，以揭示再制造成型产品寿命演变规律的科学本质。废旧零件的剩余寿命是否足够，再制造成型零件的使用寿命是否可保持一个完整的服役周期，这样一些重大问题，由于缺少理论依据，有时仅凭简单的检测设备，甚至只靠工人师傅的目测或经验判断来完成，难以保证再制造成型产品的质量。为此，必须针对再制造成型技术发展和再制造成型产品质量控制需求，探索研究更多更有效的无损检测及寿命预测理论与技术。

（2）创新再制造成型的关键技术。再制造成型技术是再制造工程的核心，是推动再制造产业化发展的技术支撑。应不断研发和创新拓展用于再制造成型的先进表面成型技术群、三维体积成型技术群等，使再制造成型零件的精度更高、性能更好、寿命更长，确保再制造产品的质量和性能。

（3）制定再制造成型技术的相关标准。技术标准是技术和产业健康、规范发展的有力保障。应尽早建立系统的再制造成型技术标准、再制造成型产品质量检测标准等体现再制造走向规范化的标准体系。我国再制造因起步较晚，再制造企业的技术积累少，再制造成型技术相关的标准缺乏，因而一定程度上阻碍了再制造成型技术的推广应用。近两年来，国内相关高等院校和再制造生产企业正在联合制定“再制造技术工艺标准、再制造质量检测标准、再制造产品认证标准”等多类标准。下一步，应深化标准内涵，制定出具有良好通用性和可操作性的标准方案。

（4）培养再制造成型技术专门人才。人才是学科发展的根本。目前，再制造工程虽然已经列入了国家学科目录，但是，由于我国再制造发展起步晚，再制造成型技术是多学科、多领域科技知识的交叉融合，我国还未能有计划地培养出再制造工程学科人才。今后，再制造产业化发展急需再制造成型技术专门人才。为此，政府、高等院校和行业应当紧密结合，加强产学研结合，加强人才培养力度。

5 结论

（1）再制造是以制造为基础，是在不改变原零件形状和性能的基础上，对其失效部分进行处理，恢复其性能，并可以根据需要进行性能提升。再制造成型技术是再制造工程的核心，其以节约资源能源、保护环境为特色，以综合利用高科技为核心，高度契合了构建循环经济、实施节能减排的战略需求。

（2）我国自主创新的先进再制造成型技术在纳米复合再制造成型技术、能束能场再制造成型技术和自动化再制造成型技术等方面取得了长足发展。先进再制造成型技术不但能恢复零件原始尺寸，并可使零件的性能得以恢复和提升，最大限度地挖掘了废旧零件中蕴含的附加值，尤其在高精度高性能表面再制造成型方面取得了显著成绩，但在再制造成型技术成套设备的研发方面还有待进一步提高。

（3）为适应再制造成型技术的发展趋势，实现再制造成型技术的智能化、复合化、专业化、微纳化和功能化，应在探索再制造成型理论基础、创新关键再制造成型技术、制定再制造成型技术标准、培养再制造成型技术专业人才等方面开展深入工作。

参考文献

[1] 徐匡迪．工程师——从物质财富的创造者到可持续发展的实践者［J］．中国表面工程，2004，17（6）：1～6.

XU Kuangdi. Engineers：From the creators of material wealth to the practitioners of sustainable development［J］. China Surface Engineering，2004，17(6)：1～6.

[2] 徐滨士．装备再制造工程的理论与技术［M］．北京：国防工业出版社，2007.

XU Binshi. Theroy and technology of equipment remanufacture engineering [M]. Beijing: National Defense Industry Press, 2007.

[3] 徐滨士. 中国再制造工程及其进展 [J]. 中国表面工程, 2010, 23(2): 1~6.
XU Binshi. Remanufacture engineering and its development in China [J]. China Surface Engineering, 2010, 23(2): 1~6.

[4] GAO Jian, CHEN Xin, ZHENG Detao, et al. Adaptive repair approach for recovering components from defects [J]. Chinese Journal of Mechanical Engineering, 2008, 21(1): 57~60.

[5] KOEHLER H, PARTES K, SEEFELD T, et al. Laser reconditioning of crankshafts: From lab to application [J]. Physics Procedia, 2010(5): 387~397.

[6] 李新亚, 王德成, 刘丰, 等. 先进成型技术与装备发展道路刍议——先进成形技术与装备发展现状与趋势 [J]. 机械工程学报, 2010, 46(17): 100~104.
LI Xinya, WANG Decheng. LIU Feng, et al. Visioning - future of advanced manufacture technology & equipment - present situation & tendency of advanced forming technique & equipment [J]. Journal of Mechanical Engineering, 2010, 46(17): 100~104.

[7] SHANMUGAM D K, MASOOD S H. Direct spray metal rapid tooling using fused deposition modeling [J]. Chinese Journal of Mechanical Engineering, 2002, 15 (Suppl.): 121~126.

[8] 宋建丽, 李永堂, 邓琦林, 等. 激光熔覆成形技术的研究进展 [J]. 机械工程学报, 2010, 46(14): 29~39.
SONG Jianli, LI Yongtang, DENG Qilin, et al. Research progress of laser cladding forming technology [J]. Journal of Mechanical Engineering, 2010, 46(14): 29~39.

[9] 雷剑波, 杨洗陈, 王云山, 等. 激光再制造快速修复海上油田关键设备 [J]. 世界制造技术与装备市场, 2006 (6): 54~56.
LEI Jianbo, YANG Xichen, WANG Yunshan, et al. Laser remanufacturing repairing key equipments in oil fields at sea [J]. World Manufacturing Engineering & Market, 2006(6): 54~56.

[10] 杨健, 黄卫东. 激光直接制造技术及其在飞机上的应用 [J]. 航空制造技术, 2009(7): 36~38.
YANG Jian, HUANG Weidong. Direct laser fabrication and its application in aircraft [J]. Aeronautical Manufacturing Technology, 2009(7): 36~38.

[11] 董世运, 徐滨士, 王志坚, 等. 激光再制造齿类零件的关键问题研究 [J]. 中国激光, 2009, 36 (增): 134~138.
DONG Shiyun, XU Binshi, WANG Zhijian, et al. Vital problems on laser remanufacturing gears [J]. Chinese Journal of Lasers, 2009, 36(Suppl.): 134~138.

[12] ZHU Sheng, MENG Fanjun, BA Dema. The remanufacturing system based on robot MAG surfacing [J]. Key Engineering Materials, 2008, 373~374: 400~403.

[13] 胡振峰, 董世运, 汪笑鹤, 等. 面向装备再制造的纳米复合电刷镀技术的新发展 [J]. 中国表面工程, 2010, 23(1): 87~91.
HU Zhenfeng, DONG Shiyun, WANG Xiaohe, et al. New development of nanocomposite electro - brush plating technique facing the equipment remanufactruing [J]. China Surface Engineering, 2010, 23(1): 87~91.

[14] 梁秀兵, 陈永雄, 白金元, 等. 自动化高速电弧喷涂技术再制造发动机曲轴 [J]. 中国表面工程, 2010, 23 (2): 112~116.
LIANG Xiubing, CHEN Yongxiong, BAI Jinyuan, et al. An automatic high velocity arc spraying technology applied to remanufacture engine crankshaft [J]. China Surface Engineering, 2010, 23(2): 112~116.

[15] 袁巍，张之敬，周敏．局部再制造的快速机加工系统设计与研究［J］．制造业自动化，2009，31 (1)：95 ~ 98.

YUAN Wei，ZHANG Zhijing，ZHOU Min. Design and research on the rapid machining system for remanufacture［J］. Manufacturing Automation，2009，31 (1)：95 ~ 98.

[16] 朱虎，张利芳．提高金属熔化沉积与五轴铣削复合成形效率［J］．辽宁工程技术大学学报，2009，28 (1)：106 ~ 108.

ZHU Hu，ZHANG Lifang. Improving efficiency of hybrid prototyping based on metal melting deposition and 5 - axis milling［J］. Journal of Liaoning Technical University，2009，28 (1)：106 ~ 108.

Prospects and Developing of Remanufacture Forming Technology

Xu Binshi，Dong Shiyun，Zhu Sheng，Shi Peijing

(Science and Technology on Remanufacturing Laboratory,
Academy of Armored Forces Engineering，Beijing，100072，China)

Abstract Remanufacture，taken as emerging strategic industry，greatly meets national strategy of recycling economy，and has highly valued by government and enterprises. Remanufacture forming technology (RFT) is the cored content of remanufacture engineering，which can rebuild original size of the worn pieces，and at the same time，restore or even upgrade its performance. It states current status of RFT in the world，and comparatively analyzes difference between China and other countries on aspects of RFT and its quality standards. Novel achievements on RFT in China are elaborated in the following three parts：additive RFT，the machining combined RFT，and quality control and evaluation for remanufacture forming. At last，it prospects developing trends of RFT in China，and proposes the advisable strategies of developing RFT.

Key words recycling economy，remanufacture forming technology，current status，developing trend

凹凸棒石黏土润滑油添加剂对钢/钢摩擦副摩擦学性能的影响*

摘　要　采用天然凹凸棒石黏土作为润滑油添加剂加入 150SN 润滑油中，在 Optimal SRV－Ⅳ 摩擦磨损试验机上研究了添加剂含量对钢/钢摩擦副摩擦学性能的影响，借助 SEM 及 EDS 分析了摩擦副的表面形貌及表面元素组成。结果表明：凹凸棒石黏土的浓度为 0.6% 可使平均摩擦系数较基础油润滑条件下降低 42.32%；凹凸棒石黏土的浓度为 0.4% 可使磨损体积降低 85.48%；凹凸棒石黏土的加入使得磨损表面更加光滑平整，同时磨损表面氧元素含量升高。分析认为凹凸棒石黏土层链状的晶体结构和摩擦过程中复杂的理化过程是实现减摩抗磨的原因。

关键词　凹凸棒石黏土　润滑油添加剂　摩擦磨损　自修复

油润滑介质中纳米润滑材料的自修复作用是指在摩擦过程中利用摩擦产生的机械摩擦作用、摩擦－化学作用和摩擦－电化学作用，摩擦副与润滑材料产生能量交换和物质交换，从而在摩擦表面上形成正机械梯度的金属保护膜、金属氧化物保护膜、有机聚合物膜、物理或化学吸附膜等，以补偿摩擦副的磨损与腐蚀，呈现磨损自修复效应[1]。近年来，以天然蛇纹石微粉为主要成分的润滑油自修复添加剂的研究报道较多[2~4]。

凹凸棒石黏土（attapulgite clay）简称凹土（AC），又名坡缕石（palygrosbite），是一种含水镁铝硅酸盐黏土矿物，具有与天然蛇纹石矿物相似的成分和晶体结构。与天然蛇纹石矿物不同的是，凹凸棒石黏土晶体多为针状纤维，单晶直径大多为 10～100nm，长度为 0.1～1μm，是天然的一维纳米材料[5~8]。但是将天然凹凸棒石黏土作为润滑油添加剂的研究鲜有报道。

本文将天然凹凸棒石黏土分散至润滑油中，并进行摩擦学试验，考察凹凸棒石黏土作为润滑油添加剂的摩擦学性能，并分析其减摩抗磨机理。

1　实验部分

1.1　试验材料及制备

试验用凹凸棒石黏土产自江苏盱眙，为天然凹凸棒石黏土经提纯后的市售产品。将少量凹凸棒石黏土加入酒精溶液中，超声分散，再滴在铜网上，采用 JEOL 电子公司的 JEM－1011 型透射电子显微镜（TEM）观察其显微形貌。从图 1 可以看出，凹凸棒石黏土颗粒为均匀细小的纤维状晶体，结构中包含棒晶和棒晶聚集形成的晶束，以及

---

* 本文合作者：王利民、许一、高飞、于鹤龙、乔英杰。原发表于《摩擦学学报》，2012，32(5)：493～499。国家“973”计划项目（2011CB013405 和 2007CB607601）、国家自然科学基金项目（5073506、50805146、50904072 和 51005243）和博士后科学基金（20090461452）资助。

少量杂质。采用荷兰帕纳科公司生产的X′Pert PRO DY2198型X射线衍射仪对试验用凹凸棒石黏土进行物相半定量分析，图2为试验用凹凸棒石黏土的XRD衍射图谱，图上主要为凹凸棒石特征峰，不含其他物质特征峰，可见试验用凹凸棒石黏土的纯度较高。

图1 试验用凹凸棒石黏土的TEM显微形貌

Fig. 1 TEM micrograph of the attapulgite clay powder

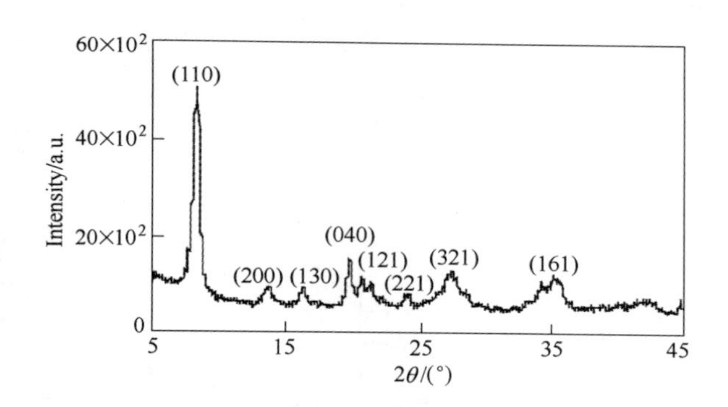

图2 试验用凹凸棒石黏土XRD图谱

Fig. 2 XRD pattern of the attapulgite clay powder

将凹凸棒石黏土按一定比例加入150SN润滑油中，并进行超声分散，分别配置成0.2%、0.4%、0.6%、0.8%（质量分数）的试验用油样。

1.2 试验方法

采用德国产Optimal SRV－Ⅳ微振动磨损试验机进行摩擦学性能研究。试验条件：行程1mm，温度50℃，球盘接触方式，上试样为ϕ10mm的GCr15钢球（HRC61～63），下试样为ϕ24mm×8mm的45号钢圆盘（HRC42～45）；载荷为20N；频率为10Hz；时间为1h。

试验结束后，试样在丙酮溶液中以超声波清洗器清洗去除表面污染物，放入烘箱烘干。在光学显微镜上测量上试样磨痕直径；采用美国ADE公司MicroXAM－3D型三维白光干涉表面形貌仪测定下试样磨损体积。

用Quanta200扫描电子显微镜（scanning electron microscope，SEM）观察试样表面磨损形貌，并用SEM附带的能量色散谱仪（energy dispersive spectrometry，EDS）测定

下试样磨损表面元素组成。

采用 ESCALAB 250Xi 型多功能光电子能谱仪（XPS）对磨损表面特征元素的化学状态进行分析。发射源采用能量为 1486.6eV 的单色 AlK 靶，能量分辨率为 ±0.2eV。谱图采集在恒定能量模式下进行，通过能量为 20eV。采用标准碳污染峰（C1s line：284.8eV）对峰位进行校正。

2 结果及讨论

2.1 摩擦学性能

由图 3a 可以看出，基础油 150SN 的平均摩擦系数较高，为 0.25。加入 0.2% 凹凸棒石黏土的油样比基础油的平均摩擦系数降低了 27.78%。随着凹凸棒石黏土添加量的增大，平均摩擦系数有继续下降的趋势。当添加量为 0.6% 时，平均摩擦系数达到最低值 0.1446，较基础油润滑条件下的平均摩擦系数降低了 42.32%。此后继续添加凹凸棒石黏土会导致摩擦系数升高。这主要是由于添加量较少时，粉体颗粒在与润滑油对摩擦副表面的吸附竞争中处于劣势，导致介入摩擦副接触区域的颗粒较少而无法形成完整的边界润滑膜；随着添加量的适当增大，越来越多的粉体颗粒吸附于摩擦表面，有利于边界润滑膜的形成；但当添加量继续增大时，就会造成颗粒在摩擦副间隙的冗余而发生聚集甚至充当磨粒，加剧磨损[9~11]。

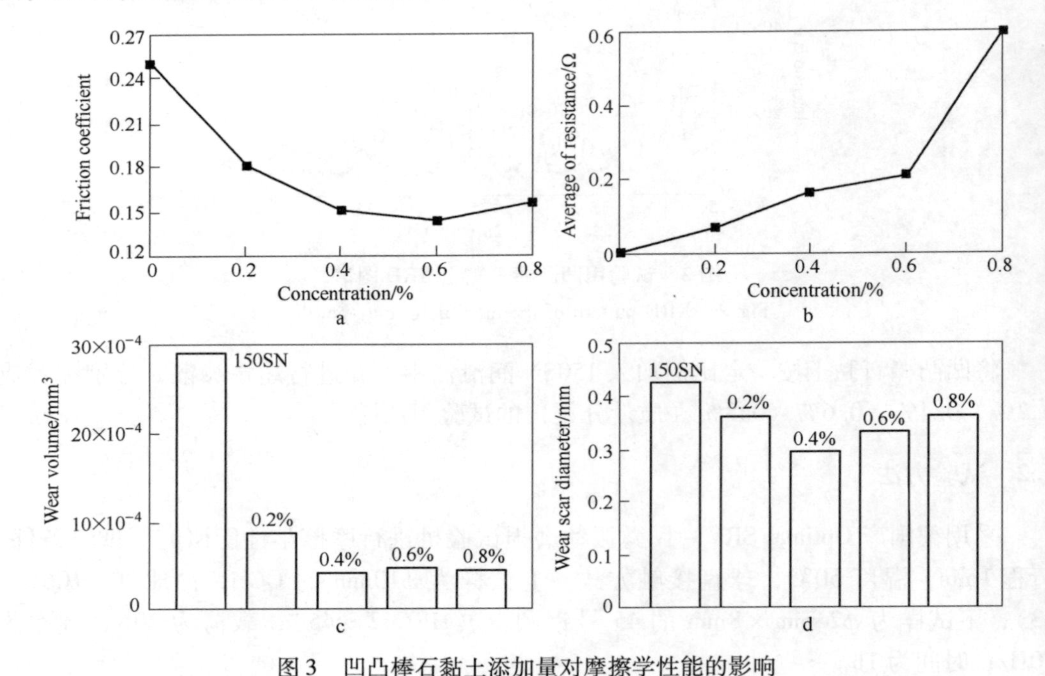

图 3 凹凸棒石黏土添加量对摩擦学性能的影响

Fig. 3 Effect of the attapulgite clay concentration on the tribological properties

a—Average of friction coefficient；b—Average of resistance；

c—Wear volume of the disk；d—Wear scar diameter of the ball

SRV－Ⅳ微振动磨损试验机可以测定上下试样之间的电阻值，并实时记录。电阻值的大小一般可以间接反映润滑油膜的厚度情况、摩擦过程氧化膜或化学转移膜（自修复膜）的厚度情况。一般来说，膜厚越大，电阻值越高。图 3b 是摩擦过程中添加量对平均电阻值的影响情况。可见随添加量的增加，电阻值呈增加的趋势，说明添加量在 0. 2% ~0. 8% 的范围内，摩擦过程中磨损表面形成的润滑油膜或氧化膜、化学转移膜厚度随添加量的增加而增加。根据 Stribeck 曲线，油膜厚度的增加可以使润滑状态从边界润滑逐渐过渡到流体润滑的方式，相应摩擦系数呈现先降低再升高的趋势[12]，这与图 3a 的趋势相符合，而氧化膜或化学转移膜厚度的变化对摩擦系数的影响不大。所以，电阻值随凹凸棒石黏土添加量升高而升高的现象，主要是油膜厚度增加引起的。

图 3c 是下试样磨损体积随添加量变化趋势图，可见下试样磨损体积变化规律与平均摩擦系数变化规律基本一致。在润滑油中添加凹凸棒石黏土之后下试样磨损体积均大幅减小。其中，添加量为 0. 4% 时，下试样磨损体积较基础油润滑条件下减少 85. 48% ，此时凹凸棒石黏土添加剂表现出最佳的抗磨性能。

上试样磨痕直径随添加量的变化规律如图 3d 所示。由图 3d 可以看出：上试样磨痕直径随添加量变化规律不如平均摩擦系数、下试样磨损体积明显，但也基本反映出类似的规律，当添加量为 0. 4% 时，上试样磨痕直径最小，为 0. 295mm，较基础油摩擦的上试样磨痕直径 0. 425mm 减小了 30. 59% 。

2. 2　磨损表面形貌

图 4 所示为下试样磨损表面的微观形貌照片。由图 4 可见，基础油润滑条件下摩擦表面的磨损较为剧烈，磨痕较深，磨损表面存在大量的犁沟和塑性变形，同时存在一定的氧化层疲劳剥落。添加量为 0. 2% 的油样磨损表面较基础油的磨损略显平滑，但是仍然存在一定量的犁沟和氧化层疲劳剥落，同时表面可见一层非连续的光滑膜。0. 4% ~0. 8% 添加量的油样磨损表面较基础油的磨损表面光滑平整，未见较深的犁沟和疲劳剥落。

2. 3　磨损表面元素分布

对不同润滑条件下磨损表面元素组成进行了 EDS 分析，结果如图 5 及表 1 所示。两种润滑条件下，摩擦表面均含有 Fe、C 和 O 等元素。当润滑油中的凹凸棒石黏土含量为 0. 4% 时，磨损表面还发现了极微量的 Mg、Al 和 Si 等元素，同时 O 元素的含量较基础油润滑时显著提高。

2. 4　磨损表面 XPS 分析

为进一步探讨磨损表面的物相组成，分析减摩抗磨机理，采用 XPS 对磨损表面进行了表征，如图 6 所示。Fe2p3/2 可拟合为 706. 9eV、708. 1eV、709. 3eV、710. 3eV 和 711. 5eV 等子峰，分别对应于 Fe、FeO、FeOOH、Fe_3O_4 和 Fe_2O_3[13,14]，当 0. 4% 凹土存在时，Fe2p3/2 在 706. 9eV 和 708. 1eV 位置的特征峰强度较基础油润滑时明显减弱，而在 709. 3eV、710. 3eV 和 711. 5eV 位置特征峰强度则明显增强（图 6a）。这说明，润滑油中加入 0. 4% 凹土之后，使得磨损表面高价态 Fe 的氧化物含量显著增加，促进了摩擦表面的氧化反应。O1s 的结合能可拟合为 529. 9eV、530. 2eV、531. 0eV、531. 6eV 和 532. 3eV 等子峰，分别对应于氧化物、羟基氧化物、硅酸盐或有机化合物[13,15,16]，添加

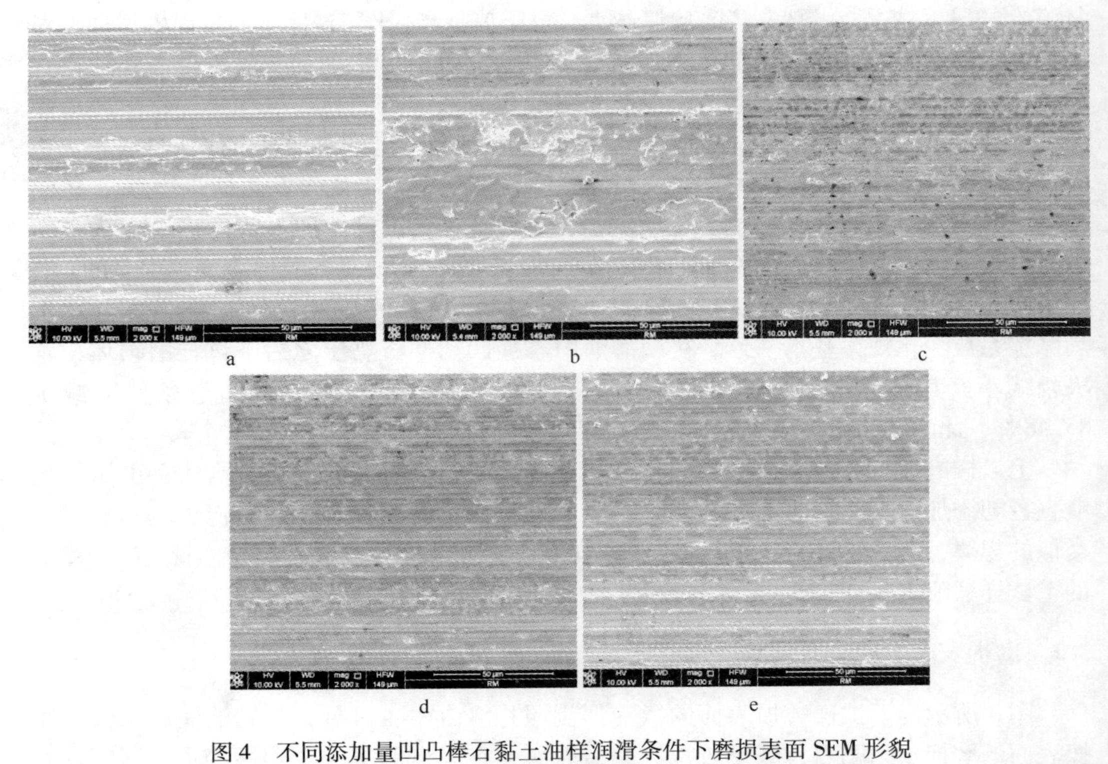

图 4　不同添加量凹凸棒石黏土油样润滑条件下磨损表面 SEM 形貌

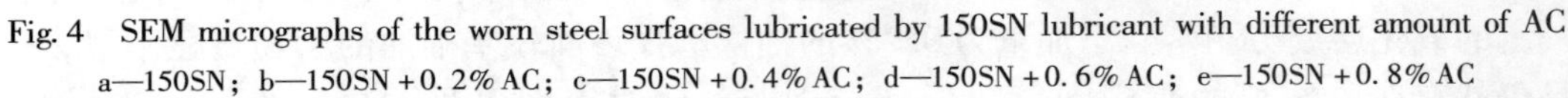

Fig. 4　SEM micrographs of the worn steel surfaces lubricated by 150SN lubricant with different amount of AC
a—150SN；b—150SN + 0. 2% AC；c—150SN + 0. 4% AC；d—150SN + 0. 6% AC；e—150SN + 0. 8% AC

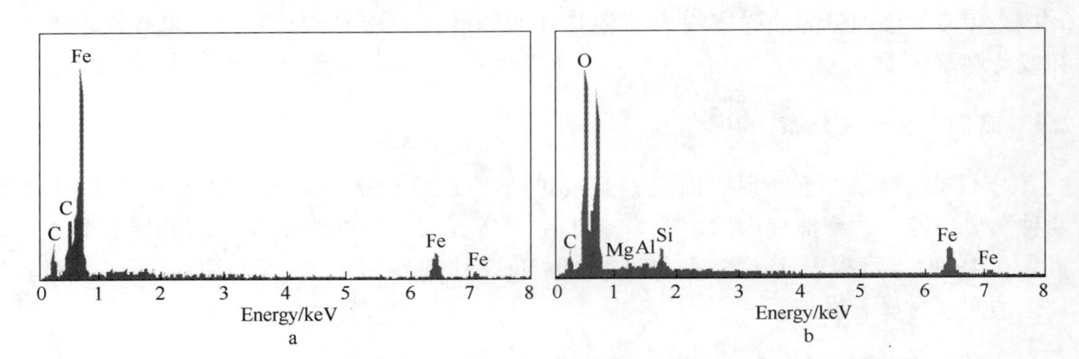

图 5　不同添加量凹凸棒石黏土摩擦试验下表面 EDS 能谱图

Fig. 5　EDS spectra of the worn steel surfaces lubricated by
150SN(a) and 150SN + 0. 4% AC(b)

了 0. 4% 凹土后磨损表面 O1s 的结合能峰出现了宽化和增强（见图 6b），这说明磨损表面除 Fe 之外其他元素的氧化物或有机物含量显著增加。基础油润滑时，C1s 可拟合为 283. 1eV、284. 4eV、284. 7eV 和 285. 4eV，分别对应于碳化物、石墨、污染碳和有机物等，而凹土存在时，C1s 未发现明显特征峰（见图 6c），分析认为磨损表面的 C 元素主要为碳钢中碳，以渗碳体（Fe_3C）的形式存在。XPS 分析时 X 射线光电子能量较低，平均自由程只有 0. 5 ~2. 5nm，而 EDS 分析时 X 射线能量较高，分析样品的深度也远大于 XPS 分析样品的深度，所以 EDS 分析中凹土存在时发现的 C 元素，主要应为深层基体碳钢中的 C 元素。当凹土存在时，Si2p 和 Mg1s 分别在 102. 0eV 和 1304. 2eV 位置上出现明显的特征峰（见图 6d、e），分析认为，磨损表面的 Si 和 Mg 元素主要以硅酸盐的形式存在[13]，其原因为凹土粉体在摩擦过程中吸附于或被碾压黏附于磨损表面。通过 Al2p 的 XPS 谱可以看出，当凹土存在时，磨损表面含有 Al 元素，但是由于其含量甚小，特征峰不是很明显（图 6f），所以很难判断其具体存在形式，但是根据 Si 和 Mg 元素是以硅酸盐的形式存在来分析，Al 元素也应该以硅酸盐（即凹土）的形式存在。

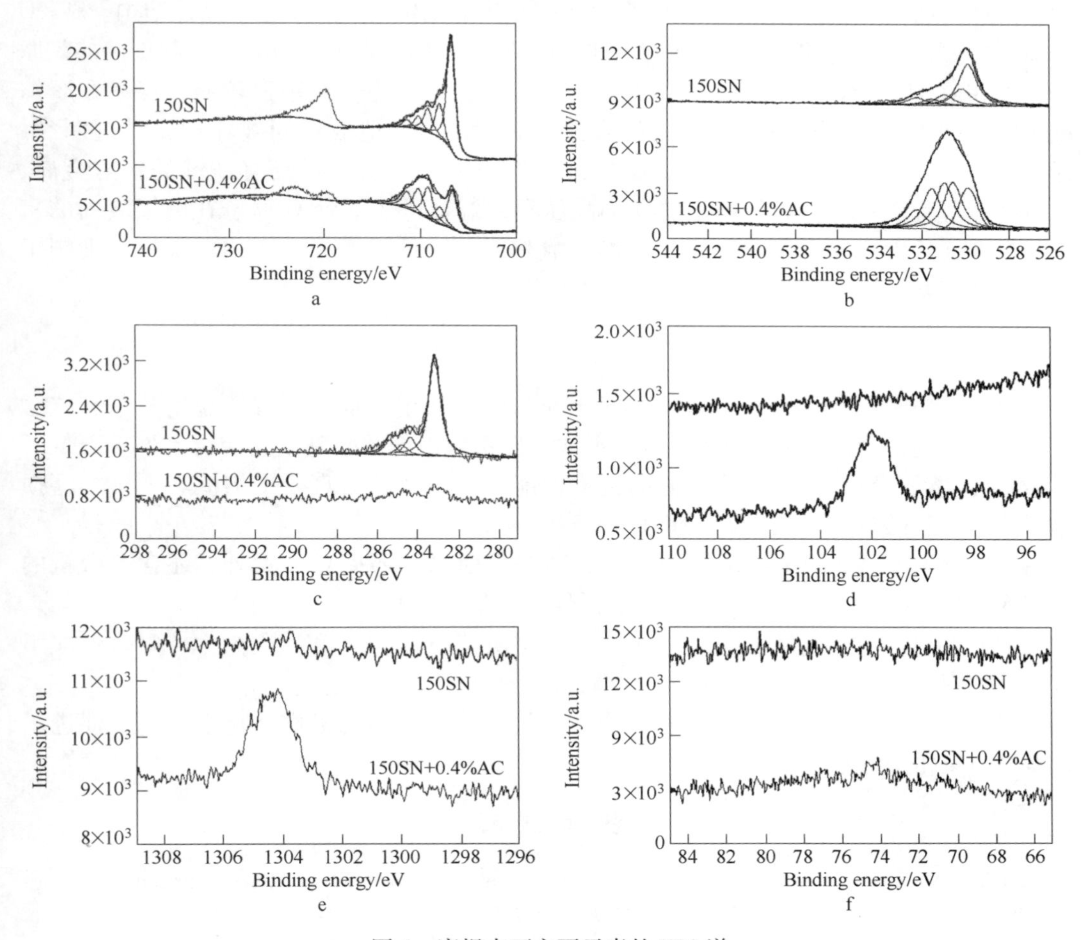

图 6　磨损表面主要元素的 XPS 谱

Fig. 6　XPS spectra of the typical elements on the worn steel surface

a—Fe2p；b—O1s；c—C1s；d—Si2p；e—Mg1s；f—Al2p

表 1　无添加和添加 0. 4%AC 的 150SN 机油润滑下钢球磨损表面的 EDS 结果

Table 1　EDS result of the worn surface lubricated by 150SN without AC and with 0. 4%AC （%）

Lubricant	Content（atomic percent）					
	Fe	C	O	Mg	Al	Si
150SN	62. 06	24. 80	13. 14	—	—	—
150SN + 0. 4%AC	46. 85	13. 50	36. 33	0. 56	0. 67	2. 09

2.5 减摩抗磨机理探讨

从EDS和XPS结果分析可以看出，凹土的加入促进了O元素在摩擦表面的富集与反应，形成了铁的氧化物和凹土复合摩擦保护膜，这一保护膜具有优异的减摩抗磨作用。这一结论与蛇纹石的自修复机理相似[2~4]。分析认为，这个以氧化物为主的摩擦保护膜的形成主要与以下几个因素有关：首先，纳米量级的凹土具有较高的比表面积，从而具有较高的活性，在摩擦过程中易于进入摩擦界面；再者，凹土与蛇纹石相似，为Si—O四面体和Mg(Al)—OH/O八面体按照2∶1比例形成的层链状晶体结构。Si—O四面体主要以共价键相连，且存在1个自由氧原子，即存在1个只有3个共价键的氧原子；而Mg(Al)—/OH/O八面体主要以离子键和氢键相连，作用力较弱，容易在摩擦过程中剪切力的作用下发生结构失稳和羟基脱除反应，造成层间破坏和化学键断裂，释放出活性氧；其三，磨损表面在摩擦力的作用下暴露出大量具有断键的新鲜金属，同时摩擦过程中在微凸体处产生的局部高压和闪温促进了活性O原子在磨损表面的富集并与表面活性金属发生化学反应，形成氧化膜，降低了磨损表面的粗糙度，抑制了磨损的持续剧烈进行，因而实现了减摩抗磨的作用。

3 结论

（1）在试验所用添加量范围内，凹土作为润滑油添加剂均可有效起到减摩和抗磨的效果，特别是当添加量为0.6%时，平均摩擦系数最低，较基础油润滑条件下降低了42.32%；添加量0.4%时具有最好的抗磨性能，上试样磨痕直径平均值和下试样磨损体积分别较基础油润滑条件下降低了30.59%和85.48%。

（2）添加0.4%凹土的油样润滑时，磨损表面除了Fe、C元素外，还含有微量的Mg、Al和Si元素，同时O元素的原子百分比较基础油润滑时显著提高。

（3）当0.4%凹土存在时，磨损表面以Fe元素的氧化物和凹土为主，复合形成了摩擦保护膜。

（4）纳米量级层链状晶体结构的凹土通过摩擦过程中的机械和化学作用，促进了磨损表面氧化层的形成。

参 考 文 献

[1] Xu B S. Nano surface engineering [M]. Beijing: Chemical Industry Press, 2003 (in Chinese).
徐滨士. 纳米表面工程 [M]. 北京：化学工业出版社，2003.

[2] Gao Y Z, Zhang H C, Xu X L, et al. Formation mechanism of self - repair coatings on the worn metal surface using silicate particles as lubricant oil additive [J]. Lubrication Engineering, 2006 (10): 39 ~ 42 (in Chinese).
高玉周，张会臣，许晓磊，等. 硅酸盐粉体作为润滑油添加剂在金属磨损表面成膜机制 [J]. 润滑与密封，2006，(10)：39 ~ 42.

[3] Guo Y B, Xu B S, Ma S N, et al. Effect of hydroxyl silicate particulates as an additive on the friction and wear behavior of mild steel/ductile cast iron pair [J]. Tribology, 2004, 24 (6): 512 ~ 516 (in Chinese).
郭延宝，徐滨士，马世宁，等. 羟基硅酸盐润滑油添加剂对45号钢/球墨铸铁摩擦副摩擦磨损性能的影响 [J]. 摩擦学学报，2004，24 (6)：512 ~ 516.

[4] Tian B, Wang C B, Yue W, et al. Influence of the wear self - repairing effect on the Crplated cylinder liner by a cermet additive in lubricating oil [J]. Tribology, 2006 (6): 574 ~ 578 (in Chinese).
田斌，王成彪，岳文，等. 陶瓷润滑油添加剂对镀铬缸套磨损自修复特性的影响 [J]. 摩擦学学报，2006 (6): 574 ~ 578.
[5] Lu Y L, Li Z, Yu Z Z, et al. Microstructure and properties of highly filled rubber/clay nanocomposites prepared by meltblending [J]. Composites Science and Technology, 2007, 67: 2903 ~ 2913.
[6] Wang L H, Sheng J. Preparation and properties of polypropylene/org - attapulgite Nanocomposites [J]. Polymer, 2005, 46: 6243 ~ 6249.
[7] Chen C H. Effect of attapulgite on the crystallization behavior and mechanical properties of poly (butylene succinate) nanocomposites [J]. Journal of Physics and Chemistry of Solids, 2008, 69: 1411 ~ 1414.
[8] Shen S Y, Yang M, Ran S L, et al. Preparation and properties of natural rubber/palygorskite composites by Co - coagulation rubber latex and clay aqueous suspension [J]. Journal of Polymer Research, 2006 (10): 469 ~ 473.
[9] Xu T, Zhao J Z, Xu K. The ball - bearing effect of diamond nanoparticles as an oil additive [J]. Phys D, 1996, 29: 2932 ~ 2937.
[10] Stempfel E M. Practical experience with highly biodegradable lubricants, especially hydraulic oils and lubricating grease [J]. NLGI Spokesman, 1998, 62: 8 ~ 23.
[11] Huang W J, Dong J X, Wu G F, et al. A study of S - [2 - (acetamido) benzothiazol - 1 - yl] N, N - dibutyl dithiocarbamate as an oil additive in liquid paraffin [J]. Tribology International, 2004, 37: 71 ~ 76.
[12] Wen S Z, Huang P. Principles of tribology [M]. Beijing: Tsinghua University Press, 2008 (in Chinese).
温诗铸，黄平. 摩擦学原理 [M]. 北京：清华大学出版社，2008.
[13] C D Wagner, W M Riggs, L E Davis, et al. Handbook of X - ray photoelectron spectroscopy [M]. Eden Prairie: Perkin - Elmer Corporation, 1979.
[14] N S McIntyre, D G Zetaruk. X - ray photoelectron spectroscopic studies of iron oxides. Anal [J]. Chem, 1977, 49: 1521 ~ 1529.
[15] Y Chen, X H Li, P L Wu, et al. Enhancement of structural stability of nanosized amorphous Fe_2O_3 powders by surface modification [J]. Mater Lett, 2007, 61: 1223 ~ 1226.
[16] K BabaU, R Hatada. Synthesis and properties of TiO_2 thin films by plasma source ion implantation [J]. Surf Coat Technol, 2001, 136: 241 ~ 243.

Tribological Properties of Attapulgite Clay as Lubricant Additive for Steel - steel Contacts

Wang Limin[1,2], Xu Binshi[2], Xu Yi[2], Gao Fei[2], Yu Helong[2], Qiao Yingjie[3]

(1. Engineering Training Center, Harbin Engineering University, Harbin, 150001, China;
2. National Key Laboratory for Remanufacturing,
Academy of Armored Force Engineering, Beijing, 100072, China;
3. College of Materials Science and Chemical Engineering,
Harbin Engineering University, Harbin, 150001, China)

Abstract The attapulgite clay powder is dispersed in 150SN mineral oil as lubricant additive. The tribological properties of steel – steel contacts under the lubrication of 150SN base oil with diversity concentration of attapulgite clay powder are investigated on a standard Optimal SRV – Ⅳ tribometer. The elemental component and morphology on worn surface are analyzed by using scanning electron microscopy coupled with energy dispersive spectroscopy. The 150SN mineral oil containing 0.6% attapulgite clay reduces friction coefficient to 57.68% and the wear of steel ball lubricated by 150SN mineral oil containing 0.4% attapulgite clay is 14.52% as low as that lubricated by 150SN mineral oil. Smooth and compact tribofilm composed of oxides is observed on the worn surface lubricated by 150SN mineral oil containing attapulgite clay. The layer – chain crystal structure of attapulgite clay and the complex physicochemical in friction process are responsible for the reduced friction and wear of the tribopair.

Key words attapulgite clay, lubricant additive, friction and wear, self – repairing

高速电弧喷涂 FeAlNbB 非晶纳米晶涂层的组织与性能*

摘　要　为了提高钢铁材料的耐磨性和硬度，利用高速电弧喷涂技术在 45 钢基体上制备了 FeAlNbB 非晶纳米晶涂层。采用扫描电镜（SEM）、能谱分析仪（EDAX），透射电镜（TEM）和 X 射线衍射仪等设备对涂层的组织结构和相组成进行了分析，研究了非晶纳米晶的形成机制。实验结果表明：FeAlNbB 非晶纳米晶涂层是非晶相、α-Fe、FeAl 纳米晶和 Fe_3Al 微晶共存的多相组织，涂层中非晶相含量约 36.2%，纳米晶尺寸约 14.1nm；涂层组织均匀，结构致密，平均孔隙率约 2.3%；非晶纳米晶涂层具有较高的硬度，其耐磨性是相同实验条件下制备的 3Cr13 涂层的 2.2 倍。

关键词　非晶纳米晶涂层　高速电弧喷涂　显微硬度　耐磨性

铁基非晶纳米晶材料具有高的强度、韧性、耐磨和耐蚀性能，且价格低廉，是当前材料研究领域的热点之一[1]。采用热喷涂技术，可以获得大面积、较大厚度的（数百微米）非晶纳米晶复合涂层[2,3]，如等离子喷涂法[4]、超音速火焰喷涂法[5]、高速电弧喷涂法[6]。与其他热喷涂方法相比，高速电弧喷涂由于设备简单，喷涂材料为粉芯丝材，易于加工等特点而被广泛应用。Borisova 等[7]用电弧喷涂工艺在 Fe-B 中加入稀土元素，成功制备了 Fe-B-RE 非晶涂层，显著提高了涂层的性能。Branagan 等人[8]利用电弧喷涂技术制备了 Fe-Cr-B-Si-Mo-W-C-Mn 非晶纳米晶涂层，涂层的组织为非晶相的基体上零星分布着尺寸为 60 ~ 140nm 的硼化物和碳化物。郭金花等[9,10]利用电弧喷涂方法制备 FeCrMoMnBCSi 铁基合金。徐滨士等[11~13]开发了多种能形成铁基非晶纳米晶涂层的高速电弧喷涂粉芯丝材，制备了 FeCrBSiMnNbY 和 FeBSiNb（Cr）系列的非晶纳米晶涂层。这些研究表明，高速电弧喷涂技术可以实现涂层非晶化，因此，开发更多的铁基高速电弧喷涂丝材制备非晶纳米晶涂层对于非晶材料在再制造工程中的应用有重要意义。

本文利用高速电弧喷涂技术在 45 钢基体上制备了 FeAlNbB 非晶纳米晶涂层，并对涂层的组织结构进行了分析，研究了非晶纳米晶涂层的力学性能和磨损行为。

1　实验

实验采用装备再制造技术国防科技重点实验室的机器人自动化高速电弧喷涂系统制备涂层，该设备采用 HAS-01 型高速电弧喷涂枪，雾化融滴的速度达到 243m/s。喷涂材料为自行开发的 ϕ2mm 粉芯丝材，化学成分为 Fe-Al-Nb-B，其组成包括了特定的原子尺寸配比和最大化的非晶形成元素。喷涂商用 3Cr13 实芯丝材作为磨损实验对比

* 本文合作者：田浩亮、魏世丞、陈永雄、刘毅。原发表于《材料科学与工艺》，2012，20（1）：108 ~ 113。国家重点基础研究发展计划（“973”计划）资助项目（2011CB013403）；国家自然科学基金资助项目（50735006；50971132）。

材料，基体材料为45钢。喷涂前对基材表面进行除锈除油等净化处理后进行喷砂预处理，完毕后立即喷涂。喷涂的主要工艺参数为：喷涂电压26V，电流140A，送丝速度2.6m/min，喷涂距离200mm，气压0.7MPa，气流速度6m^3/min。

采用D8型X射线衍射仪对涂层相结构进行分析。利用Quanta200型扫描电镜分析涂层截面的组织形貌。在JEM-2000EX型透射电镜上观察涂层的显微组织特征。使用ⅡMT－3型显微硬度计测量涂层截面的维氏硬度。在WE-10A万能材料试验机上测试涂层的结合强度。在CETR微米摩擦磨损实验机上测试了室温干摩擦条件下涂层和基体的摩擦磨损性能，磨损试验采用球-面接触方式，上试样为直径4mm的GCr15钢球，其硬度不小于60HRC。下试样为10mm×10mm×5mm的涂层方片，涂层表面抛光处理，涂层最终厚度为0.4mm。试验参数为：法向载荷F_n为15N，磨损时间30min，频率5Hz，位移幅值D为5mm。分别测试了45钢基材、3Cr13涂层和FeAlNbB非晶纳米晶涂层的摩擦系数。使用VK-9700型3D激光扫描显微镜测量了涂层和基体的二维磨痕形貌。

2 实验结果及分析

2.1 涂层的组织形貌分析

图1a为非晶纳米晶涂层截面形貌，涂层非常致密，没有裂纹，有少量的孔洞和低的孔隙率。

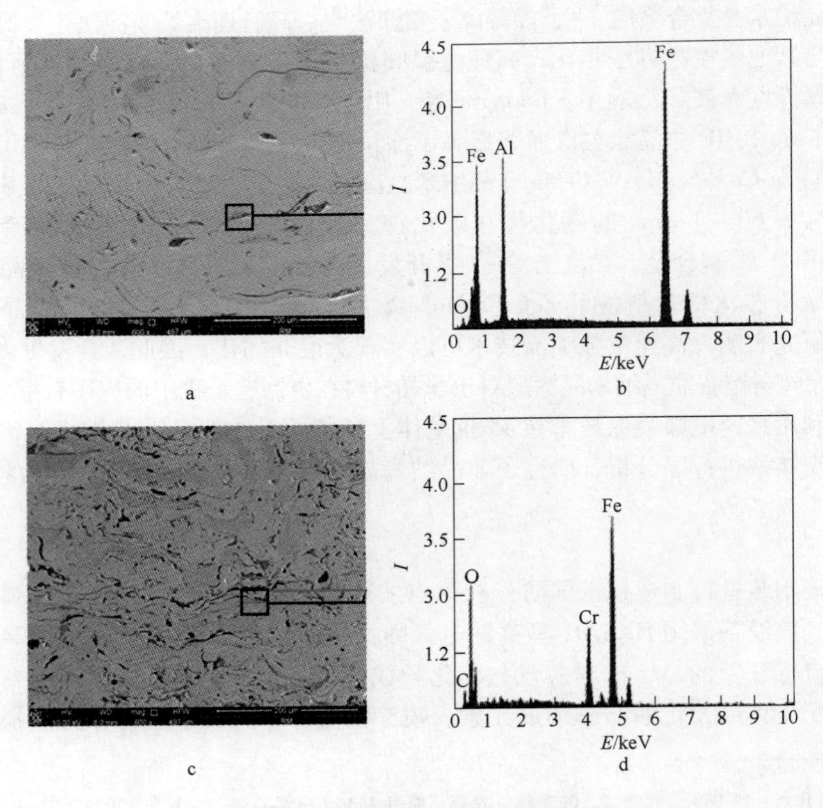

图1 涂层的截面形貌及微区能谱分析

a—FeAlNbB形貌；b—FeAlNbB微区能谱图；c—3Cr13形貌；d—3Cr13微区能谱图

利用图像分析软件测定涂层的孔隙率约为2.3%。微区能谱分析表明涂层中氧化物含量很低，约1.9%。图1c是3Cr13涂层的截面面形貌，呈明显层状结构，涂层致密度较差，存在大量空隙，经分析其孔隙率为4.3%。层与层之间存在大量的灰色氧化物，微区能谱分析氧含量约18%，这对涂层的内聚强度及耐磨性产生影响。

2.2 涂层的相结构分析

图2为涂层的X射线衍射谱，可以看出约在$2\theta=43°$处出现一个漫散射峰，说明在喷涂过程中形成了非晶相，通过Verdon方法[14]对XRD谱进行Pseudo-Voigt函数拟合，计算得到涂层中非晶相的含量为36.2%。同时图谱中还存在强度不高的晶化峰，说明涂层在沉积过程中形成了少量的晶体相，经分析主要为α-Fe、FeAl和Fe_3Al相。根据Scherrer公式[15]（系数为0.89，λ为1.540562×10^{-10}m）计算晶体相的尺寸为14.1nm，说明形成了非晶纳米晶复合结构的涂层。

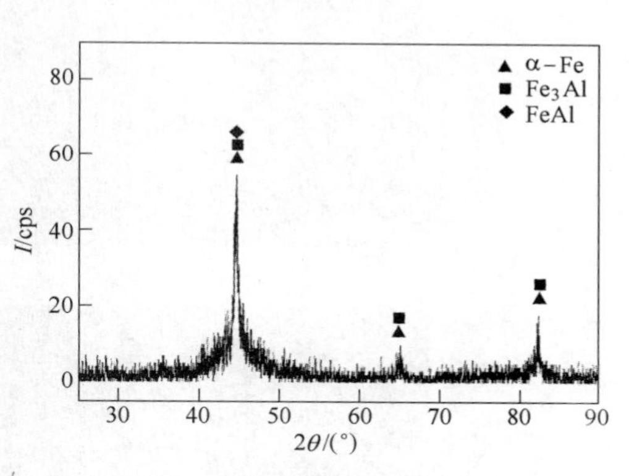

图2　涂层的XRD谱图

2.3 涂层组织结构分析

图3a是涂层微观衍射图，电子衍射花样的特点是中心有一漫散的中心斑点及漫散环，这种漫散的非晶衍射斑点的存在是非晶态的典型特征，说明涂层在沉积过程中形成了非晶相。图3b为涂层的微观组织结构形貌，可以看到涂层的内部组织比较均匀，微观组织衬度均一。

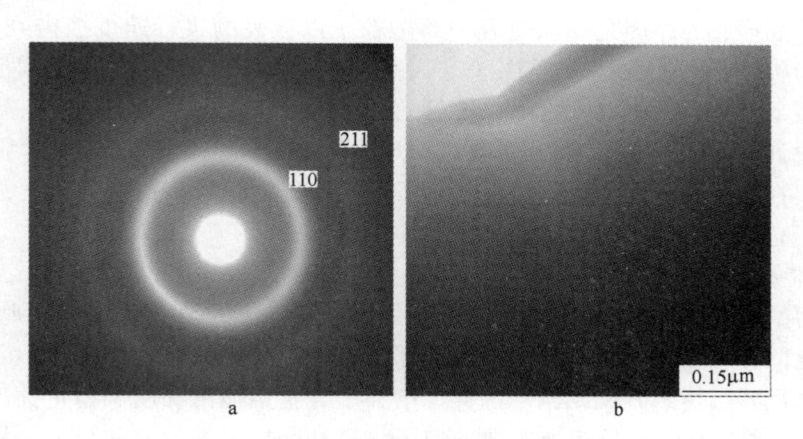

图3　非晶纳米晶涂层中非晶相衍射图

a—涂层中非晶相选取电子衍射图；b—非晶相组织形貌

图4a为非晶纳米晶电子衍射图。衍射花样由中心较宽的晕及漫散的环组成，同时在漫散的非晶衍射环上还分布着一系列小的多晶衍射斑点，经标定为α-Fe、FeAl。图4c是Fe_3Al、FeAl微晶衍射图。从图4b和d的微区微观组织形貌可以看出涂层结构是

非晶、纳米晶和微晶的混合涂层。纳米晶簇镶嵌于非晶母相上，尺寸为10～15nm，与X射线衍射峰半高宽计算得到的晶粒尺寸基本吻合。

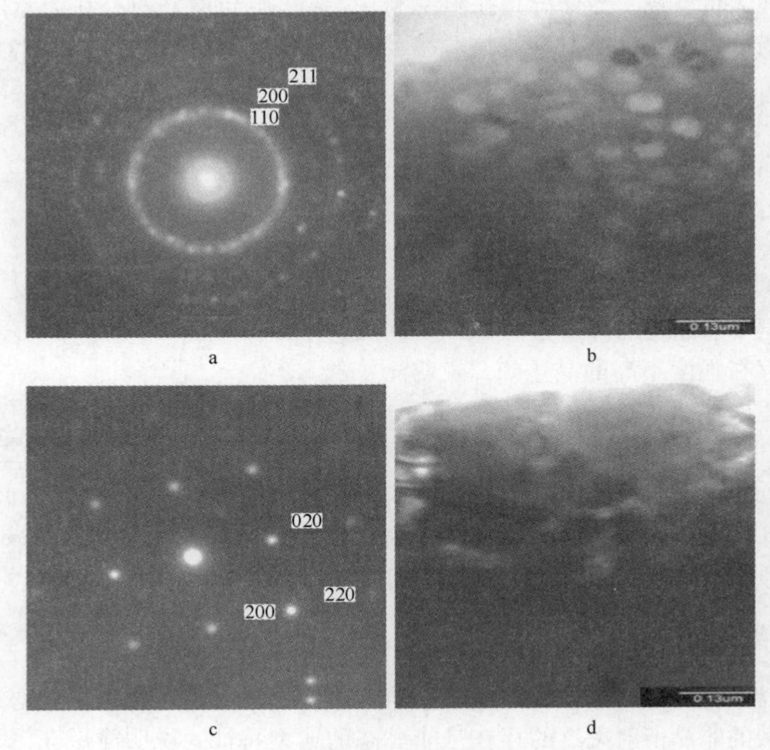

a b c d

图4 非晶纳米晶涂层中纳米晶微晶衍射图

a—涂层中非晶纳米晶选取电子衍射图；b—非晶纳米晶组织形貌；c—微晶衍射图；d—微晶组织形貌

分析非晶形成的原因是喷涂过程中熔融粒子以极高的飞行速度撞击到基体表面，并以10^5～10^7K/s的凝固速率急剧冷却[16]，快速冷却使元素长程扩散受到抑制，短程有序被保留下来形成无序堆积的凝固状态，即非晶态。另外Fe－Al－Nb－B系粉芯丝材的成分设计满足了Inoue[17]提出的高玻璃形成能力的非晶合金3大经验原则：（1）主要组元元素在3个以上；（2）主要组元原子半径差要大于12%；（3）组元之间具有较大的负的混合热。

材料体系中各元素的原子半径差较大（Nb：1.48，Al：1.43，Fe：1.27，B：0.95），大原子半径Nb和Al加入后拥有高的配位数，小的类金属原子B则占据空位，大小原子之间存在强烈的相互作用，使近邻周围彼此约束的基体原子、小原子和大原子形成类似网状结构或骨架结构。因此，增加了合金熔体的黏度和稳定性，降低了原子的可动性，抑制了晶化反应所需要的原子重排和晶体相的形核和长大，从而提高了合金的玻璃形成能力。同时，Nb元素的加入可使合金中各个组元之间负的混合热焓增大，Fe－B，Fe－Nb和B－Nb组成的混合热焓分别为－11kJ/mol、－16kJ/mol和－39kJ/mol[18]。Al的加入也能与Fe形成大的负混合热的Fe－Al原子对[19]，大的负混合热焓增大了各组元间的相互反应，促进结构的无序性，降低了原子扩散率，增加了过冷液体的黏度和体系的热稳定性，限制了结晶的发生，使涂层非晶形成能力增强。

类金属原子 B 的加入不仅可以增大原子尺寸差，使体系混乱度增强、长程无序性增加；还可以降低合金的熔点，扩大固液相线之间的距离，使合金成分在较低温度下仍保持液态，随着冷却的进行熔体的黏度增大，合金组元中长程扩散困难，从而抑制了晶态相的形核长大。可见非晶的形成是快速冷却和合理材料设计共同作用的结果。

纳米晶的产生有两方面的原因：一是非晶态在热力学上是一种亚稳状态，其自由能较高，在一定条件下，有降低能量转变成晶体的趋势。喷涂过程中后续熔融粒子对已沉积涂层的热作用或是周围粉末粒子变形凝固释放的结晶潜热，造成涂层中某些微区热量积聚超过保持非晶结构稳定的临界温度，为原子的扩散迁移提供了热量，产生较强的晶化转变趋势，生成纳米晶[20]。二是非晶态材料中存在一定程度的短程有序结构及有序原子集团[21]，非晶中的有序原子集团为非晶向纳米晶转变提供了非均质形核质点的位置，在后续涂层提供的热量和结晶潜热的作用下，部分非晶转变成了纳米晶。

Fe_3Al 和 FeAl 微晶共存的形成机理是喷涂雾化过程中高速（200～300m/s）飞行的融滴在极短的时间（<0.002s）内发生冷却、凝固，冷却速度达 $10^5 \sim 10^7$/s 数量级，这种快速冷却的非平衡凝固造成融滴内部 Fe、Al 合金化不均匀，使 FeAl→Fe_3Al 转变受到很大抑制[22]，从而导致室温下涂层中 FeAl 和 Fe_3Al 共存。

2.4 涂层的显微硬度及磨损性能分析

图 5 为从涂层表面到基体沿截面的显微硬度分布图，可以看出非晶纳米晶涂层的硬度较高，在 700～740$HV_{0.1}$变化，是 3Cr13 涂层硬度的 2.5 倍，这是由于金属间化合物 Fe_3Al 和 FeAl 均匀分布在非晶相上，对涂层起到了弥散强化作用，另外，非晶纳米晶涂层结构均匀致密、孔隙率小、也使其具有很高的硬度。

图 6 是非晶涂层、3Cr13 涂层和 45 钢基体在室温干摩擦条件下摩擦系数随滑动距离的变化曲线，非晶纳米晶涂层的摩擦系数经 5min 后进入稳定期，平均摩擦系数为 0.4，“跑合”阶段不是很明显，3Cr13 涂层的平均摩擦系数为 0.8，45 钢基体的摩擦系数较大，始终维持在 1.1 左右。说明非晶纳米晶涂层具有良好的耐磨性。

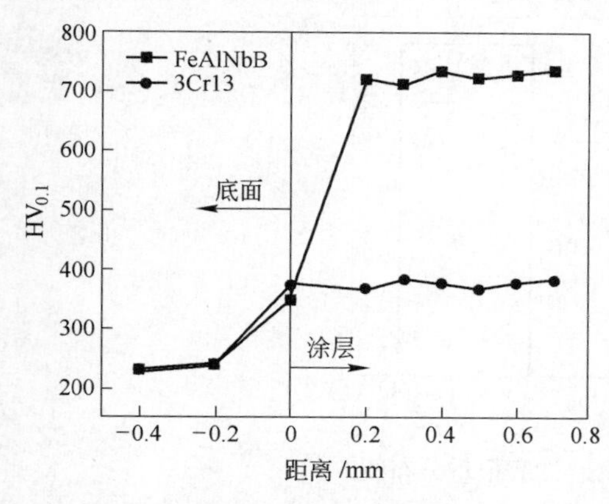

图 5　涂层与基体截面的显微硬度变化曲线

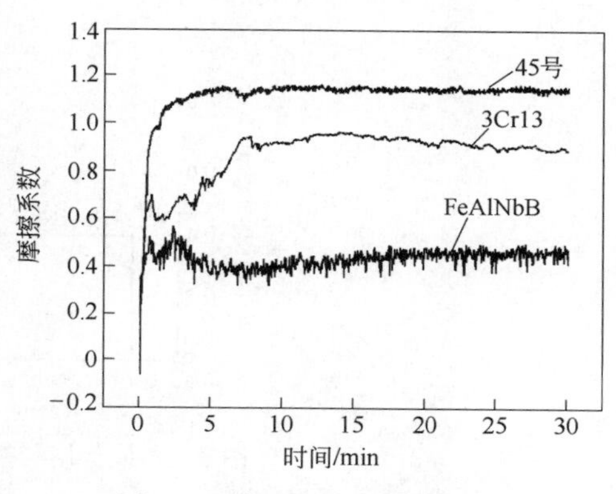

图 6　基体和涂层的摩擦系数

图 7 是 FeAlNbB 涂层、3Cr13 涂层和基体 45 钢的二维磨痕形貌，可以看出，FeAl-Nb 非晶纳米晶复合涂层的磨痕比 3Cr13 磨痕浅，磨损量也小（如图 8 所示），即相对

耐磨性是3Cr13涂层的2.2倍，是45钢基体的3.8倍。磨损量及磨损机理的不同，主要是由涂层内部结构及内聚强度的差别引起的。非晶纳米晶涂层组织结构均匀致密，孔隙率小，氧化物含量低，内聚强度高，且非晶相本身就具有良好的耐磨性，使得涂层发生剥层磨损的趋势大大减小，因此，磨损量较少。并且，非晶母相上均匀分布着Fe_3Al微晶相和α-Fe、FeAl纳米晶颗粒，涂层非晶母相中分布的纳米晶粒在一定程度上起到弥散强化作用，阻止了磨损过程中的裂纹扩展。而3Cr13涂层具有明显的层状结构，且层与层之间存在一定的氧化物，大大降低了涂层的内聚强度，较易引发剥层磨损，故磨损量较大。

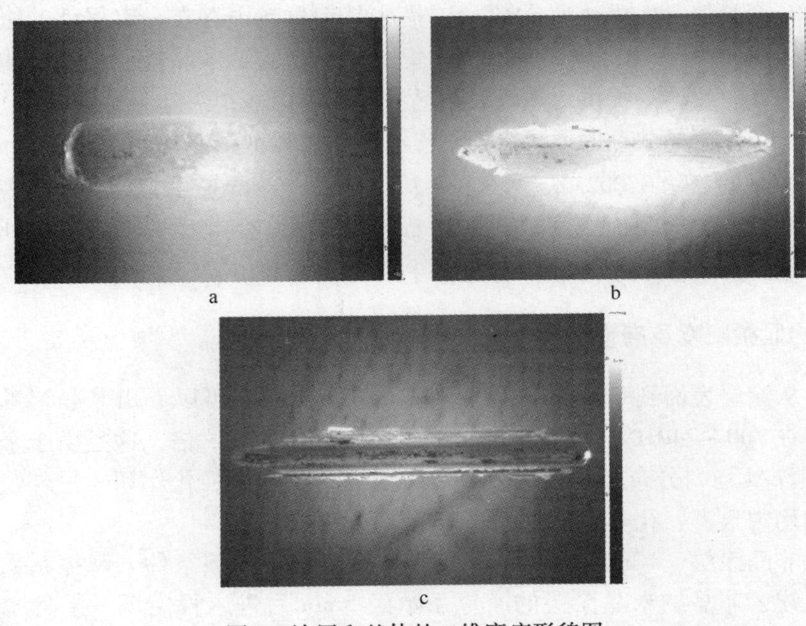

图7 涂层和基体的二维磨痕形貌图

a—FeAlNb涂层；b—3Cr13涂层；c—45钢基体

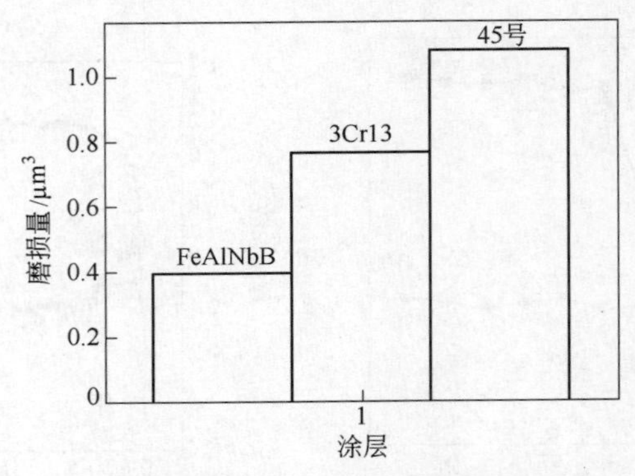

图8 不同涂层与基体材料的磨损量分布图

3 结论

（1）利用机器人自动化高速电弧喷涂技术在45钢基体上成功制备了FeAlNbB非晶

纳米晶涂层，与相同实验条件下制备的3Cr13涂层相比，非晶纳米晶涂层致密度更高，孔隙率较低，约2.3%，氧化物含量也小，约1.9%。

（2）非晶纳米晶涂层主要由非晶相和非晶母相上弥散分布的α-Fe、FeAl纳米晶Fe_3Al微晶颗粒组成。非晶相含量约36.2%，纳米晶尺寸约为14.1nm。

（3）非晶纳米晶涂层的硬度约700～740$HV_{0.1}$是3Cr13涂层的2.5倍。非晶纳米晶涂层的相对耐磨性是3Cr13涂层的2.2倍，是基体45钢的3.8倍。

参考文献

[1] SCHUH C A, HUFNAGEL T C. Mechanical behavior of amorphous alloys [J]. Acta Mater, 2007, 55: 4067～4109.

[2] WU Y, LIN P, XIE G, Formation of Amorphos and nanocrystalline phases in high velocity oxy-fuel thermally sprayed a Fe-Cr-Si-B-Mn alloy [J]. Material Science Engineering, A 2006, 430: 34～39.

[3] FAN Zishuan, SUN Dongbai, YU Hongying. Preparation of iron base amorphous and nanocrystalline alloy coatings by plasma spraying [J]. Journal of University of Science and Technology Beijing, 2005, 27(5): 582～585.

[4] LIANG Xiubing, CHENG Jiangbo, XU Binshi. Phase composition of nanostructured zirconia coatings deposited by air plasma spraying [J]. Surface and Coating Technology, 2005, 191: 267～273.

[5] SHIPWAY P H, MC CARTNEY D G, SUDAP RASRT T. Sliding wear behavior of conventional and nanostructured HVOF sprayed WC_2 Co coatings [J]. Wear, 2005, 259: 820～827.

[6] 梁秀兵，徐滨士，魏世丞，等．热喷涂亚稳态复层研究进展［J］．材料导报，2009，23(5)：1～4.

[7] BORISOVA A L, MITZ I V, PATON E O. Arc sprayed coatings of ferroalloy-base flux-cored wires [C] //Proceedings of the 1st International Thermal Spray Conference. [S. l.]: Thermal Spray-Surface Engineering via Applied Research, 2000. 705～708.

[8] BRANAGAN D J, BREISAMETER M T, MEACHAM B E. High-performance nanoscale composite coatings for boiler applications [J]. Thermal Spraying Technology, 2005, 14(2): 196～204.

[9] 郭金花，吴嘉伟，倪晓俊，等．电弧喷涂含非晶相的Fe基涂层的电化学行为［J］．金属学报，2007，43(7)：780～784.

[10] 郭金花，陆曹卫，倪晓俊，等．电弧喷涂Fe基非晶硬质涂层的组织及性能研究［J］．中国表面工程，2006，19(5)：45～48.

[11] 程江波，梁秀兵，徐滨士，等．铁基非晶纳米晶涂层组织与冲蚀性能分析［J］．稀有金属材料与工程，2009，38(12)：2142～2145.

[12] LIANG Xiubing, CHENG Jiangbo, BAI Jinyuan, et al. Erosion properties of Fe-based amorphousnanocrystalline coatings prepared by wire arc spraying process [J]. Surface Engineering, 2009 (35): 73～78.

[13] CHENG Jiangbo, LIANG Xiubing, XU Binshi. Characterization of mechanical properties of FeCrBSiMnNbY metallic glass coatings [J]. Journal of Materials Science, 2009(4): 3356～3363.

[14] VERDON C, KARIMI A, MARTIN J L. A study of high velocity oxy-fuel thermally sprayed tungsten carbide based coatings. Part 1: Microstructures [J]. Material Science Engineering, A 1998, 246: 11～24.

[15] CULLITY B D. Elements of X-ray diffraction. [M]. London: Addison-Wesley Publishing Company, 1978.

[16] NERBERY A P, GRANT P S, NEISER R A. The velocity and temperature of steel droplets during elec-

tric arc spraying [J] . Surface and Coating Technology, 2005, 195: 91 ~101.

[17] INOUE A, KOHINATA M, HTERA K. Mg-Ni-La amorphous alloys with a wide supercooled liquid region [J] . Material Transformers, 1989, 30: 378 ~381.

[18] Japan Institute of Metals. Metals Databook [M] . Tokyo: Maruzen, 2004: 139 ~141.

[19] LA P Q, YANG J, DAVID J H. Bulk Nanocrystalline Fe_3Al-Based Material Prepared by Aluminothermic Reaction [J] . Advanced Materials, 2006, 18: 733 ~737.

[20] WANG A P, WANG Z M, ZHANG J. Deposition of HVAF-sprayed Ni-based amorphous metallic coatings [J] . Journal of Alloy Compound, 2007, 440: 225 ~228.

[21] CHARLES M, FORTMAN N, ISAMU SHIMIZ U. Prospects for utilizing low temperature amorphous to crystalline phase. transformation to define circuit elements; a new frontier for very large scale integrated technology [C] //16th International Conference on Amorphous Semiconductors-Science and Technology. KOBE: [S. n.], 1995, 198 ~200: 1146 ~1150.

[22] 孙祖庆，黄原定，陈国良．急冷凝固 Fe_3Al 金属间化合物韧性研究 [J] ．北京科技大学报，1992，11(3)：328 ~332.

Microstructure and Mechanical Properties of FeAlNbB Amorphous/nanocrystalline Coatings Deposited by High Velocity Arc Spaying

Tian Haoliang[1,2], Wei Shicheng[2], Chen Yongxiong[2], Liu Yi[2], Xu Binshi[2]

(1. School of Material Science and Engineering, Beijing University of Aeronautics & Astronautics, Beijing, 100037, China;
2. National Key Laboratory for Remanufacturing, Academy of Armored Forces Engineering, Beijing, 100072, China)

Abstract To improve the hardness and the wear resistance properties of the steel materials. FeAlNbB amorphous/nanocrystalline coatings were deposited on a mild steel by high velocity arc spraying processing of which the microstructure was characterized by SEM 、EDAX 、TEM and XRD. The results show that the microstructure consists of amorphous、$\alpha-Fe$ 、FeAl nanocrystalline and Fe_3Al microcrystalline phase. The volumn fraction of amorphous phase is 36. 2% and the size of the nanocrystalline is 14. 1 nm. The coatings were fully dense with low porosity of 2. 3%. The formation mechanism of amorphous/nanocrystalline was discussed. The microhardness and wear resistance of the 3Cr13 coating fabricated under the same experiment conditions and the amorphous/nanocrystalline coatings were also analyzed. The amorphous/nanocrystalline coating has high hardness and the wear resistance is 2. 2 times better than that of 3Cr13 coatings.

Key words amorphous/nanocrystalline coating, high velocity arc spraying, microhardness, wear resistance

Research on the Microstructure and Space Tribology Properties of Electric-brush Plated Ni/MoS_2-C Composite Coating*

Abstract In order to expand the application field of MoS_2 lubricant coatings(films) and to improve their adaptability in harsh environments, a Ni/MoS_2-C composite coating composed of Ni metallic matrix and two lubricating components was prepared on AISI 52100 steel substrates with a nano electro-brush plating process. The relationship between microstructure and properties of the coating was analyzed via microstructure characterization. The tribological properties of the as-received coating samples and the samples treated in humid air(HA), high vacuum(HV), atomic oxygen(AO) erosion and ultraviolet (UV) irradiation exposure were investigated comparatively. The results show that the surface of the obtained Ni/MoS_2-C coating is smooth and clean. The micro and nano MoS_2-C grains are dispersed in the Ni matrix homogeneously. The MoS_2 and C particles also showed good synergistic effects. The added nano graphite particles not only increased the moisture resistance of the composite coating, but also reduced the residual stress and enhanced the mechanical properties of the coating. The composite coating still showed good lubricating properties after long-duration treatments in HA, HV, AO and UV environments, and the friction coefficients during the stable running periods remained at about 0.1. The influence of vacuum outgassing on the properties of the lubricants is not obvious. The coatings were oxidized slightly by AO, and thus their initial friction coefficients in AO conditions also increased, but the coating still showed good friction reduction properties after a short running-in period. The changes in microstructure and properties of the coating after UV irradiation were also inconspicuous. All test results indicate that the Ni/MoS_2-C composite coating has excellent space tribological properties and good adaptability in harsh environments. So it is promising to use Ni/MoS_2-C composite coating in a space solid lubrication area.

Key words Ni/MoS_2-C composite coating, space tribology, high vacuum, atomic oxygen, ultraviolet rays

1 Introduction

The special layered close-packed hexagonal crystal structure endows MoS_2 with excellent solid lubrication properties and it has enjoyed the reputation of "the king of lubrication" for a long time[1]. It is worth to mention that the friction coefficients of MoS_2 in vacuum are even lower than those in air. Therefore, MoS_2-base coatings is widely used to solve the lubrication problems under aerospace vacuum conditions[2]. However, the microstructure of pure MoS_2 coating is usually loose and its wear life is short[3]. In air condition (especially humid air), the pure MoS_2

* Copartner: Ma Guozheng, Wang Haidou, Wang Xiaohe, Li Guolu, Zhang Sen. Reprinted from *Surface & Coatings Technology*, 2013, 221: 142 – 149.

coating is prone to be oxidized, and thus partially loses its lubricating properties. Due to these reasons, the assembly, storage, and transportation procedures on the ground of spacecraft components that have been coated with pure MoS_2 becomes difficult, and the application of pure MoS_2 in other fields is also limited[4]. Researchers have tried their best to expand the application field of MoS_2 coatings (thin films), and to improve their adaptability in severe conditions. Adding a second phase into pure MoS_2 coatings to create composite coatings has showed good prospects. The additives usually can increase the density, bond strength to the substrate and moisture resistance of MoS_2 coatings, and thereby improve their tribological properties[5,6].

The nano-brush plating process is the combination and innovation of nano materials and the electro-brush plating technique[7]. By adding nano particles with special properties into the brush plating solution, a nano-composite coating can be obtained by the co-deposition of metal-base ions and nano particles. Our laboratory has done a lot of work on the research and application of nano-composite coatings with excellent tribological properties. We once prepared Ni/nano-Al_2O_3 and Ni/nano-SiC composite coatings having good wear resistant properties by electro-brush plating[8,9]. The Ni/PTFE and Ni/nano-graphite lubrication coatings with very low friction coefficients were prepared[10]. We also prepared a Ni matrix MoS_2 composite coating using plasma-spraying, and the Ni/MoS_2 coating exhibited excellent antifriction properties both in air and in vacuum[1].

The presented work utilizes the merits of the nano-brush plating process to overcome the flaws of pure MoS_2 coatings. In this paper, a Ni/MoS_2-C composite coating was prepared on AISI 52100 steel with the nano-brush plating process. The relationship between microstructure and properties of the coating are analyzed. Moreover, the tribological properties of the composite coating under several simulated space environments were investigated comparatively. The results would provide a reference for the space application of MoS_2-base coatings.

2 Experimental Details

The circular Q-tempered AISI 52100 steel substrate samples with the hardness of HRC58 were polished to a roughness below R_a0.3 μm. The used brush-plating Ni-base solution contains 264 g/L $NiSO_4 \cdot 6H_2O$, 56 g/L ammonium citrate, 23 g/L ammonium acetate and 105 g/L ammonia water. Then, 30 g/L MoS_2 (particle size of less than 500 nm) and 20 g/L graphite (particle size of about 40 nm) particles were added into the Ni-base solution and dispersed by high energy mechanochemical method (HEMM a patent method)[11].

The HEMM (also known as high-energy reaction ball milling) can induce mechanochemical reactions especially hydrothermal reactions in mixtures at much lower temperatures than normally required because the reaction kinetics and driving force can be increased by intensive milling[12,13]. In the present study, the uniform dispersion and suspension stabilization of micro-nano particles in the composite plating solution were realized using HEMM. The process can be described as follows: some surface active agent and dispersion stabilizer (polyacrylic nickel) were firstly added into the milling chamber together with the intermixture of Ni-base solution and micro-nano particles; and then, the intermixture was wet ball-milled with tungsten carbide balls at room temperature for 24 h. Actually, the dispersion and stabilization of the inorganic fine particles in water-base plating solution achieved by HEMM can be divided into three steps. Firstly, high-energy collisions of the grinding balls result in the depolymerization of agglomerated particles. Then, the charged ions in the solution immediately adsorb onto the fresh surfaces of micro-nano particles and form an electric double layer, causing electrostatic repulsion between the particles. At the same time, the high polymer disperser anchors and reacts with the activated fine particles, preventing the particles from approaching one another because of the steric effect. So, the HEMM exhibits an integrated effect of mechanical pulverization, electrostatic repulsion and surface modification. And then, the Ni/MoS_2-C composite coating about 100 μm thick was prepared using our automatic brush plating system. The technology procedure and equipments for preparing the composite coating can be found in our prior publications[14,15]. As a comparison, some Ni/MoS_2 coating samples containing MoS_2 lubricant only were also prepared using the same plating process.

A scanning electron microscope (SEM) equipped with energy dispersive spectroscopy (EDS), X-ray photoelectron spectroscope (XPS), X-ray diffraction (XRD), and high resolution transmission electron microscopy (HRTEM) were employed to characterize the microstructure of the Ni/MoS_2-C composite coating. Some Ni/MoS_2 and Ni/MoS_2-C coating samples were stored in humid air (80% RH) for 30 days at room temperature, and then their tribological properties were compared. The residual stress, toughness, hardness and other mechanical properties of the two coatings were also measured.

All the friction and wear tests were performed on a self-developed MSTS-1 multifunctional space tribology test system[16]. The test system not only can simulate the main space environment factors, such as high vacuum (HV), atomic oxygen (AO) erosion and high-energy ultraviolet (UV) irradiation, but also can test the tribological properties of specimens in situ. In the friction tests, the fixed upper samples were 440C steel balls with a diameter of 9.525 mm, and the circular coated samples acted as the rotating lower sample. A normal load of 12 N (about 0.68 GPa) and a sliding speed of 1.25 m/s were chosen for each test. The AO and UV irradiation sources work under low vacuum (LV, the pressure of about 8×10^{-2} Pa) conditions. Some samples were treated by AO (flux density of about 2.6×10^{15} atom/($cm^2 \cdot s$)) and vacuum UV (wavelength of 115-250 nm) beams for 10 h before tribo-testing. The wearing track morphologies and wear mechanisms were analyzed by SEM observation.

3 Results and Discussion

3.1 Microstructure of the Ni/MoS_2-C composite coating

Fig. 1 shows the surface morphology of the Ni/MoS_2-C composite coating. The composite coating is compact and smooth, showing the plating coating's typical crystal structure of cauliflower-shaped clusters. There were few cavities, heels, burrs and other macro defects in the coating. The corresponding EDS analysis results show that, though Mo, S and C elements can be detected on the coating surface, the content of Ni is relatively high.

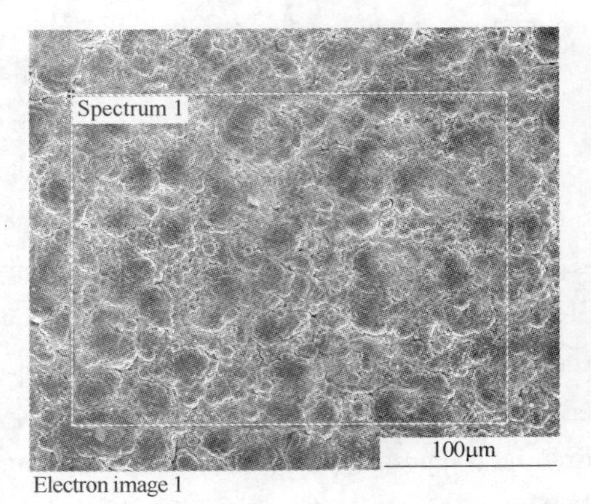

Fig. 1 Surface morphology of the Ni/MoS_2-C composite coating

In order to explore the existing form of Mo and S elements in the coating, it is necessary to do the XPS analysis of Mo3d and S2p due to the overlapping of Mo and S peaks in the EDS spectra. As shown in Fig. 2, except for a small amount of molybdic oxide, the strong peaks of Mo and S in the electron binding energy spectra correspond to Mo^{4+} and S^{2-} in MoS_2, respectively. This indicates that most MoS_2 particles co-deposited with metal Ni, and the decomposition or oxidization of MoS_2 during the plating process was slight.

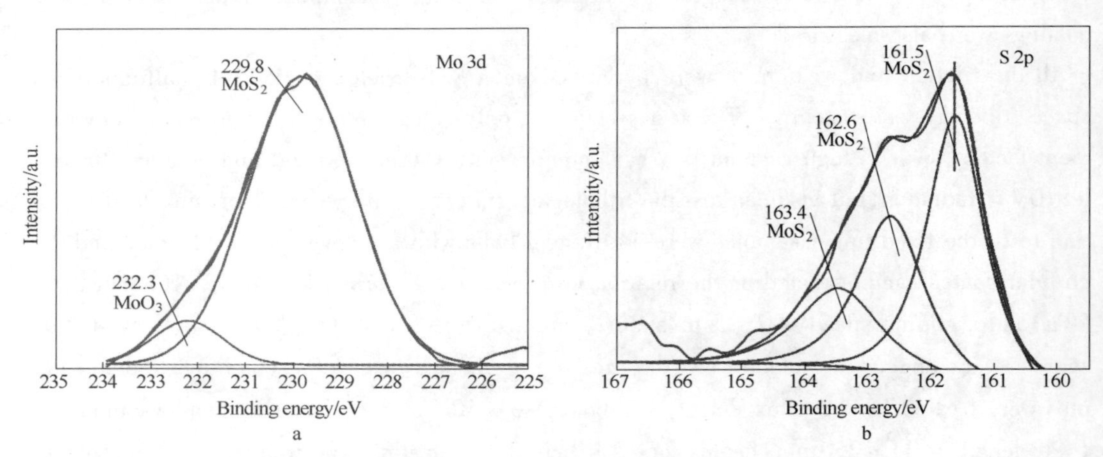

Fig. 2 XPS spectra of the Mo3d and S2p of the coating

Fig. 3 shows the XRD spectrum of the Ni/MoS_2-C coating, from which we can see that the composite coating is mainly composed of Ni, MoS_2 and graphite. Moreover, the contents of MoS_2 and graphite were relatively high. As compared with the test results by EDS, it can be seen that the MoS_2 and graphite particles were successfully deposited into the composite coating, and the matrix Ni can provide effective protection for the two lubricants. The amount of MoS_2 and graphite particles exposed directly to the outside are less, which is helpful to improve the coating's durability in harsh environments.

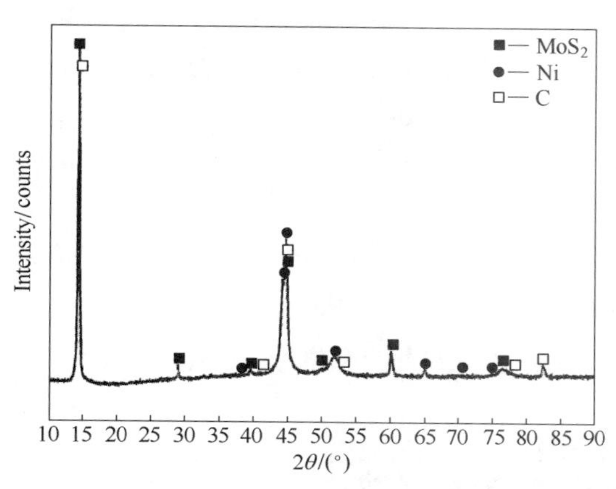

Fig. 3 Phase constituents of the Ni/MoS_2-C coating tested by XRD

The HRTEM bright field micrograph and corresponding selected area electron diffraction (SAED) pattern of the Ni/MoS_2-C coating are shown in Fig. 4. The matrix Ni showed an obvious polycrystalline structure. Most grains with random crystallographic orientations are in nano scale. The microstructure of the matrix Ni coating is compact and uniform, in which the micro and nano lubricant particles are dispersed randomly (marked areas in Fig. 4a).

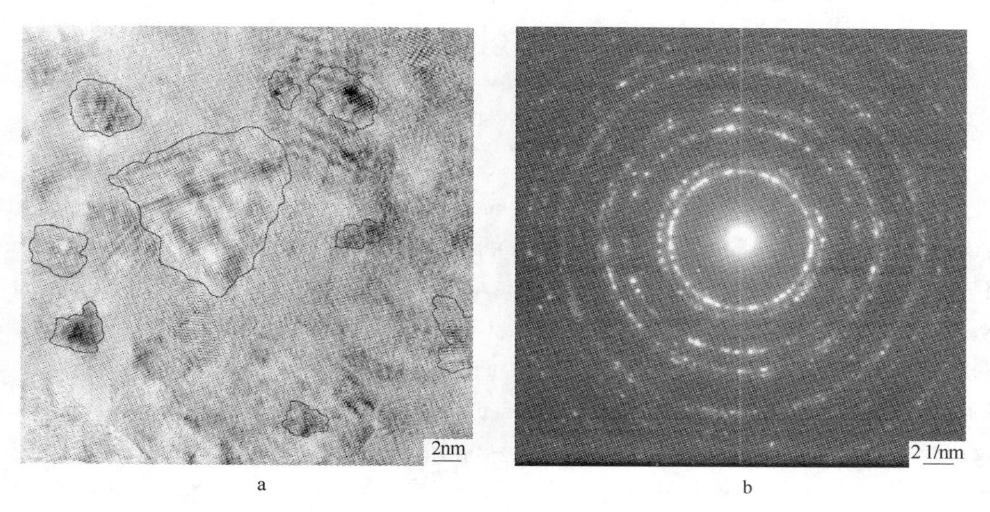

Fig. 4 HRTEM micrographs of the Ni/MoS_2-C composite coating

a—Bright field image; b—Selected area electron diffraction rings taken from a

3.2 The optimization effect of nano graphite particles on the Ni/MoS_2-C coating

Residual stress is one of the most important reasons for the cracking, peeling, delamination and other damages of coatings[17]. The inner stress generated by non-equilibrium crystallization often seriously restricts the performance of electrolytic plated coatings. The residual stress of coatings should be strictly controlled. The residual tensile stress will exacerbate the stress concentration and promote the initiation and propagation of cracks, which result in the dehiscence of coatings. Residual compressive stress not only can release the stress concentration, but also can im-

prove the fatigue resistance and bonding strength of the coating. However, excessive compressive stress may lead to bubbling or delamination of coatings. In this paper, the residual stress of the Ni/MoS_2 and Ni/MoS_2-C coatings was detected by XRD technology. Fig. 5 shows the variation of the diffraction angle (2θ) with the rocking angle ($\sin^2\psi$), the scanning diffraction angle ranged from 155° to 157°. The calculation results show that the residual stress of the Ni/MoS_2 and Ni/MoS_2-C coatings is 45 MPa and 16 MPa, respectively. Therefore, it can be concluded that the nano graphite particles can optimize the microstructure and reduce the residual stress of the coating, which is also helpful to enhance the tribological properties of the composite coating.

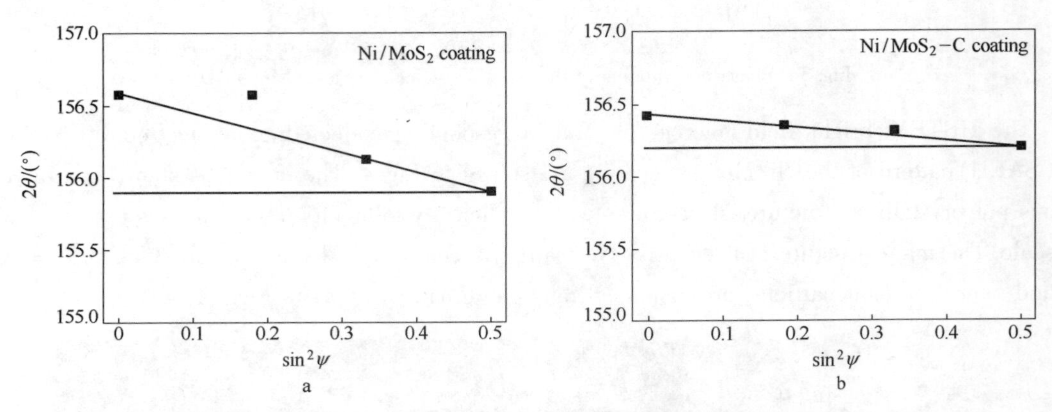

Fig. 5 Relationship between 2θ and $\sin^2\psi$ for residual stress test

a—Ni/MoS_2 coating; b—Ni/MoS_2-C coating

Fracture toughness, bond strength and other mechanical properties of a coating (film) directly affect its tribological performance. An instantaneous elastic wave will be released when a material ruptures or a coating cracks. The acoustic emission detector is sensitive to these elastic waves and can convert them to acoustic emission signals which is easy to be read and analyzed. Some mechanical properties of coatings can be characterized indirectly through analyzing the typical parameters of acoustic emission signals[18]. Many researchers tested the fracture toughness and bond strength of coatings using indentation and acoustic emission techniques[19,20]. Adopting a special indentation device[21], the mechanical properties of the two MoS_2 coatings before and after adding nano graphite particles were tested. A ball indenter of 1 mm in diameter was utilized to load the coating surface at a speed of 0.8 μm/s for 120 s (about 95 μm in indentation depth), and the final load was about 1040 N. The amplitude of the acoustic emission signal was selected as the characteristic parameter, and the threshold value was set to 55 dB. The amplitude of the acoustic emission signal directly reflects the magnitude of a material fracture event. Fig. 6 is the variation curve of the amplitudes of the acoustic emission signal during indentation. It can be seen that the signal amplitudes of the two MoS_2 coatings maintained at about 65 dB in stable period. But for the Ni/MoS_2-C coating, the vibration of the signal reduced significantly, meaning that the mechanical properties of the composite Ni/MoS_2-C coating are more uniform[22,23]. In addition, a strong acoustic emission signal was excitated as soon as the surface of Ni/MoS_2 coating was loaded, which shows that the Ni/MoS_2 coating is more brittle. At the same time, in the first 17s of loading to the Ni/MoS_2-C coating, the strength of the acoustic emission signal has not yet reached the threshold value, that is to say, the outermost layer of the Ni/MoS_2-C coating with good ductility and toughness is relatively soft. Based on the above results, it can be concluded that adding nano graphite particles into the coating not only can improve the ductility and plasticity of the composite coating, but also can enhance the homogeneity of the coating's mechanical properties.

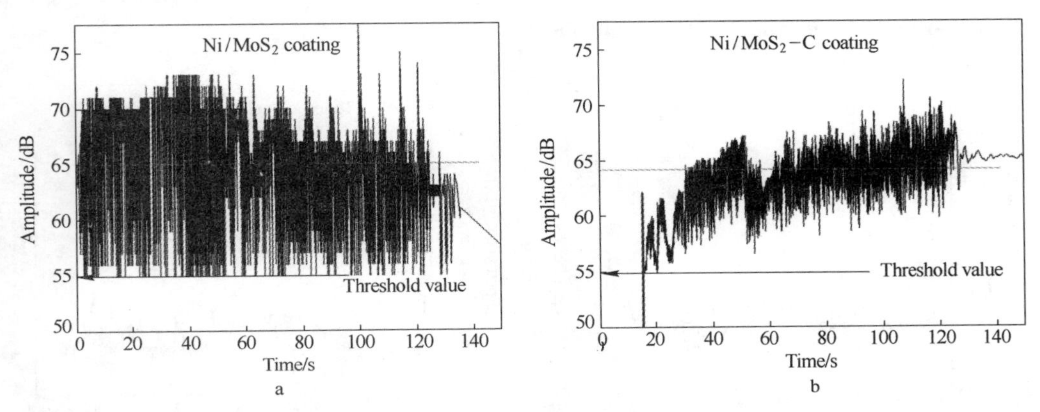

Fig. 6 Variations of the amplitude of acoustic emission signal with the indention time

Fig. 7 shows the comparison of tribological properties of the Ni/MoS_2 and Ni/MoS_2-C composite coatings after storing in humid air (80% RH) for 30 days. It can be seen that both coatings in original state had good friction reduction properties. After treating with humid air, the friction coefficients of the Ni/MoS_2 coating increased quickly and reached 0.3 when sliding for 600 s, while the friction coefficients of the Ni/MoS_2-C composite coating were low and stable during the test time.

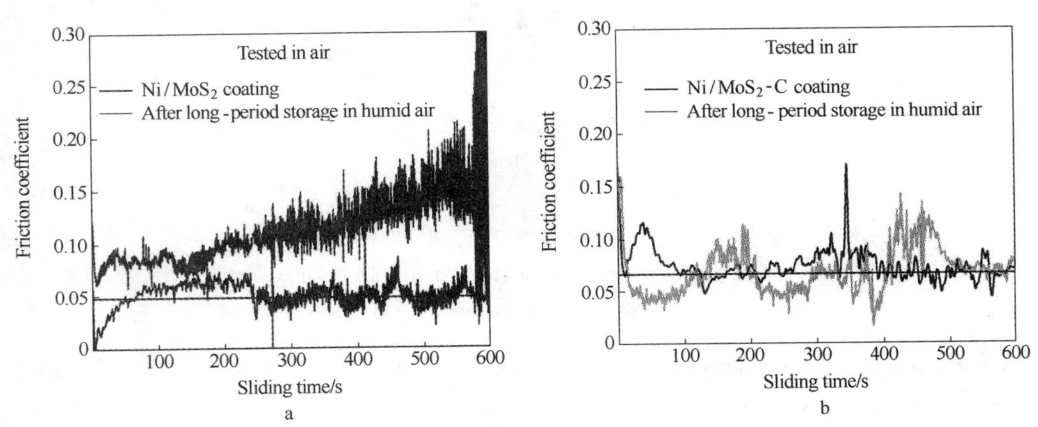

Fig. 7 Comparison of tribological properties of the coatings before and after storage in humid air

a—Ni/MoS_2 coating; b—Ni/MoS_2-C coating

It is obvious that, the Ni/MoS_2-C composite coating has good resistance to moisture environment and the two lubricants exhibit a good synergistic effect of "1 + 1 > 2". The good performance of Ni/MoS_2-C composite coating in humid air can be attributed to two reasons: one is the inherent excellent lubricating properties of graphite, the other is the graphite particles with good absorption properties could slow the oxidation of MoS_2 in humid air by absorbing excessive moisture.

3.3 The space tribology properties of Ni/MoS_2-C composite coating

Fig. 8 shows the friction coefficients of the Ni/MoS_2-C composite coating samples tested in HV, AO erosion and UV irradiation environments.

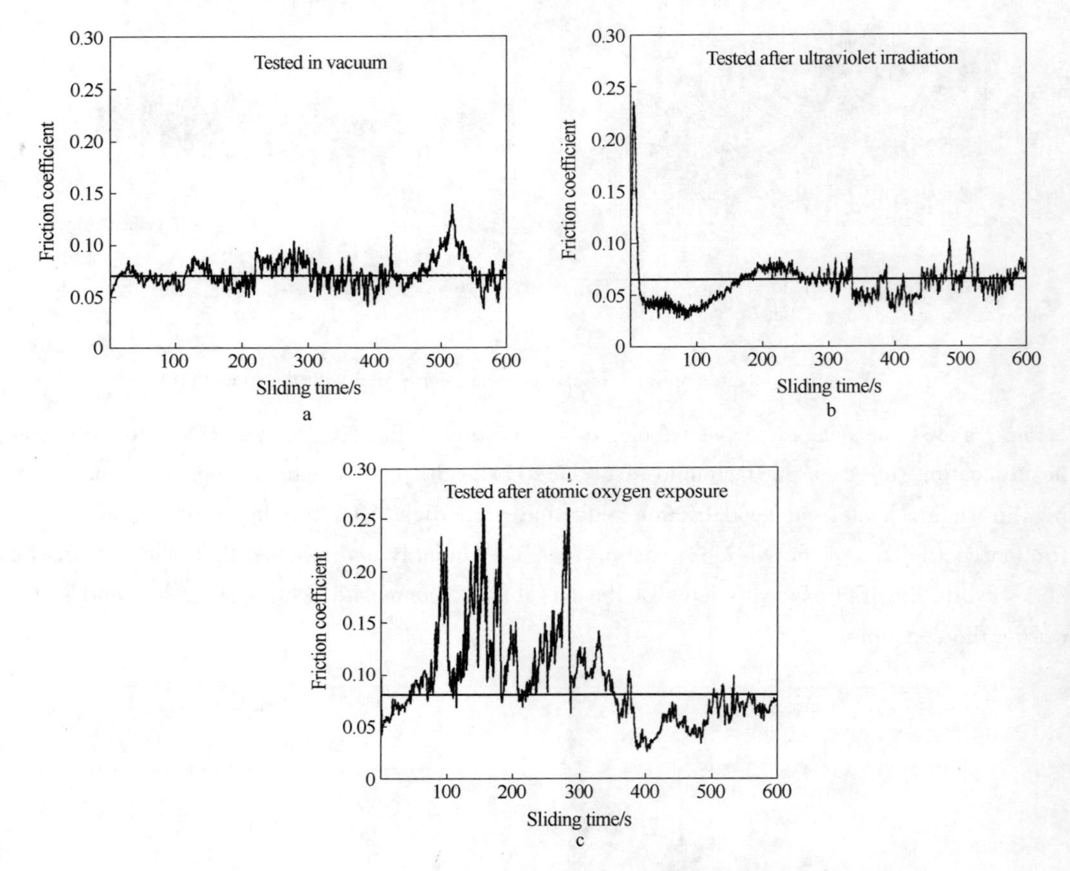

Fig. 8 Friction coefficients as a function of sliding time of the Ni/MoS_2-C coatings after treatment in different environment conditions

a—High vacuum; b—UV irradiation; c—AO erosion

By comparing with the friction coefficients of the composite coating before and after storing in humid air (as shown in Fig. 7b), it can be concluded that the Ni/MoS_2-C composite coating exhibited good friction-reduction properties in all of the presented test conditions. The friction coefficients during the stable periods maintained at less than 0.1, while that of the AISI 52100 steel substrate sample had reached as high as 0.7 to 0.8 under the same test conditions.

Nevertheless, the friction coefficient curves of the Ni/MoS_2-C composite coating also showed obvious differences when treated by various environmental factors.

As shown in Fig. 7b, the friction coefficients of the composite coating in air are low and stable, and the average value is about 0.07. The peak in the friction coefficient curve near the 350th second may result from local defects of the coating. And the coating's friction reduction properties could be recovered immediately when lubricants was transferred to the friction track. After storing in humid air, the fluctuation of the friction coefficients increased, this is because a slight oxidation of the lubricants still occurred during the long period storage in moist air.

The composite coating also showed good friction reduction properties in high vacuum, but there is a high peak around the 550th second, as shown in Fig. 8a. There are transfer films being built up frequently in the wear track and being ejected from it, and the relevant mechanical and morphological changes occurred at the contact zone. The friction coefficients remain low and stable during the dynamical equilibrium procedure, and the occasional peaks on the friction coefficient curve may result from the instantaneous break of the equilibrium.

The friction coefficients of the composite coating are still low and stable after exposure in vacuum UV beam for 10 h. The friction coefficient curve is similar to that of the samples tested in high vacuum. But there is a short running-in period with high friction coefficients during the first 20 s, which may be because the high-energy UV irradiation has a heating effect, and the initial temperature and dryness of the coating surface before sliding are higher[24]. The changes in material properties caused by UV irradiation are mainly because of the high energy UV photons. In the crystal unit layers of MoS_2 and graphite, the atoms connect with each other by strong covalent bonds. The energy of the UV photons is insufficient to cause bond breaking of the two lubricants, so the composite coating still shows excellent tribology properties after UV irradiation.

As shown in Fig. 8c, the variation of friction coefficients of the AO treated samples is quite different from other samples. The initial stage of the friction coefficient curve showed a severe fluctuation, and then decreased and stabilized gradually in the latter part of the curve. The average value of friction coefficients also exceeded 0.1. The MoS_2 and graphite particles were oxidized by the high activity AO: graphite may be oxidized to volatile oxides (CO, CO_2); MoS_2 may be oxidized to MoO_3 and other hard inclusions. All of these changes in structure resulted in the deterioration of the composite coating's tribological properties. However, the oxidization only occurred in the coating's superficial layer, the oxide layer would be removed with the continuation of sliding.

Fig. 9 shows SEM morphologies of the worn surfaces of the substrate (AISI 52100 steel) sample and Ni/MoS_2-C composite coating samples. Fig. 9a is the worn scars on the substrate sample after testing in air. It can be seen clearly that the substrate material without lubricanting coating suffered severe deformation and scuffing. There were not only deep and wide grooves, but also obvious large area adhesion speckles and material accumulations. Both the AISI 52100 steel sample and the upper 440C steel ball have high hardness surfaces, the asperity contact points bear great contact stress during wearing. The asperity peaks will be sheared off and thus form abrasive particles, then more and more large spallings and cuttings will be formed by the cyclic stress and plowing action. At the same time, with the rise of interface temperature, the surface layer of the metal substrate softened gradually and the adhesion strength of contact points increased, which ultimately resulted in severe adhesive wear.

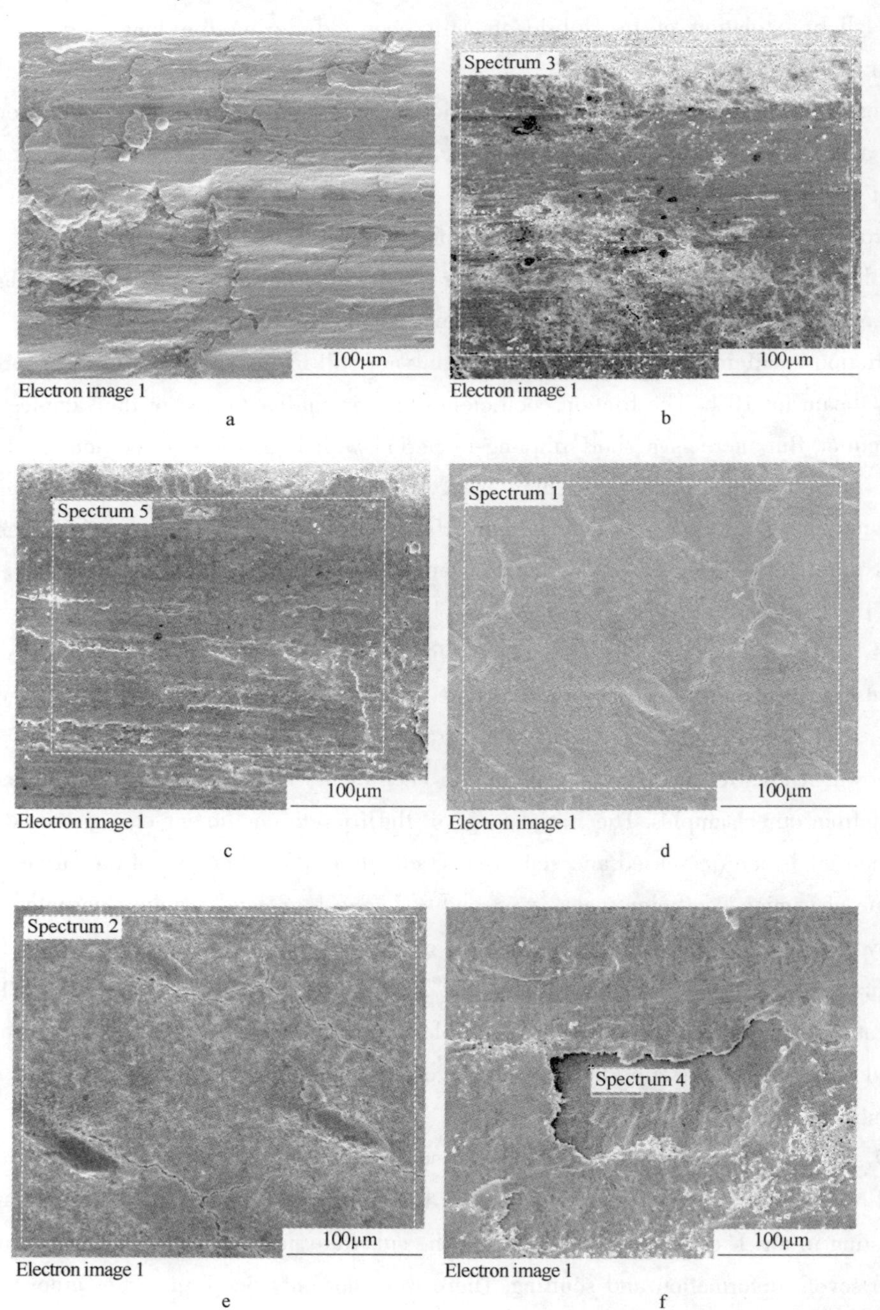

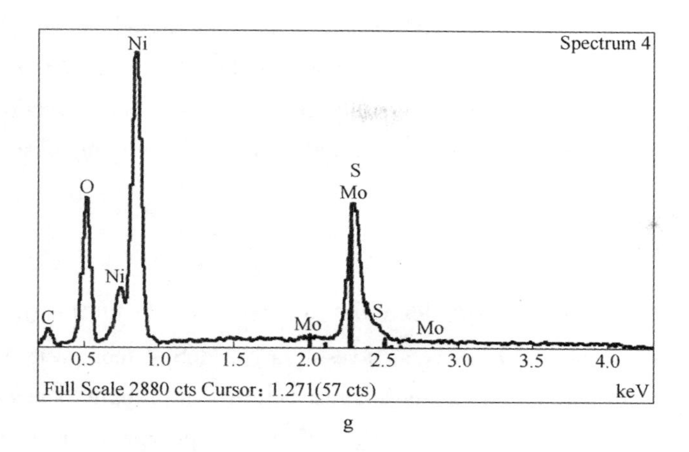

g

Fig. 9 SEM morphologies (a – f) and EDS spectra (g) of the worn surfaces after testing in different conditions
a—The substrate AISI 52100 steel after wear testing in air; b—The as-received coating in air;
c—The moisture treated coating in air; d—The as-received coating in vacuum; e—The UV treated coating in low vacuum;
f—The AO treated coating in low vacuum; g—EDS analysis corresponding to f

All the worn coated surfaces tested in the various conditions showed mild wear behaviors and were macroscopically smooth and compact, as shown in Fig. 9b-f.

Corresponding to different test conditions, the worn surface morphologies can be approximately divided into two sorts: the worn surfaces in air and that in vacuum had their respective similar morphologies.

Fig. 9b and c are the worn surfaces of the Ni/MoS_2-C composite samples before and after storing in humid air. The dominant wear damage in air conditions was caused by plowing, cutting and plastic deformation. There are some pores in the coating's microstructure and the relative strength of the soft coating is low. Under the co-action of contact stress and shearing stress, the surface layer of the composite coating would be crushed and cut. But the resulting debris particles containing abundant lubricants could be transferred continually to the contact zone and play a good role in friction reduction. Within the test duration, the spalling and cutting of the composite coating only occurred in the top layer, and the direct contact between basal metal surfaces did not occur, so the friction coefficient was always low and the wear damage was mild.

Fig. 9d, e and f shows the worn surfaces of the Ni/MoS_2-C composite coating after testing in HV, LV-AO and LV-UV environments. Though some microcracks can be observed, most worn surfaces were still covered by the composite coating. Large-area spalling of the coating only occurred on the AO treated sample, as shown in Fig. 9f. So the dominant wear mechanism of the composite coating in vacuum is fatigue wear. In vacuum, the wear debris containing lubricants can easily be transferred away by the vacuum pumping effect, and thus the subsurface layers of the coating are exposed and scratched continually. Therefore, the worn surface in vacuum is relatively smooth and clear, but the coating is easy to fatigue under the co-action of periodic scraping, plowing and milling.

The composition of the worn surfaces was similar to the test results of the as-received samples. The content of O element of the AO treated sample increased notably, but there was no peak of Fe element in the spalling region, as shown in Fig. 9g. It can be speculated that, the coating was oxidized after AO exposure and thus reduced its friction reduction and fatigue resistance properties. However, the oxidized layer and spalling pits are only observed in the superficial layer.

4 Conclusions

(1) The micro and nano MoS_2-C particles were co-deposited with the matrix Ni metal coating using nano electro-brush plating process, as a result, a Ni/MoS_2-C composite coating about 100 μm thick with a smooth surface and a fine microstructure was prepared successfully.

(2) The two lubricants in the coating, which is MoS_2 with poor moisture resistance properties and graphite with good absorption properties, showed good synergistic effects. The additive graphite particles not only can improve the moisture resistance of the composite coating, but also can reduce the residual stress and enhance the mechanical properties of the coating.

(3) The Ni/MoS_2-C composite coating still shows good friction reduction properties after long-duration treatments in humid air, high vacuum, atomic oxygen and ultraviolet irradiation environments. The dominant wear mechanism of the coating in air is abrasive wear, while its wear damage mainly resulted from plastic deformation and fatigue spalling in vacuum. The coating is oxidized by atomic oxygen and thus has a higher friction coefficient and increased wear rate. However, the oxidized layer and spalling pits are only observed in the superficial layer and the coating still shows very good tribological properties after a short running-in period.

(4) Due to its excellent tribological properties and good adaptability in harsh environments, the Ni/MoS_2-C composite coating is expected to be widely used in severe space conditions.

Acknowledgements

Authors are grateful for the financial support by the Distinguished Young Scholars of NSFC (51125023), the 973 Project (2011CB013403) and the NSF of Beijing (3120001).

References

[1] G. Z. Ma, B. S. Xu, H. D. Wang, D. X. Yang, Rare Metal Mater. Eng. 41 (2012) 331.
[2] T. Endo, T. Iijima, Y. Kaneko, M. Nishimura, Wear 190 (1995) 219.
[3] X. L. Zhang, R. G. Vitchev, W. Lauwerens, L. Stals, J. He, J. P. Celis, Thin Solid Films 396 (2001) 69.
[4] C. Donnet, Surf. Coat. Technol. 80 (1996) 151.
[5] S. Carrera, O. Salas, J. J. Moore, A. Woolverton, E. Sutter, Surf. Coat. Technol. 167 (2003) 25.
[6] M. C. Smimonds, A. Savan, E. Pfluger, H. Van Swygenhoven, Surf. Coat. Technol. 126 (2000) 15.
[7] B. S. Xu, Surf. Eng. 26 (2010) 129.
[8] X. J. Liu, B. S. Xu, S. N. Ma, Z. G. Chen, Wear 211 (1997) 151.
[9] B. S. Xu, H. D. Wang, S. Y. Dong, B. Jiang, Mater. Lett. 60 (2006) 710.
[10] J. Tan, T. T. Yu, B. S. Xu, Q. Yao, Tribol. Lett. 21 (2006) 107.
[11] B. S. Xu, China Patent: 02101196. 6.

[12] P. Bala'zˇ, E. Godocˇ1'kova', L. Kril'ova', P. Lobotka, E. Gock, Mater. Sci. Eng. , A 386(2004)442.
[13] J. Temuujin, M. Aoyama, M. Senna, T. Masuko, C. Ando, H. Kishi, J. Solid State Chem. 177(2004)3903.
[14] B. Wu, B. S. Xu, B. Zhang, X. D. Jing, C. L. Liu, Mater. Lett. 60(2006)1673.
[15] W. Y. Tu, B. S. Xu, S. Y. Dong, H. D. Wang, Mater. Lett. 60(2006)1247.
[16] B. S. Xu, H. D. Wang, G. Z. Ma, China Patent:201110106243. X.
[17] C. H. Pu, B. S. Xu, H. D. Wang, Z. Y. Piao, Tribology 29(2009)368.
[18] Z. Y. Piao, B. S. Xu, H. D. Wang, C. H. Pu, Appl. Surf. Sci. 257(2011)2581.
[19] M. K. Kazmanli, B. Rother, M. Urgen, C. Mitterer, Surf. Coat. Technol. 107(1998)65.
[20] X. L. Pang, K. W. Gao, F. Luo, Y. Emirov, A. A. Levin, A. A. Volinsky, Thin Solid Films 517(2009)1922.
[21] B. S. Xu, H. D. Wang, Y. N. Song, China Patent:201210002668. 0.
[22] H. Chang, E. H. Han, J. Q. Wang, Int. J. Fatigue 31(2009)403.
[23] S. K. Singh, K. Srinivasan, D. Chakraborty, Mater. Des. 24(2003)471.
[24] X. H. Zhao, Z. G. Shen, Y. S. Xing, J. Phys. D: Appl. Phys. 34(2001)2308.

Effect of Heat Treatment on Microwave Absorption Properties of Ni-Zn-Mg-La Ferrite Nanoparticles*

Abstract Spinel structure Ni-Zn-Mg-La ferrites have been prepared by the sol-gel route and investigated as a radar absorbing material (RAM) in a frequency range of 1 – 18 GHz. The structure and morphological studies on the nanoparticles of the ferrites have been carried out using X-ray diffraction, scanning electron microscopy and X-ray photoelectron spectroscopy. The complex permeability and complex permittivity are measured by a network analyzer. The electromagnetic wave loss and microwave absorbing property are studied as a function of frequency, annealing temperature and thickness of the absorber. The results indicate that electromagnetic wave loss of the ferrite only annealed at 850 ℃ shows two peaks. The reflection loss varies with the change of the annealing temperature. The absorber annealed at 850 ℃ exhibits the best microwave absorbing properties, which is suitable for microwave absorption materials.

Key words sol-gel growth, magnetic materials, heat treatment

1 Introduction

The production of electromagnetic wave absorbers has recently been increasing[1]. Among these absorbers, spinel ferrites, which can be used in 3 – 30 GHz band, are becoming one of the versatile magnetic materials for general use[2]. Spinel Ni-Zn ferrites have many applications in both low and high frequency devices and play an important role in many technological applications, due to their high resistivity, low dielectric loss, mechanical hardness, high Curie temperature and chemical stability[3]. Various methods have been proposed for the synthesis of sipnel Ni-Zn ferrite in the past several years, such as the sol-gel preparation[4,5], low temperature solid-state reaction[6,7], co-precipitation method[8] and high energy milling method[9]. Among these technologies, the sol-gel route is a method which can prepare pure ferrite at a relatively low temperature.

The magnetic and electrical properties of ferrites are sensitive to the preparation method and the distribution of cations[10]. Small amount of additives can be used to modify their microstructure and hence magnetic properties. Nowadays rare earth oxides are becoming promising additives to improve the magnetic properties of ferrite. Many investigations have been carried out to explore the effect of La substitution on the properties of Ni-Zn ferrites. However, in some literature conflict results are obtained[11,12]. Ahmed studied the substitution of rare earth La^{3+} into the spinel structure of Ni-Zn ferrites. In his work, a secondary phase appeared due to the substitution of

---

* Copartner: Liu Yi, Wei Shicheng, Wang Yujiang, Tian Haoliang, Tong Hui. Reprinted from *Journal of Magnetism and Magnetic Materials*, 2014 (349): 57 – 62.

La^{3+}, and dielectric constant of the ferrite showed more than one peak[13]. However, in Gable's study of the structural and magnetic properties of La substituted NiCuZn ferrites, no secondary phase was detected in XRD patterns for the calcined samples even at higher La contents[14].

Heat treatment process is an important influencing factor for the magnetic property of ferrite. Ichiyanagi investigated Mg-ferrite nanoparticles and found a clear difference in the magnetization between the quenched samples and annealed samples[15]. Pozo López studied the magnetic properties of NiZn ferrite/SiO_2 nanocomposites synthesized by ball milling and found that complete transformation of the precursor oxides into NiZn ferrite was only achieved after the as-milled powders were annealed at 1273 K in air for 1h. This heat treatment favored the formation of Ni-Zn ferrite in detriment of the precursor oxides[16]. Nevertheless, in some other literature Ni-Zn ferrite formed the spinel structure at around 350 ℃ when prepared by the sol-gel method[17]. Therefore, further studies on the effect of heat treatment and the substitution of La^{3+} on the magnetic and microwave absorption property of the ferrite still need to be done. Although researchers have already synthesized Ni-Zn-La ferrite[18] and Ni-Zn-Mg ferrite[19], Ni-Zn-Mg-La ferrite is still not studied. In this paper, Ni-Zn-Mg-La ferrite is synthesized by the sol-gel methods. The aim of our work is to describe and evaluate the effect of heat treatment on magnetic and microwave absorbing properties of the ferrite.

2 Experimental

2.1 Synthesis of ferrite

Nanoparticles ferrites have been synthesized by the conventional sol-gel method. Analytical grade metal nitrates and citric acid are used as raw materials. The molar ratios of Ni^{2+} : Zn^{2+} : Mg^{2+} : La^{3+} : Fe^{3+} are 0.5 : 0.4 : 0.1 : 0.01 : 1.99, which gives a composition of $Ni_{0.5}Zn_{0.4}Mg_{0.1}La_{0.01}Fe_{1.99}O_4$. The nitrates and citric acid are weighed in desired stoichiometric proportions and dissolved separately in the minimum amount of distilled water. The reactants are mixed together and ammonia solution is added to the solution drop by drop to adjust the pH value of the mixture to 6. The solution is slowly heated and stirred using a hot plate magnetic stirrer till it turns into a dark vicous liquid and then it is dried at 120 ℃ for 24 h. The dried gels are ignited in order to obtain loose powders. These as burnt ferrite powders are labeled as No. 1. The heat treatment process is as follows: the powders are annealed at 650 ℃, 750 ℃ and 850 ℃ for 2 h and labeled as No. 2, No. 3 and No. 4, respectively. Then these powders are cooled down to room temperature naturally.

2.2 Characterization

The crystal structure of the obtained particles is recorded by X-ray diffraction (XRD) using a Rigaku model D/max 2500 system with $\lambda = 0.154$ nm (Cu K_α radiation). The morphology is analyzed by a HITACHI S – 5500 field emission scanning electron microscope (SEM). The composition analysis of the ferrites is performed by energy dispersive spectra (EDS, OXFORD Feature Max). The valence states of elements are analyzed by X-ray photoelectron spectroscopy

(XPS, VG Scientific ESCALAB 220i – XL, USA). The complex permeability and complex permittivity are measured in the range of 1-18 GHz by an HP8722ES network analyzer. For this purpose, the Ni-Zn-Mg-La ferrite powders are homogeneously dispersed into the wax matrix and compacted into rings for the permeability and permittivity measurement. The size of the ring is 7 mm in outer diameter, 3 mm in inner diameter and 2 mm in thickness. The ferrite-wax composites contain 60% of ferrite (wt%). Static magnetic properties are studied using a Lake Shore 7410 vibrating sample magnetometer (VSM) with a maximum applied magnetic field of 10 kOe.

3 Results and Discussion

3.1 Structure and morphological study

The X-ray diffraction patterns of the powders are shown in Fig. 1. The existence of (311) peak around 35° confirms the formation of spinel structure in the prepared samples. It is found that all the peaks could be indexed to a spinel phase. Lima reported that Ni-Zn ferrite calcined in argon atmosphere at 1000 ℃/3 h showed single spinel phase structure[20]. In this work, the Ni-Zn-Mg-La ferrite powders annealed at 650 ℃/2 h already have single spinel phase. In some literature[21], secondary phase $LaFeO_3$ formed upon La substitution for Fe in the ferrite. However, in our research, the secondary phase is not found in all the samples. The sharp and strong diffraction peaks also confirm the good crystallization of the products. The average crystallite sizes of the samples are calculated using Scherer's relation for the strongest peak of the (311) plane[22]:

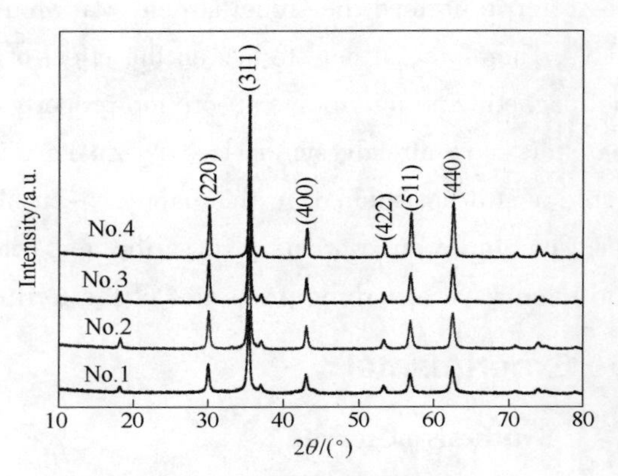

Fig. 1 X-ray diffraction of Ni-Zn-Mg-La ferrites

$$D = 0.9\lambda/(\beta\cos\theta) \tag{1}$$

where D is the crystallite size of the particle in nanometer; β is the half-maximum line width; θ is Bragg angle of diffraction; λ is the wavelength of radiation. The calculated crystallite size of the particles is exhibited in Table 1. It is observed that the average crystallite size increases with adding temperature. This result is in accordance with the study of Yousefi[23]. This is due to the grain growth of the ferrites.

Table 1 Calculated grain size of Ni-Zn-Mg-La ferrite

Sample	No. 1	No. 2	No. 3	No. 4
Crystallite size/nm	20	20	23	26

The microstructures of magnetic Ni-Zn-Mg-La ferrites annealed at different temperatures are displayed in Fig. 2a-d. SEM observations show that nanocrystallites of the ferrites are spherical in morphology. The average particle sizes of the ferrites are in the range of 40 – 80 nm. The grain size is larger than data estimated by the Scherer formula because the instrument errors are not taken into account. Moreover, the difference is indicative of the fact that every particle is formed by the aggregation of a number of crystallites or grains[24]. The surface composition of Ni-Zn-Mg-La ferrite is distinctly determined with EDS. Fig. 2e shows the composition of the Ni-Zn-Mg-La ferrite. The predominant composition is made up of iron and oxygen. In order to determine the valence states of the elements, surface/near surface of the ferrite is analyzed by XPS within a range of binding energies of 0 – 1400 eV. Core levels of Ni2p, Zn2p, Mg1s, and Fe2p can be identified in Fig. 3a. The fine spectra of the Fe2p peaks are displayed in Fig. 3b. The $Fe2p_{3/2}$ spectrum and $Fe2p_{1/2}$ spectrum obtained from the present study generally

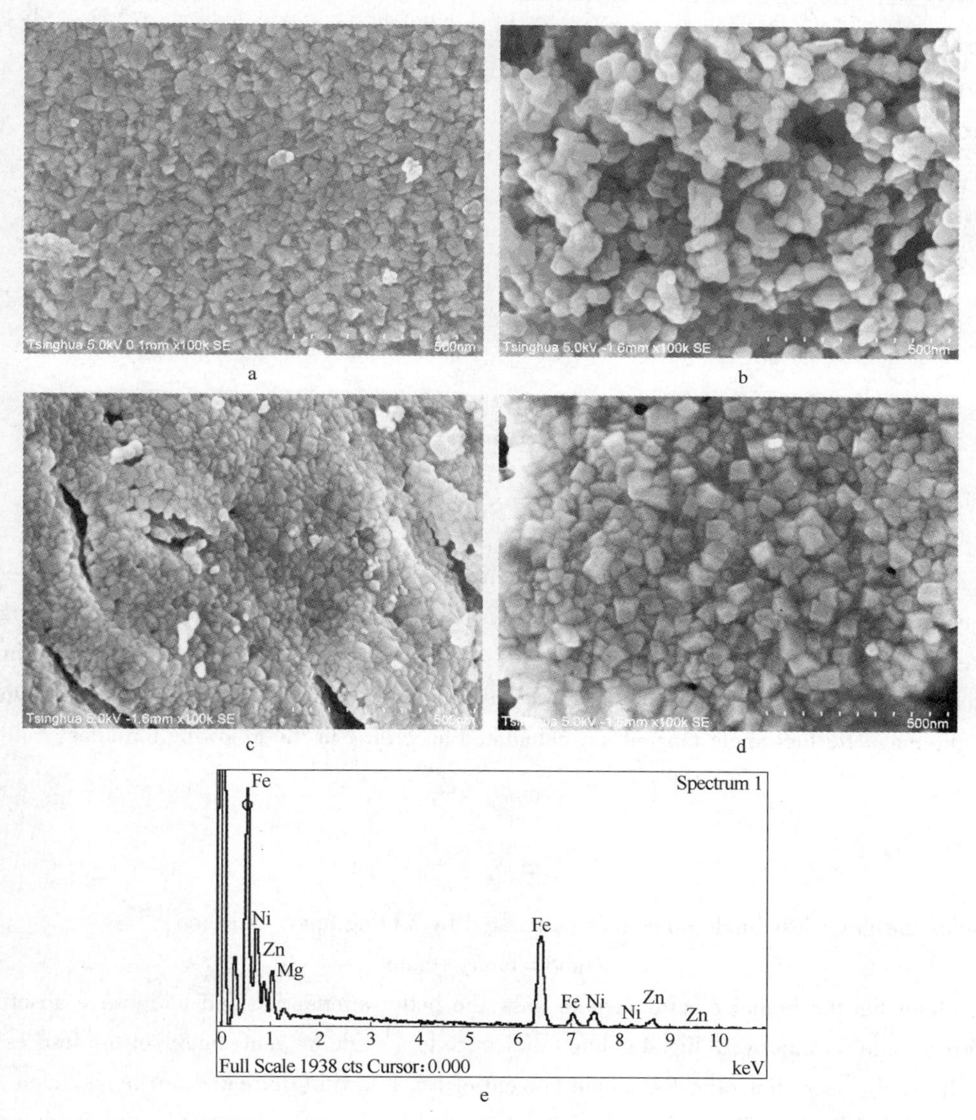

Fig. 2 SEM morphology and EDS composition analysis of Ni-Zn-Mg-La ferrites

a—No. 1 ferrite; b—No. 2 ferrite; c—No. 3 ferrite; d—No. 4 ferrite; e—EDS analysis of No. 1 ferrite

show two distinguishable main peaks of around 710.4 eV and 723.5 eV, respectively, which demonstrates the presence of Fe^{3+} cation. Meanwhile, the presence of the peak around 713.9 eV indicates that Fe^{3+} species exist in more than one chemical state. The two chemical states may be related to the different coordination environments of Fe^{3+} – the tetrahedral (A) environment and octahedral (B) environment of Fe^{3+} cations in spinel structure: ${Fe_A}^{3+}$ at higher binding energy and ${Fe_B}^{3+}$ at lower binding energy[25]. In this ferrite, the number of ${Fe_B}^{3+}$ cations is much more than that of ${Fe_A}^{3+}$ cations. This result is in accordance with Priyadharsini's report[24], which used X-ray diffraction to detect the lattice constant and cation distribution of $Ni_xZn_{1-x}Fe_2O_4$. In his report the percentage of ${Fe_A}^{3+}$ is only 27% when $x = 0.6$.

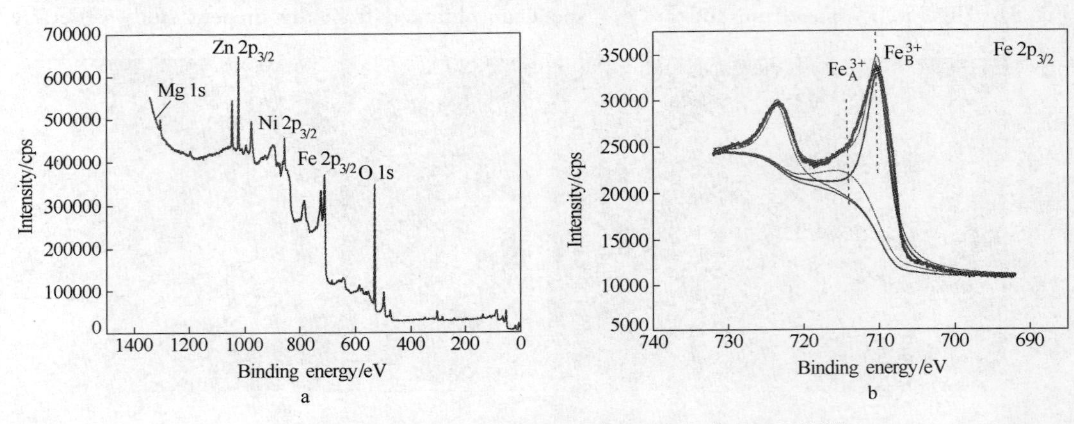

Fig. 3 XPS spectra of Ni-Zn-Mg-La ferrite annealed at 850 ℃

a—Survey scan; b—Fe 2p

3.2 Microwave absorption properties of Ni-Zn-Mg-La ferrite

It is well known that the basic principle of microwave absorption property is to consume electromagnetic wave energy by increasing the energy conversion when the electromagnetic wave signals move to the microwave absorption material. The electromagnetic wave loss is characterized by the dielectric loss angle tangent and the magnetic loss angle tangent. The dielectric loss angle tangent and the magnetic loss angle tangent are calculated according to the following formulas:

$$\tan\delta_\mu = \frac{\mu''}{\mu'} \tag{2}$$

$$\tan\delta_\varepsilon = \frac{\varepsilon''}{\varepsilon'} \tag{3}$$

Besides the gross loss angle tangent is calculated by adding $\tan\delta_\varepsilon$ and $\tan\delta_\mu$[26]:

$$\tan\delta = \tan\delta_\varepsilon + \tan\delta_\mu \tag{4}$$

It is clear that the higher electromagnetic loss, the better attenuation and microwave absorption of ferrites can be improved. Fig. 4a shows dielectric loss angle tangent ($\tan\delta_\varepsilon$) of the ferrites system. It can be seen dielectric loss angle tangent of No. 1 ferrite (untreated ferrite) is much more than annealed ferrites. The value of $\tan\delta_\varepsilon$ is similar to annealed ferrites in the frequency of 1 – 10 GHz. However, there is some difference for the annealed ferrites in the frequency of 10-

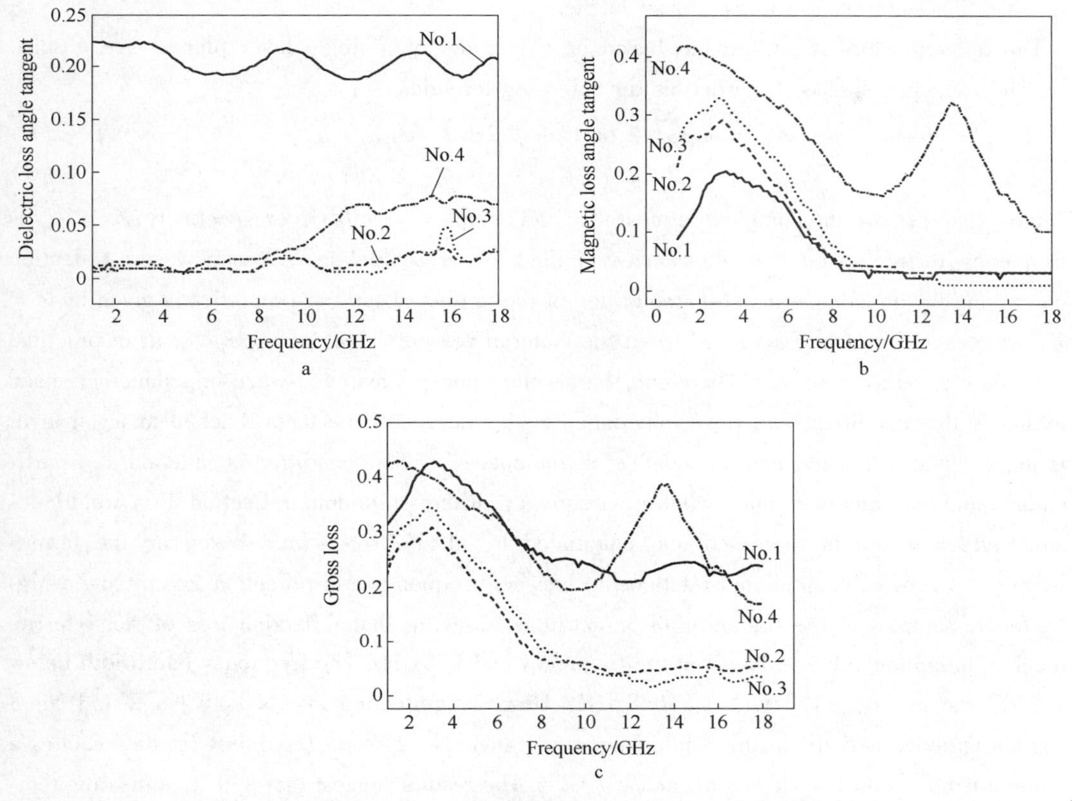

Fig. 4 The electromagnetic wave loss of Ni-Zn-Mg-La ferrites

a—Dielectric loss angle tangent; b—Magnetic loss angle tangent; c—Gross loss

18 GHz. Dielectric loss angle tangent of No. 4 ferrite (annealed at 850 ℃) increases much more than the ferrite annealed at 650 ℃ and 750 ℃. Fig. 4b exhibits the magnetic loss angle tangent of Ni-Zn-Mg-La ferrite. The mechanism of magnetic loss is the lag of domain walls with respect to the applied alternating field and imperfections in the lattice[27]. It can be seen that the magnetic loss angle tangent increases with adding temperature. It is attributed to the increase of the crystallization degree. Moreover, the ferrite annealed at 850 ℃ shows another magnetic loss peak in the frequency of 10 – 16 GHz. However, this peak is not found in No. 1, No. 2 and No. 3 ferrites. Fig. 4c shows the gross loss of Ni-Zn-Mg-La ferrites. In this picture, No. 4 ferrite (annealed at 850 ℃) shows two electromagnetic loss frequency bands: 1 – 8 GHz and 12 – 16 GHz. The peak value of gross loss is 0. 429 at the frequency of 1. 5 GHz and 0. 387 at the frequency of 13. 4 GHz, meanwhile, only one gross loss peak is found in No. 1, No. 2 and No. 3 ferrites. The peak values of gross loss of No. 1, No. 2 and No. 3 ferrites are 0. 427, 0. 308 and 0. 345, respectively. The peak of electromagnetic loss happens when the applied field frequency matches with the processing frequency of magnetic spins in ferrites. This matching leads to energy transfer from the field to the ferrite system in orienting the dipole and thus resonance occurs at the proper frequency[27,28]. Moreover, the value of magnetic loss is larger than dielectric loss for the annealed ferrite, which means magnetic loss plays a dominant role in gross loss. The

result is just contrary to the as – burnt ferrite.

The reflection loss is carried out based on the model of a single layer plane wave absorber. The reflection loss is described as the following formulas[29]:

$$Z_{in} = Z_0\sqrt{\mu_r/\varepsilon_r}\tanh[-j(2\pi f d_m/c)\sqrt{\varepsilon_r\mu_r}] \quad (5)$$

$$R = -20\lg|(Z_{in} - Z_0)/(Z_{in} + Z_0)| \quad (6)$$

where μ_r and ε_r are the complex permeability and complex permittivity, respectively; $Z_0 = 1$; f is frequency, in free space; c is the velocity of light and d is thickness. To satisfy zero-reflection condition, the impedance matching condition of the perfect absorbing properties is given by $Z = Z_0$, which can be ideally achieved when the material presents $|\mu_r| = |\varepsilon_r|$, but in practical condition, is rarely achieved. Therefore, the second concept (matched-wave-impedance) is used to satisfy this condition. The wave impedance at the surface of the metal-backed material layer is made equal to the intrinsic impedance of free space[19]. This condition is satisfied at a particular matching thickness and matching frequency, where minimum reflection loss would occur. Reflection loss is simulated and calculated by MATLAB software based on the formula. Fig. 5a shows the simulated relationship between frequency and reflection loss of Ni-Zn-Mg-La ferrite samples at the thickness of 3 mm. It is observed that reflection loss of No. 4 ferrite reaches minimum value –10 dB at the frequency of 13.7 GHz. The frequency bandwidth below –5 dB ranges from 11.2 GHz to 16.8 GHz. However, reflection loss of both No. 2 and No. 3 ferrites is above –5 dB in the whole frequency range. No. 1 ferrite (as burnt ferrite) reaches a minimum reflection loss of –8 dB at 12.7 GHz. The results suggest that 850 ℃ annealing temperature is benefit for microwave absorption property of the ferrite system. Therefore No. 4 ferrite can be used as RAM in X-band frequency and Ku-band frequency. It should be noted that NiZn ferrites were prepared through the sol-gel method in Ting's literature[30]. The ferrite powders were calcined at 900 ℃ for 2 h. The minimum reflection loss was –13 dB at 7.5 GHz with a –5 dB bandwidth in the frequency range of 2 – 12 GHz. In this work, the corresponding reflection frequency under –5dB is in the range of 11 – 17 GHz, which extends the frequency to Ku-band. If combined with NiZn ferrite, the frequency bandwidth of the composite nearly covers the whole 2 – 18 GHz.

Furthermore, microwave-absorbing property is not only related to the absorber, but also related to the thickness of the material and the frequency[31]. The above-mentioned formulas indicate there is a matching thickness at a special frequency for a magnetic material. Within the microwave region, minimum reflection loss of ferrite materials usually occurs at around a quarter-wavelength thickness of the material. There is a proper loss factor for this particular thickness. The minimum reflections can be obtained at a given frequency if the thickness of the absorber satisfies the following equation[32]:

$$d = \frac{nc}{4f_0\sqrt{|\mu_r||\varepsilon_r|}} \quad (n = 1,3,5,\cdots) \quad (7)$$

where ε_r, μ_r, f_0, c and d are the complex permittivity, complex permeability, frequency, light velocity and absorber thickness, respectively. Fig. 5b exhibits the reflection loss at different thick-

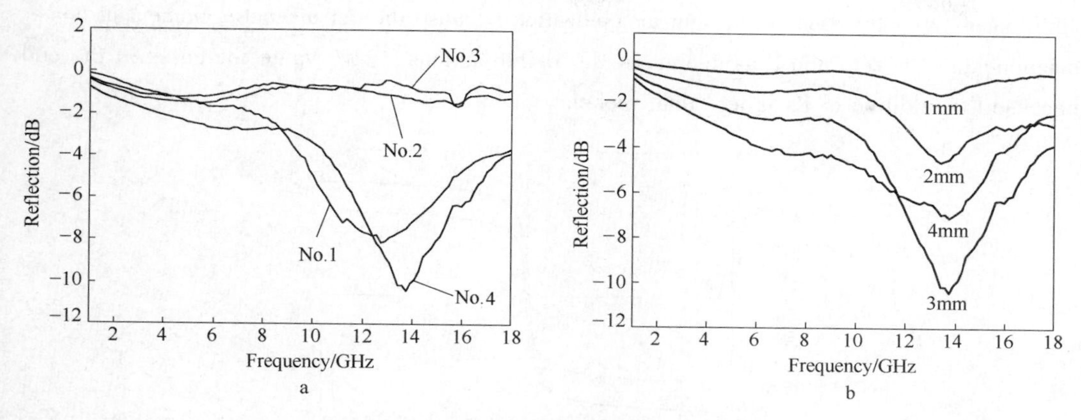

Fig. 5 Reflection loss of Ni-Zn-Mg-La ferrites

a—The ferrite annealed at different heat treatment temperature; b—No. 4 ferrite at different thickness

nesses of No. 4 ferrite. The ferrite achieves minimum reflection loss at the thickness of 3mm. If the thickness increases continually, on the contrary, the reflection loss would decrease. It is an unrealistic goal to add the weight of RAM in aircraft. In this work, the minimum reflection loss increases and subsequently decreases with adding thickness. According to the microwave absorption theory, when an electromagnetic wave strikes the surface of an absorber, it can be partially transmitted and partially reflected. The partially transmitted wave is reflected on the surface of the conductive layer, generating multiple internal reflections. These internal reflections produce multiple waves that emerge from the absorber surface, out-of-phase by 180°. Considering the frequency of interest, if the absorber's thickness is equal to d, the sum of the emerging wave and the reflected electromagnetic wave is canceled out at the air – interface on the surface of absorber. If the thickness is not equal to d, complete canceling does not occur and the microwave absorption is lower[19].

3.3 Magnetic study

The hysteresis loops are measured in order to determine parameters such as saturation magnetization (M_s), remnant magnetization (M_r) and coercivity (H_c). Fig. 6 gives the magnetic hysteresis loops of the present ferrite system at room temperature. This curve is typical of a soft magnetic material. The M_s values of both No. 1 and No. 4 ferrites tend to be saturated. The M_s of No. 4 ferrite reaches 63 emu/g while M_s of No. 1 ferrite is only about 48 emu/g. These physical properties are related to the size increment due to heat treatment. The M_s value of our $Ni_{0.5}Zn_{0.4}Mg_{0.1}La_{0.01}Fe_{1.99}O_4$ ferrite is larger than $Ni_{0.45}Zn_{0.45}Mg_{0.1}Fe_2O_4$ ferrite synthesized by Peng[33] and $Ni_{0.4}Zn_{0.5}Mg_{0.1}Fe_2O_4$ ferrite synthesized by Singh[34]. The M_s values of $Ni_{0.45}Zn_{0.45}Mg_{0.1}Fe_2O_4$ ferrite and $Ni_{0.4}Zn_{0.5}Mg_{0.1}Fe_2O_4$ ferrite are 42.85 emu/g and 36 emu/g, respectively. The difference may be related to higher nickel content in our ferrite than the other two ferrites. It can be explained that the addition of nickel favors the A-B exchange energy and dilutes the B – B super exchange energy because Ni^{2+} prefers to occupy B – site[24]. However, the addition of La^{3+} decreases saturation magnetization because the net magnetic moment of non – magnetic La^{3+} is zero and thus decreases B – B interactions[27]. M_s value increases in the end because the addition of La is less than nickel.

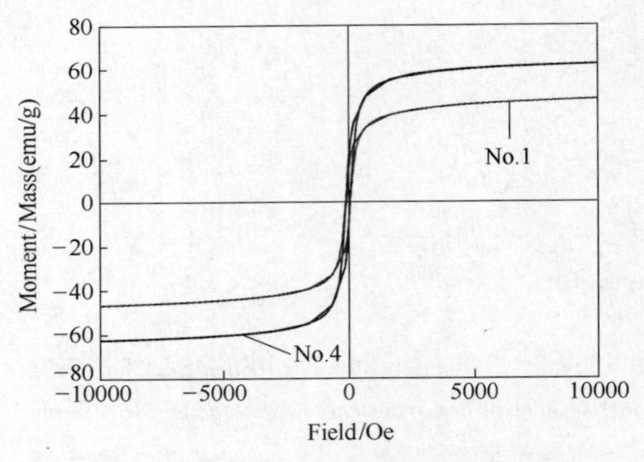

Fig. 6 Hysteresis loops of Ni-Zn-Mg-La ferrite

4 Conclusions

Nanocrystalline particles of Ni-Zn-Mg-La ferrites have been successfully synthesized by the sol – gel auto combustion route. The microstructure and microwave absorbing properties of the ferrites clearly show a dependence on the heat treatment. The present samples exhibit single – phase spinel structure and no second phase is found in this study. XRD analysis reveals the average crystallite size of these particles is in the range of 20 – 26 nm. The crystallite size increases with the increase of the annealing temperature. It is found that electromagnetic loss and reflection loss of the ferrite system are greatly improved by increasing the annealing temperature. The ferrite annealed at 850 ℃ shows microwave absorption property. Electromagnetic wave loss of the ferrite exhibits two microwave absorption peaks: 0.429 at the frequency of 1.5 GHz and 0.387 at the frequency of 13.4 GHz, respectively. The matching thickness of the ferrite is 3 mm. The hysteresis loops demonstrate the ferrite is typical of a soft magnetic material and M_s value of the ferrite annealed at 850 ℃ is 63 emu/g.

Acknowledgements

The paper is financially supported by 973 Project (2011CB013403) and National Natural Science Foundation of China (No. 51222510). The authors gratefully extend thank to the support of all members of the project working group.

References

[1] M. Begard, K. Bobzin, G. Bolelli, Thermal spraying of Co, Ti – substituted Ba – hexaferrite coatings for electromagnetic wave absorption applications, Surface and Coatings Technology. 203 (2009) 3312 – 3319.

[2] Martha Pardavi – Horvath. Microwave applications of soft ferrites, Journal of Magnetism and Magnetic Materials. 215 – 216 (2000) 171 – 183.

[3] K. H. Wu, F. C. Yang, Synthesis and characterization of organically modified silicate/NiZn ferrite hybrid

coatings, Acta Materialia. 55(2007)507 – 515.

[4] X. H. He, G. S. Song, J. H. Zhu, Non – stoichiometric NiZn ferrite by sol – gel processing, Materials. Letters. 59(2005)1941 – 1944.

[5] A. Verma, T. C. Goel, R. G. Mendiratta, M. I. Alam, Dielectric properties of NiZn ferrites prepared by the citrate precursor method, Materials Science and Engineering. B60(1999)156 – 162.

[6] J. Hu, M. Yan, W. Luo, Preparation of high – permeability NiZn ferrites at low sintering temperatures, Physica B368(2005)251 – 260.

[7] Amiri Gh. R., M. H. Yousefi, M. R. Abolhassani, S. Manouchehri, M. H. Keshavarz, S. Fatahian, Magnetic properties and microwave absorption in Ni – Zn and Mn – Zn ferrite nanoparticles synthesized by low – temperature solid – state reaction, Journal of Magnetism and Magnetic Materials. 323(2011)730 – 734.

[8] B. Parvatheeswara Rao, O. Caltun, W. S. Cho, Chong – Oh KimCheol Gi Kim. Synthesis and characterization of mixed ferrite nanoparticles, Journal of Magnetism and Magnetic Materials. 310(2007)e812 – e814.

[9] L. M. Yu, J. C. Zhang, Y. S. Liu, C. Jing, S. X. Cao, Fabrication, structure and magnetic properties of nanocrystalline NiZn – ferrite by high – energy milling, Journal of Magnetism and Magnetic Materials. 288(2005)54 – 59.

[10] S. F. Mansour, Structural and magnetic investigations of sub – nano Mn – Mg ferrite prepared by wet method, Journal of Magnetism and Magnetic Materials. 323(2011)1735 – 1740.

[11] M. A. Ahmed, N. Okasha, M. M. El – Sayed, Enhancement of the physical properties of rare – earth – substituted Mn – Zn ferrites prepared by flash method, Ceramics International. 33(2007)49 – 58.

[12] E. Rezlescu, N. Rezlescu, P. D. Popa, Fine – grained MgCu ferrite with ionic substitutions used as humidity sensor, Journal of Magnetism and Magnetic Materials. 290 – 291(2005)1001 – 1004.

[13] M. A. Ahmed, E Ateia, L. M. Salah, A. A. Ei – Gamal, Structural and electrical studies on La^{3+} substituted Ni – Zn ferrites, Materials Chemistry and Physics. 92(2005)310 – 321.

[14] M. A. Gabal, A. M. Asiri, Y. M. Al Angari, On the structural and magnetic properties of La – substituted NiCuZn ferrites prepared using egg – white, Ceramics International. 37(2011)2625 – 2630.

[15] Y. Ichiyanagi, M. Kubota, S. Moritake, Y. Kanazawa, T. Yamada, T. Uehashi, Magnetic properties of Mg – ferrite nanoparticles, Journal of Magnetism and Magnetic Materials. 310(2007)2378 – 2380.

[16] G. Pozo López, S. P. Silvetti, S. E. Urreta, A. C. Carreras, Structure and magnetic properties of NiZn ferrite/SiO_2 nanocomposites synthesized by ball milling, Journal of Alloys and Compounds. 505(2010)808 – 813.

[17] X. H. He, Q. Q. Zhang, Z. Y. Ling, Kinetics and magnetic properties of sol – gel derived NiZn ferrite – SiO_2 composites, Materials Letters 57(2003)3031 – 3036.

[18] Q. L. Li, Y. F. Wang, C. B. Chang, Study of Cu, Co, Mn and La doped NiZn ferrite nanorods synthesized by the coprecipitation method, Journal of Alloys and Compounds. 505(2010)523 – 526.

[19] A. R. Bueno, M. L. Gregori, M. C. S. Nóbrega, Microwave – absorbing properties of $Ni_{0.50-x}Zn_{0.50-x}Me_{2x}Fe_2O_4$ (Me = Cu, Mn, and Mg) ferrite – wax composite in X – band frequencies, Journal of Magnetism and Magnetic Materials. 320(2008)864 – 870.

[20] U. R. Lima, M. C. Nasar, M. C. Rezende, J. H. Araugo. Ni – Zn nanoferrite for radar – absorbing material, Journal of Magnetism and Magnetic Materials. 320(2008)1666 – 1670.

[21] P. K. Roy, B. B. Nayak, J. Bera. Study on electro – magnetic properties of La sbstituted Ni – Cu – Zn ferrite synthesized by auto – combustion method, Journal of Magnetism and Magnetic Materials. 320(2008) 1128 – 1132.

[22] H. T. Zhao, X. D. Sun, C. H. Mao, J. Du, Preparation and microwave – absorbing properties of $NiFe_2O_4$ –

polystyrene composites, Physica B404(2009)69 – 72.

[23] M. H. Yousefi, S Manouchehri, A Arab, M Mozaffari, Gh. R. Amiri, J Amighian. Preparation of cobalt – zinc ferrite($Co_{0.8}Zn_{0.2}Fe_2O_4$) nanopowder via combustion method and investigation of its magnetic properties, Materials Research Bulletin. 45(2010)1792 – 1795.

[24] P. Priyadharsini, A. Pradeep, G. Chandrasekaran. Novel combustion route of synthesis and characterization of nanocrystalline mixed ferrites of Ni – Zn, Journal of Magnetism and Magnetic Materials. 321(2009) 1898 – 1903.

[25] F. Li, X. F. Liu, Q. Z. Yang, J. J. Liu, David G. Evans, D. Xue. Synthesis and characterization of $Ni_{1-x}Zn_xFe_2O_4$ spinel ferrites from tailored layered double hydroxide precuresors, Materials Research Bulletin. 40(2005)1244 – 1255.

[26] Y. Q. Li, Y. Huang, S. H. Qi, F. F. Niu, L. Niu, Preparation and magnetic and electromagnetic properties of La – doped strontium ferrite films, Journal of Magnetism and Magnetic Materials. 323(2011) 2224 – 2232.

[27] Pawan Kumar, S. K. Sharma, M. Knobel, M. Singh, Effect of La^{3+} doping on the electric, dielectric and magnetic properties of cobalt ferrite processed by co – precipitation technique, Journal of Alloys and Compounds. 508(2010)115 – 118.

[28] A. K. Subramani, N. Matsushita, T. Watanabe, M. Tada, M. Abe, K. Kondo, M. Yoshimura, Spin – sprayed ferrite films with high resistivity and high – frequency magnetic loss for GHz conducted noise suppressors, Materials Science and Engineering. B148(2008)136 – 140.

[29] Harun Bayrakdar, Complex permittivity, complex permeability and microwave absorption properties of ferrite – paraffin polymer composites, Journal of Magnetism and Magnetic Materials. 323(2011)1882 – 1885.

[30] T. H. Ting, R. P. Yu, Y. N. Jau, Synthesis and microwave absorption characteristics of polyaniline/NiZn ferrite composites in 2 – 40 GHz, Materials Chemistry and Physics. 126(2011)364 – 368.

[31] H. Zou, S. H. Li, L. Q. Zhang, S. N. Yan, et al. Determining factors for high performance silicone rubber microwave absorbing materials, Journal of Magnetism and Magnetic Materials. 323(2011)1643 – 1651.

[32] X. G. Huang, J. Zhang, L. X. Wang, Q. T. Zhang. Simple and reproducible preparation of barium hexagonal ferrite by adsorbent combustion method, Journal of Alloys and Compounds. 540(2012)137 – 140.

[33] C. H. Peng, C. C. Hwang, J. Wan, J. S. Tsai, S. Y. Chen, Microwave – absorbing characteristics for the composites of thermal – plastic polyurethane(TPU) – bonded NiZn – ferrites prepared by combustion synthesis method, Materials Science and Engineering. B117(2005)27 – 36.

[34] N. Singh, A. Agarwal, S. Sanghi, P. Singh, Effect of magnesium substitution on dielectric and magnetic properties of Ni – Zn ferrite, Physica B406(2011)687 – 692.

高速电弧喷涂 $NiCrBMoFe/BaF_2 \cdot CaF_2$ 涂层的摩擦磨损性能研究*

摘　要　利用高速电弧喷涂技术制备了 NiCrBMoFe 和 $NiCrBMoFe/BaF_2 \cdot CaF_2$ 两种复合涂层。两种涂层均含非晶相与纳米晶相，非晶相的质量分数分别达 45% 和 33%。NiCrBMoFe 涂层中纳米晶相弥散分布在非晶母相中；$NiCrBMoFe/BaF_2 \cdot CaF_2$ 涂层纳米晶相以团聚形态存在。检测了两种涂层的摩擦磨损性能。在常温下，NiCrBMoFe 和 $NiCrBMoFe/BaF_2 \cdot CaF_2$ 两种涂层的摩擦系数分别为 0.5 和 0.375，后者比基体 $18Cr_2Ni_4WA$ 的摩擦系数下降了 25%，耐磨性能优异。在高温摩擦磨损条件下，$NiCrBMoFe/BaF_2 \cdot CaF_2$ 涂层也具有良好的耐磨性能。在 450℃左右，涂层中 $BaF_2 \cdot CaF_2$ 固体润滑相会析出涂层表面，形成一层低摩擦系数的润滑转化膜，使涂层具有良好的减摩润滑作用。

关键词　Ni 基非晶纳米晶　涂层　摩擦磨损　$BaF_2 \cdot CaF_2$

非晶纳米晶合金材料具有较高的强度和优良的耐磨性能，成为 21 世纪具有广阔应用前景的材料之一[1]。然而，制备大块非晶纳米晶三维合金技术难度很大，以涂层形式应用的性能独特的非晶纳米晶材料[2]是 21 世纪初备受瞩目的热点研究领域。目前，国内外非晶纳米晶涂层的制备多采用等离子喷涂、高速火焰喷涂或冷喷涂[3~8]非晶粉末的方法。在其制备涂层前通常要获得高质量的喷涂粉末，使制备非晶涂层的费效比很高。高速电弧喷涂技术与上述喷涂技术相比，不用预制非晶粉末，喷涂材料为粉芯丝材，丝材经过熔化、高速冷凝（冷却速度高达 $10^5 \sim 10^7 K/s$）后很容易获得非晶纳米晶涂层，在制备大面积非晶纳米晶涂层方面具有独特优势。本文利用高速电弧喷涂技术在 $18Cr_2Ni_4WA$ 基体上制备了 NiCrBMoFe 和 $NiCrBMoFe/BaF_2 \cdot CaF_2$ 两种镍基非晶纳米晶涂层。重点研究了两种涂层的常温摩擦磨损性能和高温摩擦磨损性能。

1　实验部分

1.1　试样制备

试验采用机器人自动化高速电弧喷涂系统。该系统由 Motoman HP20 机器人控制 HAS-02 型高速喷枪和 CMD AS 3000 电源系统组成。喷涂材料为自制的 Ni 基粉芯丝材，其中各组分的质量分数为 NiCr 60% ~75%，BFe 10% ~18%，Mo 1% ~10%，以及 $BaF_2 \cdot CaF_2$ 1% ~10%。基体材料为 18Cr2Ni4WA，喷涂前先用丙酮进行超声波清洗，除油除锈。喷砂粗化后进行喷涂。优化喷涂工艺参数为：电压 34V，电流 180A，空气压力 0.7MPa，喷涂距离 180mm。

* 本文合作者：梁秀兵，商俊超，郭永明，陈永雄。原发表于《摩擦学学报》，2013，33(5)：469~474。国家自然科学基金项目（51105377）和北京市优秀博士学位论文指导老师科研项目（20129003401）资助。

1.2 摩擦磨损试验

涂层的常温摩擦磨损试验在 CETR UMT－3 型（美国）摩擦磨损试验机上进行。磨损试验采用球－面接触方式，上试样为直径 ϕ4.0mm 的烧结 GCr15 球，下试样为 ϕ25.4mm×7.9mm 的涂层圆片，试验时，上试样做微小振幅往复运动，行程 4.0mm。摩擦试验条件为载荷 5N，频率 30Hz，时间 30min。

涂层高温摩擦磨损试验在球盘式高温摩擦磨损试验机（THT，CSM，Switzerland）上进行。摩擦副采用直径 6mm 的 Al_2O_3 球。摩擦磨损试验分别在 25℃、300℃、450℃、600℃的条件下进行。摩擦条件为载荷 10N，摩擦线速度 25cm/s，摩擦距离 450m。

摩擦磨损试验前都将试样进行抛光预处理，使其粗糙度达到 R_a 1.0μm。

2 结果与分析

2.1 涂层组织结构

图 1a 和 b 分别为 NiCrBMoFe 和 NiCrBMoFe/BaF_2 · CaF_2 涂层的截面形貌照片。从图中可以看出，涂层厚度大约为 500μm，涂层比较致密，与基体结合良好，有少量孔隙。电弧喷涂典型的层片状结构不明显，这是由于涂层中 B 元素的添加，起到脱氧净化作用，涂层中没有形成层片状的氧化物夹杂[9]。

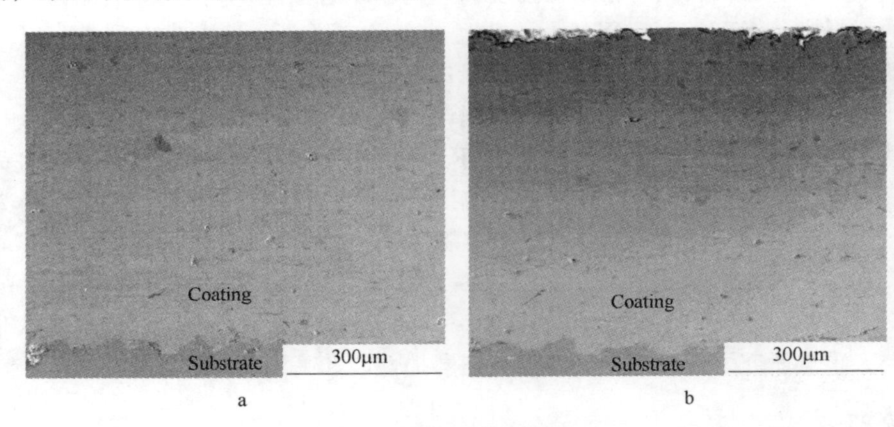

图 1 涂层的截面形貌

Fig. 1 SEM micrographes of the coatings

a—NiCrBMoFe；b—NiCrBMoFe/BaF_2 · CaF_2

图 2 为两种涂层的 XRD 图谱分析。从图 2 中可以看出：两种涂层在 2θ 为 44°附近有较宽化的漫散射峰，漫散射峰上叠加相对尖锐的晶体衍射峰[10]，说明两种涂层的相结构均由非晶相与晶体相复合而成。非晶相主要有 Ni、Cr、B 等元素组成。晶体峰主要为 Ni_3Fe 和 Ni(Cr) 固溶体。在高速电弧喷涂快速冷凝的过程中，合金的凝固温度远低于其平衡熔点 T_m，即可获得大的过冷度。根据非经典形核理论[11]，合金具有较大的过冷度时，其结晶形核率高而原子扩散能力低，可得到很高的形核率和较低的生长率，因而易于得到纳米晶。BaF_2 · CaF_2 的晶体衍射峰不明显，其原因为在晶化过程中，由于晶料长大和空洞的出现，造成大量的晶体缺陷，对平衡态固溶度很小的异质原子具

有很大的固溶度[12]，$BaF_2 \cdot CaF_2$ 会固溶在 NiCr 等粒子的晶体缺陷中，不容易检测。而且 $BaF_2 \cdot CaF_2$ 所占含量少、密度低[13]，在喷涂过程中损失比较严重。对 XRD 图谱进行 Pseudo - Voigt 函数[14]拟合，计算得到涂层中非晶相的体积分数分别达 45% 和 33%。

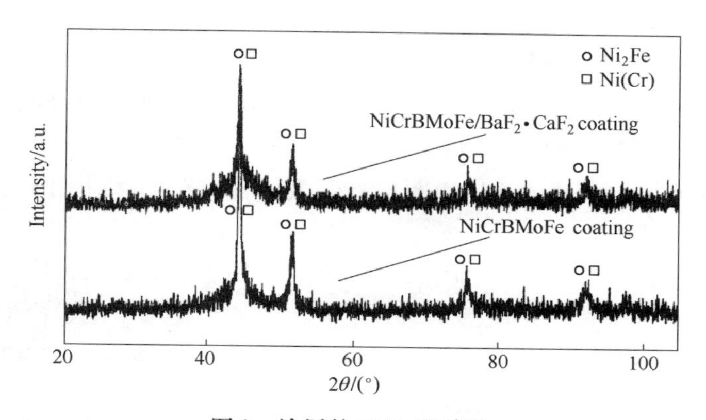

图 2　涂层的 XRD 图谱图

Fig. 2　XRD pattern of the coatings

图 3a 和 b 分别为 NiCrBMoFe 和 NiCrBMoFe/$BaF_2 \cdot CaF_2$ 两种涂层的透射电镜明场像形貌照片及选区电子衍射花样。从图 3a 可以看出，电子衍射花样由中心一漫散的中心斑点及漫散的环组成，这是非晶态的典型特征，说明涂层在成型过程中形成了完全的非晶区域。图 3b 为典型的非晶纳米晶微观组织形貌。衍射花样由中心较宽的晕和漫散的环组成，同时在漫散的非晶衍射环上分布着一系列小的多晶衍射斑点，经标定为面心立方结构的 γNi 相。NiCrBMoFe 涂层中的纳米晶（图 3a 中灰黑色部分尺寸相对均匀，约为 10nm），弥散分布在非晶母相中。NiCrBMoFe/$BaF_2 \cdot CaF_2$ 涂层中的纳米晶以团聚形态存在，涂层晶粒尺寸在 20 ~ 75nm，$BaF_2 \cdot CaF_2$ 固溶于 NiCr 等粒子的晶体缺陷中，使涂层中的纳米晶相以团聚的形态出现，形成图 3b 中所示的纳米晶区域（黑色区域）。

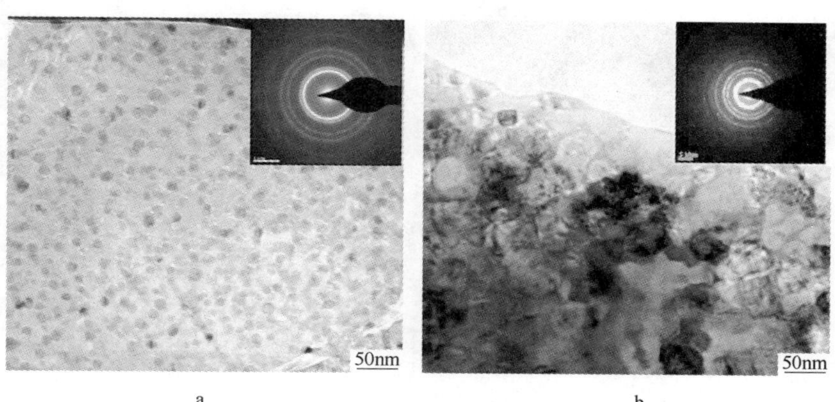

图 3　涂层的微观组织结构形貌及相应的微区衍射图

Fig. 3　TEM micrographs and electronic diffraction patterns of the coatings

a—NiCrBMoFe；b—NiCrBMoFe/$BaF_2 \cdot CaF_2$

2.2 涂层的常温摩擦磨损性能

涂层与基体的摩擦系数变化曲线如图4所示，基体的摩擦系数最高，约为0.57，NiCrBMoFe涂层约为0.5，NiCrBMoFe/BaF_2·CaF_2涂层仅为0.375，比基体摩擦系数下降25%。图5和表1为两种涂层高倍磨损形貌及EDS能谱分析结果。从图4中可以看出：NiCrBMoFe涂层为剥层磨损，NiCrBMoFe/BaF_2·CaF_2涂层为剥层磨损与轻微的磨粒磨损的混合形式。NiCrBMoFe涂层中深色区域（A区）中氧和Fe、Cr元素水平较高，主要为Fe、Cr的氧化物。在摩擦过程中产生的摩擦热使涂层产生氧化，出现Fe、Cr元素富集的深色氧化区域。氧化物薄层脆性大，在载荷持续作用下发生脆性断裂而剥落，形成图5中所示的片状剥落坑。NiCrBMoFe/BaF_2·CaF_2涂层的片状氧化剥落区分散无规律，多为扁平粒子（B-Cr）（B区）边界在持续载荷作用下产生裂纹而剥落。部分BaF_2·CaF_2在摩擦过程中脱落填入剥落坑中，使粗化不平的表面平滑化[15]，使摩擦系数降低；而且，BaF_2·CaF_2固溶在NiCr晶体缺陷中，具有抗粘着磨损的作用。细小的BaF_2·CaF_2同时诱发磨粒磨损，形成剥层磨损加轻微磨粒磨损的混合机制。

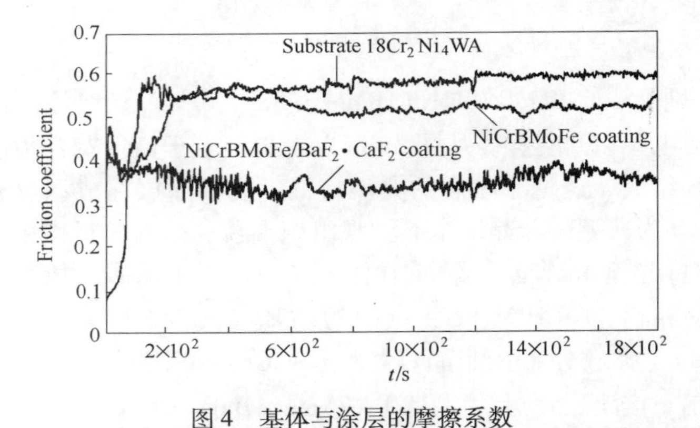

图4 基体与涂层的摩擦系数

Fig. 4 Friction coefficients of coatings and substrate

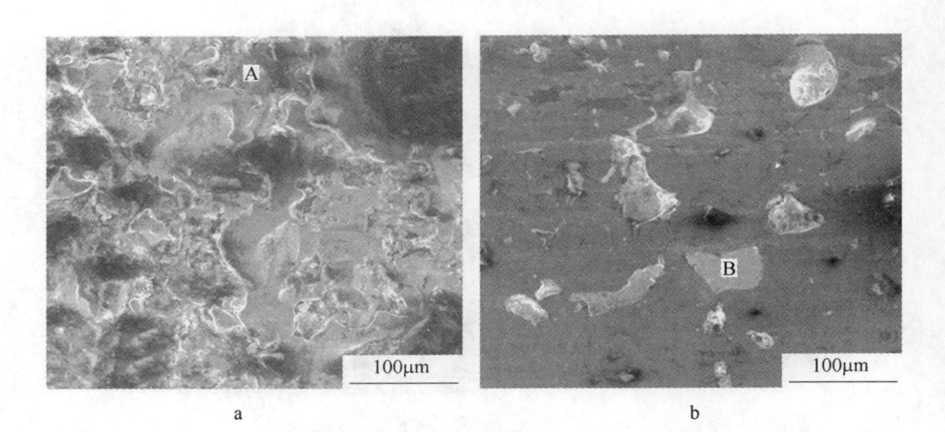

图5 涂层的高倍磨损形貌照片

Fig. 5 SEM micrographs of the coatings

a—NiCrBMoFe coating；b—NiCrBMoFe/BaF_2·CaF_2 coating

表 1　A 区和 B 区的 EDS 能谱分析结果
Table 1　EDS result of region A and B

（wt%）

Element	O	Cr	Fe	Ni	Mo	B	F	Ba
A	42.45	6.22	25.83	21.61	3.89	0	0	0
B	0	11.11	8.60	39.84	3.74	35.95	0.43	0.33

2.3　NiCrBMoFe/BaF_2 · CaF_2 涂层的高温摩擦磨损性能

图 6a 为 NiCrBMoFe 涂层在不同温度下的摩擦系数曲线。涂层在 25℃时的平均摩擦系数为 0.68，300℃与 450℃时摩擦系数为 0.55，600℃时为 0.5。随着温度升高，材料氧化趋势越来越严重，摩擦系数也不断下降；另外，高温摩擦产生大量的热量使涂层中的 NiCr 软质相间原子力减小，被挤压在涂层表面发生形变产生润滑膜[16]，起到一定的润滑作用。图 6b 为 NiCrBMoFe/BaF_2 · CaF_2 涂层在不同温度下的摩擦系数曲线。在 25℃时的摩擦系数为 0.51，随着温度升高，涂层的摩擦系数呈先下降后上升，再下降的趋势。300℃时的摩擦系数约为 0.52，450℃时为 0.44，600℃时又升至 0.49。NiCrB-MoFe/BaF_2 · CaF_2 涂层在 450～600℃区间摩擦系数先下降后上升，与其微观组织变化及涂层中自润滑相 BaF_2 · CaF_2 的作用有关。BaF_2 · CaF_2 共晶体在温度高于 400℃时，BaF_2 · CaF_2 脆性相会逐渐软化析出涂层表面，发生由脆性相向韧性相的转变[17]，降低了涂层剪切强度，使其易于形成润滑转移膜，阻碍涂层的磨损[18]，保证了涂层在高温状态下仍具有较好的润滑减摩作用。

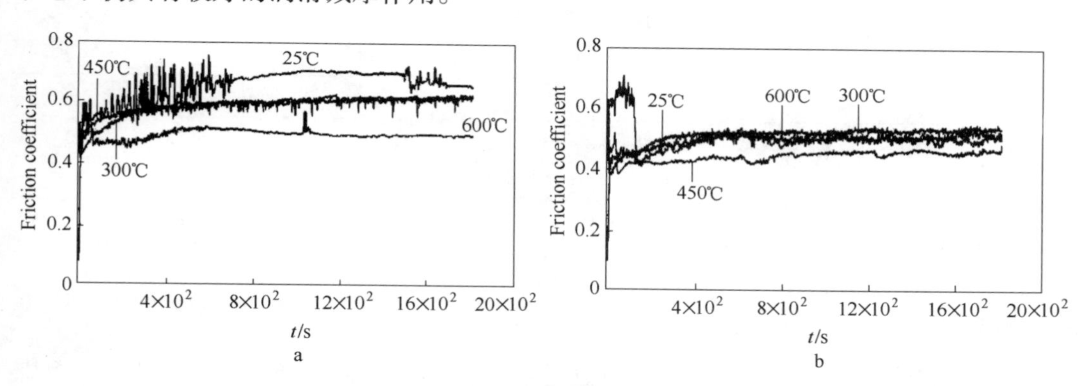

图 6　两种涂层在不同温度下的摩擦系数
Fig. 6　Friction coefficient of coatings at elevated temperature
a—NiCrBMoFe coating；b—NiCrBMoFe/BaF_2 · CaF_2 coating

图 7 和表 2 为 NiCrBMoFe/BaF_2 · CaF_2 涂层在 300℃、450℃和 600℃时的高倍 SEM 磨痕形貌及微区 EDS 分析结果。在 450℃时，涂层氧化较轻，氧的质量分数低（小于 10%）（A 区），涂层主要因未熔化颗粒（BaF_2 · CaF_2）萌生的裂纹源及涂层氧化（Cr 元素偏析富集）产生的氧化物脆断引起的疲劳剥落。在 450℃以上温度时，涂层磨痕氧化严重，氧的质量分数大于 30%。从表 2 可以看出：300℃时 BaF_2 · CaF_2 共晶体在涂层表面含量变化不大，当摩擦副之间接触温度高于 450℃时，BaF_2 · CaF_2 共晶相会逐渐析出涂层表面，并发生由脆性相向韧性自润滑相的转变。对 450℃时涂层磨痕表面进

行能谱分析（B 区），可以明显看到 $BaF_2 \cdot CaF_2$ 含量的增加，韧性相的 $BaF_2 \cdot CaF_2$ 在摩擦表面形成一层低摩擦系数的连续的润滑转移膜，对摩件 Al_2O_3 陶瓷球不断挤压和切削润滑转移膜，使其减薄和脱落，脱落的润滑转移膜不断被新转移出的润滑膜所覆盖[19]。形成的这种连续稳定的润滑转移膜阻碍了涂层磨损，具有良好的减摩润滑作用。在 600℃时，在涂层氧化磨损严重的区域已经看不到 $BaF_2 \cdot CaF_2$（C 区），这可以解释 600℃时涂层的摩擦系数比 450℃时略有升高的现象。

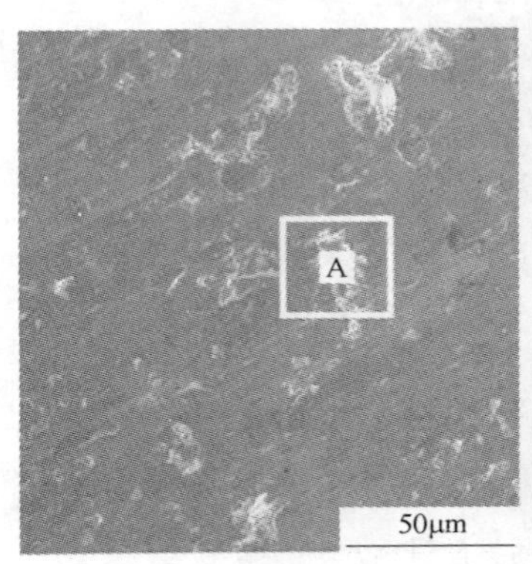

a

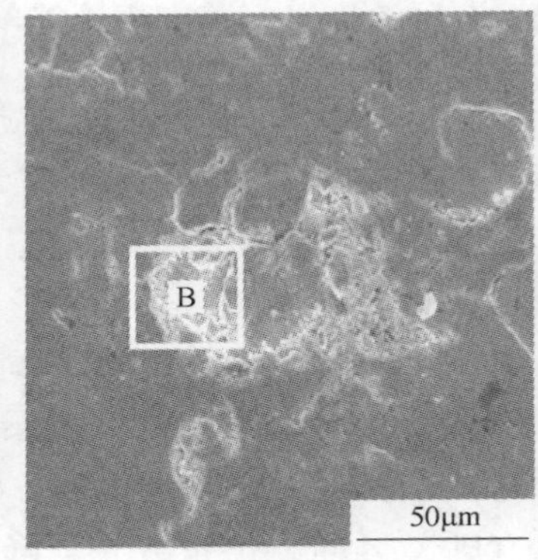

b

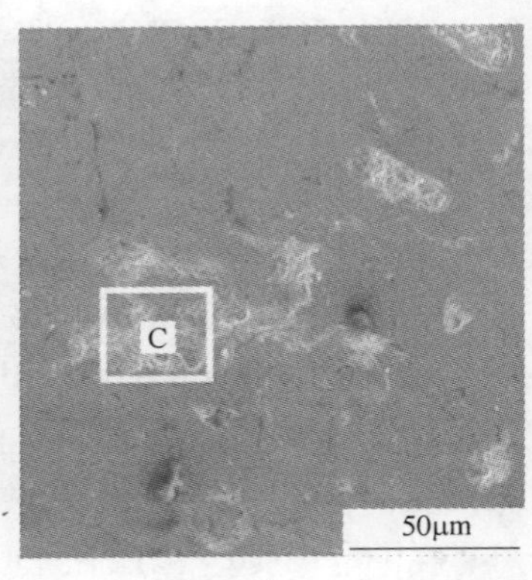

c

图 7　NiCrBMoFe/$BaF_2 \cdot CaF_2$ 涂层 SEM 照片和 EDS 分析区域

Fig. 7　SEM micrographs of NiCrBMoFe/$BaF_2 \cdot CaF_2$ coating

a—300℃；b—450℃；c—600℃

表 2　A、B 和 C 区的 EDS 能谱分析结果

Table 2　EDS result of region A，B and C　　（wt%）

Element	O	F	Ca	Cr	Fe	Ni	Mo	Ba
A	1. 22	4. 88	0. 88	13. 43	11. 44	55. 07	5. 23	0
B	19. 47	7. 93	2. 69	10. 67	8. 84	41. 89	0	7. 07
C	27. 47	0	0	11. 10	8. 30	42. 43	5. 47	0

3　结论

（1）利用高速电弧喷涂制备了 NiCrBMoFe 和 NiCrBMoFe/$BaF_2 \cdot CaF_2$ 两种 Ni 基非晶纳米晶涂层。涂层均由非晶相和纳米晶相组成，非晶相质量分数分别达 45% 和 33%。NiCrBMoFe 涂层中纳米晶尺寸均匀，弥散分布在非晶母相中；NiCrBMoFe/$BaF_2 \cdot CaF_2$ 涂层中纳米晶主要以团聚形态存在。

（2）固体自润滑剂 $BaF_2 \cdot CaF_2$ 的添加显著降低了涂层的摩擦系数。在常温情况下，团聚形态的 $BaF_2 \cdot CaF_2$ 具有抗黏着磨损的作用，使涂层的摩擦系数降低 25% 左右；在高温情况下，$BaF_2 \cdot CaF_2$ 由脆性相转变为韧性相，使得涂层的摩擦系数显著下降。

参 考 文 献

［1］Schuh C A，Hufnagel T C，Ramamurty U. Mechanical behavior of amorphous alloys［J］. Acta Mater，2007，55：4067～4109.

[2] Wei Q, Lu L Z, Li H, et al. Preparation and Properties of Amorphous Fe – Based Coating [J]. Materials for Mechanical Engineering, 2010, 34 (1): 52 ~ 54 (in Chinese).
魏琪，卢兰志，李辉，等．含非晶相铁基涂层的制备及性能 [J]．机械工程材料，2010，34 (1)：52 ~ 54.

[3] Sordelet D J, Besser M F. Oxygen Effects on Glass Formation of Plasma arc Sprayed Cu47Ti33Zr11Ni8Si1 surface coatings [J]. Materials Science and Engineering A, 2004, 375 ~ 377: 625 ~ 629.

[4] Gang J, Elkedim O, Grosdidier T. Deposition and Corrosion Resistance of HVOF Sprayed Nanocrystalline Iron Aluminium Coatings [J]. Surface & Coatings Technology, 2005, 190: 406 ~ 416.

[5] Parker F T, Spada F E, Berkowitz A E, et al. Thick Amorphous Ferromagnetic Coatings Via Thermal Spraying of Spark – eroded Powder [J]. Materials Letters, 2001, 48: 184 ~ 187.

[6] Kobayashi A, Yano S, Kimur H, et al. Fe – based Metallic Glass Coatings Produced by Smart Plasma Spraying Process [J]. Materials Science and Engineering B, 2007, 9(3): 1 ~ 35.

[7] Dent A H, Horlock A J, McCartney D G, et al. Microstructural characterisation of a Ni – Cr – B – C Based Alloy Coating Produced by High Velocity oxy – fuel Thermal Spraying [J]. Surface Coating & Technology, 2001, 13(9): 244 ~ 250.

[8] Choi H, Yoon S, Kim G, et al. Phase Evolutions of Bulk Amorphous NiTiZrSiSn Feed Stock During Thermal and Kinetic Spraying Processes [J]. Scripta Materialia, 2005, 53(1): 125 ~ 130.

[9] Ma X Q, Roth J, Gandy D W, et al. Cold gas dynamic spraying of iron – base amorphous alloy [J]. J. Therm. Spray technol., 2006, 15(4): 670 ~ 675.

[10] Cheng J B, Liang X B, Xu B S, et al. Microstructure and erosion resistance of Fe – based amorphous/nanocrystalline coatings [J]. Rare Metal Materials and Engineering, 2009, 38 (12): 2141 ~ 2145 (in chinese).
程江波，梁秀兵，徐滨士，等．铁基非晶纳米晶涂层组织及耐冲蚀性能的研究 [J]．稀有金属材料与工程，2009，38(12)：2141 ~ 2145.

[11] Lu K. Nanocrystalline Metals Crystallized from Amorphous Solids: Nanocrystallization, Structure, and Properties. Mater Sci Eng Reports, 1996, 16: 161 ~ 221.

[12] Ding C H, Wang Y P, Zhou J E. A NiCr – based High Temperature Self – lubricating Coating PM304 Prepared by Powder Metallurgy [J]. Acta Metallurgica Sinica, 2006, 42(11): 1212 ~ 1216 (in Chinese).
丁春华，王亚平，周敬恩．粉末冶金法制备 NiCr 基高温自润滑合金 PM304 涂层 [J]．金属学报，2006，42(11)：1212 ~ 1226.

[13] Wang C L, Li W G, Lu Feng, et al. Microstructure and Properties of $BaF_2 + CaF_2 + Cr_3C_2/Ni - Cr$ Coating Prepared by Detonation Gun Spraying [J]. Equipment Environmental Engineering. 2008, 5 (5): 25 ~ 28 (in Chinese).
王长亮，李伟光，陆峰，等．爆炸喷涂制备 $BaF_2 + CaF_2 + Cr_3C_2/Ni - Cr$ 涂层的组织及性能 [J]．装备环境工程，2008，5(5)：25 ~ 28.

[14] Verdon C, Karimi A, Martin J L. A Study of high velocity oxy – fuel thermally sprayed tungsten carbide based coatings. Part 1: microstructures [J]. Mater. Sci. Eng., A1998, 246: 11 ~ 24.

[15] Huang C B, Du L Z, Zhang W G. Preparation and Tribological of $NiCr/Cr_3C_2 - BaF_2 \cdot CaF_2$ High Temperature Self – lubricating Wear – resistant Coating [J]. Tribology, 2009, 29(1): 68 ~ 73 (in Chinese).
黄传兵，杜令忠，张伟刚．$NiCr/Cr_3C_2 - BaF_2 \cdot CaF_2$ 高温自润滑耐磨涂层的制备与摩擦磨损特性 [J]．摩擦学学报，2009，29(1)：68 ~ 73.

[16] Zhang Y C, Jia C C, He L M, et al. Dry Sliding Tribological Performances of Detonation Sprayed NiCr – Cr_2O_3 – BaF_2/CaF_2 Coating [J]. Lubrication Engineering, 2010, 35 (3): 57 ~ 61 (in Chinese).
张有茶，贾成厂，何利民，等．NiCr – Cr_2O_3 – BaF_2/CaF_2 爆炸喷涂涂层干滑动摩擦磨损性能[J]．润滑与密封，2010，35(3)：57 ~61.

[17] Deadmore D L, Sliney H E. Hardness of CaF_2 and BaF_2 Solid Lubricants at 25 to 670℃ [J]. NASA TM – 88979. Springfield, VA: National Technical Information Service, 1987.

[18] Cao T K, Deng J X, Sun J L. Study on Self – lubricating Al_2O_3/TiC/CaF_2 Ceramic Composites [J]. Journal of Materials Engineering, 2005, 3: 37 ~39 (in Chinese).
曹同坤，邓建新，孙军龙．Al_2O_3/TiC/CaF_2 自润滑陶瓷材料的研究 [J]．材料工程，2005，3：37 ~39.

[19] Wang H M, Yu Y L, Li S Q. Microstructure and tribological properties of laser clad CaF_2/Al_2O_3 Self – lubrication Wear Resistant ceramic matrix composite coatings [J]. Scripta Materialia, 2002, 47: 57 ~61.

Wear Properties of NiCrBMoFe/$BaF_2 \cdot CaF_2$ Coating by High Velocity Arc Spraying

Liang Xiubing[1], Shang Junchao[2], Guo Yongming[2], Chen Yongxiong[1], Xu Binshi[2]

(1. National Engineering Research Center for Mechanical Products Remanufacturing, Academy of Armored Forces Engineering, Beijing, 100072, China;
2. National Key Laboratory for Remanufacturing, Academy of Armored Forces Engineering, Beijing, 100072, China)

Abstract NiCrBMoFe and NiCrBMoFe/$BaF_2 \cdot CaF_2$ amorphous/nanocrystalline coatings were fabricated by high velocity arc spraying process. Both coatings were composed of amorphous phase and nanocrystalline phase. The friction and wear properties of the coatings were investigated. At room temperature, the coefficient of friction of the two coatings were 0. 50 and 0. 375. Room temperature friction and wear properties of NiCrBMoFe/$BaF_2 \cdot CaF_2$ coating was far superior to the substrate and NiCrBMoFe coating. At 450℃, $BaF_2 \cdot CaF_2$ riched layer on the coating surface was responsible to the low friction and wear.

Key words Ni – based amorphous/nanocrystalline, coating, frictional wear, $BaF_2 \cdot CaF_2$

电流密度对自动化电刷镀 Ni 镀层组织结构和性能的影响*

摘　要　为提高电沉积镀层质量和自动化程度，开发了新型的自动化电刷镀技术。利用扫描电镜、X 射线衍射仪、透射电镜、X 射线应力测试仪、数显显微硬度计和显微磨损试验机考察了不同电流密度下制备的电刷镀 Ni 镀层的组织结构和性能，并与电镀 Watts Ni 对比。结果表明，应用电刷镀制备的镍镀层组织平整致密，无针孔、麻点等缺陷；随着电流密度从 $4A/dm^2$ 增加 $16A/dm^2$，电刷镀 Ni 镀层的（111）面择优取向逐渐降低，（200）面择优取向逐渐增加，镀层的晶粒尺寸和应力逐渐增大，硬度约在 500～600HV 之间波动，镀层磨损失重由 6.8mg 降低到 5.2mg。

关键词　电刷镀　自动化　电流密度　组织结构　性能

1　引言

再制造工程是废旧机电产品高技术维修的产业化[1~3]，因此，它需要多种表面工程技术进行支撑，电沉积技术就是其关键技术之一，其中电刷镀已广泛地应用于再制造工程中[4,5]。

由于传统电刷镀技术多采用手工操作，劳动强度大，镀层质量不稳定，镀笔阳极多以石墨包裹棉花，浪费比较严重，刷镀工序较多，涉及镀液种类较多，不易实现机械化和自动化。因此，传统电刷镀技术已远远不能满足现代制造业和再制造业的强烈发展需求。为了克服传统手工电刷镀存在的不足，机械化和自动化电刷镀成为研究的一个热点，例如文献［6］利用计算机、传感器等多种控制设备开发了自动化电刷镀技术，劳动强度有所减低，镀层的质量有所提高[7~9]，但未对镀液体系和镀笔阳极进行改进，浪费较大。本文开发了一种新的自动化电刷镀技术，通过设计新的镀液体系和刷镀专用镀笔，使柔性介质对电刷镀层产生周期定向外力的作用，并通过数控系统控制电刷镀工序切换、刷镀参数和镀液的供给，实现了自动化电刷镀。

电流密度是电沉积的重要参数之一[10~15]，对镀层的组织结构和性能有着重要影响，本文以废旧的发动机缸套为研究对象，研究电流密度对电沉积 Ni 镀层的表面形貌、组织结构、硬度以及耐磨性的影响，以期为再制造产业的进一步发展提供技术支持和理论指导。

2　实验

2.1　实验装置

自动化电刷镀技术是在传统电镀和电刷镀基础上发展起来的一种新的工艺方法。

---

* 本文合作者：胡振峰、汪笑鹤、吕镖。原发表于《功能材料》，2013，44(17)：2507～2510。国家自然科学基金资助项目（51005244）；武器装备预研基金资助项目（9140C850201110C8501）；国家重点基础研究发展计划（“973”计划）资助项目（2011CB013403）。

该装置主要包括自动化电刷镀专用镀笔、驱动装置、镀液循环过滤系统、专用电源和控制系统。自动化电刷镀专用镀笔由安装于耐腐蚀支架上面的可溶性阳极和柔性介质组成，通过电机驱动镀笔运动，从而使支架上安装朝外的柔性介质在阴极（工件）表面上不间断地摩擦和搅拌。阴极采用内孔型金属工件，通过专用的密封装置组成一个顶部开口的容器；柔性介质采用不导电、弹性好且耐酸碱的猪鬃制成，并保证其接触距离大于内孔类工件内径 2 ~ 6mm。图 1 为自动化电刷镀装置的局部示意图。

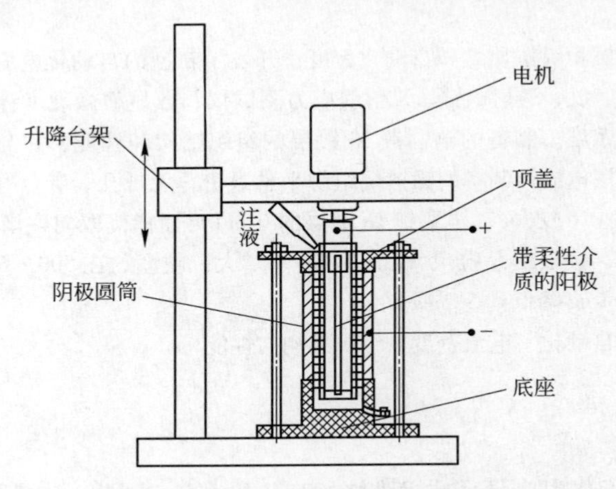

图 1　自动化电刷镀装置局部示意图

Fig. 1　Part schematic diagram of automatic brush electroplating experimental apparatus

2.2　实验方法

阴极材料采用废旧的发动机缸套，基体材质为球墨铸铁，内孔尺寸为 ϕ100mm × 210mm。自动化电刷镀的镀液采用常规的 Watts 镀液改进而成，其组成为：硫酸镍（$NiSO_4 \cdot 6H_2O$）300g/L，氯化镍（$NiCl_2 \cdot 6H_2O$）40 ~ 80g/L，硼酸（H_3BO_3）40g/L，不加添加剂。基本工艺流程为：打磨→电净→2 号活化→3 号活化→电沉积 Ni 镀层。作为对比实验，使用常规的 Watts 的 Ni 配方制备了电镀 Ni 镀层。电刷镀的工艺参数选择：镀液温度 50℃，镀笔旋转速度 100r/min，电流密度依次选择 4A/dm^2、8A/dm^2、12A/dm^2、16A/dm^2。

镀层的表面形貌测试采用 Philips Quanta200 型扫描电子显微镜（SEM）观察；镀层的晶粒尺寸采用 D8Advance 型 X 射线衍射仪测定；镀层的应力采用 X – 350A 型 X 射线应力测试仪测试；镀层的显微硬度采用 HVS – 1000 数显显微硬度计测定，载荷 100g，加载时间 15s。采用 CETR – UTM 型显微磨损试验机研究镀层的摩擦磨损性能，并用失重法确定镀层的磨损量。与镀层对磨的试样为 GCr15 钢球，直径 4mm，硬度 63HRC，磨损实验均在室温下进行。微动实验参数为载荷 10N、振幅 5mm、频率 5Hz、时间 20min。

3　结果与讨论

3.1　电流密度对镀层形貌的影响

图 2 和图 3 分别是传统电镀得到的 Ni 镀层和采用自动化电刷镀技术在不同电流密度下制备的镀层的表面形貌。对比图 2 和图 3 可以看出，自动化电刷镀技术制备的镀

层的表面形貌明显不同于传统电镀 Ni 镀层。传统电镀 Ni 镀层表面凸凹不平，有许多明显的球状微凸体，这样的镀层会产生虚尺寸，在过盈配合时会导致松动和过盈量不足。在滑动配合中会产生磨粒，导致不正常磨损。因此，该类镀层一般要进行抛光处理后才能使用。而自动化电刷镀技术得到的镀层虽有柔性介质留下的刷痕，但表面较为平整，无法分辨出晶体构成物大小。这是因为在电沉积过程中，柔性介质的摩擦作用可以有效阻止氢气泡和吸附杂质在阴极表面长期滞留，从而避免在镀层表面形成凹坑、针孔、麻点等缺陷。此外，柔性介质摩擦还可以增加阴极表面活性生长点，提高形核速率，抑制镀层的尖端放电，并对已经形成的积瘤等微观突起起到机械磨削和抛光作用，从而实现镀层表面整平的效果。比较不同电流密度下的表面形貌，在电流密度为 12A/dm^2 时，镀层的表面几乎没有划痕。可以认为，电流密度为 12A/dm^2 时，镀层的表面形貌最好。

图 2　传统 Ni 镀层的表面形貌

Fig. 2　Surface morphology of traditional Ni electroplated

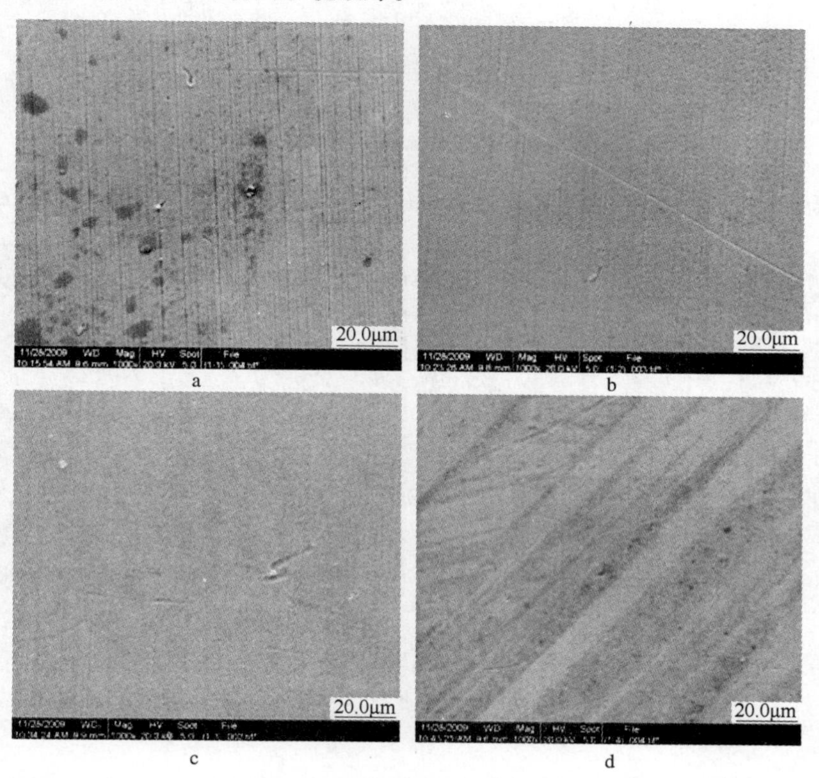

图 3　电刷镀 Ni 镀层的表面形貌

Fig. 3　Surface images of brush electroplated Ni

a—4A/dm^2；b—8A/dm^2；c—12A/dm^2；d—16A/dm^2

3.2 电流密度对镀层结构的影响

图 4 是采用不同的电沉积技术制备的镀 Ni 层的 XRD 图谱。对比可以看出，自动化电刷镀技术制备的镀层的微观结构明显不同于传统电镀 Ni 镀层，传统电镀得到的 Ni 镀层在（200）面的衍射强度明显强于（111）面，而自动化电刷镀技术得到的镀层恰恰相反，（111）面的衍射强度要高于（200）面的衍射强度。这说明柔性介质的摩擦作用会在一定程度上影响晶体的生长方式，使（111）面的生长速度变慢，而（200）面的生长速度加快，（111）面的择优程度增大，（200）面的择优程度降低。对比不同电流密度的镀层结构可以看出，镀层（200）面的择优取向随着电流密度的增加而逐渐增大。这表明电流密度的增加在一定程度上降低了电刷镀的效果，但降低程度较小。

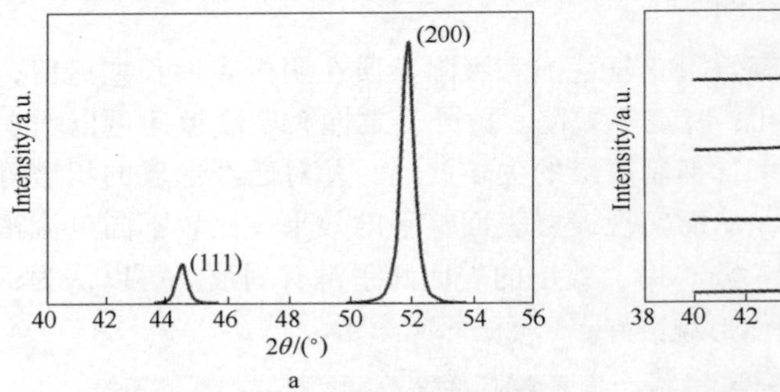

图 4 不同方法制备的镍镀层的 XRD 谱

a—传统电镀 Ni 镀层；b—自动化电刷镀 Ni 镀层

Fig. 4 XRD patterns of Ni electroplated with different methods

图 5 为电刷镀 Ni 镀层的 TEM 像及其电子衍射花样。

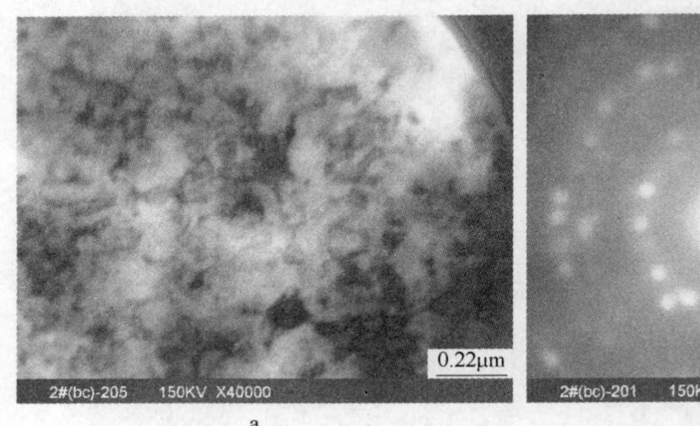

a

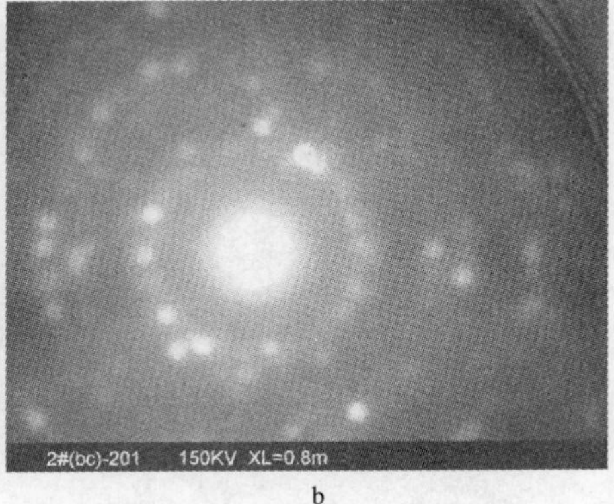

b

图 5 电刷镀 Ni 镀层的 TEM 像及其电子衍射花样

a—TEM 图像；b—衍射花样图

Fig. 5 TEM image and diffraction pattern of brush electroplated Ni

从衍射花样图可以看出，镀层是 Ni 的多晶体，并呈面心立方结构。而从 TEM 像可以看到镀层由很多形状不规则的晶粒组成，晶粒尺寸都在 200nm 以下。而每个晶粒又由许多的亚晶粒组成，如图 5a 中的很多黑色的颗粒，这是镀层中某个（或几个）取向的亚晶粒的图像，由于衬度的差别在照片上呈现黑色，这些晶粒的大小处于10～50nm之间。而Watts Ni镀层组织为粗大的柱状晶，其晶粒尺寸在0.5μm以上（甚至数个微米）[16]。

3.3 电流密度对镀层硬度的影响

图6是采用自动化电刷镀技术在不同电流密度下制备的Ni镀层的硬度。从图6可以看出，自动化电刷镀技术制备的镀层的维氏硬度普遍在500HV以上，明显高于传统电镀Ni镀层的硬度250HV。这是由于自动化电刷镀Ni镀层的晶粒细小，已经达到纳米晶，因此镀层的硬度明显高于传统电镀的硬度。对比不同电流密度的镀层硬度可以看出，镀层的硬度先降低后增高。当电流密度为4A/dm^{2}时，镀层的硬度最高，达到625HV；电流密度从8A/dm^{2}增大到16A/dm^{2}时，镀层硬度从496HV依次增大到610HV。图7给出了电流密度对镀层晶粒尺寸和镀层的应力的影响，从图7可以看出，随着电流密度的增加，镀层的晶粒尺寸增大，镀层的应力也增大，根据Hall－Petch关系，镀层的硬度和晶粒尺寸是负相关关系，即镀层的硬度随着晶粒的增大而减小。镀层的硬度和应力是正相关的关系，即镀层的硬度随着应力增大而增大。对比镀层硬度和电流密度的关系，分析认为，镀层的硬度是晶粒尺寸和应力协同作用的结果。当电流为4A/dm^{2}时，镀层的晶粒尺寸起主导作用，镀层的硬度较大；电流密度从8A/dm^{2}增大到16A/dm^{2}时，镀层的应力起主要作用，镀层的硬度逐渐增大。

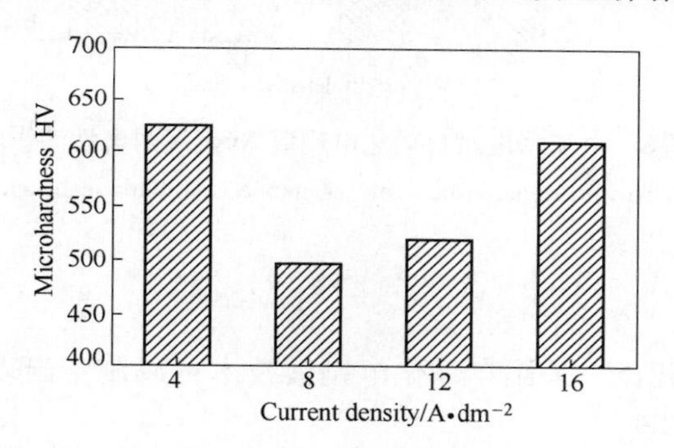

图6　电流密度对自动化电刷镀Ni镀层硬度的影响

Fig. 6　Effect of current density on microhardness of automatic brush electroplated Ni

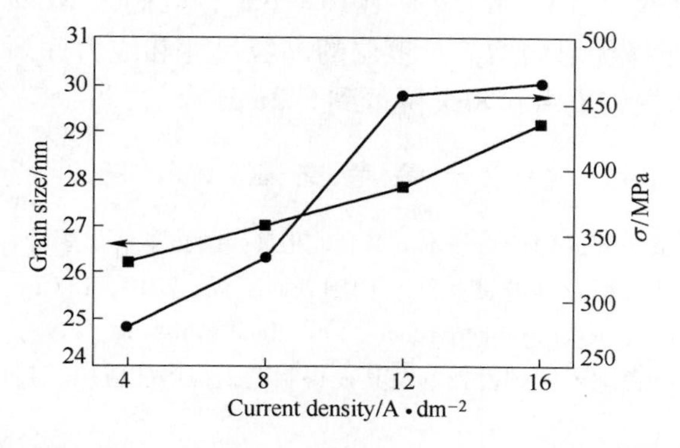

图7　电流密度对自动化电刷镀Ni镀层晶粒尺寸和应力的影响

Fig. 7　Effect of current density on grain size and stress of automatic brush electroplated Ni

3.4 电流密度对镀层耐磨性的影响

图8为不同电流密度下自动化电刷镀镀层的磨损失重图。由图8可知，当电流密度为4A/dm^{2}时，镀层磨损失重最大，为6.8mg，约占相同条件下传统电镀Ni镀层的磨损失重的35%。因此，自动化电刷镀层的耐磨性大大增强，这与镀层晶粒细化导致的硬度提高以及组织致密相关。从图8中的变化趋势还可以看出，随着电流密度的增加，电刷镀层的磨损失重逐渐减小，即镀层的耐磨性能增强。电流密度为4A/dm^{2}时，虽然镀层的硬度较高，但由于其表面粗糙度也较高，所以镀层的磨损失重依然较大；电流密度提高到8A/dm^{2}时，虽然镀层的硬度有所降低，但是由于其表面粗糙度也降低了，因此其磨损失重也降低。随着电流密度的继续增大，镀层的表面粗糙度变化不大，但其硬度和应力却逐渐升高，因此镀层的磨损失重也随之减小。

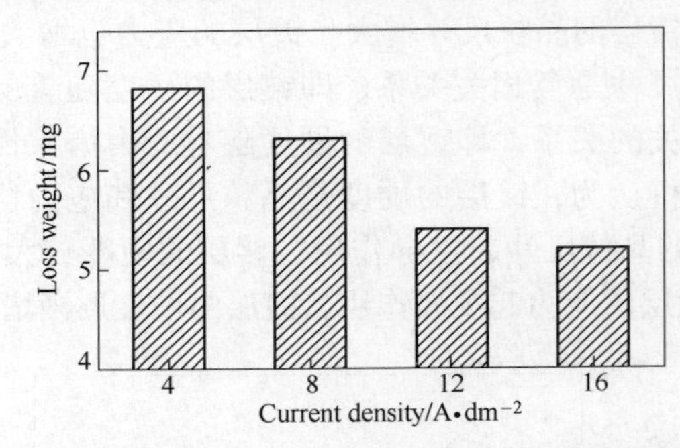

图8 电流密度对自动化电刷镀Ni镀层耐磨性的影响

Fig. 8 Effect of current density on wear resistance of automatic brush electroplated Ni

4 结论

（1）与传统电镀相比，采用自动化电刷镀技术可以制备出表面平整光滑、晶粒细小、组织致密的Ni镀层。

（2）自动化电刷镀技术制备的Ni镀层为面心立方结构，由很多形状不规则的晶粒组成，而每个晶粒又由许多的亚晶粒组成，亚晶粒的大小处于10~50nm之间。

（3）随着电流密度从4A/dm^{2}增加到16A/dm^2，电刷镀Ni镀层的（111）面择优取向降低，（200）面择优取向增大，镀层的晶粒尺寸和应力逐渐增大，硬度在500~600HV之间波动，磨损失重由6.8mg降低到5.2mg。

参考文献

［1］徐滨士．装备再制造工程的理论与技术［M］．北京：国防工业出版社，2007，1.

［2］徐滨士．中国再制造工程及其进展［J］．中国表面工程，2010，23(2)：1~6.

［3］Dini J W. Brush plating：recent property data［J］. Metal Finishing，1997，(6)：88~93.

［4］徐滨士，胡振峰．绿色纳米电刷镀技术及其在再制造工程中的应用［J］．新技术新工艺，2008，(11)：7~11.

［5］胡振峰，董世运，汪笑鹤，等．面向装备再制造的纳米复合电刷镀技术的新发展［J］．中国表面工程，2010，23(1)：87~91.

[6] 张斌，徐滨士，董世运，等．自动化电刷镀技术工艺控制专家系统的设计与应用［J］．机械工程学报，2008，44(4)：128～132.

[7] 张斌，徐滨士，董世运，等．自动化 n－SiC/Ni 复合电刷镀层的高温摩擦性能研究［J］．材料工程，2008，(1)：54～57.

[8] 张斌，徐滨士，董世运，等．自动化 n－Al_2O_3/Ni 复合电刷镀层的组织及高温耐磨性能［J］．中国表面工程，2006，(z1)：263～266.

[9] Wu Bin，Xu Binshi，Zhang Bin，et al. Automatic brush plating：An update on brush plating［J］. Materials Letters，2006，(60)：1673～1677.

[10] Wu Bin，Xu Binshi，Zhang Bin，et al. The effects of parameters on the mechanical properties of Ni－based coatings prepared by automatic brush plating technology［J］. Surface & Coatings Technology，2007，(201)：5758～5765.

[11] Li Yundong，Jiang Hui，Huang Weihua，et al. Effects of peak current density on the mechanical properties of nanocrystalline Ni－Co alloys produced by pulse electrodeposition［J］. Applied Surface Science，2008，254(21)：6865～6869.

[12] Chung C K，Chang W T. Effect of pulse frequency and current density on anomalous composition and nanomechanical property of electrodeposited Ni－Co films［J］. Thin Solid Films，2009，517(17)：4800～4804.

[13] 袁铁锤，李瑞迪，刘红江，等．电沉积制备 Ni－S 涂层电极的研究［J］．功能材料，2009，40(7)：1121～1123.

[14] 康进兴，赵文轸，徐英鸽，等．工艺参数对电沉积纳米晶镍沉积速率的影响［J］．材料热处理学报，2008，(3)：152～155.

[15] 杨余芳，文朝晖，李强国，等．Fe－Co 合金箔电沉积工艺研究［J］．功能材料，2011，42(S1)：85～88.

[16] Watanabe T. Nano－plating［M］. Amsterdam：Elsevier Ltd.，2004. 41.

Effect of Current Density on Microstructure and Properties of Automatic Brush Electroplated Nickel

Hu Zhenfeng[1]，Wang Xiaohe[1]，Lü Biao[1,2]，Xu Binshi[1]

(1. Science and Technology on Remanufacturing Laboratory，Academy of Armored Force Engineering，Beijing，100072，China；
2. School of Materials and Metallurgy，Northeastern University，Shenyang，110819，China)

Abstract A new automatic brush electroplating technology was developed to improve the quality and automatic level of electro－deposition. Effect of current density on microstructure and properties of automatic brush electroplating nickel were investigated by SEM，XRD，TEM，X－ray stress tester，micro hardness tester and CETR，and the plating was compared with Watts Ni plated. The results show that automatic brush electroplated nickel has compact and smooth microstructure，without pinholes and nodules etc. defect. With current density from $4A/dm^2$ increase to $16A/dm^2$，the degree of (111) preferential orientation decreases and that of (200) preferential orientation increases by degrees；the grain size and internal stress gradually increases；the microhardness fluctuate in 500－600HV，and the loss weight from 6. 8 to 5. 2mg.

Key words brush electroplating，automation，current density，microstructure，property

Fe314 合金熔覆层残余应力激光冲击消除机理*

摘　要　针对激光熔覆层残余应力过大导致变形、开裂的问题，采用激光冲击技术对Fe314 合金熔覆层进行了表面冲击处理，分析了熔覆层残余拉应力分布形式及消除机理。结果显示，激光熔覆时采用相对较大的激光比能量，即慢扫描速度、小光斑直径和低送粉速率工艺可有效降低熔覆层残余拉应力。而激光冲击大幅降低了熔覆层残余拉应力，随着冲击次数提高，熔覆层拉应力减小，但拉应力降低幅度呈逐渐减弱趋势。冲击波力学效应引发的极大应变率使熔覆层表层发生微塑性变形，形成压应力场，大幅抵消熔覆层初始态残余拉应力。同时，材料压缩变形时在γ-Fe 晶粒内萌发大量位错线，位错发生多系滑移并相互缠结形成位错墙，引发细晶强化作用。

关键词　激光技术　激光熔覆　激光冲击　残余应力　晶粒细化

1　引言

激光熔覆技术目前被广泛地应用于高性能复杂结构、致密金属零件的直接成型和表面磨损、腐蚀等失效零件的再制造等领域[1~3]。但激光熔覆时局部热输入导致的不均匀温度场引起了局部热效应，导致在金属熔池凝固结晶过程中形成较大的热应力，凝固结束后在熔覆层中形成残余拉应力和热变形，对成型件的尺寸精度、静载强度、疲劳强度、抗开裂等性能造成较大的影响。拉应力过大时将引发裂纹缺陷[4~6]，严重危害零件的成型或再制造质量。因此，研究激光熔覆层残余应力分布和形成机理不仅对控制残余应力具有实际意义，而且对预防熔覆层开裂提供了工艺指导。

传统消除熔覆层残余应力措施以退火热处理为主，而对于一些大型再制造零部件而言，尺寸过大，退火操作困难；同时，零件制造时经过某些特殊热处理的基体在退火处理时或将发生组织转变，导致原有工艺附加值流失。而激光冲击强化（LSP）技术凭借高压（10^9 ~ 10^{12}Pa）、极快（10 ~ 100ns）、高应变率（10^7 ~ $10^8 s^{-1}$）、非接触、可控性强等优点被广泛地应用于重要零件的表面强化与应力消除[7~9]。该技术是采用短脉冲高峰值功率密度的激光束轰击材料表面诱导高压冲击波，冲击波使材料表层发生压应力场，改善材料表层残余应力状态[10~12]。当前，针对 LSP 与激光熔覆技术复合进行熔覆层表面残余应力状态转变处理的研究较少，缺乏对熔覆层残余应力消除效果评估及消除机理的探讨。因此，本文针对激光熔覆层成型后残余应力过大的问题，采用 LSP 进行了应力消除工艺及机理研究。

2　试验材料与方法

激光熔覆试验材料为 Fe314 合金粉末，成分如表 1 所示。该粉末具有韧性好、硬度

* 本文合作者：闫世兴、董世运、王玉江、肖爱民、鲁金忠。原发表于《中国激光》，2013，40（10）：1003004-1 ~ 1003004-6。总装“十二五”预研项目（51327040401）和国家 973 计划（2011CB013403）资助。

低、抗开裂性能好和成型性能优异的特点，基体为45钢。熔覆过程采用侧向同步气动送粉方式。设备为大功率连续式Nd：YAG激光器。激光熔覆工艺参数为：激光功率1.2kW，扫描速度58mm/s，送粉速率16.2～18.0g/min，光斑直径2～3mm，载气流量200L/min，保护气为氩气。激光冲击熔覆层时采用锡箔作为吸收层，预先贴附于熔覆层表面，采用流动水作为约束层。LSP设备为脉冲式Nd：YAG激光器。LSP工艺参数为脉冲能量6J、脉宽10ns。激光熔覆与LSP工艺过程[13]分别如图1、图2所示。两种试验用Nd：YAG激光器输出光斑均为圆形，沿圆形光斑径向激光能量呈现高斯分布特征。

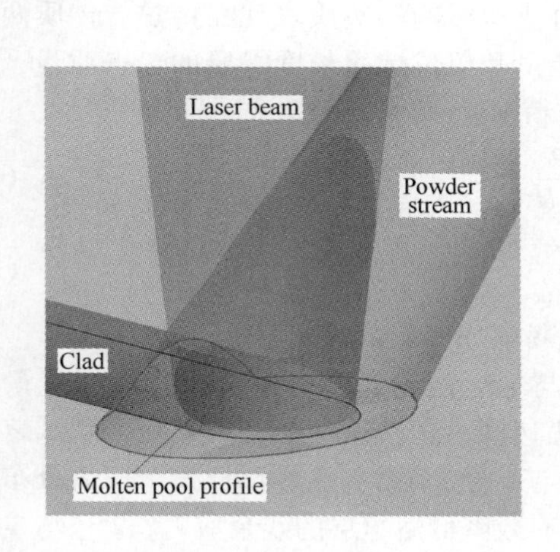

图1　侧向同步送粉式激光熔覆过程示意图
Fig. 1　Schematic view of laser cladding with a lateral synchronous powder feed

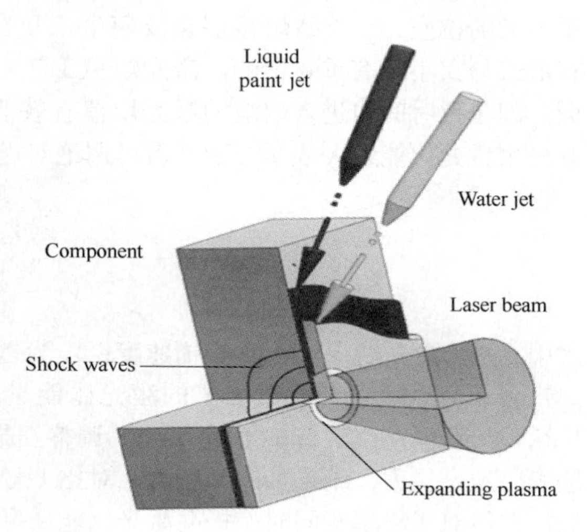

图2　LSP示意图
Fig. 2　Schematic view of LSP

采用侧倾固定Ψ法测量Fe314合金激光熔覆层LSP前后的表层残余应力，仪器为X-350A型X射线应力检测仪，测量误差率为10%。测试参数为：管电压22kV，管电流6mA，钴靶K_α特征辐射，准直管直径$\phi=3$mm，阶梯扫描步进角0.100rad，时间常数1s。对冲击后Fe314合金熔覆层试样分别切割，研磨抛光表层与截面层，采用王水（体积比为3∶1的浓盐酸和浓硝酸的混合溶液）腐蚀，在Philips XL30型扫描电镜下观察熔覆层表层及深度方向组织特征。切取表层熔覆试样，经研磨、离子减薄制备直径$\phi=3$mm的透射电镜（TEM）分析试样，采用JEM-2000型TEM观察熔覆层晶粒内部及晶界形态；采用D8型X射线衍射仪（XRD）分析熔覆层物相组成。

表1　Fe314合金粉末成分（质量分数）

Table 1　Components of Fe314 powder (mass fraction)　(%)

Element	C	Cr	Ni	B	Si	Fe
Content	0.1	15	10	1	1	bal

3 结果与讨论

3.1 激光熔覆工艺对残余应力分布的影响

激光熔覆过程中极大的加热、冷却速度使熔覆层各部位温度分布严重不均，从而使变形受到约束，在熔覆层内形成较大的宏观残余应力；同时，由于极大的熔池温度梯度和凝固速度，熔池金属结晶时在极小的范围内发生平面晶、柱状晶、细小树枝晶、等轴晶一系列晶体形式转变，组织在短程内分布严重不均，导致在熔覆层内形成较大组织应力，即微观残余应力。这两类残余应力叠加作用使激光熔覆层成型后内部残余了极大的拉应力，导致熔覆层裂纹萌生、开裂。而激光熔覆工艺对熔覆层残余应力分布形式与大小具有重要影响。激光熔覆工艺对成型影响的实质表现在传热、传质两方面。即单位时间内进入熔池的激光熔覆有效能量和单位熔覆长度需要的粉末质量，前者可用激光比能量 λ 表示，而后者可用绝对送粉率 V_g 表示，表达式分别为[14]：

$$\lambda = \frac{P}{V_s D} \tag{1}$$

$$V_g = \frac{V_f}{V_s} \tag{2}$$

式中，P 为激光功率；V_s 为扫描速度；V_f 为送粉速率；D 为光斑直径。

表 2 为不同激光熔覆工艺的激光比能量与绝对送粉速率分布，保持激光能量为 1.2kW，光斑直径为 2mm 恒定不变。调整扫描速度、送粉速率及光斑直径，测试熔覆层表层残余应力，结果如图 3 所示。对比 1 号与 4 号、2 号与 3 号试样残余应力分布可见，在绝对送粉率相同时，提高激光比能量可有效降低熔覆层的残余应力，2 号与 3 号试样残余应力差值较大，通过提高激光比能量，残余应力由 540MPa 降低到 393MPa。在激光比能量基本相同时，改变激光熔覆的绝对送粉率，如 2 号与 4 号试样，绝对送粉率由 37.5×10^{-3}g/mm 升至 54.0×10^{-3}g/mm，熔覆层残余应力由 393MPa 下降至 370MPa。但相对于改变激光比能量，控制绝对送粉率变化，熔覆层应力降低幅度较小。因此，可判断影响熔覆层残余应力分布的主要因素为激光比能量，即熔覆过程中传热影响较大。

表 2 不同激光熔覆工艺的激光比能量与绝对送粉速率分布

Table 2 Specific energy and absolute powder feed rate with different laser cladding parameters

No.	Scanning speed V_s/mm · s^{-1}	Powder feed rate V_f/g · min^{-1}	Spot size D/mm	Specific energy λ/J · mm^{-2}	Absolute powder feed rate V_g/g · mm^{-1}
1	5	16.2	2	120.0	54.0×10^{-3}
2	8	18.0	2	75.0	37.5×10^{-3}
3	8	18.0	3	50.0	37.5×10^{-3}
4	5	16.2	3	80.0	54.0×10^{-3}

根据凝固结晶原理，提高激光比能量，单位时间进入熔池的能量增加，使熔池最高温度、保温时间均增加，进而凝固时熔池相对于外界环境的温度梯度减小，熔池金属过冷度相对降低，使熔覆层底部至顶部的组织转变较均匀，组织应力相对较低。另外，熔覆层高温状态时间延长，熔覆层受热不均部位热变形拘束相对减小，有利于冷

却后期残余应力的松弛。因此，适当提高激光比能量，对于熔覆层残余应力控制具有重要作用。

将激光比能量指标细化至具体工艺参数，在扫描速度、送粉速率和光斑直径3个参数中，为保持熔覆过程较高的激光比能量，采用慢扫描速度、小光斑直径和低送粉速率工艺，有利于控制熔覆层残余拉应力。

3.2 LSP对熔覆层残余应力影响

图3表明，成型后未热处理的熔覆层内部存在较大残余拉应力，该应力状态对熔覆层抗开裂性影响较大。采用LSP对Fe314熔覆层进行处理，利用爆轰冲击波在材料表面诱发压应力效应，达到削弱或消除熔覆层残余拉应力、改善熔覆层表面组织与力学性能的目的。

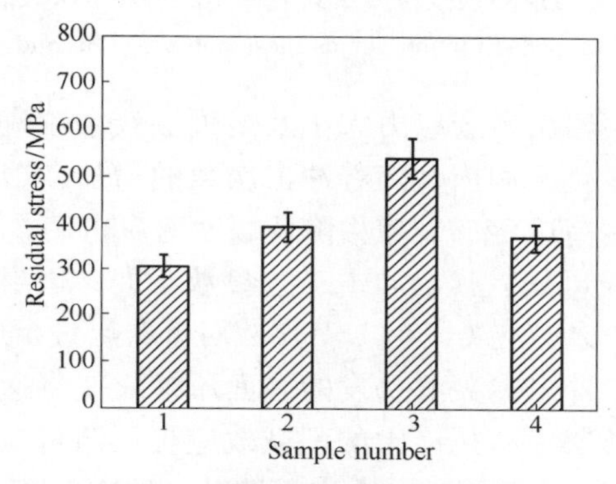

图3 不同工艺参数下熔覆层表层残余应力分布

Fig. 3 Residual stress distribution of cladding layers with different parameters

图4所示为冲击前后激光熔覆层表面不同部位的残余应力分布。可见，搭接熔覆层表面中间残余拉应力较高。而两侧边部位置拉应力较低，拉应力范围为200~800MPa。经5次LSP后（其余冲击参数为脉冲能量6J、脉宽10ns、搭接率50%），表面残余应力范围为-150~400MPa。在熔覆层边部冲击后产生了残余压应力，压应力范围为0~-150MPa，其他熔覆区域残余拉应力均大幅度降低。残余拉应力平均下降幅度为70%，可见LSP能够有效消除激光熔覆表层的残余拉应力。但由于拉应力过大，冲击后熔覆层的残余拉应力虽然大幅度降低，但未得到完全消除。另一方面，从图4可以看出冲击5次后，残余拉应力显著降低，但没有完全消除。进一步研究增加冲击次数对改善表面残余应力状态及改性区域大小的影响。测试结果如图5所示。

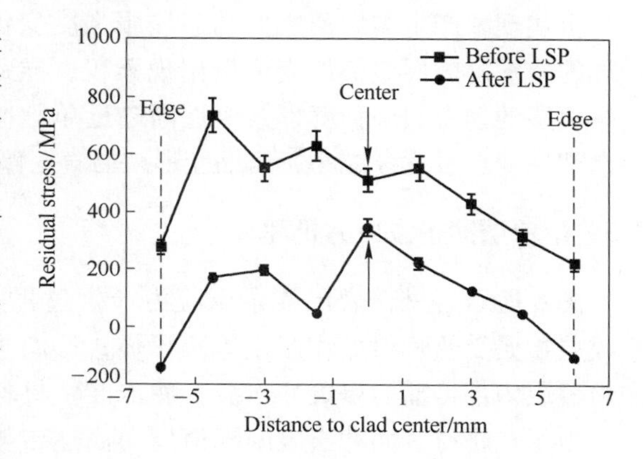

图4 冲击前后试样表面残余应力对比

Fig. 4 Comparison of residual stress distribution between layer before LSP and after LSP

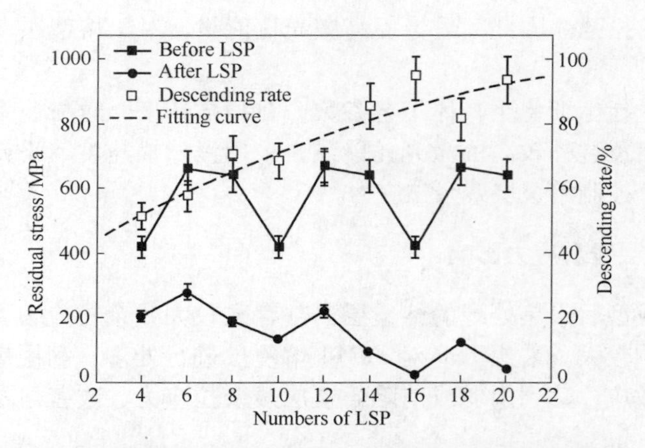

图5　熔覆层残余拉应力及其下降比例与冲击次数的关系

Fig. 5　Residual stress distribution versus the number of LSP and its desending rate

图5为冲击次数与熔覆层残余应力大小及熔覆层残余拉应力下降比率的关系。可见，在冲击4~20次的变化区间内，随着冲击次数的增加，熔覆层残余拉应力总体呈下降趋势，最大下降幅度高达94%，应力降低幅度达600MPa，根据图5中残余应力分布及应力下降率拟合曲线可见，其余冲击工艺参数不变，适当提高冲击次数，有利于大幅削弱熔覆层表层的残余拉应力。但LSP次数对熔覆层残余应力状态影响存在上限，即达到一定冲击次数后，冲击爆轰波诱发的表面压缩应变将达到饱和，压缩变形趋势逐渐减小至停止，对熔覆层原有拉应力的削减效应也逐渐将减小。如图5所示，随着冲击次数逐渐增大，熔覆层表面应力分布出现波动，拉应力减小的幅度呈现逐渐降低、平缓的趋势。

上述现象产生与熔覆层的动态屈服强度变化有关。随着冲击次数的增加，冲击波诱导的熔覆层表层微塑性变形量出现累积，导致变形层出现应变硬化效应，大幅提高了熔覆层的动态屈服强度，其值逐渐趋近单脉冲的冲击波压力，导致后续冲击时熔覆层微塑性变形量逐渐减小，压缩变形减弱直至停止。

3.3　熔覆层冲击去应力机理

图6所示为材料表层残余压应力产生过程的示意图。在激光辐照吸收层表面时，吸收层物质受热喷射爆炸产生等离子体，等离子体膨胀受到约束层的抑制，从而形成向熔覆层内部传播的爆轰冲击波，冲击波使材料表层发生三轴应力压缩而产生一维应变。由于在靶材表面冲击波的峰值压力高达吉帕量级，可使靶材表面一定深度的材料沿轴向产生弹塑性变形，同时这部分材料在平行于材料表面的平面内产生伸长变形，其周围的材料受到了挤压，如图6a所示。当冲击波压力消失后，该部分材料塑性变形无法恢复，发生塑性变形的材料与周围材料保持几何相容性，周围材料存在将已发生塑性变形的材料压缩回原始形状的趋势，最终在平行于靶材表面的平面内产生压缩变形效应，如图6b所示。冲击形成的压缩变形可部分抵消甚至消除熔覆层初始残余拉应力，从而达到减弱或消除成型熔覆层表层拉应力的技术目的。

选取2号熔覆试样，对其表面进行5次LSP（其余冲击参数为：脉冲能量6J、脉宽

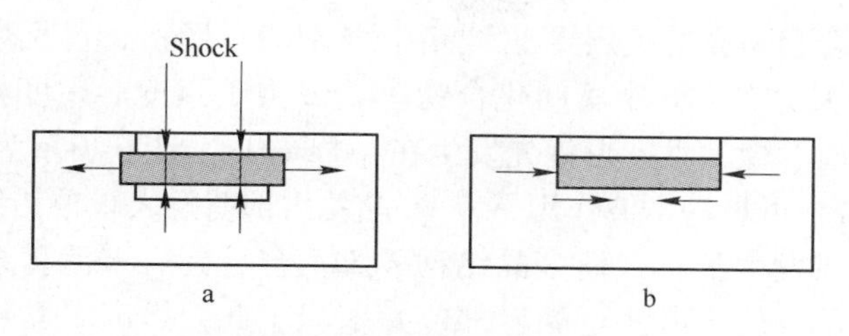

图 6　激光冲击诱导材料压应力场示意图

a—冲击过程；b—冲击后

Fig. 6　Schematic of compression stress field induced by LSP

a—During LSP；b—After LSP

10ns、搭接率 50%）。图 7 为 LSP 后熔覆层表面及深度方向上的显微组织形貌。由图 7a 可见在表面激光光斑冲击区域（A 区），熔覆层发生了微塑性变形，晶粒在冲击波的压应力作用下发生变形，组织出现致密化特征。晶粒尺寸略小于相邻区域（B 区）的晶粒尺寸，呈现为细小致密的胞状晶形貌，而附近未冲击部位（B 区）熔覆层晶体主要为尺寸相对较大的交叉树枝晶。由于 YAG 激光光斑能量的时空分布特征，光斑中心区域（约 $\phi600\mu m$ 的 A 区）的晶粒显著细化，而光斑周围晶粒尺寸呈现逐渐增大的过渡特征。由图 7b 可见，沿着冲击深度方向，可分为组织细化区和原始组织区，组织细化区内晶粒细化特征越加明显，而原始组织区熔覆层晶粒尺寸变化幅度较小，组织细化区深度约为 $200\mu m$。图 7b 中熔覆层深度方向组织变化特征表明，随着至熔覆层表面的距离增加，组织细化区逐渐缩小，组织细化区底部呈现弧形特征，反映出随熔覆层深度增加 LSP 力学效应逐渐衰弱特征。同时，由于 Nd：YAG 激光器光斑能量呈高斯分布，即光斑中心能量密度高，边缘低，导致激光诱发的冲击波压力在表层分布不均，深度方向上形成了细化区底部弧形特征。

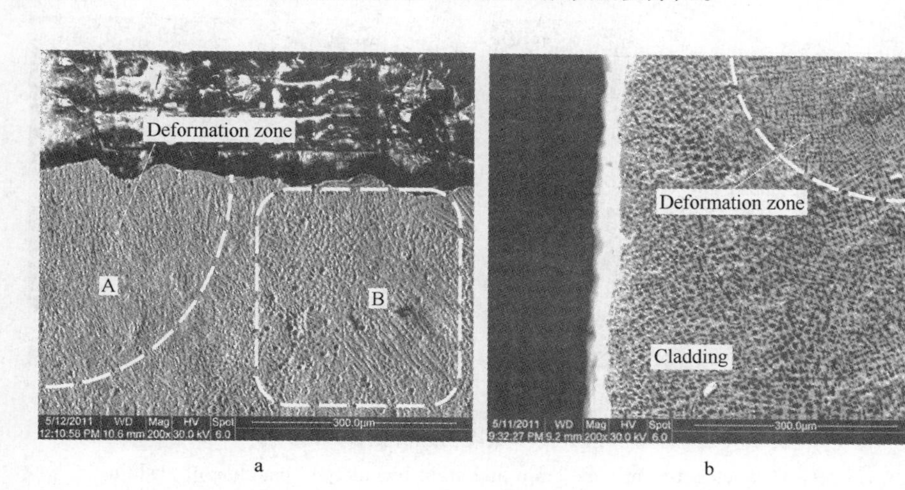

图 7　激光冲击强化层显微组织

a—表面；b—截面

Fig. 7　Microstructure of the cladding layer with LSP

a—Surface；b—Cross - section

图 8 为 Fe314 合金激光熔覆层 X 射线衍射（XRD）图谱，可见熔覆层物相主要由 γ-Fe 组成，同时少量分布 CrB 金属间化合物。γ-Fe 由于强度、硬度相对较低，塑性较好，受冲击波作用时，微塑性变形优先发生在 γ-Fe 相内。图 9 所示为熔覆层 γ-Fe 晶内与晶界的 TEM 形貌，可见冲击波作用下 γ-Fe 晶粒内部萌发大量位错线，随着微塑性变形过程的进行，位错发生滑移，而多晶结构 Fe314 合金具有多滑移系，位错滑移方向相互交叉，导致一部分位错滑移至晶界部位并塞积于此，提高了晶粒塑性变形能的临界值；而另一部分位错滑移在 γ-Fe 晶粒内形成位错缠绕结构。这些位错缠绕结构把原始晶粒细分成更小的位错胞块。当变形应变量达到一定程度时，即位错缠绕结构内的位错密度达到一定值时，柏氏矢量不同的位错开始湮灭、重排，形成位错墙，从而将 γ-Fe 晶粒细分成各个不同的亚结构，形成细化的晶粒。

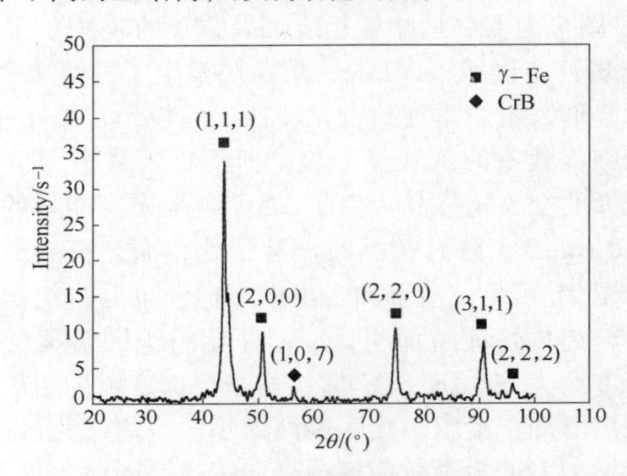

图 8　Fe314 合金激光熔覆层 XRD 图谱

Fig. 8　XRD of Fe314 alloy cladding layer

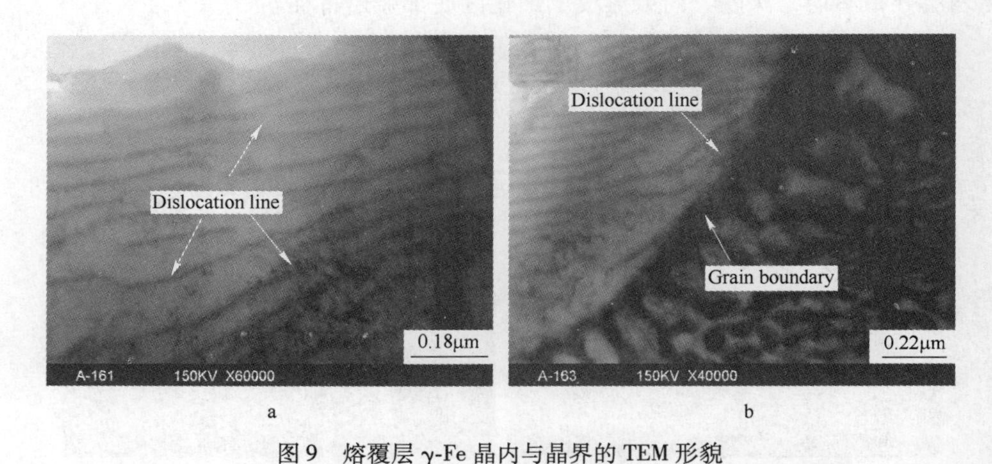

图 9　熔覆层 γ-Fe 晶内与晶界的 TEM 形貌

Fig. 9　Characteristics of γ-Fe grain and grain boundary of cladding of cladding

4　结论

（1）激光比能量与绝对送粉率影响熔覆层残余应力分布，而激光比能量大小的作用尤为显著，提高激光比能量可有效降低熔覆层残余应力。

（2）LSP 可大幅降低熔覆层残余拉应力，随着冲击次数提高，熔覆层拉应力减小，但冲击次数对 Fe314 熔覆层残余应力状态的影响存在上限，随着冲击次数逐渐增大，拉应力削弱幅度呈现逐渐减小、平缓趋势。

（3）冲击波力学效应引起的极大应变率使熔覆层表层形成压缩变形效应，大幅抵消了熔覆层的残余拉应力。熔覆层受冲击区域表层与一定深度的组织出现细化特征。

（4）Fe314 合金激光熔覆层主相由 γ-Fe 软质相组成，冲击波诱导材料压缩变形时优先在 γ-Fe 晶粒内萌发大量位错，多系滑移的位错线缠结、湮灭、重排形成位错墙，将晶粒分割为多个亚结构，形成细晶强化作用。

参 考 文 献

[1] Xu Binshi，Liu Shican，Dong Shiyun，et al. Theory and Technology of Remanufactured Equipments Engineering［M］. Beijing：National Defense Industry Press，2007：294 ~ 300.
徐滨士，刘世参，董世运，等. 装备再制造工程的理论与技术［M］. 北京：国防工业出版社，2007：294 ~ 300.

[2] Zhong Minlin，Liu Wenjin. Leading areas and hot topics on global laser materials processing research［J］. Chinese J. Laser，2008，35（11）：1653 ~ 1659.
钟敏霖，刘文今. 国际激光材料加工研究的主导领域与热点［J］. 中国激光，2008，35(11)：1653 ~ 1659.

[3] J. Dowden. The Theory of Laser Materials Processing［M］. Dordrecht：Springer Science and Business Media B V，2000：235 ~ 256.

[4] G. A. Websteret，A. N. Ezeilo. Residual stress distributions and their influence on fatigue lifetimes［J］. International J Fatigue，2001，23(s1)：S375 ~ S383.

[5] U. de Oliveira，V. Ocelík，J. Th. M. De Hosson. Residual stress analysis in Co – based laser clad layers by laboratory X – rays and synchrotron diffraction techniques［J］. Surface & Coatings Technology，2006，201(3 ~ 4)：533 ~ 542.

[6] P. Bendeich，N. Alam，M. Brandt，et al. Residual stress measurements in laser clad repaired low pressure turbine blades for the power industry［J］. Materials Science and Engineering A，2006，437(1)：70 ~ 74.

[7] P. V. Yasnii，P. O. Marushchak，Yu. M. Nikiforov，et al. Influence of laser shock – wave treatment on the impact toughness of heat – resistant steels［J］. Materials Science，2010，46(3)：425 ~ 429.

[8] T. Elperin，G. Rudin. Thermal stresses in functionally graded materials caused by a laser thermal shock［J］. Heat and Mass Transfer，2002，38(7 ~ 8)：625 ~ 630.

[9] Guan Haibing，Ye Yunxia，Wu Zhong，et al. Effects of long pulse – width stray light on shock wave induced by laser［J］Chinese J Laser. 2011，38(7)：0703007.
管海兵，叶云霞，吴忠，等. 长脉冲杂光对激光诱导冲击波的影响［J］中国激光. 2011，38(7)：0703007.

[10] C. Michaut，E. Falize，C. Cavet，et al. Classification of and recent research involving radiative shocks［J］. Astrophys Space Sci，2009，322(1 ~ 4)：77 ~ 84.

[11] B. S. Yilbas，S. B. Mansoor，A. F. M. Arif. Laser shock processing：modeling of evaporation and pressure field developed in the laser – produced cavity［J］. Int J Adv Manuf Technol，2009，42(3 ~ 4)：250 ~ 262.

[12] Luo Xinmin，Zhang Jingwen，Ma Hui，et al. Dislocation configurations induced by laser shock processing

of 2A02 Aluminum Alloy [J]. Acta Optica Sinica, 2011, 31(7): 0714002.
罗新民，张静文，马辉等.2A02 铝合金中强激光冲击诱导的位错组态分析 [J]. 光学学报，2011，31(7)：0714002.

[13] G Singh, R V. Grandhi, D S. Stargel. Modified particle swarm optimization for a multimodal mixed - variable laser peening process [J]. Struct Multidisc Optim, 2010, 42(5): 769 ~ 782.

[14] Dong Shiyun, Yan Shixing, Xu Binshi, et al. Processing optimization and properties of laser cladding Fe90 coating [J]. Journal of Functional Materials. 2011, 42(s): 15 ~ 18.
董世运，闫世兴，徐滨士，等.Fe90 合金激光熔覆工艺优化及性能研究 [J]. 功能材料，2011，42(s)：15 ~ 18.

Mechanics of Removing Residual Stress of Fe314 Cladding Layers with Laser Shock Processing

Yan Shixing[1], Dong Shiyun[1], Xu Binshi[1], Wang Yujiang[1], Xiao Aimin[2], Lu Jinzhong[2]

(1. National Key Laboratory for Remanufacturing, Academy of Armored Forces Engineering, Beijing, 100072, China;
2. School of Mechanical Engineering, Jiangsu University, Zhenjiang, Jiangsu, 212013, China)

Abstract Due to the thermal deformation and crack of laser cladding layers induced by residual stress, a treatment of laser shock processing (LSP) is performed to remove the residual stress of Fe314 alloy cladding layer, its mechanics and residual stress distribution are investigated as well. Results show that residual stress can be removed significantly when a high specific energy of laser cladding comprised of slow scanning speed, small spot size and low powder feed rate is adsorbed by melt pool. Furthermore, LSP decreases the residual stress, and with the number of LSP increasing, the residual stress of cladding layer descends gradually while the removing capability attenuates. The mechanics of variation show that a micro - plastic deformation appears on the Fe314 cladding layer surface, which is induced with a huge strain rate of shock wave. A compression stress field is formed and removes the tensile stress of cladding layer. It is observed that mass dislocation lines emerge in the γ-Fe grain due to LSP. Then a dislocation wall is formed and separates one grain into several subgrains. Eventually, the effect of grain refinement is occurred in the Fe314 alloy cladding layer.

Key words laser technique, laser cladding, laser shock processing, residual stress, grain refinement

高速电弧喷涂再制造曲轴 FeAlCr/3Cr13 复合涂层的性能研究*

摘　要　高速电弧喷涂再制造曲轴的使用寿命与涂层的结合强度和耐磨性有很大的关系。为了提高涂层的性能，研究了新型 FeAlCr/3Cr13 复合涂层的组织结构和力学性能，采用扫描电镜（SEM）、能谱仪（EDAX）和 X 射线衍射仪分析了涂层的微观结构、相组成和残余应力。利用显微硬度计和 CETR 微动摩擦磨损试验机等试验设备对涂层的力学性能进行了分析。结果表明，喷涂 FeAlCr 粉芯丝材作为打底层，合金元素反应充分，复合涂层组织均匀、致密，空隙率约 9.87%。复合涂层相主要由韧性相 α - Fe 和硬质相 $Cr_{23}C_6$、（Fe，Cr）固溶体组成，氧化物含量较低，约 3.2%，复合涂层平均残余应力较小，约 67.6MPa，平均显微硬度 $HV_{0.1}$ 为 4000MPa，结合强度约 46.6MPa，油润滑高载摩擦条件下复合涂层表现出较好的耐磨性能。

关键词　高速电弧喷涂　曲轴　FeAlCr/3Cr13 复合涂层　结合强度　耐磨性

曲轴是汽车发动机中价值最高的零件之一，工作过程中由于磨损，主轴颈和连杆轴颈出现锥度和失圆，导致发动机不能正常工作，其造价昂贵，整根曲轴的报废会带来较大的经济损失[1]。因此，对报废曲轴进行再制造有很大的意义。高速电弧喷涂是以电弧为热源，以特殊的喷枪将熔化的金属丝材用高速气流雾化，气流速度比普通喷枪提高约 1 倍，高速喷射到工件表面形成涂层的一种新型热喷涂工艺[2,3]，提高了涂层性能，是一种高效的曲轴修复技术[4]。同时，以操作机或机器人为载体的自动化喷涂技术的研究实现了发动机曲轴再制造的产业化发展[5]。

为了提高再制造曲轴涂层的性能和使用寿命，制备高性能的涂层对提高再制造曲轴的使用寿命有很大的意义。通常提高涂层的结合强度，是在粗化处理后的工件表面先喷涂一层打底层。打底层提高涂层结合强度的机理是某些特殊材料在喷涂时熔滴携带大量的热能，过热的熔滴与空气中的氧发生氧化反应又释放出大量的热量，使熔滴的温度进一步升高，在熔滴与基体碰撞时发生冶金结合，形成自结合层[6]。目前已开发的热喷涂自结合材料有镍包铝粉、铝包镍粉、镍铝复合丝、铝青铜丝[7,8]等。本研究采用高速电弧喷涂技术，在曲轴表面制备了一种新型打底层与工作层的复合涂层，在实验室条件下研究了复合涂层的组织结构、微区成分组成和力学性能。

1　实验

实验选用的基体材料是 45 号钢，打底层喷涂材料选取 ϕ2.0mm 的 FeAlCr 系粉芯丝

* 本文合作者：田浩亮、魏世丞、陈永雄、童辉、刘毅。原发表于《稀有金属材料与工程》，2014，43(3)：727 ~ 732。国家“973”项目（2011CB013403），国家自然科学基金（51105377，50971132），国家科技支撑项目（2011BAF11B07）资助。

材，丝材外皮选用0.4mm×10mm 08F优质低碳钢，粉芯丝材是在药芯焊丝生产线上采用多辊连续轧制和多道连续拔丝减径方法制造的。工作层3Cr13马氏体不锈钢丝材及新型打底丝材的化学成分如表1所示。

喷涂设备为装备再制造技术国防科技重点实验室自行研究设计的机器人自动化高速电弧喷涂系统，该设备采用HAS－01型高速电弧喷涂枪，雾化熔滴的最大速度达到243m/s[9]。曲轴在喷涂前先清洗，除去表面油污等，将轴颈磨削掉0.3mm的疲劳层，曲轴轴颈表面进行喷砂处理。喷砂工艺为：砂料为棕刚玉，粒度为1mm，气压为0.7MPa，喷砂角度为90°，喷砂距离为100mm。喷涂工艺参数为：喷涂电压26～28V，喷涂电流120～140A，压缩气体压力0.7MPa，喷涂距离180～200mm。

表1　实验选用涂层材料的化学成分（质量分数）

Table 1　Chemical composition of the cored wires　（%）

3Cr13	Cr	Mn	Si	C	O	Fe
	12～14	≤0.6	<1.0	0.1	<0.18	Bal.
FeAlCrNi	Al	Ni	Cr	Si	B	Fe
	17～20	10～14	18～20	<1	<1	Bal.

采用Quanta 200型扫描电镜分析涂层截面的组织形貌，并利用其配套的能谱仪对涂层微区进行了元素分析。采用D8型X射线衍射仪分析涂层的相结构和残余应力分布。采用孔隙率图像处理软件对复合涂层截面进行孔隙率测定。采用对偶件拉伸试验法，按照GB 9796—88标准在WE－10A万能材料试验机上测试了涂层的结合强度。使用ⅡMT－3型显微硬度计测量涂层截面的维氏硬度。在CETR微动摩擦磨损实验机上测试了室温载荷为50N油润滑摩擦条件下涂层和基体的摩擦磨损性能，磨损试验采用球－面接触方式，上试样为直径ϕ4mm的GCr15钢球，硬度（HRC）不小于60。下试样为10mm×10mm×5mm的涂层方片，涂层表面抛光处理，涂层最终厚度为0.4mm。实验采用球－面接触模式，试验参数为：磨损时间15min，频率5Hz，位移幅值D为5mm。材料的摩擦系数由试验机实时记录。使用VK－9700型3D激光扫描显微镜测量了涂层和基体摩擦后的二维和三维磨痕形貌及磨损体积损失。

2　结果与分析

2.1　复合涂层的相分析

图1为FeAlCr/3Cr13复合涂层的XRD图谱。涂层相组成有：α－Fe、（Fe，Cr）固溶体和碳化物$Cr_{23}C_6$。喷涂过程中丝材在电弧热及压缩气体的作用下熔化并雾化，熔滴温度约2000℃[10]，丝材能完全熔化并充分进行冶金反应。同时，存在比较微弱的氧化物衍射峰，主要是Fe_3O_4、Cr_2O_3，涂层中氧化物含量较少，通过Verdon方法[11,12]对

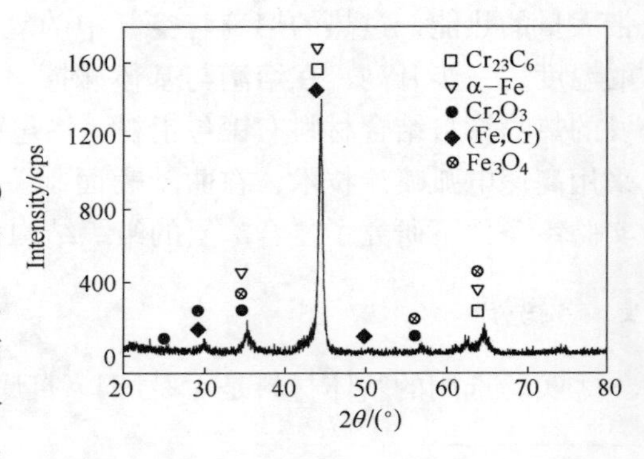

图1　FeAlCr/3Cr13复合涂层的XRD图谱

Fig. 1　XRD pattern of FeAlCr/3Cr13 composite coating

XRD 图谱进行 Pseudo - Voigt 函数拟合，计算得到涂层中氧化物的含量为 3.2%。这是因为熔滴在压缩空气的作用下，高速（≈243m/s）撞击到基体表面，瞬时冷却，空气中停留时间短，发生氧化的可能小。

2.2 涂层/基体界面形貌及成分分析

高速电弧喷涂 FeAlCr/3Cr13 复合涂层的截面微观形貌如图 2a 所示，涂层由变形良好的浅灰色条带组织和少量分布其间的深灰色氧化物组成，呈典型的层状结构，组织均匀致密，只有很少的空洞，空隙率约 9.87%。如图 2a 所示。涂层高致密度的原因是“熔滴”粒子以约 243m/s[9] 的速度撞击到基体或已沉积层表面，均匀地铺展开来，层片之间紧密结合。

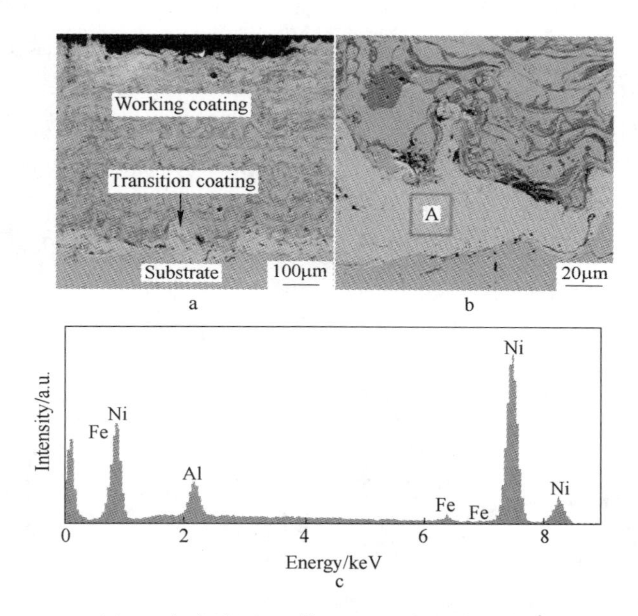

图 2　复合涂层/基体 SEM 照片及成分分析

Fig. 2　Cross - section SEM images and EDS spectrum of composite coating

a—Composite coating cross - section；b—Transition coating cross - coating；

c—Chemical composition of A zone in Fig. 2b

打底层/基体界面的微观形貌如图 2b 所示，FeAlCr 打底层厚约 50μm，与基体结合紧密没有孔洞和裂纹出现，也与工作层紧密地“咬合”在一起。同时，对打底层进行微区成分分析，主要有 Fe、Ni、Al，没有检测到 O 元素，如图 3c 所示。涂层低氧化物含量的机理在于：虽然电弧喷涂过程中，高温（≈2000℃）飞行的熔融粒子不可避免地要发生氧化，但合金元素发生的氧化反应可能性和稳定性与各元素氧化反应产生的自由能有关，FeAlCr 系丝材中含有 Fe，Al，Ni，Cr，Si，B 等元素，1727℃时 Fe，Al，Ni，Cr，Si，B 各元素生成氧化物的自由能分别为 -33.8kcal、-82.0kcal、-13kcal、-49.1kcal、-61.9kcal 和 -61.9kcal[13]，根据热力学第二定律，Al、Si 和 B 元素对氧元素的亲和力要远高于 Fe、Ni、Cr，因此容易发生氧化反应。但合金与气体能发生氧化反应还必须要求氧气压大于氧化反应的平衡分压，如 B 的氧化反应：

$$2B + \frac{2}{3}O_2 \longrightarrow B_2O_3$$

其平衡常数为：

$$K = \frac{\alpha_{B_2O_3}}{\alpha_B^2 P_{O_3}^{\frac{3}{2}}}$$

根据 $\Delta G_1 = -RT\ln K$ 可以确定1400℃时，B氧化的平衡氧分压 P_{O_2}（B/B_2O_3）$=1.67\times10^{-45}$，同理，可以得出Si和Al的平衡氧分压分别为 2.23×10^{-21} 和 1.2×10^{-3}[14]，这说明相同条件下B更容易氧化。而且，文献指出，实验温度大于1000℃时，B氧化生成 B_2O_3 以气态形式挥发，并且随温度升高挥发量增大。另外，B_2O_3 和少量的 SiO_2 发生反应生成 $B_2O_3-SiO_2$ 氧化膜，附着在熔融粒子表面，防止熔滴内部氧化[15]。这解释了为什么打底层中没有检测到氧元素存在。这对提高打底层内聚强度和其与基体的结合强度有很大的作用。

涂层内部主要由浅灰色组织组成，如图3a所示，成分分析如图3c所示，以Fe，Cr为主，结合涂层XRD分析，浅灰色区域是（Fe，Cr）固溶体和α-Fe，说明喷涂过程中熔融液滴冶金反应充分。没有检测到氧元素，这是因为喷涂过程中Fe与Cr发生氧化，在熔融粒子表面形成一薄层连续、致密的复合氧化膜[16]，对内部金属起保护作用，降低喷涂过程中涂层氧化程度。同时，铁素体作为韧性相，（Fe，Cr）固溶体是硬质相，起到弥散强化的作用，可以提高涂层的硬度和耐磨性。

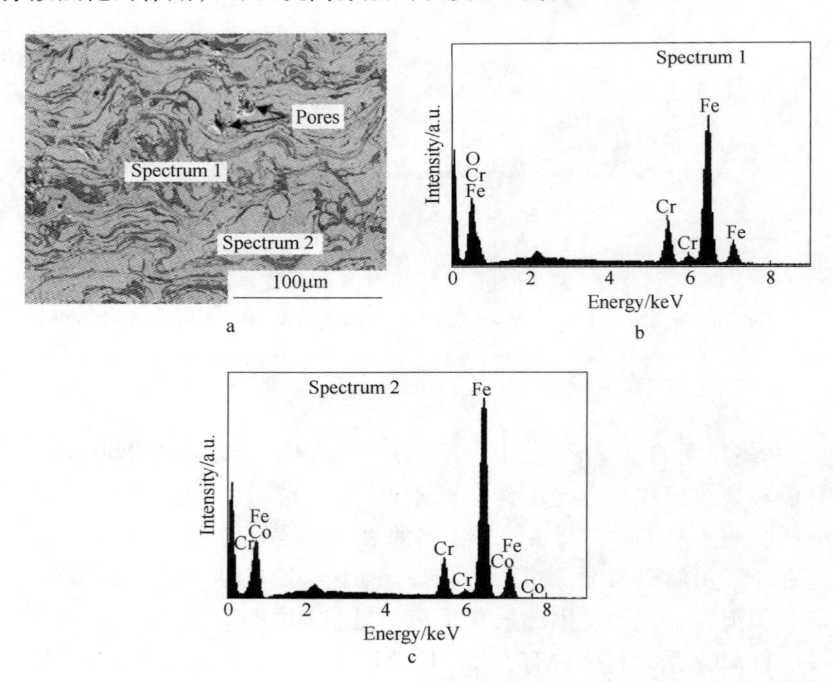

图3 3Cr13工作层的SEM照片及EDS成分分析

Fig. 3 SEM images and EDS spectra of 3Cr13 working coating

a—3Cr13 coating cross-section；b—Chemical composition of dark gray zone（spectrum 1）；
c—Chemical composition of light gray zone（spectrum 2）

另外，深灰色的断续条状组织致密地分布在浅灰色组织之间，成分分析含有 Cr、Fe、O，如图 3b 所示，与 XRD 相分析佐证，深灰色组织为 Fe_3O_4，Cr_2O_3 氧化物，喷涂过程中，高温熔滴在飞行过程中氧化不可避免，但表面薄层的复合氧化膜也可以对内部金属起到保护作用，降低了涂层整体的氧化物含量。氧化物区域存在少量微裂纹和孔隙，这是高速飞行的"熔滴"撞击到基体表面，在热应力的作用下，脆性氧化物发生开裂和破碎[16]。另外，少量氧化物 Cr_2O_3、固溶体（Fe，Cr）硬质相的存在可以提高涂层的耐磨性能。

2.3 复合涂层残余应力分析

高速电弧喷涂再制造曲轴涂层的残余应力过大，运行过程中，再制造曲轴在高载荷作用下，涂层应力集中加剧，引起涂层开裂、剥落，使曲轴报废失效。因此，有必要研究复合涂层的残余应力分布情况。涂层内部残余应力测试结果如图 4 所示。

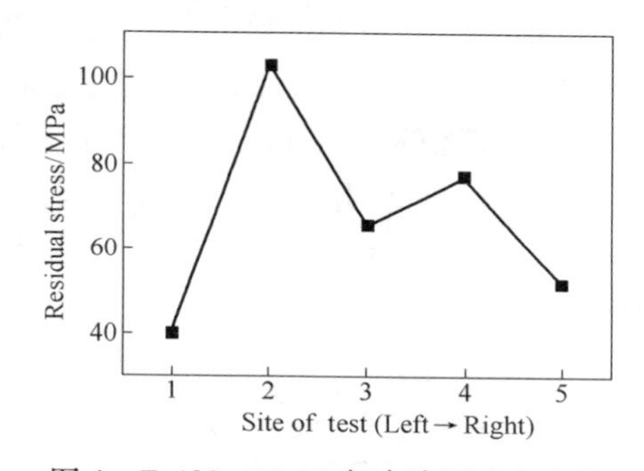

图 4 FeAlCr/3Cr13 复合涂层残余应力

Fig. 4 Residual stress of FeAlCr/3Cr13 composite coating

可以看出，涂层靠近边缘处（图 4 中横坐标的位置 1，5）残余应力值较小，在 40～50MPa 之间，中间残余应力值（图 4 中横坐标的位置 2，3，4）较高，在 65～106MPa 之间。平均残余应力约 67.6MPa。远远低于电弧喷涂 Fe 基非晶/纳米晶涂层的残余应力[17]（≈160MPa）。热喷涂产生的残余应力有两种[18]：一是高温熔滴撞击到基体，凝固收缩，但温度较低的基体或已沉积层对熔滴冷却收缩有一定束缚作用，产生收缩拉应力；二是喷涂后的冷却过程中，涂层和基体的线膨胀系数不匹配，产生热失配应力。

FeAlCr/3Cr13 复合涂层残余应力低的机理在于：（1）电弧喷涂高碳钢丝和 3Cr13 马氏体不锈钢丝材过程中，合金相会发生组织转变[19]，碳被大量烧损，同时生成贝氏体和马氏体，最终的相变产物诱发体积膨胀，产生了压应力，会部分抵消喷涂过程中的残余拉应力，使得涂层总体残余应力较低。但是，马氏体转变是在较低温度[20]（≈220℃）发生的，因此，基体或已沉积层表面温度不能太高。而涂层边缘残余应力低于中心部位的原因即涂层沉积过程中边缘散热速度大于中心部位，更易诱发马氏体转变，马氏体转变量越大，残余应力越低。（2）涂层中的微裂纹和少量空洞也能释放部分残余应力。涂层中空洞和微裂纹的分布不均匀一定程度上也解释了涂层残余应力的分布不规律性。个别点应力值较高的原因是喷涂时丝材的抖动使得沉积斑的形状发生变化，这一随机因素会影响特定位置的分布结果。（3）由于氧化物与周围金属具有不同的线膨胀系数，冷却过程中会产生热失配应力，而复合涂层氧化物含量少，尤其 FeAlCr 打底层未检测到氧元素，相应减小了热失配应力的产生[19]，这些都很好地解释了复合涂层具有低残余应力的原因。

2.4 复合涂层力学性能分析

高速电弧再制造曲轴的使用寿命很大程度上取决于涂层与基体的结合强度和涂层的内聚强度，曲轴是在交变载荷作用下工作的，涂层如果结合不好，长时间工作后，涂层会发生剥落，一些微小颗粒混合在润滑油里，造成曲轴表面的磨粒磨损，发生划伤，严重时会造成曲轴轴颈失圆报废。因此，在基体表面先喷涂新型 FeAlCr 粉芯丝材作为打底层，可以提高涂层的结合强度。FeAlCr/3Cr13 复合涂层/基体界面的结合强度测试值见表 2。可以看出复合涂层具有较高的结合强度值，约 46.6MPa，和喷涂 Ni95Al 涂层的结合强度[21]（≈49.38MPa）相当，远高于喷涂 3Cr13 不锈钢涂层的结合强度[22]（≈34.2MPa）。这是因为在喷涂过程中，高速高温熔融粒子撞击到粗化的基体表面，在冷凝时收缩与基体喷砂面或已沉积层面的“凹凸”点咬合，紧紧地结合在一起，涂层的内聚强度较高。另外，已有研究表明[23]，涂层中氧化物的存在会减小喷涂粒子的润湿性，降低涂层的结合强度，且氧化物与周围金属的线膨胀系数不匹配，导致脆性增大，涂层的内聚强度降低。

表 2 复合涂层的结合强度

Table 2 Adhesive strength of the FeAlCr/3Cr13 composite coating (MPa)

Coating	Test values					Average value
FeAlCr/3Cr13	52	41	48	50	42	46.6

喷涂 FeAlCr 打底层提高了复合涂层结合强度的原因在于：丝材中添加了自净化元素 Si 和 B，对打底层的成分分析表明，打底层不含氧元素，这就避免了氧化物降低涂层结合强度的因素。同时，丝材中含有 Ni 和 Al 两种元素，大量文献报道[24~26]，喷涂时，Al 和 Ni 会与空气中的氧气发生氧化反应放出大量的热量，使熔融的镍铝合金粒子的温度进一步升高，高温高速粒子撞击基体表面后，因放热反应而不会迅速冷却，粒子到达基材表面时温度较高，提高“熔滴”的流动性和铺展性，提高了喷涂时粉料熔化的体积分数，降低未熔粉料颗粒含量，能改善“叠层”的整体性和铺展性，使熔滴在基体表面能更好地润湿与铺展，减少了未熔颗粒、氧化物和孔洞等缺陷，改善涂层与基体界面的结合状态，从而提高涂层与基材的咬合能力。

另外，工作层与打底层之间也具有较好的结合强度，这是因为喷涂 3Cr13 工作层在已沉积层表面，相比基体喷砂粗化面，其表面粗糙度更大，也没有污染物，这些因素的共同作用促使了复合涂层具有较高的结合强度。

涂层的硬度与耐磨性有一定的关系，也能反映涂层抗承载能力[27]。硬度测试结果为：基体硬度 $HV_{0.1}$ 约 2000MPa，复合涂层平均硬度值 $HV_{0.1}$ 约 4000MPa，高于基体的硬度。涂层硬度值波动较小，表明涂层组织均匀、致密，孔隙率小，涂层的抗承载能力较高。

2.5 复合涂层摩擦磨损性能分析

FeAlCr/3Cr13 复合涂层和基体在 50N 载荷下室温油润滑过程中的摩擦系数变化曲线如图 5 所示。可以看出复合涂层的摩擦系数低于基体，基体的摩擦系数约为 0.5，且在后期（10 ~ 15min）摩擦系数有上升趋势，说明基体耐磨性能随着摩擦时间的延长而降低，基体的摩擦系数不稳定，这是因为随着摩擦过程的进行，摩擦温度上升加快[28]，摩擦表面温度升高，润滑油的黏度减小，导致油膜厚度减小，油膜失去了应有的承载和润滑作用，导致摩擦副直接接触，基体发生塑性变形，加速了磨损速率，耐磨性能降低。

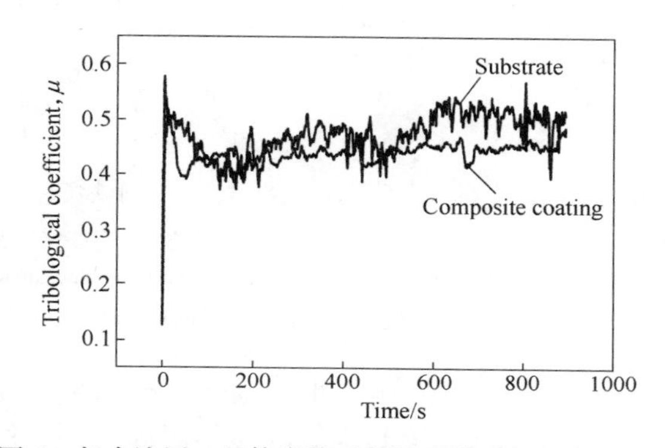

图 5 复合涂层、基体摩擦系数随摩擦时间的变化关系

Fig. 5 Changes of tribological coefficients between the composite coating and the substrate with time (Oil lubricant, Load = 50N)

复合涂层的摩擦系数在前 1min 略有上升，这是因为摩擦副（GCr15 球）硬度较高，摩擦方式为球 - 面接触摩擦，初期接触强度较高，摩擦系数波动较大，1min 后摩擦系数迅速下降，这与文献[29]的研究吻合，摩擦系数的变化与摩擦过程中的实际接触面积有关。复合涂层的平均摩擦系数约为 0.43，整个摩擦过程中也有略微波动，这正是因为涂层中存在硬质碳化物和固溶体弥散分布在 α - Fe 软基体上，硬质相的耐磨性高于周围基体，当摩擦副与硬质点接触摩擦时，润滑油的存在使摩擦过程中摩擦副之间始终保持一层油膜，被先磨掉的韧性相形成一些微小的凹点，有储油功能，因此始终在摩擦副之间有一层油膜存在，减缓了磨损速率，避免了涂层与摩擦副的接触磨损失效。复合涂层的摩擦系数在整个摩擦过程中变化平稳，说明复合涂层的耐磨性较好。

复合涂层和基体的二维和三维磨痕形貌对比如图 6 所示，复合涂层的磨痕深度和宽度都小于基体，定性分析见表 3。对比图 6b 和 d 可以看出，基体摩擦后磨痕周围挤出严重，说明基体硬度低，抗承载能力差。而复合涂层表面硬度较高，承载能力强，磨痕表面比较平整，仅有少量挤出现象，且磨痕底部有一定的粗糙度，说明软质相金属基体先磨损，而硬质相耐磨性高于基体，可以起到“骨架”支撑作用[30]。随着摩擦时间的延长，韧性相先被磨掉，露出起硬质点，导致摩擦面硬度提高，则硬度（H）升高，而弹性模量（E）是涂层抵抗变形的能力，是涂层的固有物理参数，在摩擦过程中变化不大。因此，摩擦过程中 H/E 增大，耐磨性增强。

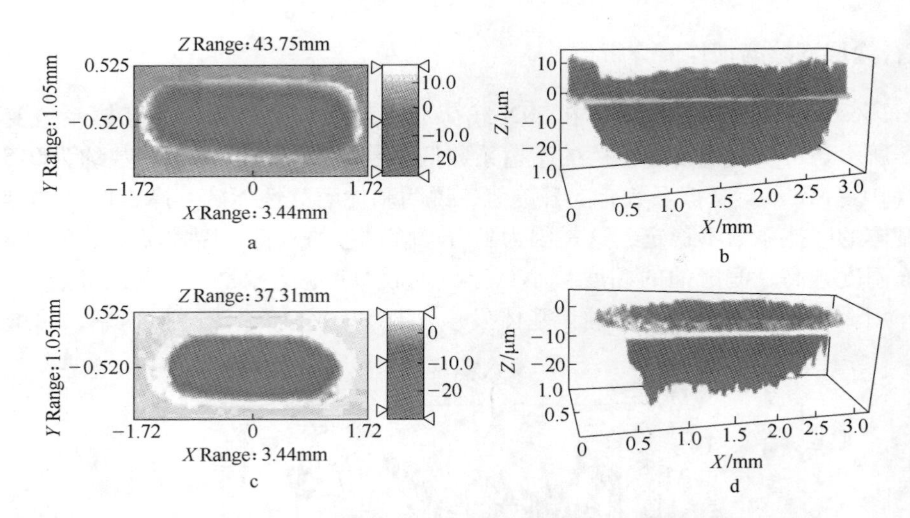

图6　载荷为50N时室温油润滑条件下磨痕形貌

Fig. 6　Two/three - dimensional morphology of wear track at room temperature and load = 50N

a—Two - dimensional morphology of the substrate；b—Three - dimensional morphology of the substrate

c—Two - dimensional morphology of the coating；d—Three - dimensional morphology of the coating

从表3中还可见，复合涂层的磨损体积小于基体，相同载荷作用下，材料的磨损量与磨损距离和硬度有关[31]，复合涂层和基体的摩擦时间相同，磨损距离相同，而复合涂层的硬度高于基体，复合涂层在摩擦过程中随着硬质点的凸出，涂层硬度提高，磨损率会呈减小的趋势，这使得再制造曲轴的使用寿命有很大的提高。

表3　复合涂层、基体磨痕参数（载荷50N，油润滑）

Table 3　Parameters of oil lubricate wear track for the composite coating and substrate（load = 50N）

Parameter	Width/mm	Depth/μm	Volume/μm^3
FeAlCr/3Cr13	2.43	37.31	1.983×10^{-6}
Substrate	2.94	43.75	2.429×10^{-6}

3　结论

（1）高速电弧喷涂制备的FeAlCr/3Cr13复合涂层组织均匀、致密。相成分主要由α - Fe、（Fe，Cr）固溶体、碳化物$Cr_{23}C_6$和少量氧化物组成。喷涂FeAlCr粉芯丝材作为打底层，丝材中Ni和Al发生放热反应，对熔融粒子有二次加热作用，促进了粒子间的致密结合，同时添加的B和Si元素有脱氧和净化基体表面的作用，涂层间和基体的结合强度较高，平均结合强度达46.6MPa，远大于喷涂3Cr13涂层结合强度，和喷涂Ni_3Al涂层的结合强度相当。

（2）复合涂层沉积过程中发生了马氏体相变，体积膨胀部分抵消了热喷涂产生的固有拉应力，涂层整体平均残余应力较低，约67.6MPa。由于涂层致密度高，含有大量固溶体和碳化物硬质相，平均显微硬度较高，$HV_{0.1}$约4000MPa，从基体界面到涂层硬度的平缓过渡提高了涂层的抗承载能力。

（3）涂层在油润滑高载摩擦条件下的耐磨性高于基体，机理在于：复合涂层中

(Fe，Cr）固溶体、碳化物 $Cr_{23}C_6$ 硬质相均匀分布在韧性相 $\alpha-Fe$ 基体上，起到弥散强化作用，有效的抵抗了黏着磨损；韧性相是易磨损部位，有储油功能，提高了涂层在油润滑条件下的摩擦稳定性，减缓了磨损率。

参 考 文 献

[1] Gao Yixin（高一新），Ren Weixia（任伟霞），Li Quanan（李全安）. *Heat Treatment*（热处理），2004，19(4)：46.

[2] Xu Binshi（徐滨士），Ma Shining（马世宁），Liu Shican（刘世参）. *Journal of Mater Protect*（材料保护），2000，33（1）：1.

[3] Chen Yongxiong（陈永雄），Xu Binshi（徐滨士），Xu Yi（许一）. *China Surface Engineering*（中国表面工程），2006，19(5)：169.

[4] Gao Weiguo（高为国），Liu Jinwu（刘金武），Yi Jiming（易际明）. *Journal of Hunan Engineering Academic*（湖南工程学院报），2004，14(4)：27.

[5] Liang Xiubing（梁秀兵），Chen Yongxiong（陈永雄），Bai Jinyuan（白金元）. *China Surface Engineering*（中国表面工程），2010，23(2)：112.

[6] Lei Hong（雷宏），Xiao Yeping（肖业平）. *China Surface Engineering*（中国表面工程），2003，59(2)：36.

[7] Takanori Takeno，Hiroyuki Shiota，Toshifumi Sugaware，et al. *Diamond and Related Materials*，2009，18：406.

[8] Martin Silber，Martin Wenzelburger，Rainer Gadow. *Surface and Coating Technology*，2008，202：4525.

[9] Bobzin K，Ernst F，Richardt K，et al. *Surface and Coating Technology*，2008，202：4438.

[10] Tamura H. *Journal of Thermal* [J]，1998，7(1)：87.

[11] Verdon C，Karimi A，Martin J L. *Material Science Engineering*，1998，246：11.

[12] Arabi Jeshvaghani R，Shamanian M，Jaberzadeh M. *Materials and Design*，2011，32：2028.

[13] Wang Fang（王方）. Shanghai：Shanghai Jiaotong University，2008.

[14] Cheng Jiangbo（程江波）. Shanghai：Shanghai Jiaotong University，2009.

[15] Luo Laima（罗来马）. Hangzhou：Zhejiang University，2010.

[16] Xu Weipu（徐维普）. Shanghai：Shanghai Jiaotong University，2004.

[17] Liu Yan（刘燕）. Beijing：Academy of Armored Forces Engineering，2010.

[18] Chen Yongxiong（陈永雄），Liang Xiubing（梁秀兵），Liu Yi（刘毅）. *Material and Design*（材料设计），2010，3：43.

[19] Rayment T，Hoile S. *Metalicll Material Transaction*，2004，35B(6)：1113.

[20] Cui Zhongqi（崔忠圻）. *Metallurgical and Heat Treatment*（金属学与热处理）[M]. Beijing：Mechanical Industry Press，1980：250.

[21] Song Yongli（宋永利），Hao Yanping（郝延平），Zhang Zhongli（张忠礼）. *Journal Jilin Engineering Technology Teacher*（吉林工程技术师范学院学报），2003，19(3)：1.

[22] Yang Hui（杨晖），Wang Hangong（王汉功），Liu Xueyuan（刘学元）. *Mechanical Engineering*（机械工程），1999，23(3)：14.

[23] Hu Junzhi（胡军志），Chen Xuerong（陈学荣），Ma Shining（马世宁）. *Surface Technology*（表面技术），2003，32(4)：18.

[24] Cezary Senderowski，Zbigniew Bojar. *Surface & Coating Technology*，2008，202：3538.

[25] Edrisy A, Alpas A T. *Thin Solid Films*, 2002, 338: 420.
[26] Mohanty M, Smith R W. *Wear*, 1996, 198: 251.
[27] Kashani H, Amadeh A, Ghasemi H M. *Wear*, 2007, 262: 800.
[28] Rayment T, Hoile S, Grant P S. *Metallic and Material Transaction*. 2004, 35B: 1113.
[29] Bowden F P, Tabor D. *Friction and Lubrication of Solid* [M] . Oxford: Clarendon Press, 1954.
[30] Fukumoto M J. *Thermal Spraying Technology*, 2008, 5: 17.
[31] Archard J F. *Journal Applied Physical*, 1953, 24: 981.

Properties of the FeAlCr/3Cr13 Composite Coating Applied in Remanufacturing Crankshaft by High Velocity Arc Spraying

Tian Haoliang[1,2], Wei Shicheng[2], Chen Yongxiong[2], Tong Hui[2], Liu Yi[2], Xu Binshi[2]

(1. Beijing University of Aeronautics & Astronautics, Beijing, 100037, China;
2. Academy of Armored Forces Engineering, Beijing, 100072, China)

Abstract The working life of a crankshaft remanufactured by robot - based automatic high velocity arc spraying has the relation with wear resistance and adhesive strength of its coating. This paper studied the characteristics and properties of a newly - designed composite coating FeAlCr/3Cr13 aimed at improving the martensitic stainless 3Cr13 coatings. Microstructures, residual stress and wear resistance of the composite coating were investigated by SEM, XRD and CETR wear tester, respectively. The results show that when the FeAlCr cored wire is used as prime coating the elements react fully. The composite coating compact and uniform and its porosity is about 9.87%. The phases of the coating are $\alpha-Fe$, $Cr_{23}C_6$, (Fe, Cr) solid solution and Ni_3Al, among which the content of oxide is as low as 3.2%. The average residual stress, microhardness and adhesive strength of the coating are 67.6MPa, $HV_{0.1}$ 4000MPa and 46.6MPa, respectively. The FeAlCr/3Cr13 composite coatings under the dry or lubricating abrasion have higher wear resistance than the substrate.

Key words high velocity arc spraying, crankshaft, FeAlCr/3Cr13 composite coating, adhesive strength, wear resistance

再制造工程管理

着眼实践寻蹊径，联想创新攀高峰*

摘　要　从个人成长经历阐述了知识的积累过程和世界观的形成过程，以及对后来发展进步所起到的基础性作用。剖析了自己在选择课题、科学研究、推广应用、理论创新、人才培养等方面取得成就的基本经验，即：选准方向、勤奋不懈，突出特色、跨越发展，实践奠基、理论升华，资源重组、综合创新。展望了自己研究领域今后的发展。

关键词　科研思维　表面工程　创新　展望

1　思维特色形成背景

1931 年我出生时，正值日本帝国主义发动九一八事变，大规模武装入侵我国东北，使东三省人民沦为亡国奴，从此，东北老百姓被奴役、被宰割，过着牛马不如的悲惨生活。我上小学时，受到的是奴化教育，只让学伪满洲国语和日语，只让知道日本的天照大神和日本的历史，不让知道自己是炎黄子孙，不准讲自己是中国人。然而，不甘做亡国奴的父母私下里告诉我：我们是中国人，老家在山东，日本帝国主义是侵略者，从此在我幼小的心灵里种下了痛恨日本侵略者和伪满洲国，要为赶走侵略者，使国家富强起来而奋斗的种子。

"八·一五"苏联红军解放哈尔滨，尚未公开的党组织已经活跃在各个阶层，我念书的四中校长和不少老师都是共产党员。他们传播革命道理，宣传新民主主义能够救中国，使我受到启蒙教育。后来在党的教育下，我懂得了共产党和国民党的不同；知道了以蒋介石为首的蒋、宋、孔、陈四大家族的腐败；知道了蒋介石的不抵抗政策才造成了东北的沦陷，使东北人民饱尝了 14 年亡国奴之苦。我也由此建立起这样的信念：只有共产党才能救中国。因此，我坚决拥护共产党，决心跟着共产党把落后的旧中国建设成为富强民主的新中国。当时有少数人盲目相信"正统"，跑到长春国民党统治区去了，我坚定地留下来，并考入了共产党领导的哈尔滨工业大学。那时，家里很穷，供不起我上大学，是党实行的供给制，使我读完了大学。我家祖祖辈辈没有大学生，我是家里第一个大学生，是党培养了我。

在哈工大，学校请来的前苏联教授介绍了前苏联的情况，特别提到建设社会主义需要大量的科技人员。对比我们国家当时落后的状况，科技人才的缺乏，我深感学习科学技术的重要，因此特别努力发愤苦读。从那时起我就下定决心，为国家的富强贡献自己的一生。

哈工大不仅具有光荣的革命传统，而且具有严谨优良的校风。学校非常重视所培养人才的质量。老校长高铁经常讲："学校水平的重要标志是培养人才的质量"。为了保证人才的质量，学校始终强调和坚持严格要求、规范化训练，使学生在校时学得扎

* 原发表于《院士思维》，安徽教育出版社，2000：558～575。

实，基础打得牢，“规范、严格、功夫到家”。学校制定了系统完整的教学大纲，使学生从一年级到五年级，通过一系列课程的训练，逐步具备一个工程师所需要的基本技术素质。在哈工大本科毕业后，我又进入了研究生班，但是未读完一年，因国防建设需要，把我抽调到中国人民解放军军事工程学院任教。

军事工程学院是我党创建的第一所高等军事工程学校，她既是我军高等技术人才的教育中心，又是尖端武器装备的研究中心。毛泽东主席为学院亲题训词。我经常背诵毛主席在训词中的一句话：“今天我们迫切需要的就是要有大批能够掌握和驾驭技术的人，并使我们的技术得到不断的改善和进步”，它激励着我勇攀科技高峰。

经过哈工大 5 年的学习，我具备了工程师最基础、最基本的素质和能力。调到军事工程学院以后，长期的军旅生涯，使我树立了正确的世界观、人生观，学习和继承了人民军队的光荣传统，坚定了为国防建设献身的决心。平时比较注意学习党的路线、方针和政策，从中掌握正确的立场、观点和方法，培养自己以马克思主义理论为指导、运用唯物辩证法不断解决新的实际问题的能力。在各种学术交流中，一方面学习别人科技方面的新成果、新知识，另一方面注意剖析别人获得成功的思路和方法。所有这些为我后来在国防科技上不断创新奠定了基础。

2　思维亮点

2.1　选准方向，勤奋不懈

一个科技工作者，最大的成就是为国家、为科技发展做出自己的研究成果，因此，从事研究工作首先要确定的是把握好研究方向，选择好研究课题。科研方向的选择有两种情况：一种是始终如一；一种是灵活变换。在科技史上采用这两种思维路线，都有不胜枚举的成功事例。

我选择的是始终如一的思维路线，为什么会选择机械维修工程作为我终生为之努力的研究方向和研究课题呢?

1954 年，服从组织需要调到军事工程学院装甲兵工程系，从事坦克维修专业的教学工作，开始我对这一工作认识并不足，认为只有搞设计、制造才是高水平，才大有作为。后来有两件事对我震动很大，一件是我给苏联坦克维修专家当翻译时得知，苏联坦克、机械化部队维修部门在卫国战争中抢修了 43 万辆次坦克和装甲车辆，相当于苏联战时最高年产量的 15 倍。这对保证装甲部队的持续战斗力、战胜法西斯起到了极其重要的作用。后来又从一份材料中得知，在第四次中东战争中，参战坦克数量多，损伤率高，但以色列军队快速修复能力强，损伤坦克的修复率高达 86%，而埃及、叙利亚军队快速抢修能力差，修复率只达到 34%，结果使以军摆脱了战争中的被动局面。这个例子又说明了装备维修在现代战争中的重要作用。另一件是下部队调查时，看到因维修设备和技术落后，许多局部磨损的坦克零部件不能修复，整个部件只好报废，既造成很大的浪费，又严重影响正常训练和战备任务的完成，干部战士们急得团团转。这两件事让我的心情久久不能平静，深深感到维修工作无论在战时还是在平时都具有极为重要的意义和作用，于是暗下决心：一定要用所学的知识改变部队维修技术的落后状况。从此，我选择了机械维修专业作为自己毕生奋斗的事业。

20 世纪五六十年代，我军坦克薄壁零件的修复技术相当落后，被称为坦克修理中的顽症。我的重要成果就是从修复坦克薄壁零件开始的。当时我和助手们既缺乏资料，又没有经验，仅在一本苏联杂志上找到了一条关于振动电弧堆焊修复薄壁零件的简要报道，而这一设备是什么样子，我们谁也没见过。为了将资料变成实际的设备，我就和助手们按照杂志上介绍的原理，一步一步地摸索着干，面对每次失败，我们都认真总结经验，相互鼓励，重新开始。经过100多个日日夜夜的苦干，终于在国内首次研制成功了振动电弧堆焊设备，摸索出了新工艺，解决了薄壁零件修复的难题，突破了部分坦克薄壁零件不能修复的禁区。1958年该成果参加了全国科技成果展览会，受到党中央和中央军委的表扬，组织上还给我记了二等功。不久，我的课题组又研制成功了“水蒸气保护振动电弧堆焊”，薄壁零件的修复质量进一步提高。在部队推广过程中，我们又通过改造老设备，研制出了两种新的振动堆焊头（该设备获得了全国科学大会奖），使坦克零件的修复范围逐步扩大。看到一辆辆“趴窝”的坦克又驰骋训练场，心里感到无比欣慰。

振动电弧堆焊虽然解决了坦克薄壁零件的修复问题，但修复后的零件质量只能接近新品，为了提高易损零件的耐磨性，我四处寻找解决的办法。1973年，我利用休假机会到哈尔滨锅炉厂参观，受到了采用等离子堆焊技术制造高压阀门来提高零件耐磨性的启发。返回学院，我就提出将等离子喷涂技术用于坦克零件维修并开始进行试验。这时，有的同志持怀疑态度；有的同志不断提醒我，一旦试验失败，将会造成坦克车辆损坏的大事故，会给自己带来什么后果；个别领导还要我深刻检查，把试验停下来。我想，搞科研是部队建设的需要，试验不能停，压力再大也要顶住。经过教研室全体同志的不懈努力，终于试验成功了等离子喷涂修复坦克零件的技术。正当第二天上级机关要来检查等离子喷涂应用于坦克修理的进展情况时，当天下午学院召开了悼念周恩来总理大会，经院领导批准，我化悲痛为力量，全力投入到设备的调试工作直到深夜，当我回到办公室时，发现门上有一张大字报，其内容是批判我没有参加追悼大会。我认为把科技搞上去，把国防建设搞上去是周总理的遗愿，在特殊情况下可以用不同的方式悼念，这张大字报使我更加坚定了为落实周总理关于我国四化建设的教导终生奋斗的决心。装甲兵首长来检查后，很快决定安排六辆坦克进行两轮各一个大修期的实车考核，考核结果证明应用等离子喷涂技术修复后的坦克零件，耐磨性比新品提高1.4~8.3倍，而成本只有新品价格的1/8，有效地延长了被修零件的使用寿命，大大提高了坦克装甲车辆的持续作战能力，节约了大量装备维修经费，为维修体制和维修制度的改革奠定了技术基础。

回顾走过的路，深感搞科技研究工作，关键是研究方向和研究课题的选择。一旦确定了方向、选准了课题，就必须有坚持不懈的决心和百折不挠的毅力，一步一个台阶地前进，否则就会一事无成。只有在长期的实践中始终如一、循序渐进地坚持对一个领域进行探索，才可能取得对这个领域较深刻的理解，把认识提高到更高的层次上去，从而取得创造性的成果。然而，研究方向和课题的选择又要紧紧与国家和部队的需要联系在一起，只有时刻想着部队，从部队的实际出发，才能不为困难压倒，不为挫折屈服，不为名利诱惑，奋勇攀登，不断进取，才能有所作为和贡献。在一定意义上可以说，国家的需要、军队建设的需要、战场的需要是我们从事科技研究永不枯竭的动力。

2.2 突击特色，跨越发展

在浩瀚的表面科学领域，不可能什么都去研究，只能有所为，有所不为。最首要的问题是要明确自己的工程对象和使命。我的工程对象是军事装备维修中的表面技术问题，使命是研究适合平时和战时装备维修需要的表面工程新技术、新材料、新工艺，以提高装备的使用寿命，缩短维修工时，降低维修成本，便于机动保障，并在此基础上促进装备制造的创新或“军转民”为国民经济建设服务。明确了自己的工程对象和使命并为之奋斗，就能逐渐形成自己研究工作的固有特色。

和固有特色相对应的就是自己的思维特色，这就是针对自己的工程对象和使命所采取的途径、方法和原则。特色的形成是一个长期的、渐进的历史过程，它既是自己成功经验的总结、优势的积累，又是在此基础上的发展和追求。跨越发展是我从事科研与教学工作的一条重要原则，也是时代的要求，因而也成为一条重要的思维创新特色。

20 世纪 70 年代末期，当国内兴起镀铁热的时候，我没有加入此项表面技术的研究，而是把适合大型设备现场修复的先进电刷镀技术作为研究方向，组织力量、协作攻关。从电刷镀设备到镀液，从工艺到应用进行了系统的研究。其中和中科院上海有机化学研究所共同研究的 200 多种电刷镀镀液，涵盖了工程需要的方方面面，为我国电刷镀技术的发展奠定了良好基础。在进行实验室系统研究的基础上，又组织了电刷镀修复坦克零件的实车考核，不断改进，逐步推广。电刷镀技术设备体积小、工艺简单、修复速度快。修复后坦克零件的耐磨性是新品的 4.3 倍，成本只有新品的 1/10，该项技术不仅解决了坦克现场修复的难题，而且还能应用于民用事业，解决了国家重点工程的多项急需，有力地支持了经济建设。例如 1983 年天津石油化纤厂的一台进口大型设备主轴损伤，被迫停产，日本专家认为中国修不了，必须购买日本制造的新轴，时间需 3 个月，仅停产一项损失就将是 2700 万元。而我们采用刷镀技术，只用了 14 天，花了 1.3 万元就修复了这台大型设备，并安全运行至今，质量良好。电刷镀技术在新装备和民用机械的维修与制造中发挥了重要作用，取得了巨大的效益。国家国民经济发展计划中，从“六五”到“八五”计划，都把电刷镀列为重点推广的先进技术，据 1995 年统计，全国电刷镀行业已创造经济效益 30 多亿元。电刷镀研究与应用在 1985 年获国家科学技术进步一等奖。最近我又组织人力把电刷镀技术的研究跨越到纳米材料的层次上，制备出多种特殊功能的复合涂层，将解决工程应用中的关键难题。

Fe_3Al 是一种抗高温冲蚀的好材料，而且成本较低，被誉为“穷人用的不锈钢”。但是过去只能用铸造的方法来获取。最近我采用高速电弧喷涂的方法制备出了 Fe_3Al 涂层，突破了 Fe_3Al 无法应用于零件表层的难题。以 Fe_3Al 为基础再与多种硬质粉末相复合，可以制备出抗高温氧化、硫化及抗冲蚀磨损的涂层，在军用装备和电站锅炉管道上有广阔的应用前景。

最近几年，在热喷涂技术方面，我重点抓了市场份额不断增加有发展前途的电弧喷涂技术。我立足于对现有电弧喷涂枪的结构改进，在技术上使它跨越到超音速喷涂水平上，而不是去全面研究喷涂设备，结果是研究周期短、投入少、喷涂效果显著、喷枪价格低廉、便于推广应用。我还研究出了用于电站锅炉的抗热腐蚀材料，使性价比超过了美国的45CT丝材。我把电弧喷涂技术的研究成果及时应用到军舰甲板的防腐处理上，寿命可达15年以上，减少了军舰维修时间和费用，提高了在航率，科技形成了战斗力，1997年该成果获军队科技进步一等奖。

在研究生的培养途径上，没有采用传统的先学课程后做课题的培养模式，而是先参加课题研究，再带着问题学习课程。实践证明这种模式培养出的人才水平高，能力强。年轻的毕业研究生有的获得了军队科技进步一等奖，有的获得了很有应用价值的专利。

这些例子进一步印证了坚持跨越思维的可行性和必要性。在跨越二字上下工夫，就能逼近研究前沿，就能形成自己独特的研究特色。

2.3 实践奠基理论升华

机械维修工程技术是实践性很强的工作，脱离了实践，维修技术既没有发展的源泉，也会失去其用武之地。因此，我一直注重经常深入部队，到部队的实践中去探索课题，在技术研究的过程中也时时想到部队的实践需要，从部队的实际出发，力求创造的新技术、新设备、新工艺在部队用得上，用着方便，容易掌握，能够很好地解决部队装备维修实践中的问题。

常言说："实践出真知"，我深深体会到这句话的真理性。在技术上能取得一些创新性或前沿性成果，不可能是凭空的突发奇想而一蹴而就，必须扎根于坚持不懈、脚踏实地的实践活动和长期经验、知识的积累才能成功。

然而每一次技术的创新，都经过刻苦的研究和反复的试验，都付出了大量的精力和体力，逐步逼近前沿，取得进展。但是，随着实践的深入，却在脑子中逐渐产生这样的想法，我从事的维修工程技术有没有共同的规律可循？能否有更好的方法取得更大、更快、更多的效果呢？答案就是向理论层次上提高。20世纪80年代初期，我积极组织了维修理论的研讨，建立了维修理论体系，促进了维修工程的发展。80年代后期我又探讨了表面技术的理论问题。在一段较长的时期内，我反复思考着。从振动电弧堆焊到等离子喷涂，由金属电刷镀到耐磨修复添加剂，它们的作用都是解决维修机件的表面，都是从解决修复中的表面问题出发，他们有一个共同的着眼点和基点，这就是"表面"。磨损在表面发生，腐蚀从表面开始，疲劳损伤由表面向基体延伸。我想如果在"表面"二字上狠下工夫，把我们进行的实践工作，上升到理论，找出表面维修的本质问题，找出具有规律性的东西，建立起"理论"体系，这样岂不可以在理论的指导下，从自发的研究走向自觉的研究，从感性的层次上升到理性的层次，这样就可能走捷径，取得"多、快、好、省"的成果。

从现代科技研究的角度讲，科技研究的成败和发展，常常取决于理论上是否取得突破性的进展。理论上的进展，又往往在于对研究过程和研究对象的深入分析。

我想，维修表面技术，无论是等离子喷涂也好，电镀（槽镀）也好，金属电刷镀也好，从它们的技术含量来说，都跨越了多种学科，其过程十分复杂，纷繁多变，但它们一定有共同的东西。于是我开始"去粗取精、去伪存真、由表及里、由此及彼"地进行归纳和总结，去寻找质的东西、核心的东西、共性的东西，即经过理论思维和科学抽象把感性经验的东西上升到理论高度。经过研究，发现在维修的表面中，其关键是表面的失效机理及表面物质与母体物质的物质属性、物质结构、相容性、结合性等基本因素。对这些基本因素必须从宏观和微观角度不断深入认识，要从宏观上、微观上的有序性，从分子–原子更深层次的结合原理、机制等方面加以理论解释。这期间我参阅了大量国内外有关资料，了解到20世纪80年代初期，英国伯明翰大学教授T. Bell建立了世界上第一个表面工程研究所，1985年创办国际《表面工程》杂志，同时将国际热处理联合会改名为国际热处理与表面工程联合会。我借鉴了国外的做法并根据多年的理论研究和实践经验，在1986年提出创建具有中国特色的表面工程学科的设想，并进行了一系列的组织和学术研究工作。1987年，在中国机械工程学会和同行专家的支持下，建立了我国第一个表面工程研究所，我任所长。1988年，又主持创办了我国第一本《表面工程》杂志，现已发展成为向国内外正式发行的科技期刊《中国表面工程》，受到国内外同行的重视和欢迎。1993年正式成立中国机械工程学会表面工程分会，我任副主任委员，从1988年至今已主持召开了三届全国表面工程学术会议、两届中日表面工程学术研讨会、两届国际表面工程学术会议。根据我军装备维修现代化发展的需要，于1991年我向上级建议建立了全军装备维修表面工程研究中心并组建了全军表面工程推广网，1996年又建立全军装备表面工程重点实验室。与此同时撰写论文和专著，阐述表面工程学科体系、基础理论与技术，从而使表面工程理论在我国确立并迅速发展起来，从维修表面技术开始，逐步发展到成为系统的综合性的表面工程学科，已成为先进制造技术的组成部分，在国民经济和社会生活的各个领域发挥着日益重要的作用。

上述例子生动地体现了实践和理论的辩证关系。实践是基础，但感性的、经验的东西，只有上升到理性的高度，才能把握事物的本质，才能发挥更大的作用。恩格斯曾指出："一个民族要想站在科学的最高峰，就一刻也不能没有理论思维"。这的确应该成为我们每位科技工作者牢牢记住的座右铭。

2.4 资源重组综合创新

有人说"综合就是创新"，我认为这个论断不全面，应该是"会综合就能创新"。举一个简单的例子，包肉包子先从养猪开始，养猪先从种粮食开始，这种思路要尽量避免。充分利用现有资源，买来面粉、肉馅、青菜和调味品就可以很快做出包子，这就是综合。但是，不同的原料配比，不同的拌馅顺序，做出的包子口味大不相同，这就是要善于综合。

我把我在机械维修技术创新中的思维规律归纳成八个字："交叉、综合、复合、系统"。

所谓"交叉"，就是注意多种学科和技术的交叉运用，相互借鉴和渗透，从而形成新的学科、新的技术。坦克零件等离子喷涂技术就是借鉴航空工程领域相关技术，进行创新而产生的。

所谓"综合"，突出表现在各种思维方法、科学理论和技术领域在技术创新中的综合运用。

现代科学技术发展有许多特点和趋势，科学技术一体化和多学科综合化是其最重要的特点和趋势。可以说，在现代，无论是科学上的发现还是技术上的发明创造，已经都不能离开科学理论的指导和技术手段的支持，而且往往表现为跨学科、多领域、各种技术的相互渗透和综合运用。所以在技术创新的过程中不能固守或单纯应用某一思维方式和技术方法，必须开展广泛的联想和思维的发散。表面工程理论就是在摩擦学、腐蚀学、材料学、表面物理化学、冶金学、力学和先进的制造技术理论、现代维修理论等的指导下，把多种表面维修技术综合起来，通过科学的思维过程建立起来的。

所谓“复合”，是指在技术的创新和创造过程中，注意将不同的理论、方法、技术有机地连接起来成为一个整体，在这个整体中既保持和发挥原有理论、方法、技术的相对独立性和优势，又具有新的整体优势。例如，热喷涂与激光重熔的复合，热喷涂与电刷镀的复合，胶黏技术与电刷镀的复合，表面强化与润滑技术的复合，金属材料基体与非金属材料涂层的复合等，多种表面技术在工程上的复合应用可以发挥各种技术和材料的综合优势，取长补短，取得 1 +1 >2 的复合协同效应。

所谓“系统”，是指在技术创新和创造的全过程，都必须坚持系统化原理，始终把理论、方法、技术的发展作为一个整体。经常进行系统分析，既善于分解，又要善于综合，把握系统中的各种环节和要素，突出重点，兼顾一般，推动科学技术的稳步发展。

近两年我从事的“再制造工程”研究，就是把设备的制造与运行作为一个整体，根据全寿命周期管理理论，全面考虑设备和零部件设计、制造、运行、维修和报废的全过程，运用系统思维方法，统筹考虑装备的前半生和后半生、装备的制造与再制造，从而使退役产品经过“再制造”，实现在后半生对环境的负影响最小、资源利用率最高的情况下，重新达到最佳的性能要求。“再制造工程”实际上是一个以优质、高效、节能、节材、环保为目标的系统工程。“再制造”已经不同于一般的维修，它将高新技术及其成果系统地应用于设备的后半生，使维修形成产业化。从技术的角度讲，再制造工程不是各种维修技术的简单叠加，而是发生了质的飞跃，成为先进制造技术的重要组成部分，是先进制造技术的补充和发展。

3　学科前沿

21 世纪已经来临，展望机械维修工程的未来，用得上两句老话：“前途是光明的，道路是曲折的”“机遇与挑战同在”。我对未来充满信心，但这种信心是和迎接挑战，战胜各种各样困难的决心及脚踏实地的工作联系在一起的。今后机械维修工程发展的趋势和重点大致有以下几个方面。

3.1　表面工程在 21 世纪要有大的发展，它将成为 21 世纪主导工业发展的关键技术之一

表面工程是固体材料通过表面预处理后，经表面涂覆，表面改性或多种表面技术复合处理，改变固体金属表面或非金属表面的形态、化学成分、组织结构和应力状态等，以获得所需要的表面性能的系统工程。表面工程技术从 20 世纪 80 年代形成后得到迅速的发展，至今已发展成为以多学科交叉、综合、复合、系统为特色的新兴学科。其最大优势是能够以多种方法制备出性能优异的表面功能薄层，使零件整体具有比基体材料更高的耐磨性、抗腐蚀性、抗疲劳性和耐高温性等性能。表面工程属于先进制造技术，同时他又对制造业技术创新提供了必要的工艺支持。它可以促进机械产品结构的创新、材料的创新和工艺的创新。表面工程在维修中的应用不仅保障了设备的正

常运行，较好地解决了进口设备配件问题，而且还为新一代产品的设计制造积累了丰富的经验。在电子信息产业中，表面工程是制造薄膜材料及其功能器件必不可少的手段，成为高新技术发展的主要技术基础。

近年来表面工程的发展异常迅速，尤其复合表面技术日益受到各方面的高度重视。国际上都在努力研究和应用各种提高零件表面性能的新技术和新工艺。众多知名专家预言，表面工程在21世纪要有大的发展，它将成为21世纪主导工业发展的关键技术之一。

3.2　制造与维修工程越来越趋于统一，装备的寿命周期理论将更受高度重视和普遍应用

以往的制造与维修处于相互脱节的状态，分属于不同的部门。例如，我军装备的设计与制造由国防工业部门完成，装备一旦配发部队，装备的管理、使用和维修等都由部队自己承担。也就是说，装备的前半生在工业部门，后半生却在部队。因此，工业部门往往只重视装备的设计、制造，很少考虑使用和维修，把使用和维修看做是用户即部队自己的事情，前、后半生严重脱节。这样，一方面造成许多由设计、制造所带来的使用与维修的不便；另一方面造成对现役装备的服役年限缺乏全面科学的认识，缺乏系统、科学的判定方法和模式。

随着科学技术的进步和装备的发展，人们越来越认识到必须对装备进行全寿命周期的管理及其费用分析，即不仅要考虑装备的论证、设计、制造，而且还要考虑装备的使用、维修直至退出现役的废品处理。

随着21世纪的来临，制造与维修工程越来越趋于统一。未来的制造与维修工程将越来越重视全寿命周期理论的研究与应用，将全面考虑设备和零部件设计、制造和运行的全过程，构成一个以优质、高效、节能、节材为目标的系统工程。

3.3　21世纪的机械维修将是“绿色机械维修工程”——再制造工程

机械维修的根本特征是现有设备的再利用，它具有节能、节材、节约资源，特别是减少原材料初级加工等重要特点，在保护环境和减少对环境的损害方面具有突出的作用。因此，机械维修具有天然的“绿色工程”的特性，符合可持续发展的战略。

但是，传统的机械维修还具有较明显的手工业作坊的特征，不能适应不断扩大的对机械维修的需求，不能充分发挥机械维修在“绿色工程”中的作用，因此有必要按照机械制造工程那样采用专业化的、批量的、流水作业的生产方式来改造机械维修生产，这样再制造工程则应运而生。

再制造工程是先进制造技术的一个重要组成部分和发展趋势，是一个统筹考虑产品部件全寿命周期的系统工程。再制造以优质、高效、节能、节材、环保为主要目标，利用原有废旧零件，采用高新表面工程技术（涂层与改性等）及其他先进加工技术，使零部件恢复尺寸、形状和性能。再制造产品的一个重要特征是依靠高新技术的应用使其性能、可靠性和寿命等都得到较大的提高，达到或超过原产品的质量，使退役产品在对环境的负影响最小，资源利用率最高的情况下重新达到最佳的性能要求。因此，再制造工程是一种极具潜力的新型产业。21世纪的机械维修必将是“绿色机械维修工程”，也就是说再制造工程在新的世纪中将得到更大的发展。

装备再制造工程学科的建设和发展*

摘　要　阐述了装备再制造工程的内涵、国内外研究现状以及装备再制造工程学科体系。重点从装备再制造工程特定的研究对象、坚实的理论基础、独立的研究内容、具有特色的研究方法与关键技术、国家重点实验室的建立这5个方面论述了装备再制造工程这一新兴学科的形成。指出再制造工程是先进制造工程的重要组成部分，是表面工程等先进技术在废旧机电产品上的综合应用，也是对维修工程的继承和发展。装备再制造工程学科及装备再制造技术国防科技重点实验室的建设必将促进装备维修保障的新发展，加速旧武器装备的延寿和升级改造，对军队现代化建设将起到重要作用。

关键词　装备再制造工程学科　全寿命周期管理　再制造关键技术　表面工程

1　引言

随着21世纪的到来，保护环境资源、实现可持续发展，已经成为世界各国共同关心的问题。机电产品制造业是最大的资源使用者，也是最大的环境污染源之一[1]。传统的制造模式向着可持续发展的模式转变就是从高投入、高消耗、高污染的传统发展模式向提高生产效率、最高限度地利用资源和最低限度地产出废物的可持续发展模式转变。再制造工程是节约资源、保护环境、旧装备升级改造和为发展新装备积累经验的有效途径，是维修工程的继承与发展，是符合国家可持续发展战略的一项绿色系统工程[2,3]。发展再制造工程是贯彻党的“十六大”精神，走“新型工业化”道路的重要举措。

2　装备再制造工程的内涵

与旧产品回收后以原材料形式再利用相比，再制造可以最大限度地挖掘出旧产品中蕴含的大部分资源。在多数产品中，附加价值（如劳动力价值、能源、资金等）在产品成本中占绝大部分。传统的再循环无法获取原产品的附加价值，因此也就降低了产品的基本价值。而再制造则获取了原产品的许多附加值，例如，发动机的再制造，只需生产新发动机所需能源的50%和劳动量的67%。工业专家认为再制造与初始制造的原料耗费量之比为1:（5～9），消费者购买1台再制造产品将比购买新产品少花50%～75%的钱[4]。

装备再制造工程是一个以装备全寿命周期设计和管理为指导，以优质、高效、节能、节材、环保为目标，以先进技术和产业化生产为手段，来修复、改造废旧（包括战损）装备的一系列技术措施或工程活动的总称[2]。简言之，装备再制造工程是高科技维修的产业化，是装备维修的重要发展方向。

---

* 本文合作者：朱胜、马世宁、刘世参、梁秀兵。原发表于《中国表面工程》，2003(3)：1～6。

20 世纪 80 年代以来，再制造工程在工业发达国家已经受到高度的重视[5]。美国不仅有全国性和行业性的再制造研究中心，而且在大学开设再制造工程方面的课程。据报道，至 1996 年美国专业化再制造公司数量超过 73000 个，直接雇员 48 万人，生产 46 种主要再制造产品，每年的销售额超过 530 亿美元。欧盟通过一项有利于环保的新规定，其要点是从 2000 年起所有报废车辆由汽车制造厂商免费回收，从 2002 年起废旧汽车的可再生利用率将达到 85%，到 2015 年达到 95%[4]。

再制造工程的研究已引起美国国防决策部门的重视，美国国家科学研究委员会制订了 2010 年国防工业制造技术的框架，提出未来所需制造能力的发展战略，2005 年将实现制造与维修一体化。美国军队目前是世界上最大的再制造受益者，美军的武器装备大量使用再制造部件。美国国防部已把“新的再制造技术”列入 2010 年及其以后优先发展的国防制造工业的新重点[6]。

我国的再制造工程虽然起步较晚，但已受到国家机关、军队总部、学术界和企业界的高度重视。2002 年 2 月出版的由科技部前部长朱丽兰同志主编的全国干部学习读本《21 世纪干部科技修养必读》（人民教育出版社）一书中阐述到“环境保护技术的重要发展方向：从回收利用废旧物向‘再制造’发展”。目前，清华大学、上海交通大学等高校正在开展有关绿色设计制造、面向“3R”（reuse，remanufacture，recycle）的设计、产品的全周期过程等研究[7]。济南“复强”、上海“大众”等企业正在针对汽车发动机再制造开展产品研发、生产与销售，并且已初具规模。再制造发动机的性能与新品相当，但成本却仅为新品的 50% 左右。民用航空器维修中采用了许多再制造产品。北京军区空军 93407 部队也已建设汽车发动机再制造工程的生产线。农业部农业机械化研究所开展履带式拖拉机的再制造研究与生产，取得了明显的经济效益和社会效益[8]。一个具有巨大潜力的再制造产业正在我国的民用与军用领域蓬勃兴起。

3 装备再制造工程学科建设的进展

1999 年 6 月，在西安召开的“先进制造技术”国际会议上，徐滨士等发表了《表面工程与再制造技术》的学术论文[9]，在国内首次提出了“再制造”的概念。

1999 年 12 月，在广州召开的国家自然科学基金委员会机械学科前沿及优先领域研讨会上，徐滨士应邀作了《现代制造科学之 21 世纪的再制造工程技术及理论研究》报告，经国家自然科学基金委员会工程与材料学部机械学科发展前沿研讨会讨论，同意将“再制造工程技术及理论研究”列为国家自然科学基金机械学科发展前沿与优先发展领域。标志着再制造工程技术与理论的研究受到了国家自然科学基金委员会的重视和认可[2]。

2000 年 3 月，在瑞典哥德堡召开的第 15 届欧洲维修国际会议上，徐滨士等发表了题为《面向 21 世纪的再制造工程》的会议论文[3]，这是我国学者在国际维修学术会议上首次发表“再制造”论文。

2001 年 4 月，“新世纪表面工程与再制造工程学术方向研讨会”在装甲兵工程学院召开，国内外学者近百人参加了会议，会议就再制造工程的研究现状和发展趋势进行了深入研讨，初步提出了再制造工程的未来发展方向。

2000 年 11 月 18 日，《光明日报》“为‘十五’计划献计献策征文”栏目发表了建议文章——《大力推广绿色再制造工程》（作者：徐滨士，石来德，易新乾）引起了原国家发展计划委员会的重视，作者应邀参加了国家计委组织的座谈讨论[10,11]，并获得二等奖。

2000 年 12 月，由徐滨士等 12 位院士及多名专家完成了《绿色再制造工程及其在我国应用的前景》咨询报告[12]，中国工程院将此报告呈报国务院。国务院办公厅批转政府 10 部委研究参阅。“再制造工程”的研究与发展得到了我国政府的高度重视。

2001 年 5 月，经总部批准，在装甲兵工程学院建立“装备再制造技术国防科技重点实验室”，装备再制造工程得到了我军的高度重视。2002 年 4 月，装备再制造国防科技重点实验室第一届学术委员会在北京召开，标志着实验室“边建设、边运行”阶段的开始。

2001 年 6 月，《中国表面工程》杂志刊登了以“再制造工程”为主题的系列文章，集中阐述了再制造工程的研究与发展现状、基本的学科体系框架及其产业化前景等[13]。

2002 年 6 月，在第 184 次香山科学会议上，徐滨士作了特邀大会报告，题目为《绿色再制造材料成型加工关键技术及其基础》[14]，报告引起了与会科学家的广泛重视，大家认为，再制造工程的研究与发展“利在当代，功在千秋”。

2002 年 9 月，“再制造基础理论与关键技术”被批准为国家自然科学基金重点项目（项目批准号：50235030）。此前，2001 年 9 月，“再制造设计基础与方法”已被列为国家自然科学基金面上项目（项目批准号：50075086）。标志着再制造工程的基础研究纳入国家自然科学基金优先资助领域。

4 装备再制造工程的学科体系

随着科学技术的进步和装备的发展，装备全寿命周期管理及其费用分析已受到世界各国的高度重视。装备发展的实践证明，装备全寿命周期管理不仅要考虑装备的论证、设计和制造，而且还要考虑装备的使用、维修直至退出现役的废品处理。装备再制造工程是以装备的全寿命周期理论为基础，以装备“后半生”中报废或改造等环节为主要研究对象，以如何开发和应用高新技术翻新与提升装备性能为研究内容，从而保障装备后半生的高性能、低投入和环境友好，为装备后半生注入新的活力。装备再制造工程学科是在装备维修工程、表面工程等学科交叉、综合的基础上建立和发展的新兴学科。按照新兴学科的建设和发展规律，装备再制造工程以其特定的研究对象，坚实的理论基础，独立的研究内容，具有特色的研究方法与关键技术、国家级重点实验室的建立及其广阔的应用前景和潜在的巨大效益，构成了相对完整的学科体系，体现了先进生产力的发展要求，这也是装备再制造工程形成新兴学科的重要标志。装备再制造工程的学科体系框架概括如图 1 所示。

4.1 装备再制造工程的研究对象

装备再制造工程的研究对象为废旧装备。从装备全寿命周期图[15]（图 2）可以看出，以往的装备从设计、制造、使用、维修至退役报废后，一部分是将可再生的材料进行回收，一部分将不可以回收的材料进行环保处理。维修在此过程中主要是针对装备零部件局部损坏而进行的性能恢复工作。再制造则是在整个装备报废后，对报废的

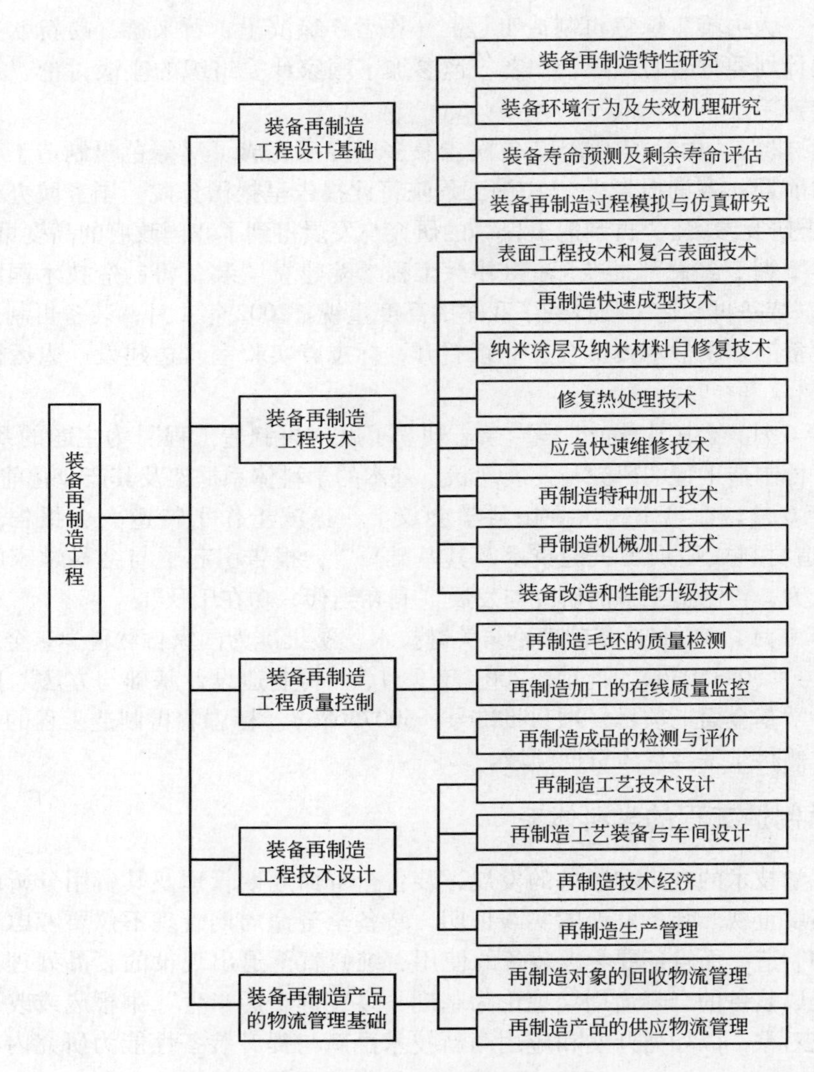

图 1　装备再制造工程学科体系框架

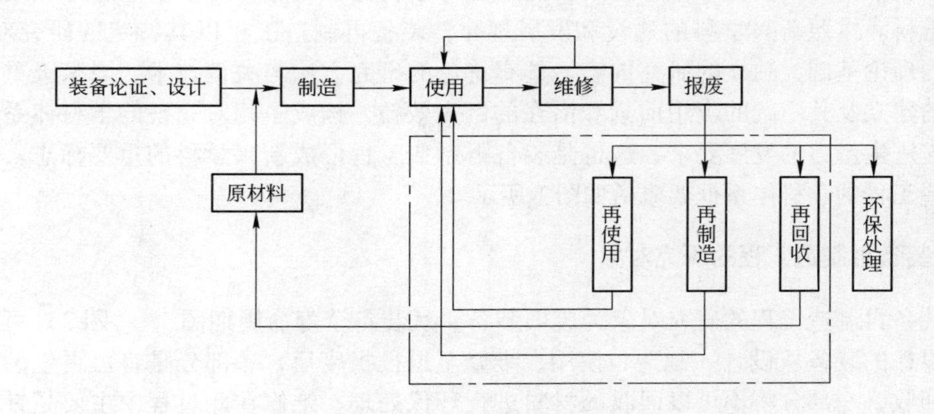

图 2　再制造对装备全寿命周期模式的拓展

装备按照规定的标准、性能指标，通过先进的技术手段进行的系统加工过程。再制造过程不但能提高装备的使用寿命，而且可以影响、反馈到装备的设计，最终使装备的全寿命周期费用最小，保证装备产生最高的效益。此外，再制造虽然与传统的回收利用有类似的环保目标，但回收利用只是重新利用它的材料，往往需要消耗大量能源并二次污染环境，其产品是低级的，属于废旧机电产品资源化处理的低级形式。再制造是一种从原部件中获取最高附加价值的好方法，可以获得等同于新品或者高于新品性能的再制造产品，属于废旧机电产品资源化处理的高级形式。

再制造与传统制造的重要区别在于毛坯的对象不同。再制造的毛坯是已经加工成型并经过服役的零部件，这种毛坯的使用性能恢复甚至提高，有很大的难度和特殊的约束条件。除一般的制造技术外，还必须应用再制造成型技术，如表面工程的各种方法和技术，一些原位修复方法和技术等。而这些方法和技术，在装备的原始制造中往往是不需要、不便用或不许用的。

再制造还是一个对旧机型升级改造的过程。科学技术的不断发展，先进技术的不断涌现，以旧机型为基础，不断吸纳先进技术、先进部件，可以使旧产品的某些重要性能大幅度提升。既具有投入少、见效快的特点，又为下一代装备的研制积累了经验。

4.2 装备再制造工程的理论基础及其研究内容

装备再制造工程是通过多学科综合、交叉和复合并系统化后正在形成中的一个新兴学科。它包含的内容十分广泛，涉及机械工程、材料科学与工程、信息科学与工程、环境科学与工程等多种学科的知识和研究成果。装备再制造工程融汇上述学科的基础理论，结合装备再制造工程实际，逐步形成了废旧产品的失效分析理论、剩余寿命预测和评估理论、再制造产品的全寿命周期评价基础以及再制造过程的模拟与仿真基础等[15,16]。此外，还要通过综合分析废旧装备技术、经济和环境三要素对恢复装备性能的影响，完成对废旧装备或其典型零部件的再制造特性研究与分析。

4.3 装备再制造工程的关键技术

废旧装备的再制造工程是通过各种高新技术来实现的。在这些再制造技术中，有很多是及时吸取最新科学技术成果的关键技术，如先进表面技术[17,18]、微纳米涂层及微纳米减摩自修复材料和技术[19]、修复热处理技术[20]、应急快速维修技术、再制造毛坯快速成型技术[17]及过时产品的性能升级技术等。再制造工程的关键技术所包含的技术种类十分广泛，其中各种表面技术和复合表面技术，主要用来修复和强化废旧零件的失效表面，是实施再制造的主要技术。由于废旧零部件的磨损和腐蚀等失效主要发生在表面，因而各种各样的表面涂敷和改性技术应用得最多；微纳米涂层及微纳米减摩自修复技术以微纳米材料为基础，通过特定涂敷工艺对表面进行高性能强化和改性，或应用摩擦化学等理论在摩擦损伤表面原位形成自修复膜层的技术，可以解决许多再制造中的难题，并使性能大幅度提高；修复热处理是一种通过恢复内部组织结构来恢复零部件整体性能的特定工艺；应急修复技术是用来对战伤装备或现场作业设备进行应急抢修的各种先进快速修复技术；再制造毛坯快速成型技术是根据零件几何信息，采用积分堆积原理和激光同轴扫描等方法进行金属的熔融堆积、快速成型的技术；过时产品的性能升级技术不仅包括通过再制造使产品强化、延寿的各种方法，而且包括产品的改装设计，特别是引进高新技术或嵌入先进的部组件使产品性能获得升级的各种方法。除上述这些有特色的技术外，通用的机械加工和特种加工技术也经常使用。

4.4 装备再制造工程的质量控制、技术设计与物流管理

再制造工程的质量控制中，毛坯的质量检测是检测废旧零部件的内部和外部损伤，从技术和经济方面分析决定其再制造的可行性。为确保再制造产品的质量，要建立全面质量管理体系，尤其是要严格进行再制造过程的在线质量控制和再制造成品的检测。再制造工程的质量控制是再制造产品性能优于或等同于新品的重要保证。

再制造工程的技术设计包括再制造工艺过程设计，工艺装备、设施和车间设计，再制造技术经济分析，再制造组织管理等多方面内容。其中，再制造的工艺过程设计是关键，需要根据再制造对象——废旧零件表面的运行环境状况，提出技术要求，选择获得零件表面性能的工艺手段和材料，编制合理的再制造工艺，提出再制造产品的质量检测标准等。再制造工程的技术设计是一种恢复或提高零件表面二次服役性能的技术设计。

再制造产品的物流管理可以简单概括为再制造对象的回收物流管理和再制造产品的供应物流管理两方面。合理的物流管理能够提高再制造产品生产效率，降低成本与提高经济效益。再制造产品的物流管理也是控制“假冒伪劣”产品冒充再制造产品的重要手段，是当前制约再制造产业发展的“瓶颈”。

5 装备再制造技术国防科技重点实验室的建立

装备再制造技术国防科技重点实验室属于国家级重点实验室，应从国家、国防科技的角度重点研究再制造工程的应用基础理论，并开发研究装备再制造工程的关键技术，为国防和经济建设服务。重点实验室从装备维修保障及装备全寿命周期管理的要求出发，以废旧装备重要部件为研究对象，以先进再制造成型加工技术为主要手段，建设成为装备寿命预测与延寿评估，零部件再制造成型加工理论与技术，装备再制造产品质量控制与检测，报废件处理技术研究等领域的装备再制造工程基础理论与关键技术的开发研究基地，解决大型装备重要部件的再制造工程理论和技术问题，使再制造装备的全寿命周期费用最低，并提高装备的性能，延长装备的使用寿命。再制造工程重点实验室的研究工作最终将带动并促进装备维修的改革和装备维修保障的建设与发展。

6 结论

（1）装备再制造工程能节省军费、提升装备性能、提高保障能力，并能节约资源、保护环境，是一项利军、利国、利民的事业。

（2）装备再制造工程是多学科的交叉与融合，又有自己特定的研究对象，坚实的理论基础，独立的研究内容与方法。

（3）再制造工程已得到我国政府、军队有关部门、学术界和企业界的广泛认同与支持。装备再制造工程学科的建设和发展将加强再制造工程基础理论和关键技术的研究，促进军用装备和民用设备的延寿和升级改造；装备再制造技术国防科技重点实验室的建设对促进经济和国防现代化建设将发挥重要作用。

参考文献

［1］清华．中国环境污染状况备忘录［J］．世界环境．1998（2）：40～42.

［2］徐滨士，张伟，等．现代制造科学之21世纪的再制造工程技术及理论研究［C］．国家自然科学基金委员会机械学科前沿及优先领域研讨论文集．广州：1999.

［3］XU Binshi，ZHANG Wei，LIU Shican，et al. Remanufacturing Technology for the 21st Century［C］. Proceedings of the 15th European Maintenance Conference，March，2000 in Gothenburg，Sweden：335～339.

［4］Robert T. Lund. The Remanufacturing Industry-Hidden Giant［R］. Research Report，Manufacturing Engineering Department，Boston University，1996.

［5］Darryl Mleynek，Kathy hammes，Helen Wong，et al. Recycling/Remanufacturing HAWAI'I an Industry Report［R］. The Clean Hawai'i Center Department of Business，Economic Development，and Tourism State of Hawai'i，1999.

［6］中国兵器工业第二一零研究所．2010年及其以后的美国国防制造工业［R］．北京：1999.

［7］机械科学研究院．先进制造技术发展前瞻研究报告集，1999.

［8］梅书文，杨金生，张福学．履带拖拉机再制造工程的质量体系和效益分析［J］．中国表面工程. 2002，15（4）：9～13.

［9］XU Binshi，ZHANG Zhenxue. Surface Engineering and Remanufacturing Technology［C］. International Conference on Advanced Manufacturing Technology'99，Xi'an. New York Press，1999：1129～1132.

［10］徐滨士，易新乾，石来德．大力推广绿色再制造工程［N］．光明日报．2000－11－18.

［11］王政．国家计委问计于民［N］．人民日报，2000－12－17.

［12］徐滨士，等．绿色再制造工程及其在我国的应用前景［R］．工程科技与发展战略咨询报告集，中国工程院，2002.

［13］徐滨士，李仁涵，梁秀兵．绿色再制造工程的进展［C］．中国表面工程．2001，14(2)：1～25.

［14］徐滨士，朱胜，马世宁，等．绿色再制造材料成型加工关键技术及其基础［C］．第184次香山科学会议论文集，2002.

［15］徐滨士，马世宁，刘世参，等．21世纪的再制造工程［J］．中国机械工程．2000（1～2）：36～39.

［16］朱绍华，刘世参，朱胜．谈绿色再制造工程的内涵和学科构架［J］．中国表面工程．2001，14(2)：5～7.

［17］徐滨士，朱绍华．表面工程的理论与技术［M］．北京：国防工业出版社，1999.

［18］徐滨士．表面工程的应用与展望［M］．1999/2000中国科学技术前沿（中国工程院版）．北京：高等教育出版社，2000.

［19］徐滨士，欧忠文，马世宁，等．纳米表面工程［J］．中国机械工程．2000，(6)：707～712.

［20］雷廷权．2010年中国的热处理［J］．金属热处理．1999（12）：1～4.

Construct and Development of Equipment Remanufacture Engineering Specialty

Xu Binshi, Zhu Sheng, Ma Shining, Liu Shican, Liang Xiubing

(National Key Laboratory for Remanufacturing, Beijing, 100072, China)

Abstract The concept, research status and discipline of remanufacturing engineering are introduced in this paper. Remanufacture engineering has special study objectives, solid theoretical foundation, independent research contents, and key technologies, which constituted the system of remanufacture engineering discipline. Remanufacture engineering is an important part of advanced manufacture engineering. It is a comprehensive application of advanced technologies used in rebuilding failed mechanical and electrical products. It is the succeeding and development of maintenance engineering. Construction and development of equipment remanufacture engineering and state key laboratory of remanufacturing will play important role in promoting the new discipline, accelerating the upgrade of old equipment and constructing military modernization.

Key words equipment remanufacture engineering, management of whole life cycle, remanufacturing key technologies, surface engineering

大力发展再制造产业*

建设节约型社会的核心是节约资源。对废旧机电产品进行再制造是节约资源的重要手段。

1 废旧机电产品中蕴含巨大财富

再好的铁矿也不如废钢，购新不如翻旧。“废旧物质”是能够再增长的物质，是“开采”成本低廉的“富矿”。

废旧机电产品中含有丰富的可再利用资源。当今世界各种物资的总量中，再生资源加工而成的钢占总产量的45%，铜为35%，铝为22%，铅为40%，锌为30%。据测算，目前我国可以回收而没有回收利用的再生资源价值高达300亿~350亿元。每年约有500万吨的废钢铁、20多万吨废有色金属、1400万吨废纸及大量的废塑料、废玻璃等没有回收利用。每回收利用1t废旧物资，可以节约自然资源4.12t，节约能源1.4t标准煤，减少6~10t垃圾处理量；每利用1t废钢铁，可炼钢850kg，相当于节约成品铁矿石2t，节能0.4t标准煤，而且用废铁炼钢周期短，如用铁矿石炼1t钢需8个工时，而用废钢铁炼1t钢只需要2~3个工时。现在，电器产品更新换代的步伐越来越快，报废电器产品中的元器件平均只用了2万小时，而元器件的平均寿命为50万小时，因此，报废电器中的元器件还具有旺盛的生命和足够长的使用寿命。

废旧机电产品中含有高附加值。以汽车发动机为例，原材料的价值只占15%，而成品附加值却高达85%。如果将发动机原始制造和再制造过程中的能源消耗、劳动力消耗和材料消耗加以对比，可以看出，再制造过程中由于充分利用了废旧产品中的附加值，能源消耗只是新品制造中的50%，劳动力消耗只是新品制造中的67%，原材料消耗只是新品制造中的11.1%~20%。在一台机器中，各部件的使用寿命不相等，每个零件的各工作表面的使用寿命也不相等，往往会因局部表面失效而造成整个机器报废。通过再制造工程对机器的局部损伤进行修复，可以最大限度地挖掘出废旧机电产品中蕴含的附加值，达到节省资金、节能、节材和保护环境的效果。

2 对废旧机电产品进行再制造是节约资源的最优途径

再制造是指以机电产品全寿命周期设计和管理为指导，以废旧机电产品实现性能跨越式提升为目标，以优质、高效、节能、节材、环保为准则，以先进技术和产业化生产为手段，对废旧机电产品进行修复和改造的一系列技术措施或工程活动的总称。简言之，再制造就是废旧机电产品高科技维修的产业化。

再制造的重要特征是再制造产品的质量和性能能够达到甚至超过新品，而成本只

* 本文合作者：刘世参、王海斗。原发表于《求是》，2005(12)：46~47。

为新品的50%，节能60%，节材70%，对环境的不良影响显著降低。再制造的对象是广义的，它既可以是设备、系统、设施，也可以是其零部件；既包括硬件，也包括软件。

高新技术的发展和应用是再制造产品在质量和性能上能达到和超过新品的根本原因。一部机电产品制造出来以后，经过若干年后才达到报废，而这期间科学技术迅速发展，新材料、新技术、新工艺不断涌现。对废旧机电产品进行再制造时，通过应用最新的研究成果，既可以提高易损零件、易损表面的使用寿命，又可以对老产品进行技术改造，使它的整体性能跟上时代的要求。

再制造工程包括以下两个主要部分：（1）再制造加工——对于达到物理寿命和经济寿命而报废的产品，在失效分析和寿命评估的基础上，把其中有剩余寿命的废旧零部件作为再制造毛坯，采用先进技术进行加工，使其性能迅速恢复，甚至超过新品。（2）过时产品的性能升级——性能过时的机电产品往往是几项关键指标落后，不等于所有的零部件都不能再使用，采用新技术镶嵌的方式对其进行局部改造，就可以使原产品跟上时代的性能要求。信息技术、微纳米技术等高科技在提升、改造过时产品性能方面有重要作用。

再制造不同于维修。维修是在产品的使用阶段为了保持其良好技术状况及正常运行而采取的技术措施。维修多以换件为主，辅以单个或小批量的零（部）件修复。而再制造是将大量相似的废旧产品回收拆卸后，按零部件的类型进行收集和检测，将有再制造价值的废旧产品作为再制造毛坯，利用高新技术对其进行批量化修复、性能升级，所获得的再制造产品在技术性能上能达到甚至超过新品。

再制造也不同于再循环。再循环是通过回炉冶炼等加工方式，得到低品位的原材料，而且回收中要消耗较多的能源，对环境有较大的影响。再制造是以废旧零部件为毛坯，通过高新技术加工获得高品质、高附加值的产品，消耗的能源少，最大限度地找回了废旧零部件中蕴含的附加值，且成本要远远低于新品。

实践证明，再制造是废旧机电产品资源化的最佳形式和首选途径。

3 废旧机电产品再制造对建设节约型社会的贡献分析

经济效益显著。1996年美国再制造工程涉及的8个工业领域中，专业化再制造公司超过7万个，生产46种主要再制造产品，年销售额超过530亿美元，接近1996年美国钢铁产业560亿美元的年销售额。其中汽车再制造是最大的再制造领域，公司总数为50538个，年销售总额365亿美元。资料表明，美国2002年资源化产业的年产值为GDP的1.6%。我国2020年GDP预计达到4万亿美元，如果以美国2002年资源化的水平作为我国2020年目标，则资源化产业年产值将达到640亿美元。

环保作用突出。废旧机电产品资源化可以减少原始矿藏开采提炼以及新产品制造过程中造成的环境污染；能够极大地节约能源，减少温室气体排放。美国环境保护局估计，如果美国汽车回收业的成果能被充分利用，大气污染水平将比目前降低85%，水污染处理量将比目前减少76%。

缓解就业压力。实施废旧机电产品再制造，将带动一批新兴产业，解决大量就业问题。美国的再制造业到2005年安排就业100万人。如果我国2020年达到美国2005

年的规模，也将会创造100万人的就业岗位。

向人民提供物美价廉的产品。由于再制造充分提取了蕴含在产品中的附加值，在产品销售时具有明显的价格优势。如再制造发动机，其质量、使用寿命、安全性能与新机相同，并有完善的售后服务，价格仅为新机的50%，可供不同收入阶层选用。

4 推进废旧机电产品再制造的建议

我国可进行再制造的资源十分丰富。据统计，2000年，我国汽车保有量为1 900万辆左右，达到报废标准的汽车210万辆。2005年，我国汽车保有量将达到3250万~3500万辆，2010年达4350万~4700万辆，每年报废的汽车将在200万辆以上。从2003年起，我国电冰箱年均报废400万台，洗衣机、电视机、电脑各在500万台以上。仅电子工业发达的东莞市，每月产生的印刷电路板、覆铜板边角料等电子产品垃圾就超过5000t，广东全省则超过8000t。我们要以建设节约型社会为契机，积极扶持再制造产业。

政策引导。政府有关部门应该制定一套行之有效的政策、法律和法规，允许再制造后的产品经标记后在市场出售；建立由国家有关部门监督的再制造企业质量认证体系；对高技术再制造企业给予贷款和税收方面的优惠。

示范试点。国家在重点抓好再制造示范企业的基础上，积极引导、建立一批专业化再制造企业群，或将一些有条件的修理企业优化组合、改造升级为再制造企业，使其采用高新技术和产业化方式生产再制造产品。大型装备制造企业应积极参加和支持自己产品的再制造，并在配件供应、销售和服务等方面提供方便。

组织关键技术攻关。发展再制造产业首先要对再制造的有关基础理论、关键技术等进行系统的研究和开发，对产品的再制造性的设计、标准及产品质量控制等方面进行研究，并及时进行成果转化。再制造企业要确保再制造产品的质量，即必须坚持严格的标准使所有再制造产品的性能等同或优于新品。应不断提高服务水平，任何环节的失误都会影响到再制造产业的健康发展。

建立制造商责任制。对一些报废量大的产品，如汽车、家用电器、电脑等，逐步建立由制造企业负责回收进行再制造处理的制度，让制造商从制造工程一开始就关注产品报废后的再制造。

唤起公众的参与意识。广大用户的理解与支持是推动再制造产业发展的重要因素。政府、企业及有关部门应通过有效的宣传工作，引导公众理解再制造的内涵和作用，消除公众对再制造的不正确认识。

发展装备再制造，提升军用装备保障力和战斗力*

摘　要　装备再制造是指废旧装备高技术修复、改造的产业化。大力发展装备再制造，可促进高技术在武器装备中的应用，推动武器装备维修保障方式的改变，是节省军费、提高装备性能的重要举措。装备再制造学科的建设是装甲兵工程学院学科建设新的增长点。通过加强宣传、组织试点、合力攻关、培养人才等综合措施，全力推进全军及学院装备再制造的发展。

关键词　装备再制造　装备维修　自主创新

2006年是“十一五”规划的开局之年。“十一五”期间，国家的经济发展将在科学发展观的指导下，将现有的以“两高两低”（即高投入、高消耗、低产出、低效益）[1]为标志的传统经济模式，逐渐转变为以“4R”（reduce 减量化、reuse 再利用、recycle 再循环、remanufacture 再制造）为原则[2]，以“两低一高”（低消耗、低排放、高效率）为特征的循环经济模式。为了与国家经济发展相适应，我军武器装备的研制与管理也应坚持“4R”原则，才能够以最低的经费投入发挥现有军事装备最佳的作战效能。

“4R”中，再制造是最活跃且最能体现自主创新和高技术含量的要素。装备再制造工程是指以装备全寿命周期理论为指导，以废旧装备性能实现跨越式提升为目标，以优质、高效、节能、节材、环保为准则，以先进技术和产业化生产为手段，进行修复、改造废旧装备的一系列技术措施或工程活动的总称[3]。简言之，装备再制造工程是废旧装备高技术修复、改造的产业化。再制造的重要特征是再制造后的装备质量和性能达到或超过新品，成本只是新品的50%，同时节能60%，节材70%，对环境的不良影响显著降低。装备再制造是先进制造的组成部分，也属于绿色制造；同时，装备再制造也是装备维修的组成部分，是维修发展的高级阶段。

1　装备再制造是节省军费、提高装备保障力和战斗力的重要举措

美军军费远远超过我军，2003年度是我军的十余倍，但是他们极为重视再制造，对装备实施再制造已成为美军为维持其庞大武器库的运转而采取的战略性措施。

1989年6月，美国国会两院通过了一项不经总统签署即可获得执行的共同决议[4]，要求国防部将更多的重点放在对现役军用武器装备的再制造上，并将再制造作为一个过渡手段，以在财政预算有限、新装备配备时间延迟以及新装备费用高昂的情况下，维持装备的战备完好率，特别是用来延长现有武器装备的服役寿命。

* 原发表于《装甲兵工程学院学报》，2006，20(3)：1～5。

隶属于美国国家科学研究委员会的“2010年及其以后国防制造工业委员会”制订了2010年国防工业制造技术的框架，提出达到未来所需制造能力的战略，并将武器系统的性能升级、延寿技术和再制造技术列为目前和将来国防制造重要的研究领域。美国波士顿大学的防御武器研究部门，专门研究包括航天飞机在内的各类武器装备再制造产品的经济性数据。

上述一系列措施使得美军成为再制造的最大受益者。

美军B－52战略轰炸机于1948年设计，1962年生产，经过1980年、1996年两次再制造技术改造，到1997年时平均自然寿命还有13000飞行小时，可服役到2030年。美军自2000年起，在5年内完成了269架阿帕奇直升机的再制造，今后10年继续完成750架的再制造。再制造后该直升机成为美国现役武装直升机中战斗力最强、性能最先进的一种机型[5]。

2000年，美陆军启动了坦克装备综合管理计划，主要目的是通过再制造将M1A1系列坦克恢复成新品状态，该计划是在军方资金不足的情况下，为维持大约7000辆“艾布拉姆”系列坦克的运转而采用的一个替代方案。美军M1A2坦克是M1系列的再制造改进型。1985年8月，美陆军将368辆M1坦克再制造升级成M1A2型，1996～2000年的5年间，又有580辆M1坦克被再制造成M1A2型[6]。

与美军相比，我军的装备采购经费更为有限，因此实施装备再制造必将成为我军装备维修技术保障的一项战略性选择。值得欣慰的是，我军在利用再制造提升装备性能方面有了喜人进展，而且具有鲜明的中国特色。美军的再制造除技术改造升级外，在恢复性再制造方面以换件为主。而我军的再制造则融入了表面工程技术，通过一系列先进表面工程技术的创新应用，使废旧发动机的旧件利用率由72%提升到90%，而且耐磨损、耐腐蚀等性能得到大幅度提升[7]。我国最大的发动机再制造公司济南复强动力有限公司，以装备再制造技术国防科技重点实验室为智力和技术支撑，通过利用先进表面工程技术，如高速电弧喷涂技术、微纳米电刷镀技术、纳米自修复添加剂技术等，革新了发动机的再制造技术，其成本只有新品的一半，质量却超过了新品，被总装备部机关列为军用发动机再制造的试点厂家。装备再制造技术国防科技重点实验室对20世纪六七十年代装备我军修理部（分）队的老旧机床进行了再制造。首先，利用各种表面工程技术对机床的磨损部位进行修复；其次，对机床进行数控化改造。再制造机床的成本只有新品的1/6 ～ 1/3，性能得到大的提升。已完成再制造机床100余台，显著提高了部（分）队的维修加工水平和加工能力[8]。

目前，我们正在“十一五”预研项目的支持下开展“装备零件战场快速再制造平台系统”研究，该平台是集信息技术、微纳米技术、快速成型技术、装备维修等为一体的武器装备先进维修系统，是一个能够在战场上靠近需要的位置迅速制造或再制造破损/战损失效零部件的机动式智能再制造系统。该系统可以实现战场装备的无备件保障和需求件的快速生成，实现装备机动性能和作战能力的快速恢复，进而为装备维修体制改革提供理论和实践指导；并将大大缩短零部件的供应链，简化后勤供给，降低维修成本，提高维修效率。

2 装备再制造可使我军装备大修的质量迈上新台阶

高技术装备要求高技术维修。我军装备的常规维修制度分为小修、中修、大修3

个层次。在总部机关的领导下，我军的维修制度改革已经取得了显著成果，“以可靠性为中心的维修”“基于状态的维修”“精确维修”“智能维修”“先进维修”“野战抢修”“智能自修复”等研究取得重要进展，但是在装备大修这个领域尚没有大的突破，存在着技术相对落后、标准不高、资源浪费等诸多问题。

美军等西方发达国家军队对装备不采用大修而采用再制造。再制造与大修的主要区别是再制造后的装备性能要达到或超过原型机的新品，而我军的大修达不到这一质量要求。以某型装备的发动机为例，新机的保质期是数百小时，大修后有一定下降。如果通过再制造能使该发动机的保质期达到新品水平，或再延长至一定时间，则军事效益、经济效益都相当巨大，也会带来维修制度的一系列变革。

随着科技的发展，新技术、新工艺、新材料大量涌现，尤其是微电子技术的发展，为装备通过再制造达到和超过新品的质量要求创造了条件。

3 装备再制造促进了新技术在武器装备上的应用

信息技术、微纳米技术和生物技术是21世纪的高新技术，对废旧装备进行再制造时利用这些新技术，既可以提高易损零件、易损表面的使用寿命，又可以对老装备进行技术改造，使它的整体性能跟上时代的要求。

装备再制造技术国防科技重点实验室在纳米表面工程技术、材料制备与成型一体化技术、快速成型应急抢修技术、纳米原位动态减摩自修复添加剂技术、无损检测技术与焊接新技术方面进行重点突破，有力推动了新技术在再制造中的应用。

在普通电刷镀技术基础上，通过添加无机陶瓷纳米颗粒，解决了纳米颗粒在多离子体系水溶液中的均匀分散与稳定悬浮，以及非导电的无机纳米颗粒与基质金属的共沉积等重大难题，进一步发展了纳米电刷镀技术[9]。利用该技术成功完成了某型进口飞机发动机压气机叶片的再制造，突破了国外对该飞机维修技术和设备的垄断，已被空军批准用于批量修复。

为了实现武器装备在运行中的自修复和再制造，通过制备出粒径可控的纳米金属及纳米化合物，研制成功具有自主知识产权的M6纳米减摩自修复添加剂[10]，经过军用吉普车和主战坦克的台架试验和实际使用，有效实现了汽车和坦克发动机运行中自修复与再制造的目标。

我军水陆两栖坦克在服役一段时间后车体会出现裂纹，直接影响军事斗争准备。通过应用先进的无损检测技术和先进材料与焊接技术，实现了两栖坦克车体裂纹的修复与再制造。首先利用最新的金属磁记忆无损检测仪快速探测被漆层覆盖的裂纹，而后利用新型焊接材料对裂纹进行焊修，再对焊缝进行超声波冲击以释放应力，最后对一些薄弱部位进行应急堵漏，防腐固化等技术实施复合处理，从而完成两栖坦克车体的再制造，节约了大量经费，保持了坦克的完好状态，为军委的重大决策提供了重要依据。

4 装备再制造促进了装备保障学科的发展

军事院校新学科点的创建与发展是一项复杂而漫长的系统工程。装备再制造是我院在全国首先倡导并全力推动的，我院现在已成为国内公认的“再制造国家队”。在这种情况下，集全院之力早日将装备再制造建设成我院新的重点学科，并将其做强，就成为我院刻不容缓的战略性任务。

建设装备再制造学科不仅是我院形势发展的需要，也是我军装备维修保障技术创新发展的需要。装备再制造，针对的是废旧及战伤的装备，要求“快速”“靠前”“高质量”地完成大规模修复或野战抢修，装备再制造关键技术必须胜任这些苛刻要求，体现了高新技术的发展前沿，如纳米表面工程技术、智能自修复技术、装备零部件战场快速制造和再制造数字化平台技术等，可在信息化战争条件下，缩短装备维修保障反应时间和物资运输时间，减少库存物资，保持装备保障力量的规模适度。

建设装备再制造学科还是我军新型装备维修保障人才培养的需要：（1）是培养军事斗争人才的需要。在军事斗争的准备中，舰艇及两栖装甲车辆等装备制造和运用中的腐蚀问题，装甲车辆的防护问题，装备战场应急抢修问题，作战中新装备研制中材料及其加工的可靠性，耐用性问题等，在以前的设计和维修中都没能给予足够的重视，都有大量的课题急需研究，对从事复杂多变技术保障工作的人才需求迫切。而普通的材料科学与工程专业培养的人员难以胜任上述要求。（2）是培养装备维修人才的需要。高技术武器装备需要高技术的维修，总装备部已批准将维修技术列入“十五”预先研究领域，与新型武器装备同步配套发展。作为维修技术的重要内容，装备维修中的材料，如应急维修材料、装备表面功能材料、战损装备维修材料等，这些都是地方院校材料专业难以涉及的领域。因此需要培养军用材料工程的专门人才。（3）是培养装备再制造工程人才的需要。在我军今后不断开展的装备再制造的理论研究及技术应用中，不断出现的新理论和新实践，如装备的全寿命周期和多寿命周期理论，废旧毛坯和再制造零件剩余寿命预测理论，基于信息化技术的装备零部件战场快速制造和再制造数字化平台，基于仿生技术的智能自修复技术，发动机再制造的工程应用等，都需要专门的人才来研究完成。因此，急需培养和造就高素质的装备再制造工程高层次人才。

5　装备再制造面临的良好发展机遇与挑战

再制造在我军得到高度重视的同时，在我国国民经济建设中也得到快速发展，并受到我国政府的重视。一个优质、高效、低耗的装备再制造正在我国兴起。

2000 年 12 月，12 位院士及 12 名专家完成了《再制造工程及其在我国应用的前景》咨询报告，中国工程院将此报告呈报国务院。国务院办公厅批转国务院 10 部委研究参阅。

2001 年 7 月，总装备部下达的“十五”武器装备预先研究计划中，“装备延寿与再制造工程技术”和“表面微损伤自修复技术”等一批关键技术被列为预先研究项目。

2003 年 7 月，总装备部通保部决定对全军车辆维修制度进行改革试验，今后将以再制造汽车发动机为基础开展汽车维修；2005 年 3 月总装通保部又启动了坦克发动机的再制造试验研究工作。

2004 年 9 月，在由国家发改委组织召开的“全国循环经济工作会议”上，笔者应邀做了《发展再制造工程，促进构建循环经济》的专题报告。

2004 年 9 月，国家发展与改革委员会下发《关于加快循环经济发展指导意见》。2005 年 11 月，国家发改委发布了全国 42 个循环经济示范试点企业名单，再制造被列为4个重点领域之一。装备再制造技术国防科技重点实验室支持的济南复强动力公司成为发动机再制造示范试点企业。

2004年11月和2005年1月，再制造研究作为一个专题，在国家自然科学基金委工程与材料学部机械学科两次召开的“十一五”前沿及优先资助领域研讨会上被深入论证。由于受到基金委及专家的一致好评，再制造研究已被国家自然科学基金委作为优先资助领域之一，将在“十一五”期间继续得到重点支持[11]。

2005年5月，中国工程院启动节约型社会战略方案研究，包括再制造在内的“4R工程”被列入研究内容。

2005年6、7月，国务院颁发的第21、22号文件《国务院关于做好建设节约型社会近期重点工作的通知》《国务院关于加快发展循环经济若干意见》中明确指出，国家将“大力支持废旧机电产品再制造”，并将“绿色再制造技术”列为“对节约资源和建设节约型社会有重大意义，且各级政府应重点支持”的关键技术之一。

2005年12月，国家发改委给装备再制造技术国防科技重点实验室下达《推进再制造产业实施方案研究》任务，要求对今后我国推广再制造业的前景进行分析和预测，并拟制再制造业在全国的实施方案及计划，提出再制造领域中需要研究开发的高新关键技术，规划再制造业的发展领域。

2005年12月中旬，由国家发改委主办的“全国首届建设节约型社会展览”上，专门安排展台向社会展示济南动力有限公司的再制造发动机。党中央、国务院和中央军委主要领导参观了现场并对再制造表现出了浓厚的兴趣。

2005年12月19日，国家发改委主任马凯在《经济日报》撰文《贯彻落实节约资源基本国策，加快建设节约型社会》。他指出，要加快技术创新，突破技术瓶颈，要组织开发包括再制造技术在内的具有重大推广意义的共性关键技术[12]。

2006年2月，《国家中长期科学和技术发展规划战略研究报告（简版）》正式公布。该报告的第三专题《制造业发展科技问题研究》中共有5处提到再制造，其中“共性关键制造技术与再制造技术”被列为制造业未来15年国家优先支持的3大重点领域之一“制造业信息化”中的重点发展主题之一，标志着再制造已成为我国未来制造业的重要组成。

2006年3月，中国工程院《建设节约型社会战略咨询研究》总报告指出，到2010年，我国废旧机械产品的再制造率达到50%；到2020年，我国将建立发动机等再制造企业超过100家，废旧机械产品的再制造率达到80%，废旧电子产品的资源化率达到95%。

6 建议和展望

结合我军武器装备再制造的发展，提出以下建议：

（1）构建装备再制造技术持续健康发展的科研体系。通盘考虑，科学谋划，把再制造的创新发展植根于整个国防科技创新体系之中；促进有利于军民结合、寓军于民的装备再制造发展机制的形成和完善，逐步建立起军民两用型重大再制造技术项目联合攻关机制；加强装备再制造技术力量建设，继续巩固和提高军用装备再制造技术力量，充分调动和发挥军工集团、高等院校等军外力量参与再制造的主动性。

（2）坚持有所为有所不为。重点突破装备再制造关键技术。适应作战方式的变化和装备发展需求，以高新技术集成应用创新为途径，以满足战场抢修为目的，研究战场损伤快速评估技术、装备损伤表面及新型材料结构的检测和修复技术、战损零部件数字化快速修复移动平台技术等，使损伤装备的应急抢修时间大大缩短，满足不同装备的要求；针对现役装备使用、存储和技术改造，研究现役主战装备及关键零部件剩余寿命预测技术，基于信息融合的装备故障综合诊断技术，解决现役装备寿命预测的关键技术，同时对新研装备寿命预测提供借鉴，现役主战装备寿命预测相对误差显著减小；以新装备研制和现役装备技术改造为需求，研究装备再制造性论证、验证、评价与增长技术，提供定性、定量的研究成果，使新研装备的再制造性不断提高。

（3）重视和加强基础研究，增强装备再制造技术发展后劲。加强研究在信息化战争条件下的战场损伤机理与模式，装备故障预测与诊断的新理论新方法，隐身、热障、电磁屏蔽、自润滑等特殊功能涂层的再制造与自修复机理，装备再制造中的表面改性原理，装备腐蚀防护与治理等；支持和鼓励优势力量进行国家队攻关，力争突破一些长期制约装备再制造技术发展的基础性难题，如装备关键零部件的剩余寿命预测技术等，为装备再制造技术的持续健康发展提供不竭动力。

（4）加强相关技术领域交流与合作，促进再制造技术的创新发展。与其他技术相比，装备再制造技术具有鲜明的复杂性和综合性，涉及的技术领域众多，还存在着很多如寿命评估、防腐、防污等基础性难题。近年来随着高技术装备的发展，又必须面临新装备中铝合金、钛合金等轻质材料及功能材料的再制造问题，信息武器中电子信息系统的再制造问题，因此急需发展适应新型装备再制造所需的新理论、新方法、新技术。

“十一五”期间，武器装备的再制造必将受到总部等领导部门的高度重视，这就要求我们积极适应这一形势，推进再制造、发展再制造，为推动我军装备维修保障事业做出更大贡献，也为建设以自主创新为核心的创新型国家发挥更大作用。

参 考 文 献

[1] 国家发改委经济体制与管理研究所《我国循环经济发展战略研究》课题组，发展循环经济是落实科学发展观的重要途径［J］. 宏观经济研究，2005，（4）：3～8.

[2] 徐匡迪. 工程师从物质财富的创造者到可持续发展的实践者［J］. 中国表面工程. 2004，17（6）：1～6.

[3] 徐滨士. 绿色再制造工程及其在我国的应用前景［R］. 中国工程院，“工程科技与发展战略”咨询报告集. 2002，2.

[4] Rep Craig. Larry E. Expressing the Sense of the Congress that the Department of Defense should Place a Greater Emphasis on the Remanufacture of Existing Mililtary Equipment as an InIerim Measure to Maintain Readiness in Light of Budget Constraints. Production Delavs and Cost Overruns of New Equipment [EB/OL]. Executive Comment Requested from DOD（H CON RES. 143）. 1989 – 6 – 7. http：//thomas. loc. gov.

[5] AH – 64D Longbow. http：//www. globalsecurith. org/military/ systems/aircraft/ah – 64d. htm. 2006 – 03 – 10.

[6] M1A1D Ahrams. http：//www. globalsecurity. org/military/systems/ground/mlald. htm. 2006 – 03 – 12.

[7] 徐滨士. 纳米表面工程［M］. 北京：化学工业出版社，2004，1.

[8] 马世宁，孙晓峰，朱胜，等，机床数控化再制造［C］. 首届机电装备再制造工程学术研讨会.

2004：24～28

[9] Xu Bin－shi，Wang Hai－dou，Dong Shi－yun，et al. Electrodepositing Nickel Silica Nano－composites Coatings [J] Electrochemistry Communications. 2005，7(6)：572～575.

[10] 徐滨士，史佩京，许一，等．纳米颗粒的摩擦学性能及其用作润滑油修复添加剂的应用研究[C] 中国工程院摩擦学工程科技论坛——润滑应用技术论文集．2004：6～14.

[11] 国家自然科学基金委员会工程与材料学部学科发展战略研究报告（2006－2010）机械与制造科学[M]．北京：科学出版社，2006.

[12] 马凯．贯彻落实节约资源基本国策，加快建设节约型社会[N]．经济日报．2005－12－19(2).

Developing Equipment Remanufacturing，Enhancing the Maintenance Performance of Military Equipment

Xu Binshi

(State Key Laboratory for Remanufacturing，The Academy of Armored Forces Engineering，Beijing，100072，China)

Abstract Equipment remanufacturing engineering is the industrialization of high－tech maintenance of the demobilizing and end-of-life equipment and products. Developing equipment remanufacturing can not only accelerate the application of high technology in the military facilities，but also promote the change of maintenance mode for the military facilities，which is the important measure to save the military expenditure and improve equipment performance. The development of equipment remanufacturing subject is the new growth point in the subject construction in our college. The integrated measures including intensified publicizing，cooperative experimenting，joint researching and talent training，could promote the progress of equipment remanufacturing in our army，as well as our college.

Key words equipment remanufacturing，equipment maintenance，independent innovation

再制造工程的发展及推进产业化中的前沿问题*

摘　要　再制造是把达到使用寿命的产品通过修复和技术改造使其达到甚至超过原型产品性能的加工过程，再制造是废旧机电产品资源化的首选途径。我国已进入到以推动再制造产业发展为中心内容的新阶段，建立生产者延伸责任制，充分发挥产品制造企业的主体作用；制定再制造企业认证法规，完善再制造产品质量体系；推广以表面工程为核心的先进技术，提高旧件利用率是当前急待研究的主要问题。

关键词　再制造工程　表面工程　产业发展　标准　关键技术

1　引言

再制造是相对于制造而言的。制造是把原材料加工成适用的产品。再制造是把达到使用寿命的产品通过修复和技术改造使其达到甚至超过原型产品性能的加工过程。再制造工程的丰富内涵在于它是以机电产品全寿命周期设计和管理为指导，以实现废旧机电产品性能提升为目标，以优质、高效、节能、节材、环保为准则，以先进技术和产业化生产为手段，对废旧机电产品进行修复和改造的一系列技术措施或工程活动的总称[1]。在贯彻科学发展观促进我国经济与社会可持续发展的进程中，再制造对节能、节材、保护环境的意义更加突出，地位和作用更加提升。前期的研究工作阐述了再制造工程在循环经济中的地位和作用、再制造工程的学科体系、再制造的国外发展情况等[2~4]。以国发［2005］21 号文件《国务院关于做好建设节约型社会近期重点工作的通知》和国发［2005］22 号文件《国务院关于加快发展循环经济若干意见》强调大力发展再制造为标志，我国的再制造工程进入到以推进再制造产业发展为中心内容的新阶段。在国家高度重视、示范试点稳步进行、法律法规逐步完善、再制造理论与技术研究取得重要成果的形势下，需要对推进我国再制造产业面临的重要问题开展深入研究。

2　我国再制造工程的发展情况

我国再制造产业发展相对较晚，仍处于起步阶段，与发达国家相比还有很大的差距。20 世纪 90 年代后期，随着国家确定了可持续发展的战略目标，再制造工程得到我国政府、学术界的高度重视，也开始得到部分企业界的认同与支持。

1999 年，中国工程院院士徐滨士教授率先在国内倡导并积极推动再制造工程的发展，并先后在 1999 年 6 月西安“先进制造技术国际会议”[5]和 2000 年 3 月瑞典“第 15 届欧洲维修国际会议”上[6]，首次提出在中国发展再制造工程的思想。

2000 年至 2006 年，中国工程院开展的咨询项目“绿色再制造工程及其在我国的应

---

* 本文合作者：刘世参、史佩京。原发表于《中国表面工程》，2008(21)：1~5。

用前景”和“废旧机电产品资源化”对再制造的内涵、意义、关键技术、国内外情况、未来发展前景以及政策建议等进行了深刻地论述；中国工程院咨询项目“建设节约型社会战略研究”将“机电产品的回收利用与再制造工程”列为支撑建设节约型社会的17项重大工程之一。中国工程院徐匡迪院长在上海“世界工程师大会”上作报告时指出“我国构建循环经济应以‘4R’原则（reduce减量化，reuse再利用，recycle再循环，remanufacture再制造）为指导”，随后又专门撰文阐述了4R原则[7]。

国家自然科学基金委将“再制造工程技术及理论研究”列为“十五”机械学科发展前沿与优先资助领域，2002年国家自然科学基金委批准“再制造设计基础与方法”和“再制造基础理论与关键技术”为面上和重点项目。2004年至2005年，国家自然科学基金委多次召开“十一五”前沿及优先资助领域研讨会，再次将“资源循环型制造与再制造”列为“十一五”机械学科发展前沿与优先资助领域，2007年“机电产品可持续性设计与复合再制造的基础研究”再次被批准为国家自然科学基金重点项目，上述项目的开展，对再制造设计和再制造过程中的多项基础理论与关键技术进行了深入细致的研究。

2003年至2004年，《国家中长期科学和技术发展规划战略研究》中第三专题《制造业发展科技问题研究》将“机械装备的自修复与再制造”列为子课题进行论证。再制造作为制造领域的优先发展主题和关键技术被列入《国家中长期科学和技术发展规划纲要（2006~2020年）》，这为再制造在中国的长远发展提供了决策性论据。为此，科技部于2006年将“汽车零部件再制造关键技术与应用”纳入“十一五”国家科技支撑计划中。

2005年，国发［2005］21号文件《国务院关于做好建设节约型社会近期重点工作的通知》和国发［2005］22号文件《国务院关于加快发展循环经济若干意见》中指出：国家将“支持废旧机电产品再制造”，并把“绿色再制造技术”列为“国务院有关部门和地方各级人民政府要加大经费支持力度的关键、共性项目之一”，提出“组织开发绿色再制造技术（等），提高循环经济技术支撑能力和创新能力”。

2005年11月，国家发改委等6部委联合颁布的《关于组织开展循环经济试点（第一批）工作的通知》（发改环资［2005］2199号）文件公布了包括七个重点行业、四个重点领域、十三个产业园区和十个省市的42个循环经济示范试点名单。其中再制造被列为四个重点领域之一，我国发动机再制造企业“济南复强动力有限公司”被列为再制造重点领域中的试点单位。

2006年4月，国家发改委资环司循环经济处向国务院递交了《关于汽车零件再制造产业发展及有关对策措施建议的报告》，曾培炎副总理作出重要批示：“同意以汽车零部件为再制造产业试点，探索经验，研发技术；同时要考虑定时修订有关法律法规”。

2006年9月，国家发改委资环司在上海主办了“汽车零部件再制造研讨会”，研讨加快推进中国再制造产业发展的对策措施，国家发改委资环司、产业政策司、工业司，以及商务部，公安部，交通部，国家质检总局等相关业务部门的领导出席了会议。

2007年1月，中国汽车工业协会受国家发改委委托，成立了以原机械部何光远部长和徐滨士院士为顾问的“汽车零部件再制造产业研究”课题组，将以国内主要汽车制造企业和现有汽车零部件再制造企业为突破口，开展汽车零部件试点研究和配套政策法规研究工作，以此来推动国内再制造产业发展。

2007 年 10 月，国家发改委向各省、自治区、直辖市下发了《关于组织申报汽车零部件再制造试点企业的通知》（发改办环资［2007］2536 号），拟在全国选择有代表性、具备再制造基础的汽车整车生产企业，或汽车零部件再制造企业开展再制造试点工作。目前全国已有 13 个省市 40 余家企业申报，国内一汽、二汽、上汽、重汽集团等大型整车制造企业，以及潍柴、玉柴等发动机制造企业越来越关注汽车零部件的再制造。

在国家积极倡导、全力推进再制造产业发展的同时，学术研究与学术交流相继开展。2003 年建成了我国第一个从事再制造领域研究的国家级重点实验室"装备再制造技术国防科技重点实验室"，目前该实验室已在再制造理论及关键技术研究方面取得一批重要成果，先后出版了一批再制造方面的专著《再制造工程基础及其应用》《再制造与循环经济》与《装备再制造工程的理论与技术》等；中国设备管理协会成立了再制造技术委员会，中国工程机械学会维修工程分会建立了再制造工程专业委员会，这两个再制造领域的学术团体组织承办了 4 次全国范围的再制造学术会议，出版了 4 部再制造学术会议论文集。

3　再制造产业发展潜力巨大

我国作为装备使用大国，设备资产已达几万亿元，其中许多大型成套机械类设备，尤其是 20 世纪 70 年代末、80 年代初引进的重大设备，从 2001 年开始将陆续到寿，面临报废，在 2010 年左右将达到最多数量。这些装备因到达技术寿命、经济寿命或环境寿命而被淘汰、报废，这为开展再制造研究提供了大量的资源。因此，大力开展再制造研究对降低装备全寿命周期费用，全面提升老旧装备性能和寿命，改善我国装备落后的状况，降低装备维修费用，节约资源、能源，减少环境污染等方面都有重要意义。

国家发改委立项的"推进再制造产业发展实施方案研究"调研结果表明，在汽车零部件、工程机械、机床、大型工业装备、铁路设备、农用机械、国防装备、医疗设备、办公设备 9 个领域可以推广再制造工程，发展再制造产业，如表 1 所示。

表 1　可实施再制造的主要领域

领　域	可用于再制造的主要零部件
汽车零部件	发动机、传动装置、离合器、转向器、启动机、水泵、空调压缩机、油泵等
工程机械	平地机、摊铺机、压路机的刮板，挖掘机、铲运机、推土机的铲斗，及其发动机、变速箱、电机
机床	车床、高精密磨床、铣床、刨床，相关设备的变速箱、齿轮、轴承、电机、转子
大型工业装备	电力设备：汽轮机缸盖、风机叶轮、锅炉"四管"； 煤炭设备：井筒、柱塞、防爆罩； 冶金设备：高炉渣口、风口、轧辊、连铸机轧辊、热轧工具等； 石化设备：储油罐、高温高压反应容器、裂解炉管柱塞、套筒； 钻井设备：泵、柱塞轴； 纺织设备：捻机锭环、加捻器、摩擦盘、导丝器、卷绕槽筒

续表1

领　域	可用于再制造的主要零部件
铁路设备	内燃机车发动机、车轮等零部件，铁路钢轨、转向架、承载鞍、铁路轴承、轴承箱，滚子轴承
农用机械	拖拉机、联合收割机、水稻插秧机、农用运输车的发动机、履带、变速器、电机
国防装备	武器装备中的发动机、变速箱等，以及适应新的作战需求的技术改造
医疗设备	核磁共振机、CT机、B超机、X光透视机
办公设备	复印机、打印机、传真机硒鼓，空调、电冰箱、照相机、通讯设备零部件，写字台、办公椅、档案橱柜、隔扇板及会议桌椅[8]

以汽车自动变速箱再制造为例，我国2000年起，大量引进自动变速箱汽车，目前有自动变速箱的汽车保有量为400万辆，根据美国的统计数据，自动变速箱的再制造量为汽车保有量的2%，我国由于新车多旧车少，再制造需求为1.5%，即6万台，而目前自动变速箱的生产能力为5 000台，市场需求旺盛，再制造产业发展潜力巨大。

预测我国汽车发动机的再制造需求量在2010至2015年间为250万~400万台，而当前我国汽车发动机再制造企业只有2家，济南复强动力有限公司和上海大众联合发展有限公司，至2010年再制造发动机的年产量不过10万余台。表1所列的九大领域中，许多领域设备的再制造工作尚未广泛开展。

上述九大领域之一的废旧机电产品是相对于自然矿产资源的第二资源，用之为宝、弃之为害。当前再制造产业规模小、范围窄，与可用于再制造的丰富资源形成巨大反差，与建设节约型环境友好型社会的要求形成巨大反差。因此，全力推进废旧机电产品再制造产业化是发展循环经济的重点工作之一。

4　发挥产品制造企业的主体作用问题

美国再制造业是一个巨大的产业，1996年，美国发布了再制造业调查报告《再制造业：潜在的巨人》[9]。报告显示，1996年美国专业化再制造公司达73000家，年销售额530亿美元，直接雇员48万人。与制药业、计算机制造业和钢铁业相比，同年美国再制造业的产值与它们基本相当（图1），但就业人数明显多于它们（图2），说明再制造业不仅能够创造巨大财富，而且能够显著解决就业问题。

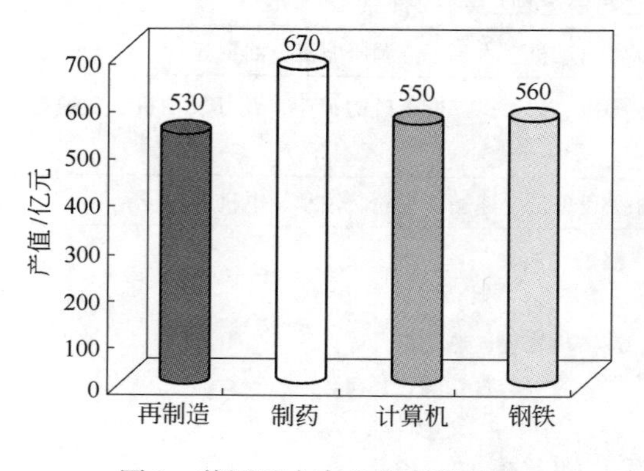

图1　美国四大产业的产值对比

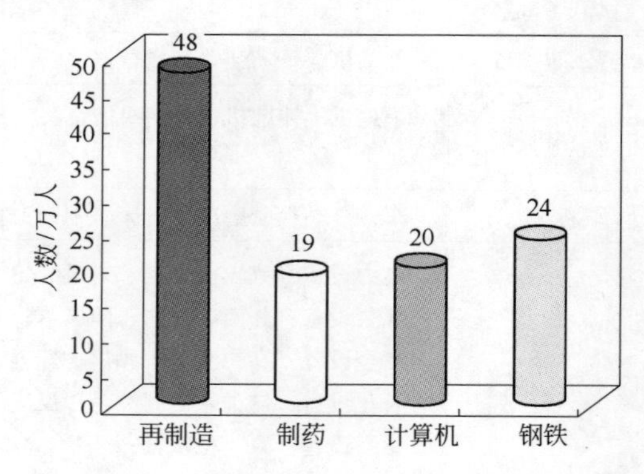

图2　美国四大产业的就业人数对比

国外产品制造企业十分重视自己产品的再制造，卡特彼勒公司把推进自己产品的再制造列为公司的发展战略，德国大众、美国福特、法国雪铁龙等汽车制造企业都委托相关的企业对自己的产品实施再制造。而我国产品制造企业对实施自己产品的再制造仍存在诸多疑虑。这就需要从理论与实践的结合上，从社会效益与企业效益的结合上阐明新品制造与产品到寿后再制造对企业发展的作用和影响，使产品制造企业提高认识、解除顾虑、明确责任、增强信心，充分发挥制造企业推进产品再制造的主体作用，推进企业综合发展。

当前工业发达国家正在建立和完善生产者延伸责任制（extended producer responsibility，EPR），在我国也是必然的发展趋势。在EPR中，生产者不仅要对自己制造的产品实施再制造负有责任，而且在新产品的设计时就要考虑未来的可拆卸性、可再制造性、可循环利用性，为自己的产品在未来的循环利用奠定良好基础[10]。

5 再制造的相关标准问题

当前准备开展再制造业务的企业很多，为保证我国再制造业的健康发展，需要一套系统的管理法规。首先，要研究和建立再制造企业的认证制度，要明确从事再制造企业的基本条件，包括再制造企业的产业基础、生产规模、加工设备、技术力量、再制造产品的质量、再制造毛坯的来源、再制造产品的销售渠道等。再制造是一种规模化的生产方式，没有规模也就没有质量和效益，没有一定规模也就不能充分吸纳各种先进技术。再制造的规模又与技术实力、旧件来源、销售状况、资金准备等密切相关。

其次是再制造产品的质量标准问题。在“再制造产品的质量和保质期应等同或超过原型产品”的总要求下，需要建立再制造产品质量标准体系，包括再制造毛坯的鉴定标准、再制造毛坯的再制造性评估标准、主要零件的寿命预测标准、整机再制造后的质量标准、各零部件再制造后的质量标准、再制造加工工艺标准、再制造加工原材料标准、再制造加工检验检测标准，以及操作者上岗标准等。这些技术标准应具有科学性和先进性，并能做到可操作性、可量化、可检测、可重复。

再制造产品的质量标准体系是建立在对再制造关键技术研究开发的基础之上的，是一项浩大的系统工程，没有完善而稳定的质量保证体系，也难以保证再制造产品的质量稳定。

6 旧件利用率问题

再制造具有潜在价值的根本原因是机器中各部件的使用寿命不相等，而且每个零件的各工作表面的使用寿命也不相等，这就为再制造的实施提供了物质基础。

在一部机器中通常固定件的使用寿命长，如箱体、支架、轴承座等，而运转件的使用寿命短。在运转件中，承担扭矩传递的主体部分使用寿命长，而摩擦表面使用寿命短。与腐蚀介质直接接触的表面使用寿命短，不与腐蚀介质接触的表面使用寿命长。这种各零部件的不等寿命性和零件各工作表面的不等寿命性，往往造成机器中部分零件以及零件上局部表面的失效而使整个机器不能使用。再制造的着眼点是把没有损坏的零部件继续使用，把有局部损伤的零件采用先进的表面工程技术等手段通过再制造加工继续使用，而且可以针对失效的原因采取措施使它的使用寿命延长。这样就挖掘出了废旧机电产品中蕴含的附加值，起到节省资金、节能、节材、保护环境的效果。

在执行“再制造产品的质量和保质期等同甚至超过原型机新品”这一总要求下，对于有磨损、腐蚀、划伤、压坑等缺陷的零件，可以用更换新件的办法，也可以用修复的办法，两者相比，用旧件修复的办法对节能、节材、保护环境的贡献最大。通常情况下，旧件修复所消耗的材料为该零件本体质量的 2 %，修复费用是该零件新品价格的1/10，同时，在旧件修复时可以根据需要对不耐磨损、不耐腐蚀的短寿命件进行表面强化，使它的使用寿命与整机寿命同步，减少一个寿命周期内的维修次数，使全寿命周期内的费用最低，功效最高。

国外再制造的历史虽然较长，所使用的技术多以更换新件为主或使用尺寸修理法。更换新零部件的再制造不符合社会持续发展的总要求，应尽可能少用；修理尺寸法的应用应留有足够的尺寸以保证再制造品出厂后能使用一个寿命周期。将以表面工程为主要内容的高新技术引用到再制造产品的加工中，是中国的再制造区别于国外再制造的主要特点，这一特色已得到国际再制造同行的认可和称赞。

因此，再制造过程中，旧件利用率的高低是衡量企业再制造水平的重要标志，是考核该企业对建设节约型社会、环境友好型社会贡献大小的重要指标。提高旧件利用率的根本办法是依靠科技进步，坚持技术集成创新。

一部机电产品制造出来以后，经过若干年后才达到报废阶段，而在此期间科学技术发展迅速，新材料、新技术、新工艺、新控制装置不断涌现，对废旧机电产品进行再制造时可以吸纳最新的成果，既可以提高易损零件、易损表面的使用寿命，又可以对老产品进行技术改造，使它的整体性能跟上时代的要求。

现在表面工程技术发展非常迅速，已由传统的单一表面工程技术发展到复合表面工程技术，进而又发展到以微纳米材料、纳米技术与传统表面工程技术相结合的纳米表面工程技术阶段。纳米表面工程中的纳米电刷镀技术、纳米热喷涂技术、纳米减摩自修复添加剂技术、纳米防腐涂料技术、纳米黏接黏涂技术等已进入实用化阶段，纳米表面工程技术在再制造中的应用可使零件表面的耐磨性、耐蚀性、抗高温氧化性、减摩性、抗疲劳损伤性等力学性能大幅度提高，为解决再制造中的疑难问题提供新途径。

7 结论

（1）国家对发展再制造产业高度重视，鼓励政策和法律法规相继出台，再制造示范试点工作稳步进行，再制造理论与技术的研究已取得重要成果，在这种背景下，调动企业的积极性，充分发挥企业的创造性是今后推动再制造产业发展的重点。

（2）建立生产者延伸责任制使产品制造企业积极参与自己产品的再制造，实施再制造企业认证制度，并完善再制造产品的质量保证体系，充分吸纳高新技术提高再制造过程的旧件利用率是再制造产业发展进入新阶段需要着重研究和解决的主要问题。

参考文献

[1] 徐滨士，朱胜，马世宁，等．装备再制造工程学科的建设与发展［J］．中国表面工程，2003，16(3)：1～6.

[2] 徐滨士，刘世参，李仁涵，等. 废旧机电产品资源化的基本途径及发展前沿研究［J］. 中国表面工程，2004，17(2)：1～6.
[3] 徐滨士，刘世参，史佩京，等. 再制造工程和表面工程对循环经济贡献分析［J］. 中国表面工程，2006，19(1)：1～6.
[4] 徐滨士，马世宁，刘世参，等. 绿色再制造工程设计基础及其关键技术［J］. 中国表面工程，2001，14(2)：12～15.
[5] Xu Binshi, Zhu Sheng. Advanced remanufacturing technologies based on nano – surface engineering [C]. Proc. 3rd Int. Conf. on Advances in Production Eng, 1999, Guangzhou: 35～43.
[6] Xu Binshi, Zhang Wei, Liu Shican. Remanufacturing Technology for the 21st Century [C]. Proceedings of the 15th European Maintenance Conference, Gothenburg, Sweden, March 2000: 335.
[7] 徐匡迪. 工程师－从物质财富的创造者到可持续发展的实践者［J］. 中国表面工程，2004，17(6)：1～6.
[8] Rolf Steinhilper. Remanufacturing—the ultimate form of recycling [M]. Fraunhofer IRB Verlag, 1998.
[9] Robot T. Lund. The Remanufacturing Industry – Hidden Giant [R]. Research Report, Manufacturing Engineering Department, Boston University, 1996.
[10] 徐滨士，等. 再制造与循环经济［M］. 北京：科学出版社，2007.

The Development and Industrialization Frontal Issues of Remanufacturing Engineering

Xu Binshi, Liu Shican, Shi Peijing

(National Key Laboratory for Remanufacturing, Beijing, 100072, China)

Abstract Remanufacturing engineering is an advanced techniques and industrializing process to maintain and rebuild worn products, and it is the optimized approach of resource recovery of waste machinery and electronic products. Remanufacturing engineering is enter into the new station, which is remanufacturing industrialization. There are three new frontal issues of remanufacturing engineering, the first is establishing extended producer responsibility to exert the main function of manufacturer, the second is establishing certification and standard of remanufacturing corporation to improve the quality system of remanufactured products, the third is developing surface engineering technology to improve the utilized efficiency of old parts.

Key words remanufacturing engineering, surface engineering, industrialization development, standard, key technology

A Research on the Concepts and Estimate Methods of Remanufacturing Rate of Engine*

Abstract Based on total life cycle assessment (TLCA) and a research project on auto engine remanufacturing, this article look forward to make certain what is "Remanufacturing Rate" of engine remanufacturing and how to express it on the occasions of OEMs, independent remanufacturers and subcontracted remanufacturers in China. The concepts of "Remanufacturing rate" of engine are put forward at first, and then the estimate methods are studied based on remanufacturing environment of China.

Key words total life cycle assessment (TLCA), remanufacturing rate, engine

1 Introduction

As the consumer market has expanded greatly, the demands for autos are rising rapidly too[1]. And also the quantity of discarded autos will increase year after year. There are lots of parts, such as engines, can be remanufactured in the discarded autos[2-4].

Chinese Remanufacturing has a high level on technique, but is still in the primary stage of industrialization and that the faultiness of policies and management methods is one of the dominating reasons[5]. One of the solutions to the problem is to design and establish a certain management system which can be adapted to the requirement. This article looks forward to contributing to that and will discuss the concepts and the corresponding estimate methods of remanufacturing rate which are ambiguous till now in China.

Now, the word of "remanufacturing rate" points to the proportion of remanufacturing productions to the total productions which also including manufacturing productions, or the production rate of remanufacturing[6-10], and the production is appeared as a single part and didn't associate with auto engine which is composed of large numbers of parts.

Some parts of engine cannot be remanufactured because of being valueless or unusefulness. Some parts can be used directly after cleaning. Whether the other parts will be remanufactured depends on life evaluation or state examination[11]. So the concept of remanufacturing rate mentioned above cannot be used here to reflect the actual remanufacturing status.

Nasr has advanced the concept of remanufacturability and described the estimate methods particularly[12]. But remanufacturability does not equate to remanufacturing rate.

---

* Copartner: Liu Bohai, Yang Shanlin, Shi Peijing, Xue Shunli. Reprinted from *Key Engineering Materials*, 2010, 419 - 420: 253 - 256.

2 Concepts and estimate models for remanufacturing rate

There are three roles play on the remanufacturing market, independent remanufacturers, OEMs (original equipment manufacturers) and subcontracted remanufacturers[3]. The concepts of remanufacturing rate for independent remanufacturers and OEMs are given below and can be referenced by subcontracted remanufacturers.

Lemma 1. For independent remanufacturers, remanufacturing rate is, in each batch, the proportion of the quantity of remanufactured engine parts to the quantity of all engine parts, and which can be called Quantity Remanufacturing Rate (QRR) and denoted by RR_q.

Lemma 2. For independent remanufacturers, except the meaning mentioned in lemma 1, remanufacturing rate is also, in each batch, the proportion of the weight of remanufactured engine parts to the weight of all engine parts, and which can be called Weight Remanufacturing Rate (WRR) and denoted by RR_w.

Lemma 3. For independent remanufacturers, except the meanings mentioned in lemma 1 and lemma 2, remanufacturing rate is also, in each batch, the proportion of the value of remanufactured engine parts to the value of all engine parts, and which can be called Value Remanufacturing Rate (VRR) and denoted by RR_v.

Lemma 4. For OEMs, except the meanings mentioned in lemma 1, lemma 2 and lemma 3, remanufacturing rate is also the proportion of the quantity of remanufactured engines to the quantity of all engines which also including new engines. We can call it Quantity – Proportion Remanufacturing Rate (QPRR), and which can be denoted by RR_{qp}.

Lemma 5. For OEMs, except the meanings mentioned in lemma 4, remanufacturing rate is also the proportion of the value of remanufactured engines to the value of all engines which also including new engines. We can call it Value – Proportion Remanufacturing Rate (VPRR), and which can be denoted by RR_{vp}.

The estimate methods for remanufacturing rate of independent remanufacturers are shown below:

$$RR_q = \frac{\sum_{i=1}^{n} q_{ri}}{n \cdot Q} \tag{1}$$

$$RR_w = \frac{\sum_{i=1}^{n} w_{ri}}{n \cdot W} \tag{2}$$

$$RR_v = \frac{\sum_{i=1}^{n} v_{ri}}{n \cdot V} \tag{3}$$

where q_{ri} denotes the quantity of remanufactured parts in remanufacturing engine $i(i = 1, 2, \cdots, n)$; Q denotes the quantity of parts in a remanufacturing engine; w_{ri} denotes the weight of remanufactured parts in remanufacturing engine $i(i = 1, 2, \cdots, n)$; W denotes the weight of a remanufacturing engine, i. e. sum of weight of all parts of a remanufacturing engine; v_{ri} denotes the value of remanufactured parts in remanufacturing engine $i(i=1,2,\cdots,n)$; V denotes the value of a remanufacturing engine, i. e. sum of value of all parts of a remanufacturing engine.

For OEMs, the concepts of QRR, WRR and VRR are same to the independent remanufacturers'. The concepts of QPRR and VPRR are as follows:

(1) If the productivity of new engine and remanufacturing engine always retain an invariant at any time, then the QPRR will be a constant. And if the value of new engine and remanufacturing engine are also invariable, the VPRR is also a constant.

$$RR_{qp} = \frac{\int_0^T r(t)\,\mathrm{d}t}{\int_0^T r(t)\,\mathrm{d}t + \int_0^T m(t)\,\mathrm{d}t} \tag{4}$$

$$RR_{vp} = \frac{p_r \cdot \int_0^T r(t)\,\mathrm{d}t}{p_r \cdot \int_0^T r(t)\,\mathrm{d}t + p_m \cdot \int_0^T m(t)\,\mathrm{d}t} \tag{5}$$

where $m(t)$ denotes the productivity of new engine at time t; $r(t)$ denotes the productivity of remanufacturing engine at time t; p_m denotes the value of a new engine; p_r denotes the value of a remanufacturing engine (for distinguishing from independent remanufacturers, it is defined with a new symbol).

(2) A much more actual situation is that the productivity of new engine and remanufacturing engine are different in different times. So, QPRR and VPRR will change with the time going. The ascertainment of relations between QPRR and VPRR needs an assumption described following.

Assumption. The value of a remanufacturing engine and the value of a new engine are constants, and the proportion of which is λ, and $0<\lambda\leqslant 1$. For different remanufacturers, since the technique levels are uneven, the λ is different either. So $p_r = \lambda \cdot p_m$.

If at time T the ultimate production of engines of a remanufacturer is z, and of them the quantity of remanufacturing engines is $\mu \cdot z$, and $0\leqslant\mu\leqslant 1$, then:

$$RR_{qp} = \mu \tag{6}$$

$$RR_{vp} = \frac{p_r \cdot (\mu \cdot z)}{p_r \cdot (\mu \cdot z) + p_m \cdot [(1-\mu)\cdot z]} \tag{7}$$

According to formula (7), formula (8) can be get:

$$RR_{vp} = \frac{\lambda \cdot \mu}{\lambda \cdot \mu + (1-\mu)} \tag{8}$$

As time goes, the number of μ changes between 0 and 1. From Fig. 1 we can see the links between QPRR and VPRR on the scale of $\lambda(0<\lambda\leqslant 1)$.

If taking no account of the limitation of assumption above, i. e. the value of engines changes with time going, $\lambda(0<\lambda\leqslant 1)$ is not a constant within a remanufacturer. Then the VPRR can be described with formula (9) below:

$$RR_{vp} = \frac{\int_0^T p_r(t) \cdot r(t)\mathrm{d}t}{\int_0^T p_r(t) \cdot r(t)\mathrm{d}t + \int_0^T p_m(t) \cdot m(t)\mathrm{d}t} \tag{9}$$

where $p_m(t)$ denotes the value of a new engine at time t and $p_r(t)$ denotes the value of a remanufacturing engine at time t, $\forall t \in [0,T]$.

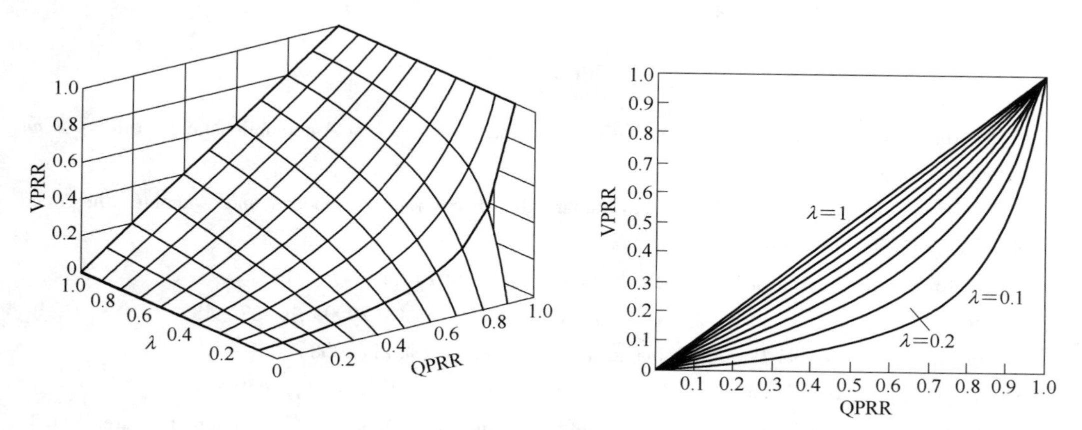

Fig. 1 Links between QPRR and VPRR on the scale of $\lambda(0<\lambda\leqslant 1)$

It is difficult to plot the links between QPRR and VPRR at such a complicated situation. As a solution, the VPRR can be calculated by separating the whole time into several segments, just as formula(10) shows. In each segment, the value of both new engine and remanufacturing engine are changeless.

$$RR_{vp} = \sum_{i=1}^{n} \frac{p_r^i \cdot \mu_i \cdot z_i}{p_r^i \cdot \mu_i \cdot z_i + p_m^i \cdot (1-\mu_i)z_i} \tag{10}$$

where p_m^i denotes the value of new engine at segment $i(i=1,2,\cdots,n)$; p_r^i denotes the value of remanufacturing engine at segment $i(i=1,2,\cdots,n)$; z_i denotes the production of engines at segment $i(i=1,2,\cdots,n$ and $\sum_{i=1}^{n} z_i = z)$; μ_i denotes the the proportion of remanufacturing engines to all engines on quantity at segment $i(i=1,2,\cdots,n)$.

3 Conclusions

(1) The remanufacturing rate of engine can be differentiated into Quantity Remanufacturing Rate(QRR), Weight Remanufacturing Rate(WRR), Value Remanufacturing Rate(VRR) and Quantity-Proportion Remanufacturing Rate(QPRR), Value-Proportion Remanufacturing Rate (VPRR). QRR, WRR and VRR aim at both independent remanufacturers and OEMs, and QPRR and VPRR aim at only OEMs.

(2) Because of the variational production and value of engines based on market research, remanufacturing rates are not constant. Also, links between QPRR and VPRR are distinguishing in different situations. They can be estimated through formula(1)-formula(5) or formula (10).

Acknowledgements

This research was supported by National Key Natural Science Foundation of China (No. 50735006), National Science and Technology Support Project (No. 2006BAF02A19 and No. 2008BAK42B03), and Open Foundation of National Key Laboratory for Remanufacturing Technology (No. 9140C8501010801).

References

[1] China Automotive Technology & Research Center, China Association of Automobile Manufacturers: *China Automotive Industry Yearbook* 2008.

[2] V. Daniel R. Guide Jr., Vaidyanathan Jayaraman, Rajesh Srivastava: Robotics and Computer Integrated Manufacturing Vol. 15 (1999), p. 221 – 230.

[3] Margarete A. Seitz: Journal of Cleaner Production Vol. 15 (2007), p. 1147 – 1157.

[4] V. Daniel R., Guide Jr.: Journal of Operations Management Vol. 18 (2000), p. 467 – 483.

[5] Xu Binshi, Liu Shican, Shi Peijing: China surface engineering Vol. 21 (2008), p. 1 – 5.

[6] Baptiste Lebreton, Axel Tuma: Int. J. Production Economics Vol. 104 (2006), p. 639 – 652.

[7] Dimitrios Vlachos, Patroklos Georgiadis and Eleftherios Iakovou: Computers & Operations Research Vol. 34 (2007), p. 367 – 394.

[8] Ruud Teunter, Konstantinos Kaparis, Ou Tang: European Journal of Operational Research Vol. 191 (2008), p. 1241 – 1253.

[9] Stefan Minner, Rainer Kleber: OR Spektrum Vol. 23 (2001), p. 3 – 24.

[10] Sergio Rubio, Albert Corominas: Computers & Industrial Engineering Vol. 55 (2008), p. 234 – 242.

[11] Xu Binshi, Liu Shican, Shi Peijing, et al: China Surface Engineering Vol. 18 (2005), p. 1 – 7.

[12] Nasr, et al: U. S. Patent 7,467,073. (2008).

再制造综合评价指标体系的设计研究*

摘　要　再制造在中国获得了良好的发展，形成了具有中国特色的再制造工程，为中国循环经济的推进及两型社会的建设做出了重要的贡献。衡量再制造的发展水平及对环境及社会的贡献，需要通过特定的评价指标对其进行描述和评价。本文介绍了中国再制造工程发展的特点以及再制造综合评价指标体系建立的原则，同时结合中国再制造的实际情况，从环境发展、经济发展及社会发展的角度提出了再制造综合评价的若干具体指标，并对提出的再制造综合评价指标体系的构建进行了分析说明。

关键词　再制造　综合评价　指标体系

1　引言

再制造（remanufacturing）是一门以先进表面工程技术（advanced surface engineering，ASE）为支撑的技术，它不仅能准确恢复尺寸，而且能显著提升性能，同时资源利用率高，能源消耗少，具有显著的节能环保特色[1]。再制造在国外已经有 80 多年的历史，从技术标准、生产工艺、加工设备、配件供应至销售和售后服务，已形成一套完整的发展体系和较大的产业规模[2]。中国的再制造仅有 10 年左右的历史，但是发展良好且迅速，形成了具有中国特色的再制造工程（Chinese characteristic remanufacturing engineering，CCRE），有力促进了循环经济的开展及两型社会（资源节约型、环境友好型）的建设。

为了促进再制造产业化的稳定发展以及衡量再制造的发展水平和对环境及社会的贡献，需要通过特定的规范及评价指标对其进行描述和评价。

2　中国再制造发展的特点

2.1　社会层面

2000 年左右，中国引入了再制造的概念，徐滨士院士指出“再制造产品的质量、性能不低于同类新产品”[3]。经过一段时间的发展，再制造显示出了显著的环境及社会效益，表 1 为一很好的再制造例子[1]。

表 1　年再制造 1 万台 WD615－67 型斯太尔发动机的节能减排效果与综合效益分析

回收附加值 /亿元人民币	直接再用金属 /t	提供就业 /人	利税 /亿元人民币	节电度 /kW·h	CO_2 减排 /kt	总能耗 /kW·h
3.23	7.65×10^3	500	0.29	1.45×10^8	0.6	1.03×10^7

* 本文合作者：刘渤海、史佩京。原发表于《检验检疫学刊》，2010，20(2)：53～56。国家自然科学基金重点项目（50735006），国家科技支撑项目（2008BAK42B03），再制造技术国家级重点实验室开放基金（9140C8501010801）资助。

再制造是循环经济的 4R（reduce，reuse，remanufacturing，recycle，即减量化、再利用、再制造及再循环）原则之一，在节能、节材、环境保护等方面做出了重大的贡献。但相比于新品，再制造产品的社会认可程度至今仍不是很高，一般的顾客不能区分再制造与维修的区别，甚至把两者等同对待，认为再制造品就是二手货，导致了再制造产品市场开拓难度的加大。

2.2 产业层面

再制造在中国的起步虽然较晚，但却受到了政府的高度重视，相关产业管理部门及科研单位为再制造的发展提供了政策支持及技术支撑，并以汽车零部件再制造为试点产业，探索再制造产业在中国的发展之路。随后，全国绿色制造标准化技术委员会再制造分技术委员会成立，负责再制造术语标准、再制造技术工艺标准、性能检测标准、质量控制标准以及关键技术标准的制定等相关工作，为再制造在中国的发展提供规范化保证。

但是，当前再制造规范的不明确，导致了再制造市场中企业水平的参差不齐，有些再制造企业开展的业务甚至不能称为再制造，其产品质量水平得不到保证，这给再制造市场的开拓及发展造成了一定的负面影响。要保证再制造产业的健康稳定发展，需要包括经济、环保、工商、公安、立法、科研等在内的相关部门的共同努力及协调，不能指望若干家企业就可以推动整个产业的发展，也不能听任部分企业只要政策而不开展再制造业务的现象存在。

2.3 企业层面

以济南复强动力有限公司为代表的一批汽车零部件生产企业开展了汽车零部件再制造业务，采用的发展模式有科研合作、平台合作及技术合作等。经过一段时间的发展，汽车零部件再制造在中国已具有一定技术基础、管理积累及市场空间。

通过对国内 14 家汽车零部件再制造试点企业的调研发现，由于一些客观存在的原因，再制造企业的发展也面临着一定的困难，主要有国家相关政策不完善、再制造回收网络不健全、再制造专业技术人员/设备的需求空白、回收产品质量状态的随机性、再制造生产计划安排的复杂性、再制造信息系统的兼容性、再制造市场的培育等，这些困难严重阻碍了再制造业务的开展，只有通过政策协调及坚持产学研结合等加以解决。

2.4 技术层面

在再制造体系中，主要的关键技术包括再制造性设计技术（remanufacturability design technology，RDT）、再制造零部件剩余寿命评估技术（remanufactured residual evaluation technology，RRET）、无损拆解与分类回收技术（nondestructive disassembling and classification recover technology，NDCRT）、绿色清洗技术（green cleaning technology，GCT）、纳米表面工程技术（nano surface engineering technology，NSET）、快速成型再制造技术（rapid forming remanufacturing technology，RFRT）、运行中的再制造技术（active remanufacturing technology，ART）以及虚拟再制造技术（virtual remanufacturing technology，VRT）等[1]，其中体现了注重环境、社会效益的特点。目前，这些技术均获得了较好的发展，已进入实用化阶段，但是离产业化应用还有一定的距离，这正是再制造企业所急需解决的问题之一。

3 再制造综合评价指标体系建立的原则

中国再制造综合评价指标体系的建立原则是既要借鉴国外再制造产业化发展经验，又必须结合中国的实际国情、自身的发展特点和目标以及参考其他相关产业的发展历程。

3.1 符合循环经济的要求

循环经济是物质循环利用、高效利用的经济发展模式，作为其支撑模式之一的再制造，运用相关先进技术，充分挖掘废旧机电产品中的高额附加值，再制造后的产品质量、性能不低于同类新品，不仅创造了良好的经济价值，且对环境保护及社会持续发展做出了重大贡献[4]。因此，再制造的综合评价指标体系不仅要反映其经济效益，还要反映其环境、社会效益，以突出再制造对循环经济的贡献程度。

3.2 遵循再制造工程发展的特点

再制造的发展在社会、产业、企业及技术层面上都面临着若干问题，为了协调解决这些问题，规范化的运营方法必不可少。在对再制造的综合评价指标体系设计过程中，必须体现再制造的节能、节材功能和物质再利用的成本优势。通过这些指标的确立，可以反映并推进先进再制造技术的应用以及规范再制造行业的准入机制及评价依据。同时，通过社会保障类指标的确立，反映再制造对社会效益的贡献以及社会对再制造的认可度改变，为再制造的发展争取更大的市场空间。

3.3 体现多层面相互之间的协调

再制造的发展包含多个层面，是一个整体的系统。根据系统工程的观点，在对其进行综合评价时，要坚持多个层面协调统一的原则，不能只偏重于某一个层面而忽略了其他层面的问题及影响。因此，评价指标的确立要不仅能反映某一层面的发展态势，还能反映不同层面之间的关系及联系，如环境保护与经济成本之间的关联等。

另外，在再制造综合评价指标体系的选取过程中还要遵循概念明确、涵盖全面、定量表达、数据获取可行、处理方法得当等原则。同时，综合评价指标在再制造的发展过程中可以进行适当的动态调整，以适应再制造的发展实情，确保对再制造健康发展的积极引导性及支持性[5]。

4 再制造综合评价指标体系的设计

4.1 再制造综合评价指标体系的主要内容

根据中国再制造发展的特点及再制造综合评价指标体系的设立原则，再制造综合评价指标体系主要包含以下 3 个方面：

（1）再制造的节能减排指标：节能、节约金属、节约用水、减少排放等方面的反映指标。

（2）再制造的社会经济效益：产值、回收附加值、税收、利润、就业等方面的反映指标。

（3）再制造的综合利用指标：资源回收与旧件利用率、工业固体废物处置量、工业固体废弃物综合利用率、废水循环利用率、生产废液/废气/粉尘排放达标率等方面的反映指标。

在设计相关环保指标时，需要注意的是再制造在减少环境污染的同时，也会产生一些新的污染物，表 2 为发动机再制造的例子。因此在对再制造综合评价指标体系的具体分解时，也要注意这一方面的体现。

表 2　发动机再制造过程中的主要污染物及其来源

工序步骤	手　段	主要污染物	来　源
回收	运输	—	—
拆解	拆机（物理方法）	废料、废油、废水、废渣	无再制造价值或因损坏严重而无法再利用的零部件；发动机拆解前的初步清洗；废旧机器里的废油、废渣等
清洗	抛丸处理、高温分解、清丸处理、干喷砂、湿喷砂、煤油清洗、超声波清洗、高压清洗、震动研磨、打磨、酸洗、碱洗等	废水、废油、废液、废气；噪声	废旧零部件的清洗用水/用油；喷砂/抛丸产生的粉尘（SO_2、Al_2O_3 等）；酸洗/碱洗后的废液；零部件表面打磨产生的锈沫及污垢；零部件高温处理后产生的废气（CO_2、SO_2 及氮化物等）；各种清洗过程带来的噪声污染等
检测	超声检测、涡流检测、磁粉检测、渗透检测、量具检测、水检等	废液、废水；噪声	各种检测过程（如水检、磁记忆检测等）中产生的废液、废水等
再制造加工	物理加工、电刷镀、电弧喷涂、激光熔覆、等离子弧熔覆、纳米粘接技术、固体润滑、冷焊等	废料、废液、废渣；噪声、燃料燃烧气体、化学污染物、电磁辐射、光电污染	一般物理加工所产生的边角料；刷镀产生的废弃镀液；激光熔覆、等离子弧熔覆及电弧喷涂产生的光电、废料、粉尘/烟雾（金属氧化物等）、废气（CO_2、SO_2 及氮化物等）及噪声等
重新装配	装机（物理方法）	废油（较少）	装配过程中使用的润滑油等
整机检验	试机/试车（台架试验）	废气；噪声	试车过程中产生的废气、噪声等
包装	木箱包装	包装废料	木料及相关捆扎设备的剩余及浪费

4.2　再制造综合评价指标体系的构成

以 1993 年联合国统计署（UNSD）提出的“环境与经济综合核算体系（system of

integrated environment and economic accounting，SEEA）”为指导，参考其他关于循环经济指标体系的研究[5~7]，提出再制造综合评价指标体系构成，见表3。

表3 再制造综合评价指标体系

目标层	系统层	状态层	要素层
再制造发展综合评价指标体系	环境发展子系统	资源节约	节约能量、节约金属量、节约用水量、减少排放量、再制造率
		资源消耗	资源材料的消耗率、万元工业增加值综合能耗、万元工业增加值取水量
		环境改善	资源回收与旧件利用率、不可再制造部分资源化比率、工业固体废弃物综合利用率、废水循环利用率、生产废气/粉尘排放达标率
	经济发展子系统	废弃物排放	万元产值废气排放量、万元产值废水排放量、万元产值废液排放量、万元产值废渣（固体废弃物）排放量
		综合经济指标	企业总产值、企业再制造部分产值、企业再制造部分资产总值、企业再制造产品出口创汇总值
		经济收益	净资产收益率、销售净利率、净利润现金流量比率、万元投资回报率、固定资产产值率
		成本	万元产值废弃物处理成本
	社会发展子系统	就业	再制造部分就业人员吸收率、就业岗位增长率
		经济	年社会公益性贡献支出、年缴纳税金、年人均工资水平
		社会保障	享有社会保障人员的数量及比重

表3显示作为目标层的再制造发展综合评价指标体系，主要由环境发展、经济发展及社会发展三个子系统组成，并进一步细分为状态层及要素层的具体指标，从总体及各相关方面详细描述了再制造的发展情况。在对具体指标进行核算时，需要较大的工作量，可以把此视为一项系统工程，需要企业内或是行业内各组成单位的全力配合，获得全面的真实数据，以更好地反映再制造发展实际情况及其对各相关方面的影响。关于各要素层指标的计算方法可参考相关文献，本文不再阐述。

在再制造综合评价中还需要考虑政策的引导及影响、生产者责任的延伸等，在完善再制造综合评价指标体系的过程中，可以考虑政策状态层的加入。

5 总结

（1）中国再制造在短时间内获得了良好的发展，对中国循环经济的发展及两型社会的建设做出了重大贡献，但是在社会、产业、企业、技术层面上都面临着若干的问题。要规范再制造行业的发展，衡量再制造的发展水平及对环境及社会的贡献，需要通过特定的评价指标对其进行描述和评价。

（2）再制造综合评价指标体系的建立要符合循环经济的要求，遵循再制造工程发展的特点，体现多层面之间相互协调的原则，从环境发展、经济发展及社会发展等方面来反映再制造的发展态势。

（3）目前，循环经济的综合评价指标体系尚不健全，因此本研究中针对再制造的综合评价指标体系的设立也具有临时性，需要不断进一步完善。

参考文献

[1] 徐滨士．再制造工程技术及其研究新进展［J］．数字制造科学，2008，6(2)：1～18.

[2] 徐滨士．基于先进表面工程技术的汽车发动机再制造工程及其产业模式［J］．数字制造科学，2008，6(2)：19～34.

[3] 徐滨士，等．装备再制造工程的理论与技术［M］．北京：国防工业出版社，2007，1～2.

[4] 徐滨士，等．再制造与循环经济［M］．北京：科学出版社，2007，7～16.

[5] 陈文晖，马胜杰，姚晓艳．中国循环经济综合评价研究［M］．北京：中国经济出版社，2009，87～90.

[6] 张颖，吴志文．循环经济与绿色核算［M］．北京：中国林业出版社，2006，69～89.

[7] 王德发．绿色GDP——环境与经济综合核算体系及其应用［M］．上海：上海财经大学出版社，2008，21～29.

[8] Liu B H，Xu B S，Yang S L，et al. A Research on the Concept and Estimate Methods of Remanufacturing Rate of Engine. Key Engineering Materials，Vols. 419～420(2010)：253～256.

A Design for Integrated Assessment Indexing System of Remanufacturing

Liu Bohai[1,2]，Xu Binshi[1]，Shi Peijing[1]

(1. National Key Laboratory for Remanufacturing，Academy of Armored Force Engineering，Beijing，100072，China；
2. Management School of Hefei University of Technology)

Abstract Remanufacturing has developed well in China and Chinese Characteristic Remanufacturing Engineering (CCRE) formed，which contributed to the development of Chinese cycle economy. In order to measure the level of economic development and the contribution to environment and society of remanufacturing，some appropriate indexes are needed to describe and assess. The development characteristics of Chinese remanufacturing are introduced at first，and the principle of establishment of integrated assessment indexing system for remanufacturing is presented too. Then combine with the practical situation of Chinese remanufacturing，some specific indexes are put forward from point of view of environment development，economy development and society development. At last，the establishment methods of integrated assessment indexing system for remanufacturing are analysed and some suggestions are given.

Key words remanufacturing，integrated assessment，indexing system

再制造认证认可发展策略思考*

再制造（remanufacturing）在国外已有80多年的历史，技术标准、生产工艺、加工设备、配件供应至销售和售后服务都比较成熟，已形成一套完整的发展体系和较大的产业规模。我国的再制造工程仅有10年左右的发展历史，在技术上获得了较好的发展，但是离产业化还有一定的距离，相关管理体系的缺乏为其主要原因之一。认证认可是国际上通行的提高产品质量和组织管理水平，促进产业经济发展的重要手段，再制造产业化进程的健康推进同样需要认证认可的保障。

1　再制造工程内涵及特征

再制造工程（remanufacturing engineering，简称RemanE）是以产品全寿命周期理论为指导，以提升废旧产品性能为目标，以优质、高效、节能、节材、环保为准则，以先进技术和产业化生产为手段，修复、改造废旧产品的一系列技术措施或工程活动的总称，既是维修发展的高级阶段，又是先进制造的新形式。再制造工程是废旧产品高技术修复、改造的产业化，其重要特征是再制造产品的质量和性能达到或超过原型新品，成本不超过新品的50%，节能60%以上，节材70%以上，具有明显的社会及环境效益。

不同于国外的换件修理法或尺寸修理法，中国特色再制造工程（CCRE）采用以先进表面工程技术为支撑的再制造关键技术，不仅准确恢复尺寸，而且显著提升性能，充分释放产品剩余寿命，最重要的是资源利用率高，能源消耗少，成本低。因此，成为我国开展循环经济“4R”原则（reduce，reuse，remanufacturing，recycle，即减量化、再利用、再制造及再循环）中十分活跃并充分体现高技术含量的要素。

2　再制造产业面临的问题

作为循环经济的4R原则之一，再制造为我国节能、节材、环境保护等方面做出了重大的贡献，但是再制造在产业化发展进程中也面临着若干的阻力及问题。要开展再制造认证认可工作及发挥认证认可对再制造产业化的推进作用，就必须要清楚地认识这些问题，主要涉及社会、产业、企业以及技术等层面。

（1）社会层面。目前，我国的再制造产品主要有汽车发动机、发电机、变速箱、起动机以及机床、轮胎及其零部件等，这些再制造产品已通过销售或售后服务等渠道进入目标顾客市场。但相比于新品，再制造产品的社会认可程度至今仍不是很高，一

---

* 本文合作者：刘渤海。原发表于《认证技术》，2010(3)：44～47。再制造项目受到国家自然科学基金重点项目（50735006）、国家科技支撑项目（2008BAK42B03）、再制造技术国家级重点实验室开放基金（9140C8501010801）资助。

般的顾客不能区分再制造与维修的区别，甚至把两者等同对待，认为再制造品就是二手货，存在隐含风险，质量水平不高，导致了再制造产品的市场开拓难度很大。

（2）产业层面。因为当前再制造规范体系的不健全及相关概念的不明确，导致了再制造市场的混乱，开展再制造业务的企业规模大小不等，水平参差不齐。有些再制造企业开展的业务甚至不能称为再制造，其产品质量得不到保证，这给再制造的发展造成了一定的负面影响。

（3）企业层面。由于一些客观存在的原因，再制造企业的发展也面临一定的困难。通过对国内14家汽车零部件再制造试点企业的调研发现，这些困难主要有国家相关政策的冲突及缺乏、再制造回收网络的建立、再制造专业技术人员/设备的需求空白、回收产品质量状态的随机性、再制造生产计划安排的复杂性、再制造信息系统的兼容性、再制造市场的培育等，这些因素都阻碍了再制造业务的开展。

（4）技术层面。我国的再制造工程是在维修工程、表面工程基础上发展起来的，主要基于表面工程、纳米表面工程和自动化表面工程技术。目前，这些技术获得了较好的发展，已进入实用化阶段，但是对再制造企业的素质要求较高，没有得到广泛的推广，离产业化应用还有一定距离，而这一方面正是再制造企业所亟需解决的问题之一。

3　再制造认证认可工作的内容

再制造认证认可工作的开展对解决再制造当前面临的问题有着积极的促进作用，再制造认证认可工作应覆盖以下三方面。

3.1　再制造产品的认证认可

再制造产品的质量、性能水平不应低于同类新品。为了保证使用者的利益，再制造产品的质量水平应有明确的规定，不能把只具有翻新水平的二手产品充当再制造品直接交付市场。因此，再制造企业在向顾客交付产品的时候应明示其产品是否为再制造品，以及应证明其质量水平达到相关要求。

认证认可在国内及国际贸易中能够保护相关方的利益，对再制造产品进行认证认可可采取的办法有以下几种。首先企业应具有相应的再制造设备、再制造工艺手段、再制造技术规范以及再制造标准指标体系及试验方法，从流程的角度来保证产品质量。二是社会相关部门应对再制造企业所生产的产品进行型式试验及质量保证，采用的方法可以是国际标准化组织（international organization for standardization，ISO）出版的《认证的原则与实践》一书中所提出的八种型式试验的前四种，即型式试验；型式试验加认证后监督——市场抽样检验；型式试验加认证后监督，供方抽样检验；型式试验加认证后监督，市场和供方抽样检验，也可以采取类似中国强制认证（CCC认证）的方式来规范及促进企业生产出合格的再制造产品。

3.2　再制造企业资质的认证认可

为了使再制造获得良好的发展环境，不仅要对再制造产品的质量进行认证，还要对再制造企业的资质进行认证。再制造的技术含量较高，对环境及社会贡献大，与传统的大修、翻新是不同的。一些没有再制造技术能力的企业打着再制造的旗号开展着实际是维修的工作，不仅对再制造的发展造成了影响，更重要的是使消费者的利益受到了极大损害。

对再制造企业的资质进行认证认可可采取的办法主要有以下几种。一是从设备、技术及人员配备上来衡量企业的再制造水平，以确保该企业具备开展再制造的能力。不同于国外的换件修理法或尺寸修理法（减法），我国的再制造技术主要是基于先进表面工程技术（加法），对企业素质要求较高，这是再制造产品质量保证的基础条件。二是从资源节约、环境保护的角度来考察企业的再制造水平，可以利用再制造率的概念来衡量再制造企业的环境及社会贡献，如利用计数再制造率、计重再制造率及计价再制造率来考察企业的资源回收再利用及价值再创造的情况，同时还可以结合再制造环境指标体系来衡量再制造的废弃物减少排放情况，以反映企业开展再制造的意义所在。

3.3 再制造企业管理体系的认证认可

为了使再制造企业的相关资源得到最大化利用，以及确保具备持续提供优质再制造产品的能力，除了对再制造设备、技术及人员的配备外，一套规范而又灵活的再制造管理体系也是必不可少的。再制造管理体系包含再制造设备管理、再制造技术/工艺管理、再制造生产管理、再制造质量管理、再制造人力资源管理以及再制造战略管理、再制造组织管理、再制造经营管理等，内容涵盖企业运营的方方面面。

对再制造企业管理体系的认证认可工作可以参考 ISO 9000 的模式，即一方内审、二方审核及三方认证相结合的形式。一方内审是指企业根据或参考相关要求建立组织管理结构、管理制度，来规范企业所开展的运营活动，在一定的时间点按照规定的指标体系进行自我评审并进行成熟度测定，以判断企业的管理水平并向顾客提供。二方审核是指顾客对再制造企业的管理体系进行评审并核实企业的再制造规范运营水平，并以此来保证所购买的再制造产品的质量状况。三方认证则是通过具有权威性且独立于供购双方的第三方组织对供应商的管理体系进行审核并给予相关证明的形式来保证贸易的正常开展，优点在于避免了大量社会重复性劳动，节约供购双方的时间及资源。

4 开展再制造认证认可工作的措施

目前，我国的再制造认证认可工作处于起步阶段，还没有形成一套适合的体系。因此，需要开展的基础性工作很多。

4.1 再制造标准体系的建立

标准为认证认可工作的开展提供了依据，再制造标准化及标准化体系建设工作是再制造实现产业化发展的基础性工作之一。再制造标准的内容较为广泛，包括再制造技术标准、再制造工作标准、再制造管理标准等。再制造技术标准包括再制造相关技术的物理规格和化学性能规范。再制造工作标准和管理标准内容广泛，包括再制造件的设计、回收、拆卸、清洗、检测、再制造加工、组装、检验、包装等操作的规范性步骤及方法等。再制造的对象产品不同（如汽车发动机再制造、机床再制造和轮胎再制造等），则对应的产品再制造标准也不同，各类再制造企业应按照所生产的产品特点来选用合适的国家、行业标准或是制定适合的再制造相关企业标准。

2008 年 10 月，全国绿色制造标准化技术委员会再制造分技术委员会成立，挂靠装甲兵工程学院装备再制造技术国防科技重点实验室，主要负责再制造术语标准、再制造技术工艺标准、性能检测标准、质量控制标准，以及关键技术标准的制定等相关工作，为再制造在我国的发展提供规范化的保证。

4.2 再制造认证认可机构的设立

再制造认证认可机构可以由国家相关监管部门指定设立，分二级组成。第一级为国家级或行业级管理部门（如中国合格评定国家认可委员会），主要工作是关注国家政策的调整对再制造的影响，并对下面层级再制造认证认可机构进行指导及监管。第二级为社会级机构，由具备认证认可资格并符合《中华人民共和国认证认可条例》要求的组织机构自愿申请，并由上一级监管部门进行资格确定，主要工作是对再制造企业的再制造产品、再制造企业资质或是再制造企业管理体系的有效性进行认证，并明示于顾客及相关方，也可以在有需求的情况下，指导再制造企业开展认证认可工作。

4.3 再制造检验检疫实验室的设立

再制造检验检疫实验室可以由第一级管理部门指定设立，也可以是第二级机构设立并由第一级部门依据《实验室和检查机构资质认定管理办法》认可并监管。再制造检验检疫实验室的主要职责首先是公平地对再制造企业所生产的再制造产品进行检测、鉴定，并进行质量、性能等方面的评定。其次是对再制造毛坯的再制造性能进行检验以判定其是否适合进行再制造，包括再制造毛坯的拆洗、受损状况检测、剩余寿命评估等内容。再制造检验检疫实验室也可以向有需要的再制造企业提供相关的技术支持。

我国的再制造技术正处于快速发展阶段，再制造检验检疫实验室设立之后，要根据再制造技术的发展对再制造检验检疫设备及技术进行更新，以更好地满足社会及再制造企业的实际需要。因此，要求再制造检验检疫实验室要具备独立的研发部门与研发人员，或是与专业研发机构进行紧密合作，以确保再制造检验检疫设备及技术的先进性及适用性。

4.4 再制造认证认可专业人员队伍的建设

上述三项工作的开展都离不开再制造专业技术人员，而再制造在中国起步较晚，人才培养及人员配备方面还不是很完善，因此，亟需建立一支包括再制造专业技术人员及认证认可技术人员在内的再制造认证认可专业人员队伍。再制造专业技术人员首先对再制造有着正确的认识，并掌握着某一项或若干项再制造专业技术，如再制造毛坯的拆卸技术、绿色清洗技术、剩余寿命评估技术等，其次对所掌握技术的相关再制造检验检疫设备操作熟练，如再制造毛坯清洗设备、疲劳试验机、电弧喷涂/电刷镀/等离子弧熔覆系统、无损检测仪等，并具备一定的研发能力。而对再制造认证认可人员则要求具备一定的再制造专业基础并熟悉认证认可的操作规范，能开展再制造相关的认证认可工作。

4.5 再制造国家政策的协调及实施

循环经济的推行涉及国家各个层面，作为其支撑之一的再制造工程如要获得其产业化的健康稳定发展，则需要包括经济、环保、工商、公安、立法、科研等在内的相关部门的继续努力及协调，给目前再制造企业予以各种形式的支持，鼓励其开展相关产品的再制造，并在此过程中建立完善的再制造市场准入制度，为再制造认证认可工作提供良好的环境。同时，加大对再制造的宣传力度，正确宣扬再制造的环境及社会效益，开发目标顾客资源，帮助再制造企业及再制造认证认可机构开拓生存及发展空间，以引导或促进再制造形成产业化规模，为中国循环经济的发展做出贡献。

废旧产品再制造性评估指标*

摘　要　通过剖析再制造生产过程，在已有模块的基础上，结合当前社会关注的问题，提出了一种全新的再制造性指标评价体系。优化了以往的模糊评价方法和离散的评估方法，用［0，1］区间上连续的函数来表示再制造性的大小，并增加了一些关于循环经济的指标。可以为再制造企业进行再制造生产提供理论依据，以及可以通过指标评价体系对现有的再制造企业进行评价，以实现资源的合理化配置。文中以济南复强动力有限公司生产斯太尔 WD615 再制造发动机为例，来说明再制造性的计算过程。

关键词　再制造性　指标体系　评估

1　引言

产品的再制造性评估有两种方式，一是对已经使用和报废的产品再制造前对其进行再制造合理性评估[1~3]；二是在设计新产品时对其进行再制造性评估，利用评估结果改进设计，增加再制造性[4]。废旧机电产品的再制造性[5]（remanufacturability）是决定其能否进行再制造的前提，是再制造基础理论研究中的首要问题。废旧产品再制造性可定义为：废旧产品在规定的条件及时间内，综合考虑技术、经济和环境因素后，通过再制造，恢复或提高原产品功能和性能的能力。随着再制造技术的发展，废旧产品的再制造性也会随着变化。产品能否再制造以及再制造的效果如何，不仅关系到再制造生产的继续进行，也对再制造产品的质量提供了一个基本保证。对产品可再制造性的评价是再制造生产的一个重要环节。可再制造性决定了对产品实施再制造的可能性和经济性。文中主要是通过对一些评价指标的量化来最终达到一个适合废旧产品再制造性的定量指标，从数量上客观反映废旧产品是否适合再制造。

2　再制造性评价

虽然 Steinhilper[1] 和 Lund[2] 给出废旧产品是否适合再制造的一个判断准则，但是这并不是一个定量的评价，而且对准则的判断很大程度上依赖人的主观因素，不能很好地对废旧产品的再制造性进行客观的评估。而随着中国再制造产业的发展与推进，人们对再制造越来越认可，再制造在环境等方面的贡献越来越受到国家政府和有关部门的重视，但在指标评价体系方面相对欠缺。目前对废旧的毛坯进行再制造性的评价主要从技术可行性、经济可行性和环境可行性三个系统层面进行考虑[6]。

2.1　技术可行性[7]（R_1）

技术可行性要求废旧产品进行再制造加工技术及工艺上可行，可以通过原产品恢

* 本文合作者：刘赟、史佩京、刘渤海。原发表于《中国表面工程》，2011，24（5）：94~99。

复或者升级恢复达到或者提高原产品性能的目的。再制造加工过程主要包括拆卸、清洗、分类检测、修复和升级、再制造零部件检测、装配、整体检测。所以对再制造技术方面的可行性评价可以从以下几个方面进行。

2.1.1 拆解合格率（R_{11}）

目前废旧产品的拆解还是以手工和半机械化拆解为主，废旧产品本身经过了长时间的服役，导致的锈蚀等因素导致了其不能够完全无损的拆解或者不能够完全分解。拆解合格率就是指可再制造件在拆解过程中未经破坏的比例，即：

$$R_{11} = \frac{Q_n}{Q_r} \times 100\% \tag{1}$$

式中，Q_n 表示无损的可再制造件数量；Q_r 表示可再制造件总数。

2.1.2 检测可靠性（R_{12}）

在进行再制造之前的毛坯检测是进行再制造加工的重要环节之一。检测会出现两种情况：一种是把不合格的毛坯错误地当成合格的毛坯，从而影响最终的产品质量；另一种是把合格的毛坯当做不合格毛坯进行回炉处理，虽然不会影响最终再制造品的质量，但是却增加了成本。其评价方式可以用以下公式表达：

$$R_{12} = \left(1 - \frac{N_r}{N}\right)\left(1 - \frac{N_i}{N}\right) \times 100\% \tag{2}$$

式中，N_r 表示合格的检验成不合格的数量；N_i 表示不合格的检验成合格的数量；N 为总的检验数量。

2.1.3 新品可靠性（R_{13}）

虽然再制造产品的质量不低于原始新品的质量，但是随着科技的进步以及人们对产品的要求逐渐严格，现有的产品质量可能会高于报废零部件生产时候的质量要求，为了保持再制造产品和现有同类产品之间的竞争优势，再制造产品的质量不能和现有同类型产品新品的质量相差太多。新品可靠性是衡量再制造产品适应新的竞争的能力（质量和价格）的一个重要的指标，用以下评估方法来评估：

$$R_{13} = \frac{n_1 + n_2\omega + n_3(1 - \omega)}{n_1 + n_2 + n_3 + n_4} \tag{3}$$

式中，n_1 表示质量比现有新品好、价格比现有新品低的数量；n_2 表示质量比现有新品差、价格比现有新品低的数量；n_3 表示质量比现有新品好、价格比现有新品高的数量；n_4 表示质量比现有新品差、价格比现有新品高的数量；ω 表示性能所促进的竞争优势比重。

2.1.4 整机性能（R_{14}）

再制造零部件虽然在单个性能上面不低于原始新品的质量，但是再制造产品很多情况下作为一个整体，作为一个系统，其性能是否不低于原始的整机性能有待核实。对于不同的再制造产品，所关心的具体内容也不同，对于再制造发动机，我们关心的是发动机的动力、扭矩、可靠性等，而再制造轮胎，所关心的就是耐磨性和寿命。所以整机性能这个指标就用达到或者超过原始新品的比例来表示，即：

$$R_{14} = \frac{Q_u}{Q} \times 100\% \tag{4}$$

式中，Q_u 表示达到原始新品数量的整机数；Q 表示整机总数。

2.1.5　相容性（R_{15}）

再制造产品作为一个整体，其中的每个零部件的使用寿命不尽相同，在使用的过程中，可能会出现零部件的损坏而需要对零部件进行更换，这要求其中的再制造零部件要与其他零部件有着良好的兼容性，所以文中用以新件代替再制造件而不影响整机性能与总的数目的比值作为评价与其他零部件相容性的方法，即：

$$R_{15} = \sum_{A_r} \frac{A_r}{A_c A_w} \tag{5}$$

式中，A_r 为用新品替换某个再制造零部件后整机性能不低于原始新品数量；A_c 为每个整机的再制造零部件数目；A_w 为整机的总数目。

2.1.6　加工效率（R_{16}）

加工效率是衡量再制造加工环节的一个结合时间性的指标，是指废旧毛坯进入再制造工厂进行清洗拆解开始到装配检测成为一个新的再制造产品所消耗的时间。

$$R_{16} = \begin{cases} 1 & t < t_0 \\ \dfrac{t_0}{t} & t > t_0 \end{cases} \tag{6}$$

式中，t 表示再制造加工平均所用时间；t_0 表示维修所需的平均时间。

2.2　环境可行性（R_2）

再制造性的环境可行性是对废旧产品再制造加工过程本身以及再制造后的产品在社会上使用时对环境影响的成本估计和预测。环境可行性的评价需要一个参照。再制造是制造的延续与补充，而且弥补了原始制造的一些缺点，其显著特点之一就是对环境友好，所以再制造的环境可行性是和制造相比得到的结果。

2.2.1　旧件利用率（R_{21}）

进入工厂的废旧整机其拆解后的零部件分成可再制造件、直接利用件和不可再制造件。旧件利用率用来衡量再制造过程对废旧产品的利用率，用可再制造件和直接利用件质量之和与整机的质量之比来表示，即

$$R_{21} = \frac{W_{reman} + W_{reuse}}{W} \times 100\% \tag{7}$$

式中，W_{reman} 表示再制造件质量；W_{reuse} 表示可利用件质量；W 表示整机质量。

2.2.2　节能（R_{22}）

原始制造出的产品经过一个寿命周期后要经过回炉变成原料再进行加工成零部件，而再制造以废旧产品为毛坯，进行加工生产，不需要经过回炉处理，因而节约大量能量，其计算公式为：

$$\begin{gathered} PW_{reman} = PW_{hreman} \times R_{21} \\ R_{22} = 1 - \frac{PW_{reman}}{PW_{md}} \times 100\% \end{gathered} \tag{8}$$

式中，PW_{reman} 表示实际再制造耗能；PW_{hreman} 表示完全再制造耗能；PW_{md} 回炉耗能。

2.2.3　减少 CO_2 排放量（R_{23}）

减少 CO_2 排放量是将废旧毛坯经回炉成原始材料与其再加工成零部件的过程相比

所减少的 CO_2 排放量；完全再制造是指再制造率为 1 的再制造过程。其计算公式如下：

$$E_r = E_{cr} \times R_{21} + E_{md} \times (1 - R_{21})$$

$$R_{23} = \frac{E_r}{E_{md}} \times 100\% \tag{9}$$

式中，E_r 为再制造 CO_2 排放量；E_{cr}为完全再制造 CO_2 排放量；E_{md}为回炉制造 CO_2 排放量。

2.3 经济可行性（R_3）

技术可行性和环境可行性决定了废旧产品能否进行再制造，而经济可行性却是最终反映进行再制造活动是否继续下去的最终判断标准和人们从事经济活动的根本出发点之一。考虑再制造活动的经济因素时，不仅要考虑回收、加工以及废弃物处理的成本，还要考虑再制造加工过程中对环境的影响所带来的环境成本。所以，引进 2 个指标：利润率和环境收益率，用来评估再制造的经济性。

利润率（R_{31}）是指净利润与销售收入之间的比值。净利润借鉴杜邦财务分析法可得其计算示意图，如图 1 所示。

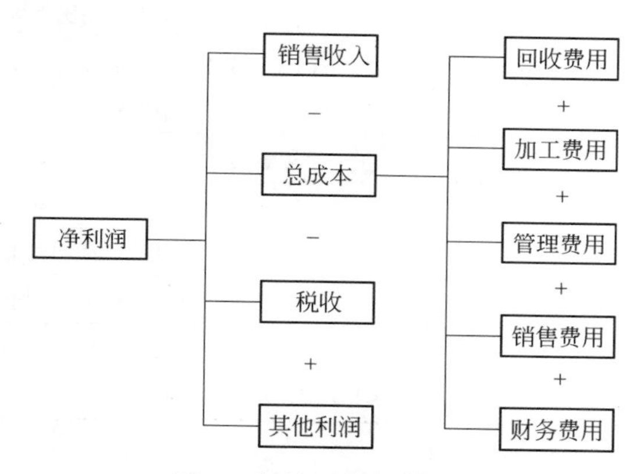

图 1　净利润计算示意图

Fig. 1　Sketch map of the net profit calculation

环境收益率（R_{32}）用来衡量再制造对环境经济的影响。废旧产品通过再制造变成了性能等同或稍高于原先新品的再制造品，这个过程不仅带来了可用的“新品”，而且避免了废旧产品的回炉所造成的原先产品附加值的丢失。环境收益率就是由再制造所引起的产品附加值的减少与新品价值的比值。

3　指标权重的确定

对于已经确定的评价指标，其权重的确定是建立再制造性指标评价模型的关键问题之一。对于这种多层次、多要素、多准则的非结构化的决策问题，层次分析法给出了很好的解决方法。

3.1 层次分析法[8]（AHP）

层次分析法是美国运筹学家 T. L. Saaty 于 20 世纪 70 年代提出的一种定性和定量相结合的决策方法。AHP 常常被用于多目标、多准则、多要素、多层次的非结构化的复杂决策问题。再制造性指标评价体系的权重的确立考虑的因素多，且指标间存在着千丝万缕的联系。利用AHP方法来确定指标的权重，不仅减少了人为因素对指标的权重的影响，增加了模型的可信度，而且建立起了模糊评价与精确评价之间的联系。

3.1.1　AHP标度

在进行层次分析法确定权重之前要进行AHP标度的确定，标度的确定一般是根据以往经验确定的，如表1所示。

3.1.2　AHP一致性检验

利用AHP标度，可以得到一个AHP判断矩阵B。通过对B的一致性检验来验证上面所定义AHP标度是否合理，其方法如表1所示。

表1　层次分析法的（AHP）标度

Table 1　Analytic hierarchy process scale

标　度	含　义
1	一个因素与另一个因素同样重要
3	一个因素与另一个因素相比稍微重要
5	一个因素与另一个因素相比明显重要
7	一个因素与另一个因素相比强烈重要
9	一个因素与另一个因素相比极为重要
2	1、3两个相邻判断的中间值
4	3、5两个相邻判断的中间值
6	5、7两个相邻判断的中间值
8	7、9两个相邻判断的中间值
倒数	因素i与因素j比较，得到的判断为$a_{ij}=1/a_{ji}$且$a_{ij}=1$（在$i=j$时）

计算判断矩阵的最大特征值$\lambda_{\max}$，并计算：

$$CI=\frac{\lambda_{\max}-n}{n-1},\quad CR=\frac{CI}{RI}$$

式中，n为判断矩阵B的阶数；RI为平均随机一致性指标（见表2）。

表2　平均随机一致性指标

Table 2　Mean indexes of stochastic consistency

阶数	4	5	6	7	8	9
RI	0.90	1.12	1.24	1.32	1.41	1.45

当$CR<0.10$时，判断矩阵有一致性，否则调整AHP标度使判断矩阵有一致性。

3.2　系统层权重的确定

对于技术、环境和经济3个系统层面来说，目前比较关心的是环境和经济，所以环境和经济相对于技术这个系统层来说稍微重要一点，根据上面的AHP标度得到相关权重，如表3所示。

表 3　AHP 法系统层权重确定

Table 3　Determination of system weight with AHP method

因素	R_1	R_2	R_3	总计	权重/%
R_1	1.0	0.5	0.5	2.0	20.0
R_2	2.0	1.0	1.0	4.0	40.0
R_3	2.0	1.0	1.0	4.0	40.0

判断矩阵

$$B=\begin{bmatrix}1 & 0.5 & 0.5\\ 2 & 1 & 1\\ 2 & 1 & 1\end{bmatrix},\ \lambda_{\max}=3,\ 可得\ CI=0,\ CR=0<0.10$$

所以 AHP 标度是有效的，且技术可行性、环境可行性和经济可行性的权重分别为20.0%、40.0%和40.0%。

3.3　指标层权重的确定

表 4 中所给出的是技术可行性指标的关系。由表 4 可看出技术可行性评价指标拆解合格率、检测可靠性、新品可靠性、整机性能、相容性和加工效率的权重分别为：16.46%、16.46%、22.78%、17.72%、17.72%和 8.86%。

根据环境可行性指标间的关系，可得到表 5 的数据，即环境可行性评价指标旧件利用率、节能和减少 CO_2 排放量的权重分别为 20.0%、40.0%和 40.0%。

根据经济可行性间的关系，可得表 6，即经济可行性评价指标利润率和环境收益率的权重分别为 50.0%和 50.0%。

表 4　AHP 法技术可行性权重确定

Table 4　Determination of technical feasibility weight with AHP method

因素	R_{11}	R_{12}	R_{13}	R_{14}	R_{15}	R_{16}	总计	权重/%
R_{11}	1.0	1.0	0.5	1.0	1.0	2.0	6.5	16.46
R_{12}	1.0	1.0	0.5	1.0	1.0	2.0	6.5	16.46
R_{13}	2.0	2.0	1.0	1.0	1.0	2.0	9.0	22.78
R_{14}	1.0	1.0	1.0	1.0	1.0	2.0	7.0	17.72
R_{15}	1.0	1.0	1.0	1.0	1.0	2.0	7.0	17.72
R_{16}	0.5	0.5	0.5	0.5	0.5	1.0	3.5	8.86

注：$CI=0.0162$，$CR=0.0131<0.10$。

表 5　AHP 法环境可行性权重确定

Table 5　Determination of environment feasibility weight with AHP method

因素	R_{21}	R_{22}	R_{23}	总计	权重/%
R_{21}	1.0	0.5	0.5	2.0	20.0
R_{22}	2.0	1.0	1.0	4.0	40.0
R_{23}	2.0	1.0	1.0	4.0	40.0

注：$CI=0$，$CR=0<0.10$。

表 6　AHP 法经济可行性权重确定

Table 6　Determination of economy feasibility weight with AHP method

因素	R_{31}	R_{32}	总计	权重/%
R_{31}	1.0	1.0	2.0	50.0
R_{32}	1.0	1.0	2.0	50.0

注：$CI=0$，$CR=0<0.10$。

4　计算实例

以济南复强动力有限公司再制造一批斯太尔 WD615（500 台）为例。发动机的总质量为 873.68kg。工厂对发动机缸体、缸盖、曲轴、连杆、气门、挺柱和凸轮轴进行再制造，其相关数据见表 7。

表 7　济南复强再制造基本数据

Table 7　Basic remanufacturing data of JiNan FuQiang Power Co. Ltd

再制造零部件	报废率/%	质量/kg
缸体	5.62	260.00
缸盖	4.48	93.60
曲轴	3.70	103.00
连杆	3.43	21.10
气门	4.12	3.15
挺柱	2.98	
凸轮轴	6.15	11.25

由式（1）和式（7）可得：拆解合格率 $R_{11}=94.92\%$，旧件利用率 $R_{21}=56.32\%$。

对该批发动机进行统计分析，拆解合格的零部件经检测后均可再制造，由于没有对报废的零部件进行二次检验，所以由式（2）可以估算出 $R_{12}\geqslant 94.92\%$。

再制造发动机的价格均低于原先新品的价格，该批 500 台再制造发动机有 14 台因质量问题被召回，通过对用户的调查，约 80% 的用户是因为再制造发动机便宜而选择使用再制造发动机的，所以，由式（3）、式（4）可得：$R_{13}=(500-14+14\times 0.2)/500=97.76\%$，$R_{14}=(500-14)/500=97.20\%$。

公司采用表面工程技术结合尺寸恢复法来再制造发动机，所得到的再制造发动机的零部件均可以和原先零部件进行互换使用，所以由式（5）可得：$R_{15}=1$。

发动机大修平均所需时间为 36h，再制造一台发动机平均所需时间为 24h，由式（6）可得 $R_{16}=1$。

每回炉 1t 钢铁、铝材、铜材的耗能分别为 1784kW · h/t、2000kW · h/t 和 1726kW · h/t，排放 CO_2 分别为 0.086t、0.17t 和 0.25t；再制造 1 台发动机耗能 102.857kW · h，排放 CO_2 0.0281t（工厂实际数据）；用回炉后的型材制造 1 台新发动机耗能 205.714kW · h，排放 CO_2 0.0562t。一台发动机含钢铁：583.668kg；铝

材：16.01kg；铜材：1.886kg。

可计算出：一台发动机从回炉到再制造所需的能量为1769.22kW·h，排放的CO_2为133.70kg；再制造一台发动机所消耗的能量为：785.68kW·h、排放的CO_2为86.50kg。利用式（7）、式（8）可以计算出节能$R_{22}=1-785.68/1769.22=55.59\%$、$R_{23}=1-86.50/133.70=35.3\%$。

单台再制造发动机的售价为29000元，其中设备费400元、材料费300元、能源费300元、新加零件费10000元、税费3400元、人力费1600元和管理费400元。可计算出获得利润为13000元，$R_{31}=13000/29000=44.82\%$，和新品的利润率27.5%相比，利润率大大提高。

每台发动机平均价格4.5万元，实行再制造后部件价格分别为：气缸体总成1件10500元、连杆总成6件共1466.5元、气门挺柱12件共386.4元、汽缸盖总成6件共5241.6元、凸轮轴1件520元。$R_{32}=(10500+1466.5+386.4+5241.6+520)/45000=40.25\%$。

综上，计算出该批废旧发动机的可再制造性为：$R=55.91\%$。

5　结论

（1）再制造过程是一个对资源进行高级利用的过程，其产品的质量不低于新品。

（2）通过对节能和节材指标的计算，可以看出再制造在能源节约和环境保护方面具有良好的效果。

（3）通过计算利润率发现，再制造的利润与销售新品发动机相比有了明显提高，使资源得到了进一步的合理化配置。

（4）济南复强再制造动力有限公司只是对斯太尔WD615系列发动机的汽缸体总成、连杆总成、气门挺柱和气缸盖总成进行再制造，对废旧发动机的剩余价值尚没有达到最大限度的利用，应积极开展其他零部件的再制造，以实现再制造对废旧产品剩余价值的充分利用。

参考文献

[1] Steinhilper R. Product recycling and eco-design：challenges，solutions and examples［R］. International Conference on Clean Electronics，1995.

[2] Robot T L. The remanufacturing industry hidden giant［R］. Research Report，1996.

[3] Daniel V R. Production planning and control for remanufacturing［J］. Journal of Operations Management，2000，18：467～483.

[4] Bras B. Towards design for remanufacturing metrics for assessing remanufacturability［C］. Proceedings of the 1 st International Workshop on Reuse，Eindhoven，The Netherlands，1996.

[5] 徐滨士，马世宁，刘世参，等. 绿色再制造工程设计基础及关键技术［J］. 中国表面工程，2001，2：12～15.

[6] 朱胜，姚巨坤. 装备再制造设计及其内容体系［J］. 中国表面工程，2011，24（4）：1～6.

[7] 再制造技术标准体系及评价体系研究［R］. 中国标准化研究院，2009，11.

[8] 赵焕臣，许树柏，和金生. 层次分析法［M］. 北京：科学出版社，1986.

Assessment Indexes of Used Products Remanufacturability

Liu Yun[1], Xu Binshi[1], Shi Peijing[1], Liu Bohai[1,2]

(1. Science and Technology on Remanufacturing Laborotory, Academy of Armored Forces Engineering, Beijing, 100072, China;
2. Management School of Hefei University of Technology, Hefei, 230009, China)

Abstract The paper raises a new system on remanufacturability assessment indexes based on the existing modules, dissecting the process of remanufacturing and considering the common focuses. The system is different from vague assessment and discrete assessment, which uses a continous function on [0, 1] to weight remanufacturability. At the same time, some economic indexes are added into the system. It can not only provide a theoretical basis for remanufacturing factory, but also can appraise remanufacturing factory's existing assessment system to rationalize the allocation of resources. Finally, taking JiNan FuQiang power Company Limited's Steyr WD - 615 engine remanufacturing for example, the system was explained how the remanufacturability works.

Key words remanufacturability, indexes system, assessment

再制造产业发展过程中的管理问题*

摘　要　再制造正处在产业发展的初期阶段，在产业化过程中面临着来自社会、政策、技术、管理等层面的压力。从再制造产业发展的实际出发，针对再制造产业化过程中的若干关键管理问题进行研究。结合企业开展再制造业务的关注点及再制造产业未来的发展趋势，重点研究企业开展再制造业务的风险管理、再制造生产管理、再制造质量管理、再制造认证认可及再制造绩效考核等内容，分析其各自特点，并提出相应的对策。

关键词　再制造　产业发展　管理问题

1　引言

随着国家促进循环经济力度的加大，作为循环经济支撑技术之一的再制造受到了越来越多的关注，并得到了各方面的认可，走向了产业化发展的进程。再制造以制造为基础，但其生产流程不同于制造，增加了废旧产品的回收、拆解、清洗、废旧零部件质量检测及寿命评估、再制造加工、再装配等工序[1~4]。再制造的工序更加复杂，新品制造没有面临此类问题，这些问题有技术方面的，也有管理方面的。另外，再制造作为一个新兴的产业，很多的政策措施没有出台作为保障，导致再制造产业的发展缓慢或受阻。因此，有必要对再制造产业发展过程中面临的管理问题做出研究并加以解决，本文主要针对再制造产业发展过程中所面临的受到普遍关注的若干关键管理问题加以研究，并给出相关的建议。

2　再制造风险管理

国家已经给予了再制造企业若干优惠政策，但产业发展仍处于初期阶段，还存在较多的已知或未知因素影响再制造企业的发展，包括政策、技术、管理、市场等方面，导致再制造企业运营的风险。对国家第一批14家汽车零部件再制造试点企业的调研分析显示，部分企业只看重技术及设备的重要性，而忽略企业风险分析，造成再制造业务发展困难重重。如能结合实际对再制造企业进行风险分析，则可以帮助企业认识当前情况，在合理的情况下开展再制造业务或是决定是否开展再制造业务。同时，开展再制造风险分析还可以有效帮助企业合理规避风险，将企业所面临的风险降到最低程度。

对于再制造生产过程中面临的风险的识别是风险管理的首要任务，再制造生产过程中的风险识别技术可以采用较为客观的流程图法、工作分解结构法（WBS）和风险分解结构法（RBS）等方法，这些方法是基于过程的具体流程及构成情况，对每一个

* 原发表于《中国工程科学》，2012，14(12)：10~14。国家重点基础研究发展计划（2011CB013400）；军队预研项目（51327040401，9140C850102）资助。

流程或结构所涉及的风险源进行识别，确认其对项目的影响程度，并采取相应的措施进行控制。根据分析，再制造生产过程的风险主要有设备风险、技术风险、管理风险、操作风险以及环境风险5类，如表1所示。

在再制造风险因素确定下来之后就可以利用一定的方法（如Delphi法、模糊综合评分法等）对其权重进行确认，其后，再制造企业则可以对照风险因素表对自身的情况进行分析，明确企业的自身能力，以助其进行风险控制或业务决策。

另外，如果将考虑范围继续扩大，即不仅只考虑再制造生产过程，而是考虑再制造运营的全过程，则再制造的风险还包括政策风险和市场风险等因素，在对其进行分析的时候可以采用的方法同上。

表1　再制造生产过程中的风险因素举例

Table 1　Risk factors in the producing process of remanufacturing

准则层	指标层
设备风险（R_1）	手工拆解的破坏性及效率低下（R_{11}）
	清洗设备不具备导致的清洗质量差及效率低（R_{12}）
	传统检测手段的落后导致质量检测准确率低（R_{13}）
	表面预处理设备不符合产品要求（R_{14}）
	没有根据产品特点选择专用加工设备（R_{15}）
	不具备具有针对性的表面涂层性能测量设备（R_{16}）
	再制造产品缺少与原型新品的匹配验证过程（R_{17}）
	再制造加工设备的投资大及通用性不高（R_{18}）
技术风险（R_2）	没有掌握产品结构导致的拆解技术或顺序错误（R_{21}）
	清洗技术及材料的错误导致清洗不干净（R_{22}）
	检测技术不完善导致产品剩余寿命评估失真（R_{23}）
	不具备合格的表面预处理技术导致表面涂层脱落（R_{24}）
	再制造加工技术及材料选择的盲目性影响涂层性能（R_{25}）
	表面涂层性能检测技术不具备或不完善（R_{26}）
	后续加工达不到原产品精度要求（R_{27}）
	整机寿命预测难度大（R_{28}）
	再制造核心技术的研发周期长及针对性强（R_{29}）
管理风险（R_3）	废旧产品回收渠道不畅通（R_{31}）
	拆解前没有对废旧产品的状态进行确认（R_{32}）
	清洗设备及材料的选择错误（R_{33}）
	不可利用件流入后续加工过程（R_{34}）
	对拆解后零部件的检测频次不够（R_{35}）
	内部损伤件没有做出标示（R_{36}）
	用于再制造产品的新件没有进行检测而直接装配（R_{37}）
	废旧产品损坏状态不一致导致再制造生产计划制定困难（R_{38}）
	对市场需求的不掌握导致再制造产品不足或剩余（R_{39}）

续表 1

准则层	指标层
操作风险（R_4）	未按规定的顺序和工具进行拆解而导致损坏（R_{41}）
	检测过程不规范致使缺陷没有被检出（R_{42}）
	预处理过程前确定的修理等级不够（R_{43}）
	再制造加工过程的不规范导致返工、返修或报废（R_{44}）
	未按要求的装配顺序及设备进行装配（R_{45}）
	未按原型新品的要求进行试机致产品质量不达标（R_{46}）
环境风险（R_5）	拆解后的报废品随意丢弃（R_{51}）
	大量采用化学清洗技术（R_{52}）
	没有对再制造生产过程中产生的废物进行处理（R_{53}）
	没有对再制造加工过程中的噪声污染进行隔绝处理（R_{54}）

3 再制造生产管理

再制造生产与新品生产的过程不同，再制造生产管理也具有其特殊性，主要有：(1) 毛坯返回在时间和质量上的不确切性；(2) 返回与需求之间平衡的必要性；(3) 返回产品需要拆卸、分解；(4) 回收材料质量水平的不一致性；(5) 需要逆向物流网络；(6) 材料匹配限制的复杂性；(7) 混乱的加工路线以及高度不同的加工时间。

因为再制造毛坯是从市场上回收的废旧产品，在产业发展初期部分再制造企业还没有形成固定的回收渠道，再制造毛坯的回收量及质量具有一定的随机性。而企业的生产计划要根据回收量的大小及质量的好坏来制定，因此具有很大的不稳定性。

根据对调研情况的分析得知，目前大多数企业的回收渠道还不是很健全，所采用的再制造生产方式属于“推动式”，即生产出的产品通过各种渠道销售给消费者，离“拉动式”的生产还有一定的差距。无论何种生产方式，再制造毛坯回收量的不确定性都会造成两种情况，一是再制造毛坯回收量较大，超出企业的生产能力及市场消费能力，另一种情况是回收量较小，满足不了企业的生产能力及市场消费能力。因此，诸多企业采用了安全库存的形式来平衡生产能力，安全库存包括拆解件库存及更新件库存，这就会给企业带来库存成本压力，而传统的库存控制方法，如 EOQ（经济订货批量）、EPL（经济生产批量）库存控制模型不再完全适用于再制造生产运营，企业的生产计划制定具有更大的复杂性。

目前的再制造生产计划制定及生产调度研究中，大多都是基于一定的假设条件，如假设已知需求量和回收量，根据库存和能力约束建立生产计划规划模型等。研究认为，目前的再制造生产计划和调度的研究还存在着缺乏描述不确定性的有效方法、缺乏有效的建模工具等难题。对于独立再制造商来说，再制造毛坯回收量的多少及可再制造率的大小决定了其生产经营行为，影响较大，因此需要企业不断地开展回收渠道拓展业务。对于原始设备再制造商来说，需要根据再制造毛坯回收量及可再制造率来决定下一周期内的再制造业务量与新品制造业务量，也需要对再制造毛坯回收量及可再制造率做出预测及判断，以合理安排生产计划。

针对再制造毛坯回收量的预测目前还比较少见，根据预测理论及方法的发展状况以及再制造毛坯回收量的分布状况，可以探讨将人工神经网络及支持向量机技术引入该领域的适用性。而针对再制造毛坯的可再制造率的预测，则可以选择针对某一数据区间预测的马尔科夫链方法，以方便企业合理地安排生产计划及资源需求计划。后续工作计划中要研究选择方法及适用性分析。

4 再制造质量管理

基于表面工程技术的中国特色再制造质量管理内容包括：再制造毛坯质量管理、再制造成型过程质量管理、再制造成品质量管理。

再制造过程中，废旧件作为“毛坯”，通过多种高新技术在废旧零部件的失效表面生成涂覆层，恢复失效零件的尺寸并提升其性能，获得再制造产品。因此，再制造产品的质量由废旧件（即再制造毛坯）的原始质量和再制造恢复涂层的质量两部分共同决定。其中，废旧件的原始质量则是制造质量和服役工况共同作用的结果，尤其服役工况中含有很多不可控制的随机因素，一些危险缺陷常常在服役条件下生成并扩展，这将导致废旧件的制造质量急剧降低；而再制造恢复涂层的质量取决于再制造技术，包含再制造材料、技术工艺和工艺设备等。再制造零件使用过程中，依靠再制造毛坯和修复涂层共同承担服役工况的载荷要求，控制再制造毛坯的原始质量和修复涂层的质量就能够控制再制造产品的质量。

质量管理是一个有序的过程，先进合理的再制造工艺技术可以从根本上保证产品的质量，但是因为任何一个生产系统都会受到5M1E因素的影响，导致产品质量问题发生的必然性，因此也需要从管理的角度对生产系统进行良好的控制。再制造质量管理的要求主要如下所述：

（1）研发再制造专用技术及装备。再制造技术及设备的研发是再制造质量管理的硬件基础，从根本上保证了再制造产品及过程的质量，在再制造过程中，应根据再制造对象的不同，选择不同的加工技术、加工工艺、成型材料等，为了提高再制造产品质量及再制造生产加工效率，也应有针对性地研发专用技术及装备。

（2）建立健全再制造技术标准及管理标准。标准化工作是质量管理的基础工作之一，标准是衡量产品质量好坏的尺度，也是开展生产制造、质量管理工作的依据。再制造标准包括再制造技术标准、工作标准和管理标准。技术标准包括再制造相关技术的物理规格和化学性能规范，用作质量检测活动依据。工作标准和管理标准内容广泛，包括再制造件的设计、回收、拆卸、清洗、检测、再制造加工、组装、检验、包装等操作的规范性步骤、方法及管理依据。2011年10月，全国绿色制造标准化技术委员会再制造分技术委员会成立，挂靠装甲兵工程学院装备再制造技术国防科技重点实验室，主要负责再制造术语标准、再制造技术工艺标准、性能检测标准、质量控制标准以及关键技术标准的制定等相关工作，为再制造在中国的发展提供规范化的保证。

（3）建立健全再制造质量管理体系。在标准化工作的基础上，以朱兰质量管理三部曲（质量策划、质量控制、质量改进）为指导思想，以通行的ISO 9000族质量管理体系为参考，建立健全再制造质量管理体系。其中要体现质量策划工作是重点，因为质量策划决定质量目标，质量策划也同样影响着产品的可再制造性及其在多个生命周期的质量情况。为了实现再制造质量策划的目的，必不可少的后续工作即是质量控制及质量改进，这里的控制不仅是技术方面的控制，也包括管理方面的控制。质量工具箱为上述各项工作提供了可用的方法和技术。

5 再制造认证认可

从不同产业的发展历史来看，认证认可是推动及规范产业发展的重要手段，处于起步阶段的再制造产业的发展同样离不开认证认可的保障。开展再制造产品体系认证、认可技术研究和再制造企业的认证示范，对保证再制造产品质量、规范再制造行业、保障消费者权益、为政府部门提供技术支撑和认证采信，从而进一步促进再制造行业的健康发展都具有重要意义。中国质量认证中心与装备再制造技术国防科技重点实验室对再制造汽车发动机和轮胎产品认证技术进行了研究，提出了适合我国国情的再制造认证认可模式，并建立了再制造汽车零部件认证示范基地，是我国再制造认证认可研究的先行者。

根据实际情况调研及分析，再制造产品认证采用“型式试验 + 工厂质量体系评定 + 认证后监督”的基本模式，初始工厂检查应含质保体系和环保体系的审核，监督检查应采用飞行检查及现场抽样检测，同时根据产品的不同选取不同的认证操作模式。

（1）三方认证结合技术可行性评审或二方评审相结合的方式。对于非通用类的再制造零部件，因产品的结构设计、材料要求等技术性能的基本要求均来自于整机企业或者 OEM 零部件工厂，而且这些再制造零部件也全部用于原配整机售后维修，二方在参数管理、制造规范上有明确的要求，并进行二方评审，而且再制造企业为不同的委托方服务时，委托方的技术、质量要求有所差别，而三方认证基本的操作模式基本一致，质量控制和质量保证要求一致，评价标准统一，认证结果公平、公正。因此，采取二方评审和三方认证的认证操作模式，能够使再制造产品认证的可操作性更强，认证有效性更能保证。认证中二方与三方的工作内容和要求，在产品认证实施规则中予以明确，适合这种方式的如发动机、变速箱等汽车零部件产品。

对于再制造产品认证，因其产品结构复杂，使用寿命周期较长，对整机的质量性能、安全环保性能等影响大，认证除有证书外还需有符合性声明——一致性证书，以便于整机售后维修企业、整机用户的选择使用和相关管理部门的管理。

未得到整机厂授权的独立再制造企业，因缺少二方评审环节和整机厂的技术支持，应在认证流程上进行严格的质量控制，在认证申请阶段进行技术可行性评审，确保其质量保证能力及生产能力达到再制造产品的要求；在工厂审核时进行全条款审查，确保质量保证体系和能力达到或超过原机的要求，型式试验要进行全项试验，包括可靠型式试验，保证再制造产品的性能和质量不低于原型新品的水平。

（2）认证过程采用产品认证和体系认证相结合的模式。因为再制造产品是用报废的零部件进行再制造的，其在报废件的拆解、清洗、废物处理及再制造的加工过程中都有二次污染的防治及监管问题，而且再制造零部件产品的质量保证体系建设不容忽视。根据产品不同、企业需求不同，可以采用产品认证结合体系认证联合进行的操作模式，即产品认证 + 质量体系认证。产品认证与体系认证一体化进行的方式有效、高效、方便，而且可以节省认证费用。再制造产品认证模式如图 1 所示。

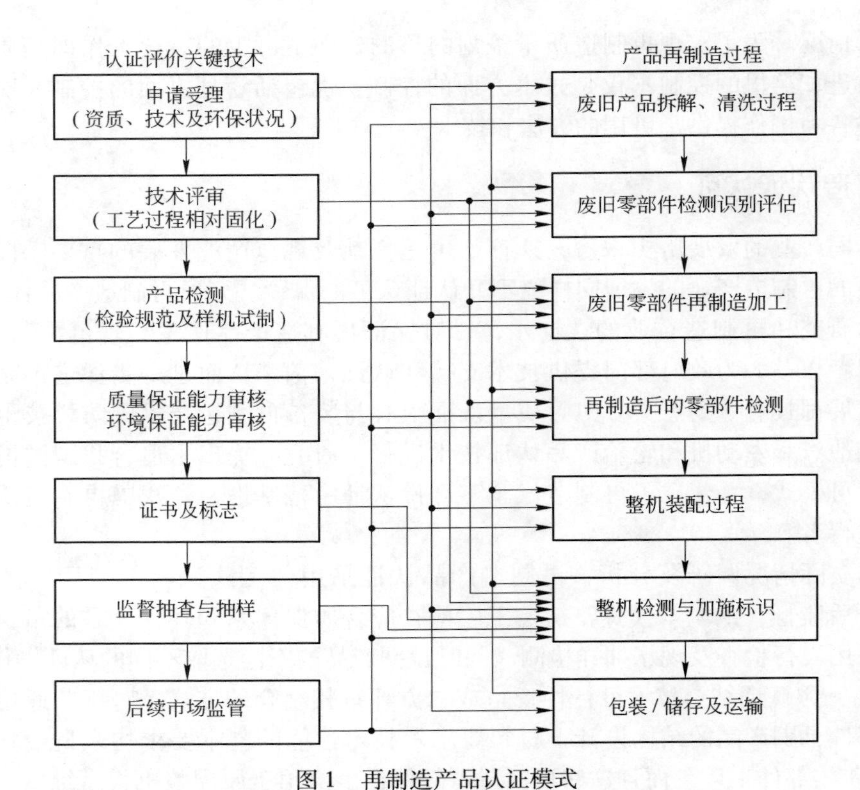

图 1　再制造产品认证模式

Fig. 1　Authentication mode of remanufacturing products

6　再制造绩效考核

再制造在中国的发展历程较短，面临着来自社会、产业、企业及技术等各级层面的压力。2008 年国家发展改革委员会批准了国内 14 家汽车零部件产品制造企业作为再制造试点，2009 年国家工业和信息化部批准了国内 35 家企业和产业园区作为机械产品再制造试点，以寻求中国再制造的产业发展途径。同时，在产业政策的不断调整及完善下，越来越多的企业开始在市场的引导下开展再制造工作。企业的再制造产品不尽相同，所采取的再制造发展途径也是各具特色，为了确定某种发展途径是否符合中国再制造产业的发展趋势，绩效考核也成为了一项必不可少的工作。绩效考核制度的建立可以不断地激励再制造企业在摸索中发现自身的不足，改正不足，最终实现再制造资源的合理配置。

绩效考核的目的是评估并掌握组织或是组织内成员的现状，作为下一步工作的依据。再制造绩效考核可以是再制造企业内部的自我考核，也可以是由上级管理部门所开展的针对性考核。再制造企业的考核方法可以采取与一般制造企业相同的方法工具或是其组合，而当管理部门或是企业自身在对再制造企业进行考核的时候，关注点则会因为企业的性质而产生一定的转变，并且要客观反映再制造本身所具有的特点和优势。与一般制造型企业相似，再制造绩效考核也分为个人绩效考核和组织绩效考核，其目标主体可以是组织整体或是组织的各个组成单位，也可以是组织内的人员。而绩效考核的实施主体可以是组织的管理方或是与之合作的第三方。

绩效考核的方式有很多，考核方法应针对被考核对象的特点来选择，这与绩效考核的成败有一定的关系。与传统制造不同，处在产业发展初期阶段的再制造还面临着种种问题，再制造企业的发展也依然是边探索边发展，因此在这种时候发展方法的绩效及发展经验的积累对再制造产业的发展非常重要。再制造企业进行绩效考核的方式及关注点必然有着自身的特点，而目前针对再制造绩效考核的研究还不多见。现有的绩效考核方法的通用性较好，在再制造中应用时，要注意与中国再制造的发展状况及环境相结合。

7 结语

（1）中国再制造产业处于初级发展阶段，在技术上取得了巨大的成就，但是在管理体系的建立和完善上还存在着一定的不足，需要加以研究解决。

（2）对目前再制造产业发展过程中受到普遍关注的关键的再制造管理问题进行了研究，包括再制造风险管理、再制造生产管理、再制造质量管理、再制造认证认可及再制造绩效考核等内容，并给出了相应的对策建议。

参 考 文 献

[1] 徐滨士．中国再制造工程及其进展［J］．中国表面工程，2010，23(2)：1～6.

[2] 徐滨士．再制造与循环经济［M］．北京：科学出版社，2007.

[3] 徐滨士，刘世参，王海斗．大力发展再制造产业［J］．求是，2005(12)：46～47.

[4] 徐滨士．装备再制造工程的理论与技术［M］．北京：国防工业出版社，2007.

Management Problems in Development Process of Remanufacturing Industry

Xu Binshi

(National Key Laboratory for Remanufacturing,
Academy of Armored Force Engineering, Beijing, 100072, China)

Abstract Remanufacturing in China is still in its early stage and faces pressures from society, policy, technology and management. Considering the actual states of remanufacturing in China, this paper researched several key management issues which were cared by various aspects from angle of view of remanufacturing players. Based on need analysis on trend of development of remanufacturing in China, six key management problems were mainly researched, and they were: risk management of remanufacturing players, remanufacturing production management, remanufacturing quality management, authentication mode of remanufacturing in China, subsidy policy of remanufacturing industry, and performance assessment of remanufacturing. The characteristics of six issues were analyzed and the corresponding countermeasures were put forward.

Key words remanufacturing, industry development, management problems

建设工程研究中心，提升再制造产业技术水平*

摘　要　再制造是中国经济战略性新兴产业，已成为循环经济建设的亮点。文中综述了机械产品再制造技术在国内外的发展现状，以及中国采取的以“尺寸恢复和性能提升”为特色的再制造模式与国外的差别。为弥补再制造产业化发展在技术体系、市场管理等方面的不足，特别是先期采取的企业试点模式存在分散化运作、产业示范能力不够强的风险，文中提出了建设机械产品再制造国家工程研究中心作为再制造产业发展科技支撑平台的发展模式，并阐述了该工程中心的定位、发展方向以及四大技术平台建设情况，以提升我国再制造产业的整体水平。

关键词　再制造　机械产品　工程研究中心　产业化

1　引言

再制造已成为我国循环经济建设的亮点，2008 年开始，在国家发改委的推动下，我国已有 14 家企业开展汽车零部件再制造试点，取得了初步成效，并于 2013 年正式启动第二批共 28 家再制造试点企业。工业与信息化部于 2010 年开始在机床、工程机械、大型工业设备、船舶、铁路机车等 8 个行业的 35 家企业实施“机电产品再制造试点”。目前，我国已形成了汽车发动机、变速箱、矿山机械等机电产品 32.5 万套的再制造能力。但与国外发达国家相比，我国机械产品再制造市场需求量大面广，但支撑企业数量很少，产业发展仍处于起步阶段，尚未形成规模，距离战略性新兴产业发展需求还有较大差距。主要表现在我国再制造关键技术体系尚不够完善，急需解决适合产业化生产的再制造质量评价体系和再制造成型加工技术体系的关键技术；缺乏技术成果转化平台，已开发的关键技术设备推广应用滞后；缺乏再制造领域的管理和研发人才；再制造质量标准不健全，缺乏市场准入机制和评价机制；社会各界对再制造的认识不统一、不深入；市场存在无序竞争现象，再制造产品质量存在隐患，影响和限制了再制造产业的健康发展[1]。

2009 年 1 月 1 日实施的国家《循环经济促进法》中明确提出要发展再制造产业，以促进循环经济向产业化、集群化方向发展。同时部署大力发展汽车、工程机械、机床、电机等重点领域的再制造；提倡加大科技投入，以科技创新为本；要求全面建设配套服务体系，为再制造企业发展创造良好的环境和提供有利支撑。2010 年国家发改委、工信部等 11 部委联合公布了《关于推进再制造产业发展的意见》，明确指出“加快建立再制造国家工程研究（技术）中心，加强再制造技术研发能力建设，形成再制造关键设备生产研发体系”。国家工程研究中心是国家发改委根据建设创新型国家和产业结构优化升级的重大战略需求，以提高自主创新能力、增强核心竞争力和发展后劲

* 本文合作者：梁秀兵、张伟、杨庆东、陈永雄。原发表于《中国表面工程》，2013，26(4)：99~104。国家自然科学基金（51105377，51005244）资助。

为目标，组织具有较强研发实力的高校、科研机构和企业等建设的研发实体，是国家创新体系的重要组成部分。因此，实施再制造国家工程研究中心建设模式，高度契合了国家构建循环经济的战略需求，为国家再制造产业的健康发展提供科技创新平台支撑，对落实生态文明国策，建设“资源节约型、环境友好型、健康和谐型”社会具有重要促进作用。

2 国内外机械产品再制造技术发展状况

2.1 国外再制造模式与技术发展现状

再制造产业在国外已经发展成为一个新兴的朝阳产业，据统计，2010 年仅全球汽车零部件再制造产业规模已达 850 亿美元以上，在美国机动车辆维修市场中，再制造配件占据了 70% 以上的市场份额[2]。世界发达国家多年的实践经验表明，再制造产业非常重要，是发展循环经济、扩大内需和环境保护的重要途径，同时，再制造也是一项复杂的系统工程，需要统筹协调发展，从技术层面来讲，它涉及回收、拆解、清洗、检测、分选、换件或修理、装配使用与检验等流程。对于单个机械零部件的再制造，国外主要采用以“减尺寸”为特征的尺寸修理法和换件修理法的再制造模式[3]。尺寸修理法是将失配的零件表面尺寸加工到可以配合的范围，如缸套 - 活塞磨损失效后，通过镗缸的方法恢复缸套的尺寸精度，再配以大尺寸的活塞完成再制造，涉及的再制造技术相对单一；换件修理法是将损伤零件整体更换为新品零件。近年来国外重点发展的再制造相关技术包括以下几个方面：

（1）再制造回收技术。美国福特公司已建立起全球最大的汽车回收中心，专门从事汽车的回收、拆卸与再制造，充分利用回收再制造的部件，一年实现 10 亿美元的营业额；而且福特与通用、克莱斯勒等大汽车公司已结成回收联盟，在密歇根建立汽车拆卸中心，专门研究开发汽车零部件的拆卸、再制造和再循环利用。美国罗切斯特理工学院有一个专门从事再制造工程研究的全国再制造和资源恢复中心[4]。日本截至 2008 年的统计数据显示，再制造的工程机械中，58% 由日本国内用户使用，34% 出口到国外，其余 8% 拆解后作为零部件配件出售。

（2）再制造设计。国外许多学校已开设再制造技术课程，而且将零部件的再制造性充分融入到工业产品设计当中，认为在设计产品时只考虑一次性使用是不合理的[5]。再制造设计是在新产品设计阶段就考虑产品的可再制造性，为更新换代打下基础，主要结合具体产品，针对再制造过程中的重要设计要素（如拆卸性能、零件的材料种类、设计结构与坚固方式等）进行研究，同时考虑在报废时如何减少环境污染等。再制造往往是在产品使用数年后实施的，在此期间许多新技术、新材料相继出现，因而通过再制造修复、改造的产品质量往往优于原产品。

（3）军用装备再制造技术研究。再制造工程的研究也已引起美国国防决策部门的重视，美国制订了 2010 年国防工业制造技术的框架，特别强调为延长武器系统的使用寿命，已经将系统性能升级、延寿技术和再制造技术列为目前和将来国防制造重要的研究领域，体现出其对再制造新技术的认识[6]。另外，美国正在实施的国家纳米技术计划中，也把研究开发强度和硬度更高、重量更轻、安全性更好，并将再制造自行修复的纳米材料列为重要方面之一[7,8]。

2.2 国内机械产品再制造技术现状

我国机械产品再制造发展还处于起步阶段，但是近几年废旧装备的数量却在不断增加。根据2010年统计数据，我国汽车年报废量超过300万辆，主要工程机械报废量超过40万台，农业拖拉机报废量超过100万台，目前机床保有量逾700万台，其中服役超过10年的机床占到60%，“十一五”期间机床报废量超过80万台，而且还将大量增加，预计到2020年工程机械的报废量将达到120万辆左右，其数字惊人。随着我国进入机械装备报废的高峰期，再制造产业在社会、资源、环境效益等方面的优势决定了发展再制造产业势在必行。

除装甲兵工程学院有国家级再制造重点实验室专门从事再制造技术研发外，还有如西安交通大学、清华大学、机械科学研究院、广州有色金属研究院等科研单位开始从事机械产品再制造技术、工艺及设备研发生产，但局限于研发再制造工程领域中的少数技术，还不能形成完善的技术体系，且工程化应用的领域较少。从事机械产品再制造生产的企业主要集中在发改委批准的42家汽车零部件再制造试点企业和工信部批准的35家机电产品再制造试点企业。其中济南复强动力有限公司是一家起步较早的具有国际水准的汽车发动机再制造公司。其他多数企业近几年才开始从事再制造生产，普遍存在技术缺乏、人才需求量大、经济效益不明显的特点。中国汽车工业协会公布的2010年我国汽车零部件再制造业产能数据显示，再制造发动机约11万台，变速器约6万台，发电机、起动机约100万台，再制造总产值低于25亿元。随着国家对再制造产业发展各项政策的落实，预计我国机械再制造产业将迎来跨越式的发展。

如前所述，国外再制造企业在进行再制造生产过程中主要采用以“减法”为特征的尺寸修理法和换件修理法。这种再制造模式导致废旧产品的再制造率很难进一步提高，由于缺少高新技术应用，该模式很难保证再制造产品性能的跨越式提升。为此，装甲兵工程学院率先提出发展以尺寸恢复与性能提升为特色的中国再制造模式和技术体系[9]。同时在再制造的基础理论与相关前沿技术进行了深入的探索研究，特别是以微纳技术、信息技术和先进表面工程技术为支撑的再制造关键技术不断有所突破，研发出具有自主知识产权的再制造关键技术与设备，可将某型号旧发动机的旧件再利用率提高到90%，使零件的尺寸精度和质量标准不低于原型新品水平，而且在耐磨、耐蚀、抗疲劳等性能方面达到原型新品水平。目前，已研发了废旧零部件缺陷、裂纹和应力集中综合检测仪器，自动化表面工程再制造专机等一系列发动机零部件再制造关键技术，以及发展了再制造零部件剩余寿命评估理论，自动化高速电弧喷涂，纳米复合表面制备与成型一体化技术，等离子熔敷再制造快速成型技术以及缸体、曲轴、连杆等零部件自动化再制造工艺等创新性成果。

2.3 中国特色再制造与国外的区别

中国虽然采用以尺寸恢复与性能提升为特色的再制造，机械产品再制造产业与国外发达国家的机械产品再制造产业相比，仍存在较大的差距，主要表现在行业范围、企业规模、销售收入、评价体系等多个方面：

（1）再制造市场需求量大、而支撑企业尚未跟进。我国作为装备使用大国，许多大型成套机械设备面临退役，同时机械设备零配件的维修需求量大，但是国内年再制造5万台（套）零部件的再制造企业数量较少。

（2）实施再制造的质量控制体系尚未完善。许多企业以前是专门从事维修或表面加工的，有的甚至是从事拆解的企业，连修理都没有从事过，其规模相对较小，并没有充分认识到再制造对象和过程的复杂性，大部分在旧件检测、再制造毛坯修复等关键环节没有建立完善的质量控制体系，缺乏先进的无损检测与质量评价技术手段；一些废旧零部件的质量检测和寿命评估技术的许多难题还未攻克，直接修复投入再制造市场后，严重影响产品的质量和可靠性，甚至使整个行业的信誉受损。而国外的质量评价体系成熟度非常高，如美国预计2015年75%的再制造公司通过ISO认证，其汽车及配件和工业设备的再制造比例达到72%以上，而且对假冒伪劣产品的惩罚力度很大。

（3）可实现产业化生产先进成熟的再制造技术还非常缺乏。大部分再制造企业主要还是采用换件法和尺寸修理法进行再制造，许多旧件毛坯作废弃处理，再利用率很低，而且，对产品的再制造加工手段还处于低水平，大多沿用制造工艺的设备和技术，导致再制造后产品非标件多，用户认可程度低，加工成本高。

（4）市场存在无序竞争现象，再制造产品质量存在隐患。各企业之间互相压价，致使产品利润日渐降低，同时部分小企业采用价格低廉、质量无保证的零件，并贴牌大厂产品标识，假冒大厂再制造产品，导致再制造产品质量难以保证。各企业生产规模相差较大，导致生产设备和生产工艺良莠不齐，工人技能和管理理念相差较大，这也是难以保证再制造产品质量的隐患之一。

3　机械产品再制造国家工程研究中心建设的意义与方向

3.1　工程中心建设的意义

我国循环经济法提出，循环经济是在生产、流通和消费等过程中进行的减量化、再利用、资源化活动的总称，它是推进可持续发展战略的一种优选模式，强调以循环发展模式替代传统的线性增长模式，表现为“资源—产品—再生资源”和“生产—消费—再循环”的模式。“再制造产业化”作为循环经济的重点工程之一，需要快速健康发展。

从国内外制造及再制造产业的发展历程看，“技术产业化”是一个普遍的发展模式，我国的再制造产业在这一环节上先期采取的是企业试点模式，其总体上是分散化运作，产业示范能力明显不足。因此，要解决机械产品再制造的共性关键技术问题，并把科研成果进行产业转化和推广，光靠单独的企业是无法实现的，迫切需要建立一个开发关键共性技术的基地；需要一个建立行业标准和为行业提供高技术服务、带动全行业技术水平提升的国家工程研究中心，以承担构建我国机械产品再制造产学研用技术创新体系的重任。

鉴于再制造产业有其独特的技术和相应的政策法规环境，机械产品再制造国家工程研究中心成了推动再制造产业发展的重要科技创新支撑平台。在此背景下，装甲兵工程学院凭借雄厚的再制造技术优势，联合山东能源机械集团有限公司和北京首科集团公司申报建设“机械产品再制造国家工程研究中心”，于2012年5月25日正式得到国家发改委批复。

自1992年原国家计委正式启动国家工程中心计划以来，经过20年的发展，总数已达130个，分布在信息、新材料、机械等国民经济重要领域。主要任务是，根据国家和产业发展需求研发关键共性技术，开展具有重要市场价值的科技成果的工程化和系统集成，为规模化生产提供先进成熟的技术、工艺及装备，提供技术验证和咨询服务，培养高层次人才。机械产品再制造国家工程研究中心的主要定位是，关键共性技术、成套工艺和装备的开发与工程化，研制相关标准和规范，建立再制造检测评价体系，培养再制造领域的创新技术人才团队，构建再制造产学研用技术创新体系，促进先进成熟的再制造技术在装备保障领域的应用，提升装备修复和再生能力，为促进再制造产业军民融合式发展提供有力的技术支撑。因此，该工程中心的建设，将集中解决再制造行业关键共性技术问题，形成和推广具有自主知识产权的创新性成果，构建一个有高技术支撑和促进国际交流与合作，并引领和服务于全行业的公共平台，加快我国机械产品再制造科技成果向现实生产力转化，将为推动我国机械再制造行业的发展和提高整体技术水平起到重要作用。

3.2 工程研究中心的发展方向

3.2.1 再制造共性和关键技术研发

对于单个零部件的再制造，我国采用的是不同于国外的再制造模式，主要发展的是以尺寸恢复与性能提升为特色的技术体系。因此，机械产品再制造国家工程研究中心将以机械零部件的再制造成型技术研发为主线，再制造共性和关键技术研发的重点方向集中在以下几方面：

（1）工程论证设计技术体系。包括机械产品再制造性设计与分析、再制造经济效益分析模型、再制造工艺流程设计和再制造逆向物流信息化管理等。

（2）寿命评估与质量检测技术设备体系。包括再制造毛坯剩余寿命评估系统、毛坯的结构与表面完整性评估技术、再制造产品使用寿命预测技术设备、再制造产品质量检测技术设备、再制造涂层检测与评估技术设备和零部件失效分析技术等。

（3）绿色拆解及清洗技术设备体系。包括快速无损拆解技术、高温分解清洗技术设备、物理喷丸清洗技术设备、清洗液循环再利用技术设备等。

（4）零部件表面修复成型技术设备体系。包括纳米电刷镀、自动化热喷涂、高能束熔覆、薄膜沉积等先进的表面修复与改性技术、工艺和设备。

（5）成型材料制备技术体系。包括再制造电沉积镀液、再制造合金丝材、再制造粉体、微纳米表面自修复材料等。

（6）零部件加工技术设备体系。包括再制造加工工艺及方法、自动控制及信息化设备、再制造精密加工设备工艺、再制造尺寸恢复与尺寸加工融合技术设备等。

3.2.2 再制造科技成果的工程化与产业化

针对我国机械产品再制造行业效率低、运作散、技术水平较低等一系列突出问题，开展具有重要市场价值的科技成果的工程化与系统集成，开发机械产品再制造亟需的重大技术产品和设备。重点开展民用和军用车辆、工程机械、矿采装备、冶金装备等行业领域的关键零部件成套再制造技术的工程化研究，包括再制造工艺流程的定型研究、再制造产品的后加工以及质量检验等工作，最终形成成熟的可产业化的再制造成套技术、工艺、材料和装备，提升再制造产业的核心竞争力。

3.2.3 再制造产业化服务

在再制造关键技术与设备推广应用过程中，注重再制造的企业标准、行业标准和国家标准的研究与制定，加强认证认可方法与制度研究。同时，围绕国家循环经济宏观发展战略，通过再制造关键技术设备在产业化推广过程中的技术验证，深入分析再制造产业化过程中的制约因素，探究促进我国再制造产业做大做强的措施与对策。

4 机械产品再制造国家工程研究中心的技术平台建设

针对国家发改委对国家工程研究中心的定位，工程研究中心整合目前国内外再制造技术领域的优势资源，依托目前“装备再制造技术国防科技重点实验室”的理论基础和技术储备优势，引进消化吸收，转化科研成果，并利用图 1 所示的技术平台，进行机械产品再制造核心共性技术的研发、系统集成和研究成果的工程化与产业化，为行业提供技术开发及科技成果工程化的试验、验证、检测、定标等所需的技术装备，最终形成我国机械产品再制造领域关键技术的展示、研究与产业化转化的基地。工程中心的技术平台主要包括 4 个方面：

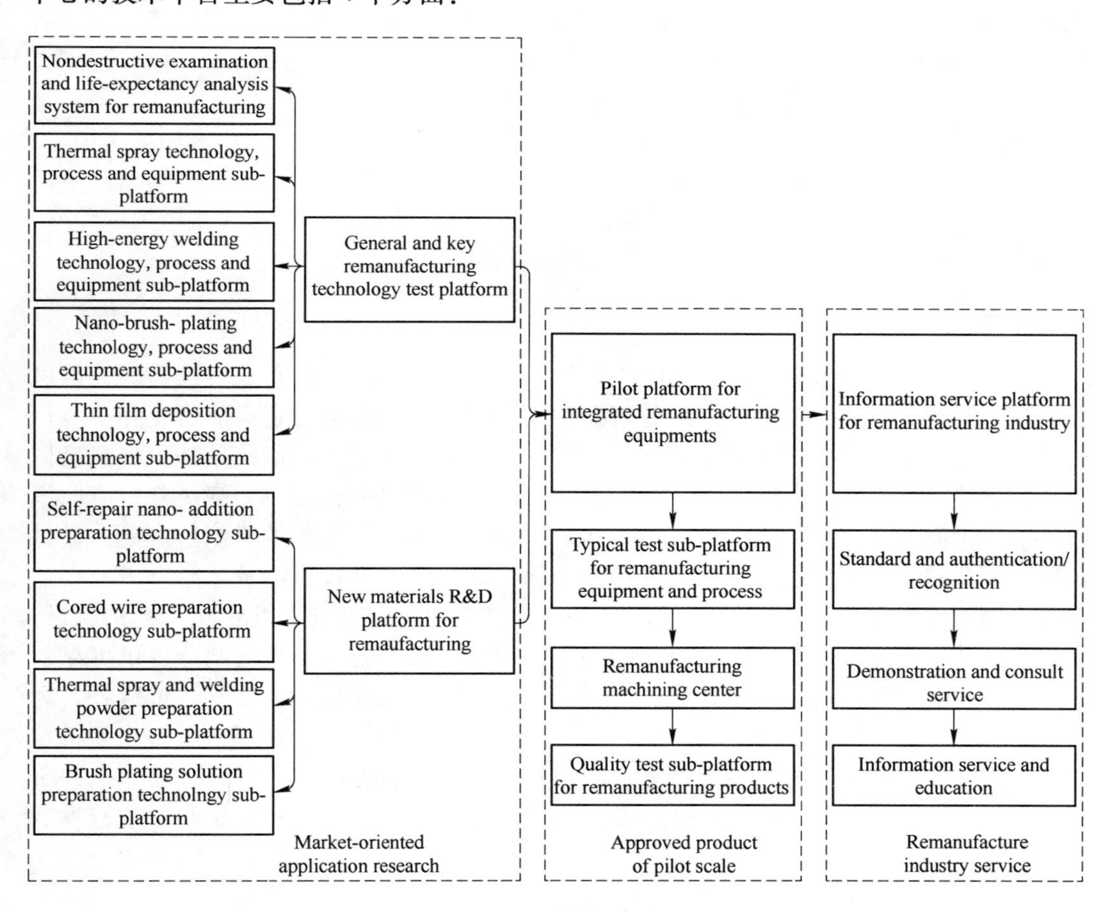

图 1 机械产品再制造国家工程研究中心的技术平台结构

Fig. 1 Structure of technology platform for NERC – MPR

（1）再制造共性关键技术试验平台。主要包括研发再制造毛坯质量检测和剩余寿命评价技术体系，解决判定废旧零部件能否再制造的难题；研发系列新型再制造修复成型技术、工艺与设备，解决量大面广和附加值高的典型零部件的再制造难题。该平台的建成为提升旧件的再制造率、保证再制造零件的尺寸精度和质量标准不低于原型新品水平提供技术保障。

（2）再制造新材料研发平台。新型材料研发是基于表面技术再制造的重要研究内容，依托装备再制造技术国防科技重点实验室的研究基础，在各种表面涂、镀、覆层制备材料及减摩自修复添加剂材料等领域拥有的多项成果，工程中心将完善这些新材料的性能、体系和制备工艺，开展工程化应用，并结合开发的再制造关键设备和工艺实施典型机械产品的再制造产业化推广，不断提升工程研究中心的创新能力和综合技术水平。

（3）再制造成套装备工程化中试平台。为将研发成果和产品转化到工程用当中，很有必要建设基于这些技术和成果的成套装备技术工程化中试平台，实现成套装备与技术产品推广应用前的定型与质量考核，为规模化再制造提供可靠的技术保障。因此，再制造成套装备技术工程化中试平台是检验机械产品再制造技术研发成果成败的关键，也是规模化推广再制造工艺和产品的必要环节，同时，基于该平台，可以针对汽车、工程机械等领域的某几种量大面广的典型废旧机械零部件，开展小规模的再制造生产，为后期标准制定和再制造分中心的建设打下基础。

（4）再制造产业信息化服务平台。再制造产业信息化服务平台主要采用虚拟现实、网络和数据库、技术成熟度评价等技术，支撑再制造生产线与运行模式规划论证、国家再制造产业发展政策法规咨询、标准制定、资质认证认可和高层次人才培训服务等，是推进和规范再制造国家战略性新兴产业发展的基础条件之一。

5 结语

机械产品再制造国家工程研究中心根据国家机械产品再制造行业发展需要，以机械产品再制造关键共性技术、再制造新材料、再制造成套技术与装备的研发和工程化为主导，通过建设高水平的技术创新、工程化研发验证和产业化示范基地，提高机械产品再制造自主创新能力。通过再制造关键共性技术试验平台建设，解决判定废旧零部件能否再制造的难题，解决量大面广和附加值高的典型零部件的再制造难题；通过再制造新材料研发平台建设，开展各种表面涂、镀、覆层制备材料及减摩自修复添加剂材料等领域的研发和工程化应用推广；通过再制造成套装备工程化中试平台的建设，实现成套装备与技术产品推广应用前的定型与质量考核，为规模化再制造提供可靠的技术保障；通过再制造产业信息化服务平台建设，提供工程技术验证和咨询服务，为行业培养工程技术研究与管理的高层次人才。因此，建设国家工程研究中心作为科技创新支撑平台发展我国的再制造产业，时机和条件成熟，方向正确。通过建立工程研究中心，将搭建再制造产业与科研之间的“桥梁”，为我国再制造产业的科学规范发展提供支撑。

参考文献

[1] 董碧娟．“中国再制造”发展正当其时［N］．经济日报，2012－4－17(15版)．

[2] Coverage 专题 [J]. 机电一体化，2012(5)：4~8.
[3] 徐滨士. 国内外再制造的新发展及未来趋势 [C]. 科技支撑科学发展——2009 年促进中部崛起专家论坛暨第五届湖北科技论坛论文集，中国科技出版社，2009.
[4] 李源. 欧美汽车回收业的发展 [J]. 世界汽车，1999，29(1)：7~8.
[5] Robert T. Lund. The Remanufacturing Industry—Hidden Giant [R]. Boston University，1996.
[6] 中国兵器工业第二一零研究所. 2010 年及其以后的美国国防制造工业 [R]. 1999.
[7] 郭廷杰. 欧盟各国报废汽车再生利用管理简介 [J]. 中国资源综合利用，2001，19(7)：36~38.
[8] 曹学军. 克林顿要实施 NNI 计划 [J]. 科学时报，2000，5.
[9] Xu B S，Zhang Z X. Surface engineering and remanufacturing technology [C]. International Conference on Advanced Manufacturing Technology，Xi'an. New York Press，1999：1129~1132.

Development of Remanufacturing Industrialization by Building Engineering Research Center

Xu Binshi[1a,1b]，Liang Xiubing[1b,1c]，Zhang Wei[1d]，
Yang Qingdong[2]，Chen Yongxiong[1b]

(1a. Science and Technology on Remanufacturing Laboratory，
1b. National Engineering Research Center for Mechanical Product Remanufacturing，
1c. Science Research Department，
1d. Engineering Department for Equipment Maintenance and Remanufacture，
Academy of Armored Force Engineering，Beijing，100072，China；
2. Shandong Energy Mechanical Co.，Ltd.，Xintai，271222，China)

Abstract Remanufacturing is one of the strategic newly industries in Chinese economic construction，becoming one of the hot topics of the recycling economic construction. In this study，the domestic and overseas developments of the mechanical product remanufacturing technology were discussed，including the differences in remanufacturing modes. In China，a remanufacturing mode featured with the concept of dimension resume and property upgrade was developed. To make up the shortcomings of technology system and market management during the remanufacturing industrialization，especially for the mode of enterprise pilot project，a development mode of establishing national engineering research center for mechanical product remanufacturing (NERC-MPR) as technical support for remanufacture industry was put forward. The registration，developing direction and technological platforms of NERC-MPR were represented additionally，and the developing trend of the engineering research center was put forward consequently.

Key words remanufacturing，mechanical product，national engineering research center，industrialization

附录

附录1　获奖情况

1. 国家奖

2013年，面向再制造的表面工程技术基础，国家自然科学二等奖

2011年，装备零件快速智能电弧喷涂抢修技术及其应用，国家科技进步二等奖

2009年，军用装备机械零部件纳米颗粒复合延寿技术，国家技术发明二等奖

2005年，院士科普丛书——《神奇的表面工程》，国家科技进步二等奖

2003年，坦克履带板材料——ZGMn8CrMo研究，国家技术发明二等奖

2003年，高效能超音速等离子喷涂技术开发与应用，国家科技进步二等奖

1998年，电弧喷涂防腐技术在我军主战装备的应用，国家科技进步三等奖

1988年，有色、贵金属和合金等新型电刷镀溶液系列（66种）的研究与应用，国家科技进步二等奖

1985年，电刷镀技术及其推广应用，国家科技进步一等奖

1985年，热喷涂技术开发、推广、应用，国家科技进步二等奖

2. 省部级及军队奖

2013年，再制造关键技术及产业化，中国机械工业科学技术一等奖

2013年，微纳米金属硫化物固体润滑技术及应用，教育部技术发明一等奖

2012年，面向再制造的表面工程技术基础，北京市自然科学一等奖

2009年，装备零件快速智能电弧喷涂抢修技术及其应用，军队科技进步一等奖

2007年，纳米微粒复合表面工程技术及其推广应用，中国机械工业科学技术一等奖

2006年，原位动态纳米减摩自修复添加剂技术，军队科技进步二等奖

2005年，《纳米表面工程》，中国石油和化学工业协会科技进步二等奖

2004年，纳米颗粒复合电刷镀技术开发研究及其在装备上的应用，军队科技进步二等奖

2002年，舰船甲板防滑技术研究与应用，军队科技进步二等奖

2002年，高效能超音速等离子喷涂技术开发与应用，军队科技进步一等奖

2000年，应急维修技术研究与应用，军队科技进步二等奖

2000年，坦克履带板换代材料研究，军队科技进步一等奖

1998年，《表面工程与维修》专著，军队科技进步二等奖

1997年，电弧喷涂防腐技术开发研究及在我军主战装备上的应用，军队科技进步一等奖

1995年，摩擦电喷镀技术研究，军队科技进步一等奖

1989年，刷镀复合技术金刚石砂轮及其制品研究，军队科技进步二等奖

1987年，高硬度装甲钢低温焊修技术研究，军队科技进步二等奖

1987年，62种系列刷镀液研制，军队科技进步二等奖

1982年，坦克零件等离子喷涂修复工艺试验研究，军队科技成果一等奖

附录2　出版的学术著作

序号	著作名称	出版日期	出版单位
1	维修焊接	1985.09	天津科学技术出版社
2	刷镀技术	1985.09	天津科学技术出版社
3	等离子喷涂与堆焊	1986.12	中国铁道出版社
4	现代机械维修	1994.07	中国铁道出版社
5	表面工程与维修	1996.06	机械工业出版社
6	表面工程的理论与技术	1999.07	国防工业出版社
7	表面工程	2000.06	机械工业出版社
8	神奇的表面工程	2000.12	清华大学出版社
9	徐滨士院士科研文选	2001.03	机械工业出版社
10	表面工程新技术	2002.01	国防工业出版社
11	纳米表面工程	2004.01	化学工业出版社
12	再制造工程基础及其应用	2005.11	哈尔滨工业大学出版社
13	材料表面工程	2005.12	哈尔滨工业大学出版社
14	《中国材料工程大典》第16、17卷 材料表面工程（上、下）	2006.01	化学工业出版社
15	装备再制造工程的理论与技术	2007.07	国防工业出版社
16	再制造与循环经济	2007.08	科学出版社
17	机械加工手册—第十三章：表面技术	2007.10	机械工业出版社
18	徐滨士院士教学、科研文选	2010.04	化学工业出版社
19	表面工程的理论与技术	2010.04	国防工业出版社
20	机械工程学科发展报告（成形制造）	2011.04	中国科学技术出版社
21	高稳定性高速电弧喷涂腐蚀防护技术	2011.02	科学出版社
22	Micro and Nano Sulfide Solid Lubrication	2011.07	Springer Press
23	中国机械工程技术路线图	2011.08	中国科学技术出版社
24	装备再制造工程	2013.12	国防工业出版社

附录3　制定的国家标准

序号	标准名称	标准号	标准类别
1	机械产品再制造　通用技术要求	GB/T28618—2012	国家标准
2	再制造　术语	GB/T28619—2012	国家标准
3	再制造率的计算方法	GB/T28620—2012	国家标准
4	机械产品再制造质量管理要求	报批	国家标准
5	再制造毛坯质量检验方法	报批	国家标准
6	再制造企业技术条件	报批	国家标准

附录4 获得授权的国家发明专利（部分）

序号	专利名称	专利号
1	自强化双相抗疲劳耐磨修复层合金	ZL88100157. 0
2	钴－镍－磷非晶态合金刷镀镀层及其方法	ZL89108740. 0
3	复合刷镀金刚石砂轮制品、装置及工艺	ZL90100944. X
4	超细硼酸盐润滑油添加剂的制备方法	ZL96109484. 2
5	含稀土元素的粉芯铁基合金的热喷涂线材	ZL96120778. 7
6	装甲车辆履带板的中锰铸钢	ZL98123882. 3
7	一种抗高温冲蚀复合涂层及其制备方法	ZL00114712. 9
8	超分散稳定剂的合成方法	ZL01125113. 1
9	舰船甲板防滑涂层用粉芯丝材、防滑涂层及其制备方法	ZL00114749. 8
10	超音速等离子喷涂枪	ZL01101077. 0
11	在金属基体上电刷镀纳米颗粒复合镀层的方法	ZL02101195. 8
12	纳米颗粒复合电刷镀镀液料浆的制备方法	ZL02101196. 6
13	纳米减摩自修复添加剂及其制备方法	ZL200310102060. 6
14	高分子聚合物/硫属化合物纳米复合材料的原位制备方法	ZL200310110836. 9
15	硫化银半导体纳米颗粒的制备方法	ZL200410049452. 5
16	自动化刷镀机	ZL2005610001480. 4
17	轮足复合蠕动式机器人	ZL200510001481. 9
18	快换滚动阳极液冷电刷镀笔	ZL200610056168. X
19	便携式逆变刷镀电源	ZL200610056169. 4
20	渗碳类重载齿类件齿面激光熔覆粉末材料及修复方法	ZL200710119899. 9
21	一种高速电弧喷涂制备非晶纳米晶涂层的粉芯丝材	ZL200810074939. 7
22	一种 Zn/ZnS 复合固体润滑薄膜及其制备方法	ZL200710175458. 0
23	铁磁性金属构件疲劳裂纹和应力集中的磁记忆检测方法	ZL200710119911. 6
24	内孔类零部件电刷镀的装置及方法	ZL200910077394. X
25	旋转式材料热震试验机	ZL200510117854. 9
26	一种电弧压力的测试装置	ZL201020123688. X
27	一种铁基体上镍镀层的无损测厚方法	ZL200610069899. 8
28	旧缸体缸筒内壁表面缺陷的自动化涡流/磁记忆检测装置	ZL201010272384. 4
29	发动机旧曲轴内部缺陷的自动化超声波检测方法及装置	ZL201010272402. 9
30	一种 WS_2/MoS_2 固体润滑多层膜及其制备方法	ZL200710175460. 8
31	一种铁铬硼硅/FeS 复合固体润滑薄膜及其制备方法	ZL200710175459. 5
32	适于连接声发射探头的标准杆部件	ZL201120184125. 6
33	等离子熔覆快速成形铁基合金粉末材料	ZL200910241851. 4

续表

序号	专利名称	专利号
34	一种金属表面强化减摩自修复剂及其制备方法	ZL200810076277.7
35	发动机缸体全自动立式电刷镀设备	ZL201010103228.5
36	连杆自动化电刷镀机床	ZL201010103144.1
37	一种高速电弧喷涂制备纳米晶金属间化合物涂层的粉芯丝材	ZL200810076758.8
38	一种焊接热效率的测试装置	ZL201020123692.6
39	内孔类零部件电刷镀用镀笔	ZL200910077204.4
40	一种坦克柴油发动机复合减摩自修复延寿方法	ZL200810076278.1
41	再制造零部件表面涂层加速磨损寿命试验机及其检测方法	ZL201010167331.6
42	一种3Cr13/FeS复合固体润滑薄膜及其制备方法	ZL200710175461.2
43	大负载真空室送样装置	ZL201120127515.X
44	提高喷涂层接触疲劳寿命的方法	ZL200810223548.7
45	制备单质钨膜的方法	ZL200710100161.8
46	MSTS-1型多功能空间摩擦学试验机测量控制软件	2012SR017469
47	再制造前废旧曲轴R角部位磁记忆信号采集装置	ZL201110003051.6
48	废润滑油再生混凝沉降复合设备	ZL201110431606.7
49	一种抗海洋腐蚀与热腐蚀的复合涂层及其制备方法	ZL201110111175.6
50	一种雷达吸波材料及制备方法以及应用	ZL201210020084.6
51	自动化高速电弧喷涂再制造发动机曲轴的工艺	ZL201210049550.3
52	CRLE-1曲轴再制造寿命检测评估系统（CRLE-1测评系统）	2013SR074670
53	纳米羟基硅酸镁粉体及其制备方法	ZL200810117289.X
54	一种基于表面超声波测定薄镀层残余应力的方法	201210106043.9
55	一种面向快速加工的数控多功能复合加工机床	ZL201010047410.3